SOME IMPORTANT GRAPHS

parabola

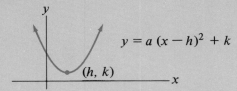

$$y = a(x-h)^2 + k$$

(h, k)

parabola

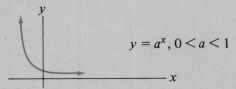

$$x = a(y-k)^2 + h$$

(h, k)

circle

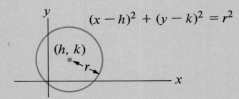

$$(x-h)^2 + (y-k)^2 = r^2$$

(h, k)

r

exponential functions

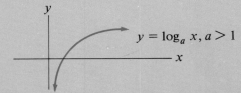

$$y = a^x, 0 < a < 1$$

logarithmic function

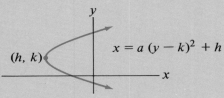

$$y = \log_a x, a > 1$$

ellipse

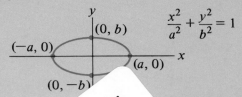

$$\frac{x^2}{a^2} + \frac{y^2}{b^2} = 1$$

$(0, b)$

$(-a, 0)$

$(a, 0)$

$(0, -b)$

hyperbola

$$\frac{x^2}{a^2} - \frac{y^2}{b^2} = 1$$

$(a, 0)$

hyperbola

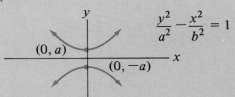

$$\frac{y^2}{a^2} - \frac{x^2}{b^2} = 1$$

$(0, a)$

$(0, -a)$

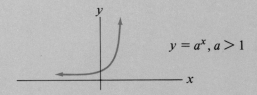

$$y = a^x, a > 1$$

College
Algebra

2nd
Edition

COLLEGE ALGEBRA

Walter Fleming
Hamline University

Dale Varberg
Hamline University

Prentice-Hall, Inc., Englewood Cliffs, New Jersey 07632

Library of Congress Cataloging in Publication Data

Fleming, Walter.
   College algebra.

   Includes index.
   1. Algebra. I. Varberg, Dale E. II. Title.
QA154.2.F53 1984   512.9   83–10935
ISBN 0–13–141630–8

Editorial/production supervision: Paula Martinac
Interior and cover designs: Jayne Conte
Manufacturing buyer: John B. Hall
Computer art created by: GENIGRAPHICS

Other titles by the same authors:
Algebra and Trigonometry, Second Edition
Plane Trigonometry
Precalculus Mathematics
Calculus with Analytic Geometry, Fourth Edition, by Edwin J. Purcell and Dale Varberg

COLLEGE ALGEBRA
Second Edition
Walter Fleming/Dale Varberg

Printed in the United States of America

10 9 8 7 6 5 4 3 2

ISBN 0-13-141630-8

Prentice-Hall International, Inc., *London*
Prentice-Hall of Australia Pty. Limited, *Sydney*
Editora Prentice-Hall do Brasil, Ltda., *Rio de Janeiro*
Prentice-Hall Canada Inc., *Toronto*
Prentice-Hall of India Private Limited, *New Delhi*
Prentice-Hall of Japan, Inc., *Tokyo*
Prentice-Hall of Southeast Asia Pte. Ltd., *Singapore*
Whitehall Books Limited, *Wellington, New Zealand*

To Mr. and Mrs. Ivar Lindgren,
who have generously supported
the Hamline Mathematics Department

CONTENTS

PREFACE
TO THE SECOND EDITION

Solving a problem is similar to building a house. We must collect the right material, but collecting the material is not enough; a heap of stones is not yet a house. To construct the house or the solution, we must put the parts together and organize them into a purposeful whole.

George Polya

As with the first edition, we begin our text with a quote from George Polya. This reflects our admiration of a great mathematician and suggests continuity with the first edition and its emphases. We have aimed to retain those features that made the first edition so successful, while incorporating a number of improvements.

Our book continues to be a *solid introduction to algebra,* especially those parts of the subject that are useful in the physical and social sciences. We write in a *free-flowing relaxed style,* avoiding the overuse of technical jargon. To motivate students and add interest, we begin each section with a *special boxed display.* This box may contain an historical anecdote, a famous quotation, an appropriate cartoon, or a diagram illustrating the key result of the section.

PROBLEM-SOLVING EMPHASIS

We continue to emphase *problem solving* both in the text discussions and in the problem sets. There are over 3000 problems for the student to work—drill problems, calculator problems, numerous word problems, and an occasional theoretical problem (at the end of a problem set). We have increased the number of *calculator problems* but they are always marked with the symbol ⓒ and can be omitted if desired. Every problem set ends with a group of problems labeled *miscellaneous problems*. These are designed to challenge the better students by testing the techniques of the section in a somewhat random manner and by offering a few hard nuts to crack.

Like all textbook authors, we use *examples* to illustrate the discussions in the text. In contrast to most, we also put examples within the problem sets. Why do we follow this novel practice? First, we are convinced that putting examples in the problem sets is good pedagogy. It allowed us to cluster a set of problems right next to an illustrative example; this should help students learn the corresponding concept. Second, it permitted shorter, more coherent, and more interesting sections. We were able to concentrate on the main ideas without illustrating every technique or elaborating every detail. Consequently, a student arrives at a problem set sooner and with a better overall perspective. In any case, we emphasize that *the problem set examples are an integral part of the text*. Note that they are always identified with a descriptive title.

CHANGES IN THE SECOND EDITION

In addition to adding more problems, especially of the calculator variety, and to rewriting scattered paragraphs too numerous to mention, we have made the following changes and improvements.

1. Square roots and cube roots are introduced early (Section 1-3) and used in examples from there on. A full treatment of radicals waits until Chapter 6 (as in the first edition).
2. Complex numbers are deemphasized, except in Chapter 7 (Theory of Polynomial Equations), where they play a necessary role. Section 1-6 introduces these numbers, but that section can be postponed until just before Chapter 7. While complex numbers are briefly referred to in a self-contained treatment of the quadratic equation in Chapter 3, they otherwise appear only in a few problems marked with the symbol ⓘ These problems can be omitted by the instructor who wishes to put all the emphasis on the real numbers.
3. The geometric treatment of the conic sections (Chapter 4) is relegated to the ends of the problem sets. The approach is now completely algebraic via the standard equations.
4. The section on exponential growth and decay has been completely rewritten.
5. Appearing for the first time are "caution boxes" placed in the margins at strategic places to warn students of commonly made errors.

AUDIENCE

This book is appropriate for all typical college algebra courses taught in American universities and colleges. It contains all the algebraic topics needed in the standard brief calculus course. Business students will welcome inclusion of material on matrices, linear programming, probability, and mathematics of investment. Well-prepared students can either omit or review quickly Chapters 1 and 2. The dependence chart below will aid instructors in designing a syllabus.

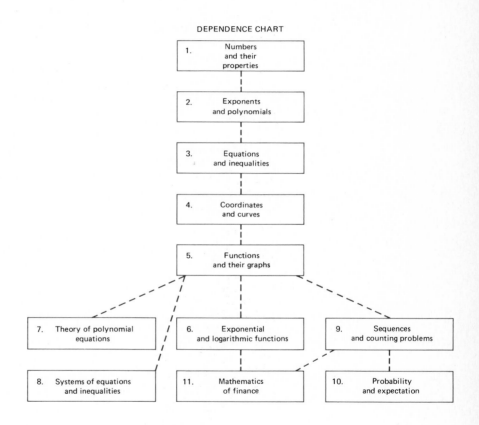

DEPENDENCE CHART

1. Numbers and their properties

2. Exponents and polynomials

3. Equations and inequalities

4. Coordinates and curves

5. Functions and their graphs

7. Theory of polynomial equations

6. Exponential and logarithmic functions

9. Sequences and counting problems

8. Systems of equations and inequalities

11. Mathematics of finance

10. Probability and expectation

SUPPLEMENTARY MATERIALS

An Instructor's Manual is available upon adoption of the text. Advice on how to use the text, answers to the even-numbered problems, four versions of chapter tests for each chapter, and an answer key for these tests are included in the manual.

ACKNOWLEDGMENTS

The second edition has profited from the warm praise and the constructive criticism of a host of reviewers. We offer our thanks to the following people:

Karen E. Barker, Indiana University, South Bend
Phillip M. Eastman, Boise State University
Gerald A. Goldin, Northern Illinois University, DeKalb
William J. Gordon, State University of New York, Buffalo
Judith J. Grasso, University of New Mexico, Albuquerque
James Jerkofsky, Benedictine College
James P. Muhich, University of Wisconsin, Oshkosh
Richard Nadel, Florida State University, Tallahassee
Stephen Peterson, University of Notre Dame
Stuart Thomas, Oregon State University, Corvallis
Paul D. Trembeth, Delaware Valley College
Robert F. Wall, Bryant College

We wish to acknowledge the hard work and general competence of the production staff at Prentice-Hall, especially that of Paula Martinac, Eleanor Henshaw Hiatt, and Jayne Conte. Our heartiest thanks go to mathematics editor, Robert Sickles, whose gentle prodding kept us at the task.

Walter Fleming
Dale Varberg

1

NUMBERS
AND THEIR PROPERTIES

Numbers are an indispensable tool of civilization, serving to whip its activities into some sort of order . . . The complexity of a civilization is mirrored in the complexity of its numbers.

Philip J. Davis

"Algebra is a merry science," Uncle Jakob would say. "We go hunting for a little animal whose name we don't know, so we call it *x*. When we bag our game we pounce on it and give it its right name."

Albert Einstein

1-1
What Is Algebra?

Sometimes the simplest questions seem the hardest to answer. One frustrated ninth grader responded, "Algebra is all about *x* and *y*, but nobody knows what they are." Albert Einstein was fond of his Uncle Jakob's definition, which is quoted above. A contemporary mathematician, Morris Kline, refers to algebra as generalized arithmetic. There is some truth in all of these statements, but perhaps Kline's statement is closest to the heart of the matter. What does he mean?

In arithmetic we are concerned with numbers and the four operations of addition, subtraction, multiplication, and division. We learn to understand and manipulate expressions like

$$16 - 11 \qquad \frac{3}{24} \qquad (13)(29)$$

In algebra we do the same thing, but we are more likely to write

$$a - b \qquad \frac{x}{y} \qquad mn$$

without specifying precisely what numbers these letters represent. This determination to stay uncommitted (not to know what x and y are) offers some tremendous advantages. Here are two of them.

GENERALITY AND CONCISENESS

All of us know that $3 + 4$ is the same as $4 + 3$ and that $7 + 9$ equals $9 + 7$. We could fill pages and books, even libraries, with the corresponding facts about other numbers. All of them would be correct and all would be important. But we can achieve the same effect much more economically by writing

$$a + b = b + a$$

The simple formula says all there is to be said about adding two numbers in opposite order. It states a general law and does it on one-fourth of a line.

Or take the well-known facts that if I drive 30 miles per hour for 2 hours, I will travel 60 miles, and that if I fly 120 miles per hour for 3 hours, I will cover 360 miles. These and all other similar facts are summarized in the general formula

$$D = RT$$

which is an abbreviation for

$$\text{distance} = \text{rate} \times \text{time}$$

PROBLEM SOLVING

Uncle Jakob's definition of algebra hinted at something else that is very important. In algebra, as in life, there are many problems. Often they involve finding a number which is initially unknown but which must satisfy certain conditions. If these conditions can be translated into the symbols of algebra, it may take only a simple manipulation to find the answer or, as Uncle Jakob put it, to bag our game. Here is an illustration.

> Roger Longbottom has rented a motorboat for 5 hours from a river resort. He was told that the boat will travel 6 miles per hour upstream and 12 miles per hour downstream. How far upstream can he go and still return the boat to the resort within the allotted 5-hour time period?

We recognize immediately that this is a distance-rate-time problem; the formula $D = RT$ is certain to be important. Now what is it that we want to know? We want to find a distance, namely, how far upstream Roger dares to go. Let us call that distance x miles. Next we summarize the information that is given, keeping in mind that, since $D = RT$, it is also true that $T = D/R$.

	GOING	*RETURNING*
Distance (miles)	x	x
Rate (miles per hour)	6	12
Time (hours)	$x/6$	$x/12$

There is one piece of information we have not used; it is the key to the whole problem. The total time allowed is 5 hours, which is the sum of the time going and the time returning. Thus

$$\frac{x}{6} + \frac{x}{12} = 5$$

After multiplying both sides by 12, we have

$$2x + x = 60$$

$$3x = 60$$

$$x = 20$$

Roger can travel 20 miles upstream and still return within 5 hours.

We intend to emphasize problem solving in this book. To be able to read a mathematics book with understanding is important. To learn to calculate accurately and to manipulate symbols with ease is a worthy goal. But to be able to solve problems, easy problems and hard ones, practical problems and abstract ones, is a supreme achievement.

It is time for you to try your hand at some problems. If some of them seem difficult, do not become alarmed. All of the ideas of this section will be treated in more detail later. As the title "What Is Algebra?" suggests, we wanted to give you a preview of what lies ahead.

Problem Set 1-1

EXAMPLE A (Writing phrases in algebraic notation)   Use the symbols x and y to express the following phrases in algebraic notation.
(a) A number divided by the sum of twice that number and another number.
(b) The sum of 32 and $\frac{9}{5}$ of a Celsius temperature reading.

Solution.
(a) If x is the first number and y is the second, then the given phrase can be expressed by

$$\frac{x}{2x + y}$$

(b) If x represents the Celsius reading, then we can write the given phrase as

$$32 + \frac{9}{5}x$$

Perhaps you recognize this as the Fahrenheit reading corresponding to a Celsius reading of x.

Express each of the phrases on page 5 in algebraic notation using the symbols x and y. Be sure to indicate what x and y represent unless this is already stated in the problem.

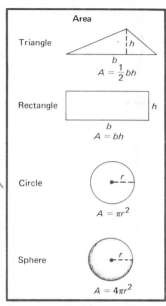

Area

Triangle
$A = \frac{1}{2}bh$

Rectangle
$A = bh$

Circle
$A = \pi r^2$

Sphere
$A = 4\pi r^2$

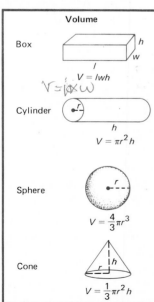

Volume

Box
$V = lwh$

Cylinder
$V = \pi r^2 h$

Sphere
$V = \frac{4}{3}\pi r^3$

Cone
$V = \frac{1}{3}\pi r^2 h$

1. One number plus one-third of another number.
2. The average of two numbers.
3. Twice one number divided by three times another.
4. The sum of a number and its square.
5. Ten percent of a number added to that number.
6. Twenty percent of the amount by which a number exceeds 50.
7. The sum of the squares of two sides of a triangle.
8. One-half the product of the base and the height of a triangle.
9. The distance in miles that a car travels in x hours at y miles per hour.
10. The time in hours it takes to go x miles at y miles per hour.
11. The rate in miles per hour of a boat that traveled y miles in x hours.
12. The total distance a car traveled in 8 hours if its rate was x miles per hour for 3 hours and y miles per hour for 5 hours.
13. The time in hours it took a boat to travel 30 miles upstream and back if its rate upstream was x miles per hour and its rate downstream was y miles per hour.
14. The time in hours it took a boat to travel 30 miles upstream and back if its rate in still water was x miles per hour and the rate of the stream was y miles per hour. Assume that x is greater than y.

In the margin are some formulas from geometry. Use them to express each of the following in algebraic symbols. In each problem, assume that all dimensions are given in terms of the same unit of length (such as a centimeter).

15. The area of a square of side x.
16. The area of a triangle whose height is $\frac{1}{3}$ the length of its base.
17. The surface area of a cube of side x.
18. The area of the surface of a rectangular box whose dimensions are x, $2x$, and $3x$.
19. The surface area of a sphere whose diameter is x.
20. The area of a Norman window whose shape is that of a square of side x topped with a semicircle.
21. The volume of a box with square base of length x and height 10.
22. The volume of a cylinder whose radius and height are both x.
23. The volume of a sphere of diameter x.
24. The volume of what is left of a cylinder of radius 10 and altitude 12 when a hole of radius x is drilled along the center axis of the cylinder. (Assume x is less than 10.)
25. The volume of what is left when a round hole of radius 2 is drilled through a cube of side x. (Assume that the hole is drilled perpendicular to a face and that x is greater than 4.)
26. The volume of what is left of a cylinder when a square hole of width x is drilled along the center axis of the cylinder of radius $3x$ and height y.
27. The area and perimeter of a running track in the shape of a square with semicircular ends if the square has side x. (Recall that the circumference of a circle satisfies $C = 2\pi r$, where r is the radius.)

28. The area of what is left of a circle of radius r after removing an isosceles triangle which has a diameter as a base and the opposite vertex on the circumference of the circle.

EXAMPLE B (Stating sentences as algebraic equations)   Express each of the following sentences as an equation using the symbol x. Then solve for x.
   (a) The sum of a number and three-fourths of that number is 21.
   (b) A rectangular field which is 125 meters longer than it is wide has a perimeter of 650 meters.

Solution.
   (a) Let x be the number. The sentence can be written as

$$x + \frac{3}{4}x = 21$$

If we multiply both sides by 4, we have

$$4x + 3x = 84$$
$$7x = 84$$
$$x = 12$$

It is always wise to make at least a mental check of the solution. The sum of 12 and three-fourths of 12—that is, the sum of 12 and 9—does equal 21.
   (b) Let the width of the field be x meters. Then its length is $x + 125$ meters. Since the perimeter is twice the width plus twice the length, we write

$$2x + 2(x + 125) = 650$$

To solve, we remove parentheses and simplify.

$$2x + 2x + 250 = 650$$
$$4x + 250 = 650$$
$$4x = 400$$
$$x = 100$$

The field is 225 meters long and 100 meters wide.

Express each of the following sentences as an algebraic equation in x and then solve for x. Start by writing down what x represents. (Note: Chapter 3 covers this kind of problem in more detail.)

29. The sum of a number and one-half of that number is 45.
30. Fifteen percent of a number added to that number is 10.35.
31. The sum of two consecutive odd numbers is 168.
32. The sum of three consecutive even numbers is 180.

33. A car going at x miles per hour travels 252 miles in $4\frac{1}{2}$ hours.

34. A circle of radius x centimeters has an area of 64π square centimeters.

Miscellaneous Problems

35. A rectangle 3 meters longer than it is wide has a perimeter of 48 meters. How wide is the rectangle?

36. The area of a rectangle is 54 square feet and its length is $1\frac{1}{2}$ times its width. Find the length.

37. A cube has side length 4 inches. Find its volume and its surface area.

38. The surface area of a certain sphere is the same as the area of a circle of radius 8 centimeters. Find the radius of the sphere.

39. If the radius of a sphere is doubled, what happens to its volume? Its surface area?

40. A boat can go 6 miles per hour in still water. If the rate of the current is 2 miles per hour, how far upstream can this boat go and return in 4 hours? (*Hint:* The rate upstream equals the rate in still water minus the rate of the current; the rate downstream is the sum of those two rates.)

41. Thomas Goodroad knows that his small plane can cruise 150 miles per hour in still air and that his tank holds enough gas for 4 hours of flying. If he takes off on patrol against a north wind of 50 miles per hour, how long can he fly in that direction and then return safely? (See the hint in Problem 40.)

42. On a 200-mile trip, Janet Creighton drove 40 miles per hour the first 100 miles and 60 miles per hour the rest of the way. What is her average speed for the trip?

43. On a 400-kilometer trip, Jonathan drove at 50 kilometers per hour the first 200 kilometers and at 70 kilometers per hour the rest of the way. What was his average speed?

44. Ken Woodward drove from Oregon City to Cozy Harbor, a distance of 200 kilometers, in $2\frac{1}{2}$ hours. What speed must he average on the return trip in order to average 90 kilometers per hour for the entire trip?

"And how many hours a day did you do lessons?" said Alice, in a hurry to change the subject.

"Ten hours the first day," said the Mock Turtle; "nine the next, and so on."

"What a curious plan!" exclaimed Alice.

"That's the reason they're called lessons," the Gryphon remarked, "because they lessen from day to day."

This was quite a new idea to Alice, and she thought it over a little before she made her next remark. "Then the eleventh day must have been a holiday?"

"Of course it was," said the Mock Turtle.

"And how did you manage on the twelfth?" Alice went on eagerly.

"That's enough about lessons," the Gryphon interrupted in a very decided tone; "tell her something about the games now."

From *Alice's Adventures in Wonderland* by Lewis Carroll, mathematician, clergyman, and storyteller

1-2
The Integers and the Rational Numbers

If algebra is generalized arithmetic, then arithmetic is going to be important in this course. That is why we first review the familiar number systems of mathematics.

The human race learned to count before it learned to write; so did most of us. The **whole numbers** 0, 1, 2, 3, . . . entered our vocabularies before we started school. We used them to count our toys, our friends, and our money. Counting forward to larger and larger numbers was no problem; counting backwards was a different matter: 5, 4, 3, 2, 1, 0. But what comes after 0? It bothered Alice in Wonderland, it bothered mathematicians as late as 1300, and it still bothers some people. However, if we are to talk about debts, cold temperatures, and even moon launch countdowns, we must have an answer. Faced with this problem, mathematicians invented a host of new numbers, -1, -2, -3, . . . , called **negative integers.** Together with the whole numbers, they form the system of integers.

THE INTEGERS

The **integers** are the numbers . . . , -5, -4, -3, -2, -1, 0, 1, 2, 3, 4, 5, ... This array is familiar to anyone who has looked at a thermometer lying on its side. In fact, that is an excellent way to think of the integers. There they label points, equally spaced along a line, as in the following diagram.

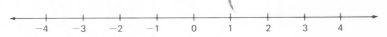

If we are talking about temperatures, -4 means 4 degrees below zero; if we are referring to money, -4 might represent a debt of 4 dollars.

Numbers, used as labels for points, debts, or football players, are significant. But what really makes them useful is our ability to combine them, that is, to add, subtract, and multiply them. All of this is familiar to you, so we shall remind you of only two properties, or rules, that sometimes cause trouble.

(i) $$(-a)(b) = (a)(-b) = -(ab)$$

(ii) $$(-a)(-b) = ab$$

These rules actually make perfectly good sense. For example, if I am flat broke today and am losing 4 dollars each day (that is, gaining -4 dollars each day), then 3 days from now I will be worth $3(-4) = -(3 \cdot 4) = -12$ dollars. On the other hand, 3 days ago (that is, on day -3), I must have been worth 12 dollars; that is, $(-3)(-4) = 3 \cdot 4 = 12$. The chart in the margin shows my worth each day starting 4 days ago and continuing to 4 days from now. Can you tell how much I was worth 6 days ago and how much I will be worth 10 days from now?

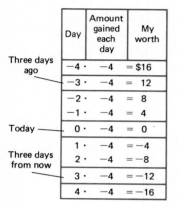

The problem set will give you practice in working with the integers. In the meantime, we note that with all their beauty and usefulness, the integers are plagued by a serious flaw: You cannot always divide them—that is, if you want an integer as the answer. For, while $\frac{4}{2}$, $\frac{9}{3}$, and $\frac{12}{2}$ are perfectly respectable integers, $\frac{1}{2}$, $\frac{3}{4}$, and $\frac{5}{3}$ are not. To make sense of these symbols requires a further enlargement of the number system.

THE RATIONAL NUMBERS

In a certain sense, mathematicians are master inventors. When new kinds of numbers are needed, they invent them. Faced with the need to divide, mathematicians simply decided that the result of dividing an integer by a nonzero integer should be regarded as a number. That meant

$$\frac{3}{4} \qquad \frac{7}{8} \qquad \frac{-2}{3} \qquad \frac{14}{-16} \qquad \frac{8}{2}$$

and all similar ratios were numbers, numbers with all the rights and privileges of the integers and even a bit more: division, except by zero, was always possible. Naturally, these numbers were called rational numbers (ratio numbers). Stated more explicitly, a **rational number** is a number that can be expressed as the ratio of two integers p/q, $q \neq 0$.

The informal use of fractions (ratios) is very old. We know from the Rhind papyrus that the Egyptians were quite proficient with fractions by 1650 B.C. However, they used a different notation and considered only ratios of positive integers.

The rational numbers are admirably suited for certain very practical measurement problems. Take a piece of string of length 1 unit and divide it into two parts of equal length. We say that each part has length $\frac{1}{2}$. Take the same piece of string and divide it into 4 equal parts. Then each part has length $\frac{1}{4}$ and two of them together have length $\frac{2}{4}$. Thus $\frac{1}{2}$ and $\frac{2}{4}$ must stand for the same number—that is,

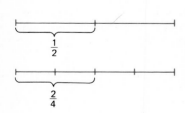

$$\frac{1}{2} = \frac{2 \cdot 1}{2 \cdot 2} = \frac{2}{4}$$

Considerations like this suggest an important agreement. We agree that

$$\frac{a}{b} = \frac{k \cdot a}{k \cdot b}$$

for any nonzero number k. Thus $\frac{1}{2}, \frac{2}{4}, \frac{3}{6}, \frac{-4}{-8}, \ldots$ are all treated as symbols for the same rational number.

We should learn to read equations backwards as well as forwards. Read backwards, the boxed equation tells us that we can divide numerator and denominator (top and bottom) of a quotient by the same nonzero number k. Or, in language that may convey the idea even better, it says that we can cancel a common factor from numerator and denominator.

$$\frac{24}{32} = \frac{\cancel{8} \cdot 3}{\cancel{8} \cdot 4} = \frac{3}{4}$$

Among the many symbols for the same rational number, one is given special honor, the reduced form. If numerator a and denominator b of the rational number a/b have no common integer divisors (factors) greater than 1 and if b is positive, we say a/b is in **reduced form.** Thus $\frac{3}{4}$ is the reduced form of $\frac{24}{32}$, and $-2/3$ is the reduced form of $50/-75$.

We call attention to another rather obvious fact. Notice that $\frac{2}{1}$ is technically the reduced form of $\frac{4}{2}, \frac{6}{3}$, and so on. However, we almost never write $\frac{2}{1}$, since the ordinary meaning of division implies that $\frac{2}{1}$ is equal to the integer 2. In fact, for any integer a,

$$\frac{a}{1} = a$$

Thus the class of rational numbers contains the integers as a subclass.

Let us go back to that horizontal thermometer, the calibrated line we looked at earlier. Now we can label many more points. In fact, it seems that we can fill the line with labels.

No discussion of the rational numbers is complete without mention of how to add, subtract, multiply, and divide them. You are familiar with these operations but perhaps you need to review them. Such a review is one of the major purposes of the following problem set.

Problem Set 1-2

EXAMPLE A (Removing grouping symbols) Simplify $-3[2 - (6 + x)]$.

Solution. Keep these two things in mind. First, always begin work with the innermost parentheses. Second, remember that a minus sign preceeding a set of parentheses (or brackets) means all that is between them must be multiplied by -1, and so the sign of each term must be changed when the parentheses are removed. Thus

$$-3[2 - (6 + x)] = -3[2 - 6 - x] = -3[-4 - x] = 12 + 3x$$

Simplify each of the following

1. $4 - 2(8 - 12)$
2. $-5 + 2(3 - 18)$
3. $-3 + 2[-5 - (12 - 3)]$
4. $-2[1 + 3(6 - 8) + 4] + 3$
5. $-4[3(-6 + 13) - 2(7 - 5)] + 1$
6. $5[-(7 + 12 - 16) + 4] + 2$
7. $-3[4(5 - x) + 2x] + 3x$
8. $-4[x - (3 - x) + 5] + x$
9. $2[-t(3 + 5 - 11) + 4t]$
10. $y - 2[-3(y + 1) + y]$

EXAMPLE B (Reducing fractions) Reduce (a) $\dfrac{24}{36}$; (b) $\dfrac{3 + 9}{3 + 6}$;

(c) $\dfrac{12 + 3x}{3x}$.

Solution. You can cancel common factors (which are multiplied) from numerator and denominator. Do not make the common mistake of trying to cancel common terms (which are added).

(a) $\dfrac{24}{36} = \dfrac{\cancel{12} \cdot 2}{\cancel{12} \cdot 3} = \dfrac{2}{3}$

(b) $\dfrac{3 + 9}{3 + 6} = \dfrac{\cancel{3} + 9}{\cancel{3} + 6} = \dfrac{9}{6}$ Wrong

$\dfrac{3 + 9}{3 + 6} = \dfrac{12}{9} = \dfrac{\cancel{3} \cdot 4}{\cancel{3} \cdot 3} = \dfrac{4}{3}$ Right

(c) $\dfrac{12 + 3x}{3x} = \dfrac{12 + \cancel{3x}}{\cancel{3x}} = 12$ Wrong

$\dfrac{12 + 3x}{3x} = \dfrac{\cancel{3}(4 + x)}{\cancel{3} \cdot x} = \dfrac{4 + x}{x}$ Right

Reduce each of the following, leaving your answer with a positive denominator.

11. $\dfrac{24}{27}$
12. $\dfrac{16}{36}$
13. $\dfrac{45}{-60}$

14. $\dfrac{63}{-81}$ 15. $\dfrac{3 - 9x}{6}$ 16. $\dfrac{4 + 6x}{4 - 8}$

17. $\dfrac{4x - 6}{4 - 8}$ 18. $\dfrac{4x - 12}{8}$

EXAMPLE C (Adding and subtracting fractions) Simplify (a) $\frac{3}{4} + \frac{5}{4}$; (b) $\frac{3}{5} + \frac{5}{4}$; (c) $\frac{8}{9} - \frac{5}{12}$.

Solution. We add fractions with the same denominator by adding numerators. If the fractions have different denominators, we first rewrite them as equivalent fractions with the same denominator and then add. Similar rules apply for subtraction.

(a) $\dfrac{3}{4} + \dfrac{5}{4} = \dfrac{3 + 5}{4} = \dfrac{8}{4} = 2$

(b) $\dfrac{3}{5} + \dfrac{5}{4} = \dfrac{12}{20} + \dfrac{25}{20} = \dfrac{37}{20}$

(c) $\dfrac{8}{9} - \dfrac{5}{12} = \dfrac{32}{36} - \dfrac{15}{36} = \dfrac{17}{36}$

Simplify each of the following.

19. $\dfrac{5}{6} + \dfrac{11}{12}$ 20. $\dfrac{8}{10} - \dfrac{3}{20}$ 21. $\dfrac{4}{5} - \dfrac{3}{20} + \dfrac{3}{10}$

22. $\dfrac{11}{24} + \dfrac{2}{3} - \dfrac{5}{12}$ 23. $\dfrac{5}{12} + \dfrac{7}{18} - \dfrac{1}{6}$ 24. $\dfrac{11}{24} + \dfrac{3}{4} - \dfrac{5}{6}$

25. $\dfrac{-5}{27} + \dfrac{5}{12} + \dfrac{3}{4}$ 26. $\dfrac{23}{30} + \dfrac{2}{25} - \dfrac{3}{5}$

EXAMPLE D (Multiplying and dividing fractions) Simplify (a) $\frac{3}{4} \cdot \frac{5}{7}$; (b) $\frac{3}{4} \cdot \frac{16}{27}$; (c) $\frac{\frac{3}{4}}{\frac{9}{16}}$; (d) $4 \div \frac{5}{3}$.

Solution. We multiply fractions by multiplying numerators and multiplying denominators. To divide fractions, invert (that is, take the reciprocal of) the divisor and multiply.

(a) $\dfrac{3}{4} \cdot \dfrac{5}{7} = \dfrac{3 \cdot 5}{4 \cdot 7} = \dfrac{15}{28}$

(b) $\dfrac{3}{4} \cdot \dfrac{16}{27} = \dfrac{3 \cdot 16}{4 \cdot 27} = \dfrac{3 \cdot 4 \cdot 4}{4 \cdot 3 \cdot 9} = \dfrac{4}{9}$

(c) $\dfrac{\frac{3}{4}}{\frac{9}{16}} = \dfrac{3}{4} \cdot \dfrac{16}{9} = \dfrac{3 \cdot 4 \cdot 4}{4 \cdot 3 \cdot 3} = \dfrac{4}{3}$

(d) $4 \div \dfrac{5}{3} = \dfrac{4}{1} \cdot \dfrac{3}{5} = \dfrac{4 \cdot 3}{1 \cdot 5} = \dfrac{12}{5}$

$$b \div \dfrac{a}{c} = \dfrac{1}{b} \cdot \dfrac{a}{c} = \dfrac{a}{bc}$$

$$b \div \dfrac{a}{c} = b \cdot \dfrac{c}{a} = \dfrac{bc}{a}$$

Simplify.

27. $\dfrac{5}{6} \cdot \dfrac{9}{15}$ 28. $\dfrac{4}{13} \cdot \dfrac{5}{12}$ 29. $\dfrac{3}{4} \cdot \dfrac{6}{15} \cdot \dfrac{5}{2}$ 30. $\dfrac{9}{11} \cdot \dfrac{33}{5} \cdot \dfrac{15}{18}$ 31. $\dfrac{\frac{5}{6}}{\frac{8}{12}}$

32. $\dfrac{\frac{9}{24}}{\frac{15}{12}}$ 33. $\dfrac{\frac{3}{4}}{2}$ 34. $\dfrac{6}{7} \div 9$ 35. $6 \div \dfrac{7}{9}$ 36. $\dfrac{3}{\frac{4}{5}} \cdot \dfrac{7}{5}$

EXAMPLE E (Complicated fractions)   Simplify (a) $\dfrac{\frac{2}{3} + \frac{1}{5}}{\frac{5}{7} - \frac{1}{2}}$;   (b) $\dfrac{\frac{5}{6} - \frac{2}{15}}{\frac{11}{30} + \frac{3}{5}}$.

Solution.   We show two methods of attacking four-story expressions.

(a) Start by working with the top portion and bottom portion separately.

$$\frac{\dfrac{2}{3} + \dfrac{1}{5}}{\dfrac{5}{7} - \dfrac{1}{2}} = \frac{\dfrac{10}{15} + \dfrac{3}{15}}{\dfrac{10}{14} - \dfrac{7}{14}} = \frac{\dfrac{13}{15}}{\dfrac{3}{14}} = \frac{13}{15} \cdot \frac{14}{3} = \frac{182}{45}$$

(b) Multiply the top and bottom portions by a common denominator of all the simple fractions.

$$\frac{\dfrac{5}{6} - \dfrac{2}{15}}{\dfrac{11}{30} + \dfrac{3}{5}} = \frac{30\left(\dfrac{5}{6} - \dfrac{2}{15}\right)}{30\left(\dfrac{11}{30} + \dfrac{3}{5}\right)} = \frac{25 - 4}{11 + 18} = \frac{21}{29}$$

Simplify.

37. $\dfrac{\frac{2}{3} + \frac{3}{4}}{\frac{7}{12}}$ 38. $\dfrac{\frac{3}{5} - \frac{3}{4}}{\frac{9}{20}}$ 39. $\dfrac{\frac{2}{3} + \frac{3}{4}}{\frac{2}{3} - \frac{3}{4}}$

40. $\dfrac{\frac{8}{9} - \frac{2}{27}}{\frac{8}{9} + \frac{2}{27}}$ 41. $\dfrac{\frac{5}{6} - \frac{1}{12}}{\frac{3}{4} + \frac{2}{3}}$ 42. $\dfrac{\frac{3}{50} - \frac{1}{2} + \frac{4}{5}}{\frac{4}{25} + \frac{7}{10}}$

Miscellaneous Problems

Perform the indicated operations and simplify.

43. $\dfrac{5}{6} - \dfrac{1}{9} + \dfrac{2}{3}$ 44. $\dfrac{3}{4} - \dfrac{7}{12} + \dfrac{2}{9}$ 45. $-\dfrac{2}{3}\left(\dfrac{5}{4} - \dfrac{1}{12}\right)$

46. $\dfrac{3}{2}\left(\dfrac{8}{9} - \dfrac{5}{6}\right)$ 47. $\dfrac{1}{3}\left[\dfrac{1}{2}\left(\dfrac{1}{4} - \dfrac{1}{3}\right) + \dfrac{1}{6}\right]$ 48. $-\dfrac{1}{3}\left[\dfrac{2}{5} - \dfrac{1}{2}\left(\dfrac{1}{3} - \dfrac{1}{5}\right)\right]$

49. $\dfrac{14}{33}\left(\dfrac{2}{3} - \dfrac{1}{7}\right)^2$ 50. $\left(\dfrac{5}{7} + \dfrac{7}{9}\right) \div \dfrac{3}{2}$ 51. $1 \div (2 + \frac{3}{4})$

52. $15(2 - \frac{3}{5})$ 53. $\dfrac{\frac{11}{49} - \frac{3}{7}}{\frac{11}{49} + \frac{3}{7}}$ 54. $\dfrac{\frac{1}{2} - \frac{3}{4} + \frac{7}{8}}{\frac{1}{2} + \frac{3}{4} - \frac{7}{8}}$

55. $2 + \dfrac{\frac{3}{4} + \frac{5}{12}}{\frac{2}{3}}$ 56. $\dfrac{11}{12} - \dfrac{\frac{2}{3} - \frac{3}{4}}{\frac{5}{6} + \frac{1}{12}}$ 57. $1 - \dfrac{2}{2 + \frac{3}{4}}$

58. $2 + \dfrac{3}{1 + \frac{5}{2}}$

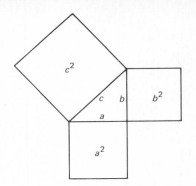

1-3
The Real Numbers

The Pythagorean theorem, that beautiful gem of geometry, came to be the great nemesis of Greek mathematics. For hidden in that elegant formula is a consequence that seemed to destroy the Greek conception of numbers. Here is a paraphrased version of an old story reported by Euclid.

"Consider," said Euclid, "a right triangle with legs of unit length and hypotenuse of length C. By the theorem of Pythagoras,

$$C^2 = 1^2 + 1^2 = 2$$

That is, $C = \sqrt{2}$. But then, as I will show by a logical argument, the quantity $\sqrt{2}$, whatever it is, is not a rational number. To put it bluntly, if only rational numbers exist, then the simplest right triangle we know has a hypotenuse whose length cannot be measured."

Euclid's argument (soon to be presented) is flawless; it is also devilishly clever. But to understand it, we need to know some basic facts about prime numbers.

A DIGRESSION ON PRIME NUMBERS

A **prime number** is a whole number with exactly two whole-number divisors, itself and 1. The first few primes are

$$2, 3, 5, 7, 11, 13, 17, 19, 23, 29$$

Prime numbers are the building blocks of other whole numbers. For example,

$$18 = 2 \cdot 3 \cdot 3 \qquad 40 = 2 \cdot 2 \cdot 2 \cdot 5$$

This type of factorization is possible for all nonprime whole numbers greater than 1.

FUNDAMENTAL THEOREM OF ARITHMETIC

Any nonprime whole number (greater than 1) can be written as the product of a unique set of prime numbers.

This theorem is important in many parts of mathematics (see the problem set) but here we want to point out one simple consequence. When the square of any whole number is written as a product of primes, each prime occurs as a factor an even number of times. For example,

$$(18)^2 = 18 \cdot 18 = 2 \cdot 3 \cdot 3 \cdot 2 \cdot 3 \cdot 3 \cdot = \underbrace{2 \cdot 2}_{\text{two 2's}} \cdot \underbrace{3 \cdot 3 \cdot 3 \cdot 3}_{\text{four 3's}}$$

$$(40)^2 = 40 \cdot 40 = 2 \cdot 2 \cdot 2 \cdot 5 \cdot 2 \cdot 2 \cdot 2 \cdot 5 = \underbrace{2 \cdot 2 \cdot 2 \cdot 2 \cdot 2 \cdot 2}_{\text{six 2's}} \cdot \underbrace{5 \cdot 5}_{\text{two 5's}}$$

THE PROOF THAT $\sqrt{2}$ IS IRRATIONAL

Suppose that $\sqrt{2}$ is a rational number; that is, suppose that $\sqrt{2} = m/n$, where m and n are whole numbers (necessarily greater than 1). Then

$$2 = \frac{m^2}{n^2}$$

and so

$$2n^2 = m^2$$

Now imagine that both n and m are written as products of primes. As we saw above, both n^2 and m^2 must then have an even number of 2's as factors. Thus in the above equation, the prime 2 appears on the left an odd number of times but on the right an even number of times. This is clearly impossible. What can be wrong? The only thing that can be wrong is our supposition that $\sqrt{2}$ was a rational number.

To let one number, $\sqrt{2}$, through the dike was bad enough. But, as Euclid realized, a host of others came pouring through with it. Exactly the same proof shows that $\sqrt{3}$, $\sqrt{5}$, $\sqrt{7}$, and, in fact, the square roots of all primes are irrational. The Greeks, who steadfastly insisted that all measurements must be based on whole numbers and their ratios, could not find a satisfactory way out of this dilemma. Today we recognize that the only adequate solution is to enlarge the number system.

> Reductio ad absurdum, which Euclid loved so much, is one of a mathematician's finest weapons. It is a far finer gambit than any chess gambit: a chess player may offer the sacrifice of a pawn or even a piece, but a mathematician offers the game.
>
> G. H. Hardy

THE REAL NUMBERS

Let us take a bold step and simply declare that every line segment shall have a number that measures its length. The set of all numbers that can measure lengths, together with their negatives and zero, constitute the **real numbers.** Thus the rational numbers are automatically real numbers; the positive rational numbers certainly measure lengths.

Consider again the thermometer on its side, the calibrated line. We may have thought we had labeled every point. Not so; there were many holes corresponding to what we now call $\sqrt{2}$, $\sqrt{5}$, π, and so on. But with the introduction of the real numbers, all the holes are filled in; every point has a number label. Because of this, we often call this calibrated line the **real line.**

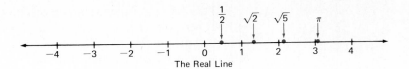

The Real Line

DECIMALS

There is another important way to describe the real numbers. It calls for review of a basic idea. Recall that

$$.4 = \frac{4}{10} \qquad .7 = \frac{7}{10}$$

Similarly,

$$.41 = \frac{4}{10} + \frac{1}{100} = \frac{40}{100} + \frac{1}{100} = \frac{41}{100}$$

$$.731 = \frac{7}{10} + \frac{3}{100} + \frac{1}{1000} = \frac{700}{1000} + \frac{30}{1000} + \frac{1}{1000} = \frac{731}{1000}$$

It is a simple matter to locate a decimal on the number line. For example, to locate 1.4, we divide the interval from 1 to 2 into 10 equal parts and pick the fourth point of division.

If the interval from 1.4 to 1.5 is divided into 10 equal parts, the second point of division corresponds to 1.42.

We can find the decimal corresponding to a rational number by long division. For example, the division in the margin shows that $\frac{7}{8} = .875$. If we try the same procedure on $\frac{1}{3}$, something different happens. The decimal just keeps on going; it is an **unending decimal.** Actually, the terminating decimal .875 can be thought of as unending if we annex zeros. Thus

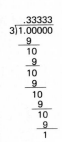

$$\frac{7}{8} = .875 = .8750000 \ldots$$

$$\frac{1}{3} = .3333333 \ldots$$

$$\frac{1}{6} = .16666 \ldots$$

Let us take an example that is a bit more complicated, $\frac{2}{7}$.

$$
\begin{array}{r}
.285714 \\
7\,\overline{)\,2.0000000} \\
\underline{14} \\
60 \\
\underline{56} \\
40 \\
\underline{35} \\
50 \\
\underline{49} \\
10 \\
\underline{7} \\
30 \\
\underline{28} \\
20
\end{array}
$$

If we continue the division, the pattern must repeat (note the circled 20's). Thus

$$
\frac{2}{7} = .285714285714285714\ldots
$$

which can also be written

$$
\frac{2}{7} = .\overline{285714}
$$

The bar indicates that the block of digits of 285714 repeats indefinitely.

In fact, the decimal expansion of any rational number must inevitably start repeating, because there are only finitely many different possible remainders in the division process (at most as many as the divisor). It is a remarkable coincidence that the converse statement is also true. A repeating decimal must inevitably represent a rational number (see Problems 25–31). Thus the rational numbers are precisely those numbers that can be represented by repeating decimals.

What about the nonrepeating, unending decimals like

$$
.12112111211112\ldots
$$

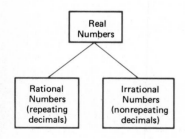

They represent the **irrational numbers.** And they, together with the rational numbers, constitute the real numbers.

We showed that $\sqrt{2}$ is not rational (that is, it is irrational). It too has a decimal expansion.

$$
\sqrt{2} = 1.414214\ldots
$$

Actually, the decimal expansion of $\sqrt{2}$ is known to several thousand places. It does not repeat. It cannot. It is a fact of mathematics.

We have supposed that you are familiar with the symbols $\sqrt{2}, \sqrt{3}, \ldots$ from earlier courses. For completeness, we now make the meaning of these symbols precise.

SQUARE ROOTS AND CUBE ROOTS

Every positive number has two square roots. For example, 3 and -3 are the two square roots of 9; this is so because $3^2 = 9$ and $(-3)^2 = 9$. If a is positive, the symbol $\sqrt{a}$ always denotes the *positive* square root of a. Thus $\sqrt{9} = 3$, and the two square roots of 7 are $\sqrt{7}$ and $-\sqrt{7}$. For the present, $\sqrt{-9}$ is a meaningless symbol since there is no real number whose square is -9. To make sense of the square root of a negative number requires a further extension of the number system, a topic we take up in Section 1-6.

In contrast, $\sqrt[3]{a}$, called the cube root of a, makes sense for any real number a. It is the unique real number whose cube is a. Thus $\sqrt[3]{8} = 2$ since $2^3 = 8$, and $\sqrt[3]{-64} = -4$ since $(-4)^3 = -64$.

We postpone the discussion of general nth roots until Section 6-1. In the meantime, we shall use square roots and cube roots in examples and problems.

Problem Set 1-3

EXAMPLE A (Prime factorization)   Find the prime factorizations of 168 and 420.

Solution.   Factor out as many 2's as possible, then 3's, then 5's, and so on.

$$168 = 2 \cdot 84 = 2 \cdot 2 \cdot 42 = 2 \cdot 2 \cdot 2 \cdot 21 = 2 \cdot 2 \cdot 2 \cdot 3 \cdot 7$$
$$420 = 2 \cdot 210 = 2 \cdot 2 \cdot 105 = 2 \cdot 2 \cdot 3 \cdot 35 = 2 \cdot 2 \cdot 3 \cdot 5 \cdot 7$$

Write the prime factorization of each number.

1. 250   2. 504   3. 200   4. 2079   5. 2100   6. 1650

EXAMPLE B (Least common multiple)   The *least common multiple* (lcm) of several positive integers is the smallest positive integer that is a multiple of all of them. Find the least common multiple of 168 and 420.

Solution.   We found the prime factorizations of these two numbers in Example A. To find the least common multiple of 168 and 420, write down the product of all factors that occur in either number, repeating a factor according to the greatest number of times it occurs in either number.

$$\text{lcm}(168, 420) = 2 \cdot 2 \cdot 2 \cdot 3 \cdot 5 \cdot 7 = 840$$

Find each of the following using the information you obtained in Problems 1–6.

7. lcm(250, 200)   8. lcm(504, 2079)   9. lcm(250, 2100)
10. lcm(504, 1650)   11. lcm(250, 200, 2100)   12. (lcm(504, 2079, 1650)

EXAMPLE C (Least common denominator)   Calculate $\frac{5}{168} + \frac{13}{420}$.

Solution.   Write both fractions with a common denominator. The best choice of common denominator is the least common multiple of 168 and

420—namely, 840—obtained in Example B. We call it the *least common denominator*.

$$\frac{5}{168} + \frac{13}{420} = \frac{5 \cdot 5}{840} + \frac{13 \cdot 2}{840} = \frac{25 + 26}{840} = \frac{51}{840} = \frac{17}{280}$$

Calculate each of the following, using the results of Problems 7–12.

13. $\dfrac{3}{250} + \dfrac{17}{200}$ 14. $\dfrac{5}{504} - \dfrac{1}{2079}$ 15. $\dfrac{7}{250} - \dfrac{1}{2100}$

16. $\dfrac{13}{504} + \dfrac{13}{1650}$ 17. $\dfrac{3}{250} - \dfrac{17}{200} + \dfrac{11}{2100}$ 18. $\dfrac{13}{504} + \dfrac{13}{1650} - \dfrac{17}{2079}$

EXAMPLE D (Rational numbers as repeating decimals)   Write $\frac{68}{165}$ as a repeating decimal.

Solution.

$$
\begin{array}{r}
.412 \\
165\overline{)68.0000} \\
660 \\
\hline
200 \\
165 \\
\hline
350 \\
330 \\
\hline
200
\end{array}
$$

Answer: $\dfrac{68}{165} = .41212 \ldots = .4\overline{12}$

Find the repeating decimal expansion for each number. Use the bar notation for your answer.

19. $\frac{2}{3}$ 20. $\frac{3}{5}$ 21. $\frac{5}{8}$ 22. $\frac{13}{11}$ 23. $\frac{6}{13}$ 24. $\frac{4}{13}$

EXAMPLE E (Repeating decimals as rational numbers)   Write $.\overline{24}$ as the ratio of two integers.

Solution.   Let $x = .\overline{24} = .242424 \ldots$. Then $100x = 24.242424 \ldots$. Subtract x from $100x$ and simplify.

$$100x = 24.2424 \ldots$$
$$x = .2424 \ldots$$
$$99x = 24$$
$$x = \frac{24}{99} = \frac{8}{33}$$

We multiplied x by 100 because x is a decimal that repeats in a two-digit group. If the decimal had repeated in a three-digit group, we would have multiplied by 1000.

Write each of the following as a ratio of two integers.

25. $.\overline{7}$ 26. $.\overline{123}$ 27. $.2\overline{35}$ 28. $.875$

29. $.3\overline{25}$ 30. $.5\overline{21}$ 31. $.3\overline{21}$

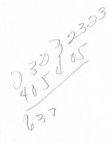

Miscellaneous Problems

32. Find the prime factorization of 420 and 750.

33. Find the least common multiple of 420 and 750. Use the result of Problem 32.

34. Calculate each of the following.

(a) $\dfrac{17}{420} + \dfrac{11}{750}$ (b) $\dfrac{17}{420} \div \dfrac{11}{750}$

35. Find the repeating decimal expansion of each of the following.

(a) $\dfrac{5}{16}$ (b) $\dfrac{3}{11}$

36. Write as a ratio of two integers.
(a) $.\overline{27}$ (b) $.2\overline{07}$

37. Write the sum $.\overline{23} + .\overline{405}$ as a repeating decimal.

38. Is the sum of two rational numbers necessarily rational?

39. Is the sum of two irrational numbers necessarily irrational?

40. Answer the questions 38 and 39 with the word *sum* replaced by *product*.

41. Show that $\sqrt{2} + \frac{3}{4}$ is irrational (*Hint:* Let $\sqrt{2} + \frac{3}{4} = r$. By adding $-\frac{3}{4}$ to both sides, show that it is impossible for r to be rational.)

42. Show that if x is irrational and y is rational, then $x + y$ is irrational.

43. Show that $\frac{3}{4}\sqrt{2}$ is irrational.

44. Show that if x is a nonzero rational number and y is an irrational number, then xy is irrational.

45. Which of the following are rational numbers?
(a) $\sqrt{\frac{8}{18}}$ (b) $(3\sqrt{2})(\sqrt{2})$
(c) $\sqrt{2}(\sqrt{2} + 1)$ (d) $(3 + 2\sqrt{2}) + (4 - 2\sqrt{2})$
(e) $\dfrac{4\sqrt{2} + 2}{2\sqrt{2} + 1}$ (f) $(.12)\sqrt{2}$
(g) $.\overline{12}$ (h) $.12112111211112\ldots$

46. What is the smallest positive integer?

47. Write a positive rational number that is smaller than .00000001. Is there a smallest positive rational number?

48. The number $(1/10{,}000{,}000)\sqrt{2}$ is irrational (see Problem 44 of this set). Write a smaller positive irrational number. Is there a smallest positive irrational number?

49. Mimic the proof that $\sqrt{2}$ is irrational (see the first part of this section) to show that $\sqrt{3}$ is irrational.

50. Show that the square root of any prime number is irrational.

51. It is known that π is irrational. What does this mean about its decimal expansion?

52. Show that $\sqrt{\pi}$ is irrational.

53. The beginning of the decimal expansion of π is

$$\pi = 3.14159 \ldots$$

Is $\pi - \frac{22}{7}$ positive, negative, or zero?

Playing by the Rules

Chess has rules; so does mathematics. You can't move a pawn backwards; you must not divide by zero. Chess without rules is no game at all. Numbers without definite properties are worthless curiosities. A good chess player and a good mathematician play by the rules.

1-4 Fundamental Properties of the Real Numbers

Now we face an important question. Are the real numbers adequate to handle all the applications we are likely to encounter? Or will some problems arise that will force us to enlarge our number system again? The answer is that the real numbers are sufficient for most purposes. Except for one type of problem, to be described in Section 1-6, we will do our work within the context of the real numbers. From now on, when we say number with no qualifying adjective, we mean real number. You can count on it.

All of this suggests that we ought to pay some attention to the fundamental properties of the real numbers. By a fundamental property, we mean something so basic that we must understand it, and to understand means more than to memorize. Understanding a property means to see the purpose the property serves, to recognize its implications, and to be able to derive other things from it.

ASSOCIATIVE PROPERTY

Addition and multiplication are the fundamental operations; subtraction and division are offshoots of them. These operations are binary operations, that is, they work on two numbers at a time. Thus $3 + 4 + 5$ is technically meaningless. We ought to write either $3 + (4 + 5)$ or $(3 + 4) + 5$. But, luckily, it really does not matter; we get the same answer either way. Addition is **associative.**

$$a + (b + c) = (a + b) + c$$

Thus we can write $3 + 4 + 5$ or even $3 + 4 + 5 + 6 + 7$ without ambiguity. The answer will be the same regardless of the order in which the addition is done.

Addition and multiplication are like Siamese twins: What is true for one is quite likely to be true for the other. Thus multiplication, too, is associative.

$$a \cdot (b \cdot c) = (a \cdot b) \cdot c$$

If we wish, we can write $a \cdot b \cdot c$ with no parentheses at all.

COMMUTATIVE PROPERTY

It makes some difference whether you first put on your slippers and then take a bath or vice versa—the difference between wet and dry slippers. But for addition or multiplication, order does not matter. Both operations are **commutative.**

$$a + b = b + a$$

$$a \cdot b = b \cdot a$$

Thus

$$3 + 4 = 4 + 3$$
$$3 + 4 + 6 = 4 + 3 + 6 = 6 + 4 + 3$$

and

$$7 \cdot 8 \cdot 9 = 8 \cdot 7 \cdot 9 = 9 \cdot 8 \cdot 7$$

NEUTRAL ELEMENTS

To be neutral is to sit on the sidelines and refuse to do battle. A neutral party can be ignored; the outcome is not affected by its presence. Thus we call 0 the **neutral element** for addition; its presence can be ignored in an addition.

$$a + 0 = 0 + a = a$$

then $0 = 0 \cdot p$, which is true for any number p. We choose to avoid such ambiguities by excluding division by zero.

Problem Set 1-4

EXAMPLE A (Simple calculations)   Calculate $31.9 + 45 + 68.1 + 155 + 43.2$.

Solution.   Intelligent use of the associative and commutative properties makes this a breeze.

$$(31.9 + 68.1) + (45 + 155) + 43.2 = 100 + 200 + 43.2 = 343.2$$

Find the following sums or products using the basic properties we have introduced. If you can do it all in your head, that will be fine.

1. $420 + 431 + 580$
2. $99,985 + 67 + 15$
3. $983 + 400 + 300 + 17$
4. $8.75 + 14 + 36 + 1.25$
5. $\frac{11}{13} + 43 + \frac{2}{13} + 17$
6. $\frac{15}{8} + \frac{5}{6} + \frac{1}{6} + \frac{9}{8}$
7. $6 \cdot \frac{3}{4} \cdot \frac{1}{6} \cdot 4$
8. $99 + 98 + 97 + 3 + 2 + 1$
9. $5 \cdot \frac{1}{3} \cdot \frac{2}{5} \cdot 6 \cdot \frac{1}{2}$
10. $(.25)(363)(400)(\frac{1}{3})$

EXAMPLE B (Number properties)   What properties justify each of the following?
(a) $6 + [5 + (-6)] = [6 + (-6)] + 5$
(b) $-\frac{1}{3}(\frac{6}{7} - \frac{9}{11}) = -\frac{2}{7} + \frac{3}{11}$

Solution.
(a) Commutative and associative properties for addition:

$$6 + [5 + (-6)] = 6 + [(-6) + 5] = [6 + (-6)] + 5$$

(b) The definition of subtraction and the distributive property:

$$-\frac{1}{3}(\frac{6}{7} - \frac{9}{11}) = -\frac{1}{3}(\frac{6}{7} + (-\frac{9}{11})) = -\frac{1}{3} \cdot \frac{6}{7} + (-\frac{1}{3})(-\frac{9}{11}) = -\frac{2}{7} + \frac{3}{11}$$

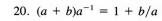 *Name the properties that justify each of the following.*

11. $\frac{5}{6}(\frac{3}{4} \cdot 12) = (12 \cdot \frac{5}{6})\frac{3}{4}$
12. $4 + (.52 - 2) = (4 - 2) + .52$
13. $(2 + 3) + 4 = 4 + (2 + 3)$
14. $9(48) = 9(40) + 9(8)$
15. $6 + [-6 + 5]$
    $= [6 + (-6)] + 5 = 5$
16. $6[\frac{1}{6} \cdot 4] = [6 \cdot \frac{1}{6}]4 = 4$
17. $-(\sqrt{5} + \sqrt{3} - 5)$
    $= -\sqrt{5} - \sqrt{3} + 5$
18. $(b/c)(b/c)^{-1} = 1$
19. $(x + 4)(x + 2)$
    $= (x + 4)x + (x + 4)2$
20. $(a + b)a^{-1} = 1 + b/a$

EXAMPLE C (More on properties)   Which of the following are true for all real numbers a, b, and c?

(a) $(a - b) - c = a - (b - c)$

(b) $(a + b) \div c = (a \div c) + (b \div c)$

Solution.

(a) Not true; subtraction is not associative. For example, if $a = 12$, $b = 9$, and $c = 5$, then

$$(a - b) - c = (12 - 9) - 5 = 3 - 5 = -2$$

$$a - (b - c) = 12 - (9 - 5) = 12 - 4 = 8$$

(b) True, provided $c \neq 0$:

$$(a + b) \div c = (a + b)\left(\frac{1}{c}\right) = a\left(\frac{1}{c}\right) + b\left(\frac{1}{c}\right)$$

$$= (a \div c) + (b \div c)$$

Which of the equalities in Problems 21–32 are true for all choices of a, b, and c? If false, provide an example. If true, provide a demonstration similar to that in part (b) above.

21. $a - (b - c) = a - b + c$
22. $a + bc = ac + bc$
23. $a \div (b + c)$
    $= (a \div b) + (a \div c)$
24. $(a + b)^{-1} = a^{-1} + b^{-1}$
25. $ab(a^{-1} + b^{-1}) = b + a$
26. $a(a + b + c) = a^2 + ab + ac$
27. $(a + b)(a^{-1} + b^{-1}) = 1$
28. $(a + b)(a + b)^{-1} = 1$
29. $(a + b)(a + b) = a^2 + b^2$
30. $(a + b)(a - b) = a^2 - b^2$
31. $a \div (b \div c) = (a \div b) \div c$
32. $a \div (b \div c) = (a \cdot c) \div b$

33. We claimed that we could show that $a \cdot 0 = 0$ for all a using only the properties displayed in boxes. Justify each of the following equalities by one of those properties.

$$0 = -(a \cdot 0) + a \cdot 0$$

$$= -(a \cdot 0) + a \cdot (0 + 0)$$

$$= -(a \cdot 0) + (a \cdot 0 + a \cdot 0)$$

$$= [-(a \cdot 0) + a \cdot 0] + a \cdot 0$$

$$= 0 + a \cdot 0$$

$$= a \cdot 0$$

34. Demonstrate that $(-a) \cdot b = -(a \cdot b)$ and $(-a) \cdot (-b) = a \cdot b$ using only the properties displayed in boxes and the result of Problem 33.

35. Let # represent the exponentiation operation, that is, $a \# b = a^b$. Is # commutative? Is # associative?

"I wonder, sir, if you would be willing to move to the end?"

1-5 Order and Absolute Value

We may question the bartender's tact, but not his mathematical taste. He has a deep feeling for an important notion that we call order; it is intimately tied up with the real number system. To describe this notion, we introduce a special symbol $<$; it stands for the phrase "is less than."

We begin by recalling that every real number (except 0) falls into one of two classes. Either it is positive or it is negative. Then, given two real numbers a and b, we say that

$$a < b \text{ if } b - a \text{ is positive}$$

Thus $-3 < -2$ since $-2 - (-3)$ is positive. Similarly, $3 < \pi$ since

$$\pi - 3 = 3.14159 \ldots - 3 = .14159 \ldots$$

which is a positive number.

Another and more intuitive way to think about the symbol $<$ is to relate it to the real number line. To say that $a < b$ means that a is to the left of b on the real line.

The symbol $<$ has a twin sister, denoted by $>$, which is read "is greater than." If you know how $<$ behaves, you automatically know how $>$ behaves. Thus there is no need to say much about $>$. It is enough to note that $b > a$ means exactly the same thing as $a < b$. In particular, $b > 0$ and $0 < b$ mean the same thing; both say that b is a positive number. Relations like $a < b$ and $b > a$ are called **inequalities.**

PROPERTIES OF $<$

If Homer is shorter than Ichabod and Ichabod is shorter than Jehu, then of course Homer is shorter than Jehu. This and other properties of the "less than"

Homer   Ichabod   Jehu

relation seem almost obvious. The following is a formal statement of the three most important properties.

PROPERTIES OF INEQUALITIES

1. (Transitivity). If $a < b$ and $b < c$, then $a < c$.
2. (Addition). If $a < b$, then $a + c < b + c$.
3. (Multiplication). If $a < b$ and $c > 0$, then $a \cdot c < b \cdot c$
   If $a < b$ and $c < 0$, then $a \cdot c > b \cdot c$

Property 2 says that you can add the same number to both sides of an inequality. It also says that you can subtract the same number from both sides, since c can be negative. Property 3 has a catch. Notice that if we multiply both sides by a positive number, we preserve the direction of the inequality; however, if we multiply by a negative number, we reverse the direction of the inequality. Thus

$$2 < 3$$

and

$$2 \cdot 4 < 3 \cdot 4$$

but

$$2(-4) > 3(-4)$$

These facts are shown on the following number line.

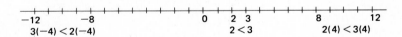

Division, which is equivalent to multiplication by the reciprocal, also satisfies Property 3.

THE ≤ RELATION

In addition to its twin sister $>$, the symbol $<$ has a half sister. It is denoted by $\leq$ and is read "is less than or equal to." We say

$a \leq b$ if $b - a$ is either positive or zero

For example, it is correct to say $2 \leq 3$; it is also correct to say $2 \leq 2$. This new relation behaves very much like $<$. In fact, if we put a bar under every $<$ and $>$ in the properties of inequalities displayed above, the resulting statements will be correct. Naturally, $b \geq a$ means the same thing as $a \leq b$.

INTERVALS

We can use the order symbols $<$ and $\leq$ to describe intervals on the real line. When we write $-1.5 < x \leq 3.2$, we mean that x is simultaneously greater than -1.5 and less than or equal to 3.2. The set of all such numbers is the interval shown below.

The small circle at the left indicates that -1.5 is left out; the heavy dot at the right indicates that 3.2 is included. The following diagram illustrates other possibilities.

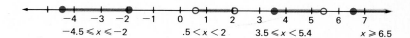

Sometimes we use set notation to describe an interval. For example, to denote the interval at the far left above, we could write $\{x : -4.5 \leq x \leq -2\}$, which is read "the set of all x such that x is greater than or equal to -4.5 and less than or equal to -2."

We should not write nonsense such as

$$3 < x < 2 \quad \text{or} \quad 2 > x < 3$$

The first is simply a contradiction. There is no number both greater than 3 and less than 2. The second says that x is both less than 2 and less than 3. This should be written simply as $x < 2$.

ABSOLUTE VALUE — *always positive*

Often we want to describe the size of a number, not caring whether it is positive or negative. To do this, we introduce the concept of absolute value, symbolized by two vertical bars $|\ \ |$. It is defined by

$$|a| = \begin{cases} a & \text{if } a \geq 0 \\ -a & \text{if } a < 0 \end{cases}$$

This two-pronged definition can cause confusion. It says that if a number is positive or zero, its absolute value is itself. But if a number is negative, then its absolute value is its additive inverse (which is a positive number). For example,

$$|7| = 7$$

since $a = 7$ is positive. On the other hand,

$$|-7| = -(-7) = 7$$

since $a = -7$ is negative. Note that $|0| = 0$.

We may also think of $|a|$ geometrically. It represents the distance between a and 0 on the number line. More generally, $|a - b|$ is the distance between a and b. The number $|b - a|$ represents this same distance.

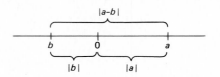

You should satisfy yourself that this is true for arbitrary choices of a and b, for example, that $|7 - (-3)|$ really is the distance between 7 and -3.

The properties of absolute values are straightforward and easy to remember.

PROPERTIES OF ABSOLUTE VALUES

1. $|a \cdot b| = |a| \cdot |b|$

2. $\left|\dfrac{a}{b}\right| = \dfrac{|a|}{|b|}$

3. $|a + b| \leq |a| + |b|$

There is an important connection between absolute values and square roots, namely,

$$\sqrt{a^2} = |a|$$

For example, $\sqrt{6^2} = |6| = 6$ and $\sqrt{(-6)^2} = |-6| = 6$.

Problem Set 1-5

In Problems 1–12, replace the symbol # by the appropriate symbol: $<$, $>$, or $=$.

1. $1.5 \ \# \ -1.6$ 2. $-2 \ \# \ -3$

3. $\sqrt{2} \ \# \ 1.4$ 4. $\pi \ \# \ 3.15$

5. $\frac{1}{5} \ \# \ \frac{1}{6}$ 6. $-\frac{1}{5} \ \# \ -\frac{1}{6}$

7. $5 - \sqrt{2} \ \# \ 5 - \sqrt{3}$ 8. $\sqrt{2} - 5 \ \# \ \sqrt{3} - 5$

9. $-\frac{3}{16}\pi \ \# \ -\frac{3}{17}\pi$ 10. $\left(\frac{16}{17}\right)^2 \ \# \ \left(\frac{17}{18}\right)^2$

11. $|-\pi + (-2)| \ \# \ |-\pi| + |-2|$ 12. $|\pi - 2| \ \# \ \pi - 2$

13. Order the following numbers from least to greatest.

$$\frac{3}{4}, \ -2, \ \sqrt{2}, \ \frac{-\pi}{2}, \ -\frac{3}{2}\sqrt{2}, \ \frac{43}{24}$$

14. Order the following numbers from least to greatest.

$$5 - 5, .37, -\sqrt{3}, \frac{14}{33}, -\frac{7}{4}, -\frac{49}{35}, \frac{3}{8}$$

Use a real number line to show the set of numbers that satisfy each given inequality.

15. $x < -4$ 16. $x < 3$ 17. $x \geq -2$

18. $x \leq 3$ 19. $-1 < x < 3$ 20. $2 < x < 5$

21. $0 < x \leq 3$ 22. $-3 \leq x < 2$ 23. $-\frac{1}{2} \leq x \leq \frac{3}{2}$

24. $-\frac{7}{4} \leq x \leq -\frac{3}{4}$

Write an inequality for each interval.

25.   (number line from -2 to 3, closed dots at -2 and 3)   26. (number line from -1 to 3, closed dots at -1 and 1)

27.   (number line from -2 to 3, closed dot at 1)   28. (number line from -1 to 3, open dot at 1)

29.   (number line from -2 to 3, open dot at -2, closed dot at 2)   30. (number line from -1 to 3, open dots at -1 and 1)

31.   (number line from -2 to 3, open dots at -1 and 2)   32. (number line from -1 to 3, open dot at -1, closed dot at 2)

EXAMPLE (Removing absolute values) Write each of the following without the absolute value symbol. Then show the corresponding interval(s) on the real number line.

(a) $|x| < 3$ (b) $|x - 2| < 3$ (c) $|x - 4| \geq 2$

Solution.

(a) Since $|x|$, the distance between x and 0 on the number line, is less than 3, x must be between -3 and 3. We write $-3 < x < 3$.

(b) Here $x - 2$, instead of x, must be between -3 and 3, that is,

$$-3 < x - 2 < 3$$

Adding 2 to each quantity gives

$$-1 < x < 5$$

(c) The distance between x and 4 is greater than or equal to 2. This means that $x - 4 \leq -2$ or $x - 4 \geq 2$. That is,

$$x \leq 2 \quad \text{or} \quad x \geq 6$$

In set notation, we may write the solution set for the inequality in part (c) as $\{x : x \le 2 \ \text{or} \ x \ge 6\}$ or equivalently as $\{x : x \le 2\} \cup \{x : x \ge 6\}$.

Write each of the following without the absolute value symbol. Show the corresponding interval(s) on the real number line.

33. $|x| \le 4$ 34. $|x| \ge 2$ 35. $|x - 3| < 2$

36. $|x - 5| < 1$ 37. $|x + 1| \le 3$ 38. $|x + \frac{3}{2}| \le \frac{1}{2}$

39. $|x - 5| > 5$ 40. $|x + 3| < 3$

Miscellaneous Problems

In Problems 41–48, show on the number line the set of numbers that satisfy the given inequality (or inequalities).

41. $x \le -2$ 42. $x > 9$ 43. $-4 < x \le 2$

44. $2 \le x < 7$ 45. $|x + 3| < -4$ 46. $|x - 2| \le 3$

47. $|x - 5| > 2$ 48. $|x - 2| \ge 3$

49. If $a < b$, does it follow that $a^2 < b^2$?

50. If $0 < a < b$, does it follow that $a^2 < ab$? That $ab < b^2$? That $a^2 < b^2$?

51. If $0 < a < b$, how are $1/a$ and $1/b$ related to each other?

52. If $0 < a < b$, how are $\sqrt{a}$ and $\sqrt{b}$ related to each other?

53. Note that $-1 < x < 7$ can be rewritten as $|x - 3| < 4$. Rewrite each of the following double inequalities as a single inequality involving an absolute value.
    (a) $2 < x < 10$ (b) $-6 \le x \le 4$
    (c) $3 < x < 6$ (d) $-3 < x < 6$

54. Use Property 1 of absolute values to show that $|-a| = |a|$. (*Hint:* $-a = (-1)(a)$.)

55. Use Property 1 of absolute values to show that $|a|^2 = a^2$.

56. Show that $|a| < 1$ implies $a^2 < |a|$, but $|a| > 1$ implies $a^2 > |a|$.

57. Show that $|a + b + c| \le |a| + |b| + |c|$.

58. Show that if $|a| \le 1$, then $|a + 2a^2 + 3a^3| \le 6$.

59. Under what conditions is $|a + b| = |a| + |b|$?

60. Show that $\sqrt{a^2 + b^2} \le |a| + |b|$.

$$i = \sqrt{-1}$$

"The Divine Spirit found a sublime outlet in that wonder of analysis, that portent of the ideal world, that amphibian between being and not-being, which we call the imaginary root of negative unity."

Gottfried Wilhelm Leibniz
1646–1716

1-6
The Complex Numbers

As early as 1550, the Italian mathematician Raffael Bombelli had introduced numbers like $\sqrt{-1}$, $\sqrt{-2}$, and so on, to solve certain equations. But mathematicians had a hard time deciding whether they should be considered legitimate numbers. Even Leibniz, who ranks with Newton as the greatest mathematician of the late seventeenth century, called them amphibians between being and not being. He wrote them down, he used them in calculations, but he carefully covered his tracks by calling them imaginary numbers. Unfortunately that name (which was actually first used by Descartes) has stuck, though these numbers are now well accepted by all mathematicians and have numerous applications in science. Let us see what these new numbers are and why they are needed.

Go back to the whole numbers 0, 1, 2, 3, ... We can easily solve the equation $x + 3 = 7$ within this system ($x = 4$). On the other hand, the equation $x + 7 = 3$ has no whole number solution. To solve it, we need the negative integer -4. Similarly, we cannot solve $3x = 2$ in the integers. We can say that the solution is $\frac{2}{3}$ only after the rational numbers have been introduced. To their dismay, the Greeks discovered that $x^2 - 2 = 0$ had no rational solution. We conquered that problem by enlarging our family of numbers to the real numbers. But there are still simple equations without solutions. Consider $x^2 + 1 = 0$. Try as you will, you will never solve it within the real number system.

By now, our procedure is well established. When we need new numbers, we invent them. This time we invent a number denoted by i (or by $\sqrt{-1}$) which satisfies $i^2 = -1$. However, we cannot get by with just one number. For after we have adjoined it to the real numbers, we still must be able to multiply and add. Thus with i, we also need numbers such as

$$2i \qquad -4i \qquad (\tfrac{3}{2})i, \ldots$$

which are called pure imaginary numbers. We also need

$$3 + 2i \qquad 11 + (-4i) \qquad \tfrac{3}{4} + \tfrac{3}{2}i, \ldots$$

and it appears that we need even more complicated things such as

$$(3 + 8i + 2i^2 + 6i^3)(5 + 2i)$$

Actually, this last number can be simplified to $1 + 12i$, as we shall see later. In fact, no matter how many additions and multiplications we do, after the expressions are simplified we shall never have anything more complicated than a number of the form $a + bi$. Such numbers, that is, numbers of the form $a + bi$ where a and b are real, are called **complex numbers.** We refer to a as the **real part** and b as the **imaginary part** of $a + bi$. We shall agree that $0 \cdot i = 0$. Then every real number is a complex number, for $a = a + 0 = a + 0i$.

ADDITION AND MULTIPLICATION

We cannot say anything sensible about operations for complex numbers until we agree on the meaning of equality. The definition that seems most natural is this: $a + bi = c + di$ if and only if $a = c$ and $b = d$. That is, two complex numbers are equal if and only if their real parts and their imaginary parts are equal.

Now we can consider addition. Actually we have already used the plus sign in $a + bi$. That was like trying to add apples and bananas. The addition can be indicated but no further simplification is possible. We do not get apples and we do not get bananas; we get fruit salad.

When we have two numbers of the form $a + bi$, we can actually perform an addition. We just add the real parts and the imaginary parts separately—that is, we add the apples and we add the bananas. Thus

$$(3 + 2i) + (6 + 5i) = 9 + 7i$$

and more generally,

$$\boxed{(a + bi) + (c + di) = (a + c) + (b + d)i}$$

When we consider multiplication, our desire to maintain the properties of Section 1.4 leads to a definition that looks complicated. Thus let us first look at some examples.

(i) $\qquad 2(3 + 4i) = 6 + 8i$ (distributive property)

(ii) $\qquad (3i)(-4i) = -12i^2 = 12$ (commutative and associative properties and $i^2 = -1$)

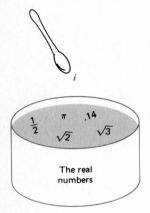

The real numbers

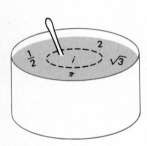

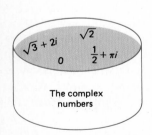

The complex numbers

(iii) $(3 + 2i)(6 + 5i) = (3 + 2i)6 + (3 + 2i)5i$ (distributive property)

$$= 18 + 12i + 15i + 10i^2 \qquad \text{(distributive, commutative, and associative properties)}$$

$$= 18 + (12 + 15)i - 10 \qquad \text{(distributive property)}$$

$$= 8 + 27i \qquad \text{(commutative property)}$$

The same kind of reasoning applied to the general case leads to

$$(a + bi)(c + di) = (ac - bd) + (ad + bc)i$$

which we take as the definition of multiplication for complex numbers.

Actually there is no need to memorize the formula for multiplication. Just do what comes naturally (that is, use familiar properties) and then replace i^2 by -1 wherever it arises, as in the following example.

$$(2 - 3i)(5 + 4i) = (10 - 12i^2) + (8i - 15i)$$

$$= (10 + 12) + (-7i)$$

$$= 22 - 7i$$

Consider the more complicated expression mentioned earlier. After noting that $i^3 = i^2i = -i$, we have

$$(3 + 8i + 2i^2 + 6i^3)(5 + 2i) = (3 + 8i - 2 - 6i)(5 + 2i)$$

$$= (1 + 2i)(5 + 2i)$$

$$= (5 + 4i^2) + (2i + 10i)$$

$$= 1 + 12i$$

SUBTRACTION AND DIVISION

Subtraction is easy; we simply subtract corresponding real and imaginary parts. For example,

$$(3 + 6i) - (5 + 2i) = (3 - 5) + (6i - 2i)$$

$$= -2 + 4i$$

and

$$(5 + 2i) - (3 + 7i) = (5 - 3) + (2i - 7i)$$

$$= 2 + (-5i)$$

$$= 2 - 5i$$

Division is somewhat more difficult. We first note that $a - bi$ is called the **conjugate** of $a + bi$. Thus $2 - 3i$ is the conjugate of $2 + 3i$ and $-2 + 5i$ is the conjugate of $-2 - 5i$. Next, we observe that a complex number times its conjugate is a real number. For example,

$$(3 + 4i)(3 - 4i) = 9 + 16 = 25$$

and in general

$$(a + bi)(a - bi) = a^2 + b^2$$

To simplify the quotient of two complex numbers, multiply the numerator and denominator by the conjugate of the denominator, as illustrated below.

$$\frac{2 + 3i}{3 + 4i} = \frac{(2 + 3i)(3 - 4i)}{(3 + 4i)(3 - 4i)} = \frac{18 + i}{9 + 16} = \frac{18}{25} + \frac{1}{25}i$$

The effect of this multiplication is to replace a complex denominator by a real one. Notice that the result of the division is a number in the form $a + bi$.

A GENUINE EXTENSION

We assert that the complex numbers constitute a genuine enlargement of the real numbers. This means first of all that they include the real numbers, since any real number a can be written as $a + 0i$. Second, the complex numbers satisfy all the properties we discussed in Section 1-4 for the real numbers. The order properties of Section 1-5, however, do not apply to the complex numbers.

The diagram below summarizes our development of the number of systems.

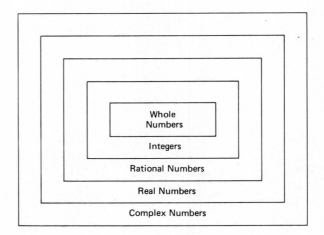

Is there any need to enlarge the number system again? The answer is no, and for a good reason. With the complex numbers, we can solve any equation that arises in algebra. Right now, we expect you to take this statement on faith.

Problem Set 1-6

Carry out the indicated operations and write the answer in the form a + bi.

1. $(2 + 3i) + (-4 + 5i)$
2. $(3 - 4i) + (-5 - 6i)$
3. $5i - (4 + 6i)$
4. $(-3i + 4) + 8i$
5. $(3i - 6) + (3i + 6)$
6. $(6 + 3i) + (6 - 3i)$
7. $4i^2 + 7i$
8. $i^3 + 2i$
9. $i(4 - 11i)$
10. $(3 + 5i)i$
11. $(3i + 5)(2i + 4)$
12. $(2i + 3)(3i - 7)$
13. $(3i + 5)^2$
14. $(-3i - 5)^2$
15. $(5 + 6i)(5 - 6i)$
16. $(\sqrt{3} + \sqrt{2}i)(\sqrt{3} - \sqrt{2}i)$
17. $\dfrac{5 + 2i}{1 - i}$
18. $\dfrac{4 + 9i}{2 + 3i}$
19. $\dfrac{5 + 2i}{i}$
20. $\dfrac{-4 + 11i}{-i}$
21. $\dfrac{(2 + i)(3 + 2i)}{1 + i}$
22. $\dfrac{(4 - i)(5 - 2i)}{4 + i}$

EXAMPLE A (Multiplicative inverses)   Find $(5 + 4i)^{-1}$, the multiplicative inverse of $5 + 4i$.

Solution.   The multiplicative inverse is just the reciprocal. Thus

$$(5 + 4i)^{-1} = \frac{1}{5 + 4i} = \frac{1(5 - 4i)}{(5 + 4i)(5 - 4i)} = \frac{5 - 4i}{25 - 16i^2}$$

$$= \frac{5 - 4i}{25 + 16} = \frac{5 - 4i}{41} = \frac{5}{41} - \frac{4}{41}i$$

Find each value.

23. $(2 - i)^{-1}$
24. $(3 + 2i)^{-1}$
25. $(\sqrt{3} + i)^{-1}$
26. $(\sqrt{2} - \sqrt{2}i)^{-1}$

Use the results above to perform the following divisions. For example, division by $2 - i$ is the same as multiplication by $(2 - i)^{-1}$.

27. $\dfrac{2 + 3i}{2 - i}$
28. $\dfrac{1 + 2i}{3 + 2i}$
29. $\dfrac{4 - i}{\sqrt{3} + i}$
30. $\dfrac{\sqrt{2} + \sqrt{2}i}{\sqrt{2} - \sqrt{2}i}$

EXAMPLE B (High powers of i)   Simplify (a) i^{51};   (b) $(2i)^6 i^{19}$.

Solution.   Keep in mind that $i^2 = -1$, $i^3 = i^2 i = -i$, and $i^4 = i^2 i^2 = (-1)(-1) = 1$. Then use the usual rules of exponents.
(a) $i^{51} = i^{48} i^3 = (i^4)^{12} i^3 = (1)^{12}(-i) = -i$
(b) $(2i)^6 i^{19} = 2^6 i^6 i^{19} = 64 i^{25} = 64 i^{24} i = 64i$

Simplify each of the following.

31. i^{94}
32. i^{39}
33. $(-i)^{17}$
34. $(2i)^3 i^{12}$
35. $\dfrac{(3i)^{16}}{(9i)^5}$
36. $\dfrac{2^8 i^{19}}{(-2i)^{11}}$
37. $(1 + i)^3$
38. $(2 - i)^4$

EXAMPLE C (Complex roots)   We say that a is a 4th root of b if $a \cdot a \cdot a \cdot a = b$ (more briefly, $a^4 = b$). Thus 3 is a 4th root of 81 because $3 \cdot 3 \cdot 3 \cdot 3 = 81$. Show that $1 + i$ is a 4th root of -4.

Solution.

$$(1 + i)(1 + i)(1 + i)(1 + i) = (1 + i)^2(1 + i)^2$$
$$= (1 + 2i - 1)(1 + 2i - 1)$$
$$= (2i)(2i)$$
$$= 4i^2$$
$$= -4$$

39. Show that i is a 4th root of 1. Can you find three other 4th roots of 1?
40. Show that $-1 - i$ is a 4th root of -4.
41. Show that $1 - i$ is a 4th root of -4.
42. Show that $-1 + i$ is a 4th root of -4.

Miscellaneous Problems

43. Perform the indicated operations and write your answers in the form $a + bi$.
    (a) $(5 - 3i) - (7 + 2i) + (6 - 7i)$
    (b) $(5 - 3i)(5 + 3i) - (4 + 2i)(3 - 4i)$
    (c) $\dfrac{5 - 2i}{3 + 4i} + i^{31}$

44. Simplify, giving your answer in the form $a + bi$.
    (a) $\dfrac{(3 + 2i)(2 - 3i)}{5 - i}$ (b) $\dfrac{2}{3 - i} - \dfrac{5}{2 + 2i} + \dfrac{1}{i^{49}}$

45. Show that $1 + \sqrt{3}i$ is a cube root of -8. (You need to demonstrate that $(1 + \sqrt{3}i)(1 + \sqrt{3}i)(1 + \sqrt{3}i) = -8$.)

46. Show that $1 + 2i$ and $2 - i$ are both 4th roots of $-7 - 24i$. Can you guess two more 4th roots of $-7 - 24i$? Check your answers.

47. Solve the equation $3ix - 7i = 2i + 1$.

48. Show that $2 + i$ and $-2 - i$ are solutions of $x^2 - 3 - 4i = 0$.

CHAPTER SUMMARY

Algebra is generalized arithmetic. In it, we use letters to stand for numbers and then manipulate these letters according to definite rules.

The number systems used in algebra are the **whole numbers,** the **integers,** the **rational numbers,** the **real numbers,** and the **complex numbers.** Except for the complex numbers, we can visualize all these numbers as labels for points along a line, called the **real line.** Rational numbers can be expressed as ratios of integers. Real numbers (and therefore rational numbers) can be expressed as **decimals.** In fact, the rational numbers can be expressed as **repeating decimals** and the **irrational** (not rational) numbers as **nonrepeating decimals.**

The fundamental properties of numbers include the **commutative, associative,** and **distributive** laws. Because of their special properties, 0 and 1 are called the **neutral elements** for addition and multiplication. Within the rational number system, the real number system, and the complex number system, all numbers have **additive inverses** and all but 0 have **multiplicative inverses.**

The real numbers are **ordered** by the relation $<$ (is less than); its properties should be well understood as should those of its relatives $\leq$, $>$, and $\geq$. The symbol $|\ |$, read **absolute value,** denotes the magnitude of a number regardless of whether it is positive or negative.

CHAPTER REVIEW PROBLEM SET

1. An open box with a square base of length x centimeters has height y centimeters. Express its volume V and surface area S in terms of x and y.

2. Suppose that an airplane can fly x miles per hour in still air and that the wind is blowing from east at y miles per hour. Express in terms of x and y the time it will take for the plane to fly 100 miles due east and then return.

3. Perform the indicated operations and simplify.

   (a) $\dfrac{13}{24} - \dfrac{7}{12} + \dfrac{2}{3}$

   (b) $\dfrac{9}{24} \cdot \dfrac{15}{18} \cdot \dfrac{8}{5}$

   (c) $\dfrac{1}{3}\left(\dfrac{1}{2} - \left(\dfrac{5}{6} - \dfrac{2}{3}\right)\right)$

   (d) $(\tfrac{11}{30})/(\tfrac{33}{18})$

   (e) $\dfrac{\frac{3}{4} - \frac{1}{12} + \frac{3}{8}}{\frac{3}{4} + \frac{5}{12} - \frac{7}{8}}$

   (f) $3 + \dfrac{\frac{3}{4} - \frac{7}{8}}{\frac{5}{12}}$

4. Find the prime factorizations of 1000 and 180.

5. Find the least common multiple of 1000 and 180.

6. Express $\frac{5}{13}$ and $\frac{11}{7}$ as repeating decimals.

7. Express $.\overline{257}$ and $1.\overline{23}$ as ratios of two integers.

8. Is $(a \div b) \times c = (a \times c) \div b$? Is $(a + b)^{-1} = a^{-1} + b^{-1}$?

9. State the associative law of multiplication and the commutative law of addition.

10. Order the following numbers from least to greatest: $\frac{29}{20}$, 1.4, $1.\overline{4}$, $\sqrt{2}$, $\frac{13}{8}$.

11. Write the inequalities $|x + 2| < 6$ and $|2x - 1| \leq 3$ without using the absolute value symbol. Then show the corresponding intervals on the real line.

12. Is $|-x| = x$? Explain.

13. Write each of the following in the form $a + bi$.
    (a) $(3 - 2i) + (-7 + 6i) - (2 - 3i)$
    (b) $(3 - 2i)(3 + 2i) - 4i^2 + (2i)^3$
    (c) $2 + 3i + (3 - i)/(2 + 3i)$
    (d) $(2 + i)^3$
    (e) $(5 - 3i)^{-1}$

14. Assuming that $\sqrt{5}$ is irrational, show that $5 + \sqrt{5}$ and $5\sqrt{5}$ are irrational.

15. A rectangle with a perimeter of 72 centimeters is 4 centimeters longer than it is wide. Find its area.

2

EXPONENTS AND POLYNOMIALS

Algebra is the intellectual instrument for rendering clear the quantitative aspects of the world

Alfred North Whitehead

The sermon was long; the sanctuary was hot. It was even hard to stay awake. For awhile, young Franklin Figit amused himself by doodling on the Sunday bulletin. Growing tired of that, he tried folding the bulletin in more and more complicated ways. That's when a bright idea hit him. "I wonder what would happen," he said to himself, "if I took a mammoth bulletin, folded it in half, then in half again, and so on until I had folded it 40 times. How high a stack of paper would that make?"

2-1
Integral Exponents

Young Franklin has posed a very interesting question. What do you think the answer is? 10 inches? 3 feet? 500 feet? Make a guess and write it in the margin. When we finally work out the solution late in this section, you are likely to be surprised at the answer.

Let us make a start on the problem right away. If the bulletin were c units thick ($c = .01$ inches would be a reasonable value), then after folding once, it would be $2c$ units thick. After two folds, it would measure $2 \cdot 2c$ units thick, and after 40 folds it would have a thickness of

$$2 \cdot 2$$

$$\cdot 2 \cdot c$$

Nobody with a sense of economy and elegance would write a product of forty 2's in this manner. To indicate such a product, most ordinary people and all mathematicians prefer to write 2^{40}. The number 40 is called an **exponent;** it tells you how many 2's to multiply together. The number 2^{40} is called a **power** of 2 and we read it "2 to the 40th power."

In the general case, if b is any number ($\frac{3}{4}$, π, $\sqrt{5}$, i, . . .) and n is a positive integer, then

$$b^n = \underbrace{b \cdot b \cdot b \cdots b}_{n \text{ factors}}$$

Thus

$$b^3 = b \cdot b \cdot b \qquad b^5 = b \cdot b \cdot b \cdot b \cdot b$$

$$-2^4 = -16$$
$$(-2)^4 = 16$$

How do we write the product of 1000 b's? (Honey is not the answer we have in mind.) The product of 1000 b's is written as b^{1000}.

RULES FOR EXPONENTS

The behavior of exponents is excellent, being governed by a few simple rules that are easy to remember. Consider multiplication first. If we multiply 2^5 by 2^8, we have

$$2^5 \cdot 2^8 = \underbrace{(2 \cdot 2 \cdot 2 \cdot 2 \cdot 2)}_{5}\underbrace{(2 \cdot 2 \cdot 2 \cdot 2 \cdot 2 \cdot 2 \cdot 2 \cdot 2)}_{8}$$

$$= \underbrace{2 \cdot 2 \cdot 2 \cdot 2 \cdot 2 \cdot 2 \cdot 2 \cdot 2 \cdot 2 \cdot 2 \cdot 2 \cdot 2 \cdot 2}_{13}$$

$$= 2^{13}$$

$$= 2^{5+8}$$

This suggests that to find the product of powers of 2, you should add the exponents. There is nothing special about 2; it could just as well be 5, $\frac{2}{3}$, or π. We can put the general rule in a nutshell by using the symbols of algebra.

$$b^m \cdot b^n = b^{m+n}$$

Here b can stand for any number, but (for now) think of m and n as positive integers. Be careful with this rule. If you write

$$3^4 \cdot 3^5 = 3^9$$

or

$$\pi^9 \cdot \pi^{12} \cdot \pi^2 = \pi^{23}$$

that is fine. But do not try to use the rule on $2^4 \cdot 3^5$ or $a^2 \cdot b^3$; it just does not apply.

Next consider the problem of raising a power to a power. By definition, $(2^{10})^3$ is $2^{10} \cdot 2^{10} \cdot 2^{10}$, which allows us to apply the rule above. Thus

$$(2^{10})^3 = 2^{10} \cdot 2^{10} \cdot 2^{10} = 2^{10+10+10} = 2^{10 \cdot 3}$$

It appears that to raise a power to a power we should multiply the exponents; in symbols

$$(b^m)^n = b^{m \cdot n}$$

Try to convince yourself that this rule is true for any number b and for any positive integer exponents m and n.

Sometimes we need to simplify quotients like

$$\frac{8^6}{8^6} \qquad \frac{2^9}{2^5} \qquad \frac{10^4}{10^6}$$

The first one is easy enough; it equals 1. Furthermore,

$$\frac{2^9}{2^5} = \frac{2^5 \cdot 2^4}{2^5} = 2^4 = 2^{9-5}$$

and

$$\frac{10^4}{10^6} = \frac{10^4}{10^4 \cdot 10^2} = \frac{1}{10^2} = \frac{1}{10^{6-4}}$$

These illustrate the general rules.

$$\frac{b^m}{b^n} = 1 \qquad \text{if } m = n$$

$$\frac{b^m}{b^n} = b^{m-n} \qquad \text{if } m > n$$

$$\frac{b^m}{b^n} = \frac{1}{b^{n-m}} \qquad \text{if } n > m$$

In each case, we assume $b \neq 0$.

We did not put a box around these rules simply because we are not happy with them. It took three lines to describe what happens when you divide powers of the same number. Surely we can do better than that, but first we shall have to extend the notion of exponents to numbers other than positive integers.

ZERO AND NEGATIVE EXPONENTS

So far, symbols like 4^0 and 10^{-3} have not been used. We want to give them meaning and do it in a way that is consistent with what we have already learned. For example, 4^0 must behave so that

$$4^0 \cdot 4^7 = 4^{0+7} = 4^7$$

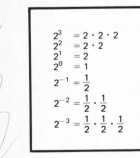

$$
\begin{aligned}
2^3 &= 2 \cdot 2 \cdot 2 \\
2^2 &= 2 \cdot 2 \\
2^1 &= 2 \\
2^0 &= 1 \\
2^{-1} &= \frac{1}{2} \\
2^{-2} &= \frac{1}{2} \cdot \frac{1}{2} \\
2^{-3} &= \frac{1}{2} \cdot \frac{1}{2} \cdot \frac{1}{2}
\end{aligned}
$$

This can happen only if $4^0 = 1$. More generally, we require that

$$\boxed{b^0 = 1}$$

Here b can be any number except 0 (0^0 will be left undefined).

What about 10^{-3}? If it is to be admitted to the family of powers, it too must abide by the rules. Thus we insist that

$$10^{-3} \cdot 10^3 = 10^{-3+3} = 10^0 = 1$$

This means that 10^{-3} has to be the reciprocal of 10^3. Consequently, we are led to make the definition

$$b^{-n} = \frac{1}{b^n} \qquad b \neq 0$$

Now the complicated law for quotients of powers can be put in a simple form, as the following examples illustrate.

$$\frac{b^4}{b^9} = \frac{b^4}{b^4 \cdot b^5} = \frac{1}{b^5} = b^{-5} = b^{4-9}$$

$$\frac{b^5}{b^5} = 1 = b^0 = b^{5-5}$$

$$\frac{b^{-3}}{b^{-9}} = \frac{1/b^3}{1/b^9} = \frac{b^9}{b^3} = b^6 = b^{-3-(-9)}$$

In fact, for any choice of integers m and n, we find that

$$\frac{b^m}{b^n} = b^{m-n} \qquad b \neq 0$$

What about the two rules we learned earlier? Are they still valid when m and n are arbitrary (possibly negative) integers? The answer is yes. A few illustrations may help convince you.

$$b^{-3} \cdot b^7 = \frac{1}{b^3} \cdot b^7 = \frac{b^7}{b^3} = b^4 = b^{-3+7}$$

$$(b^{-5})^2 = \left(\frac{1}{b^5}\right)^2 = \frac{1}{b^5} \cdot \frac{1}{b^5} = \frac{1}{b^{10}} = b^{-10} = b^{(-5)(2)}$$

POWERS OF PRODUCTS AND QUOTIENTS

Expressions like $(ab)^n$ and $(a/b)^n$ often arise; we need rules for handling them. Notice that

$$(ab)^n = \underbrace{(ab)(ab) \ldots (ab)}_{n \text{ factors}} = \underbrace{a \cdot a \cdots a}_{n \text{ factors}} \cdot \underbrace{b \cdot b \cdots b}_{n \text{ factors}} = a^n b^n$$

$$\left(\frac{a}{b}\right)^n = \underbrace{\left(\frac{a}{b}\right)\left(\frac{a}{b}\right) \cdots \left(\frac{a}{b}\right)}_{n \text{ factors}} = \frac{a \cdot a \cdots a}{b \cdot b \cdots b} = \frac{a^n}{b^n}$$

Our demonstrations are valid for any positive integer n, but the results are correct even if n is negative or zero. Thus for any integer n,

$$(ab)^n = a^n b^n$$

$$\left(\frac{a}{b}\right)^n = \frac{a^n}{b^n}$$

Thus

$$(3x^2 y)^4 = 3^4 (x^2)^4 y^4 = 81 x^8 y^4$$

and

$$\left(\frac{2x^{-1}}{y}\right)^3 = \frac{2^3 (x^{-1})^3}{y^3} = \frac{2^3 x^{-3}}{y^3} = \frac{8}{x^3 y^3}$$

We summarize our discussion of exponents by stating the five main rules together. In using these rules, it is always understood that division by zero is to be avoided.

RULES FOR EXPONENTS

In each case, a and b are (real or complex) numbers and m and n are any integers.

1. $b^m b^n = b^{m+n}$

2. $(b^m)^n = b^{mn}$

3. $\dfrac{b^m}{b^n} = b^{m-n}$

4. $(ab)^n = a^n b^n$

5. $\left(\dfrac{a}{b}\right)^n = \dfrac{a^n}{b^n}$

THE PAPER FOLDING PROBLEM AGAIN

It is now a simple matter to solve Franklin Figit's paper folding problem especially if we are satisfied with a reasonable approximation to the answer. To be specific, let us approximate 1 foot by 10 inches, 1 mile by 5000 feet, and 2^{10} (which is really 1024) by 1000. Then a bulletin of thickness .01 inch will make a stack of the following height when folded 40 times ($\approx$ means "is approximately equal to").

$$(.01)2^{40} = (.01) \cdot 2^{10} \cdot 2^{10} \cdot 2^{10} \cdot 2^{10} \text{ inches}$$

$$\approx \frac{1}{10^2} \cdot 10^3 \cdot 10^3 \cdot 10^3 \cdot 10^3 \text{ inches}$$

$$= 10^{10} \text{ inches}$$

$$\approx 10^9 \text{ feet}$$

$$\approx \frac{10 \cdot 10^8}{5 \cdot 10^3} \text{ miles}$$

$$= 2 \cdot 10^5 \text{ miles}$$

$$= 200{,}000 \text{ miles}$$

That is a stack of paper that would reach almost to the moon.

Problem Set 2-1

Use the rules for exponents to simplify each of the following. Then calculate the result.

1. $\dfrac{3^2 3^5}{3^4}$ 2. $\dfrac{2^6 2^7}{2^{10}}$ 3. $\dfrac{(2^2)^4}{2^6}$ 4. $\dfrac{(5^4)^3}{5^9}$ 5. $\dfrac{(3^2 2^3)^3}{6^6}$ 6. $6^6 \left(\dfrac{2}{3^2}\right)^3$

EXAMPLE A (Removing negative exponents)   Rewrite without negative exponents and simplify.

(a) -4^{-2} (b) $(-4)^{-2}$ (c) $[(\tfrac{3}{4})^{-1}]^2$ (d) $2^5 3^{-2} 2^{-3}$

Solution.

(a) The exponent -2 applies just to 4.

$$-4^{-2} = -\frac{1}{4^2} = -\frac{1}{4 \cdot 4} = -\frac{1}{16}$$

(b) The exponent -2 now applies to -4.

$$(-4)^{-2} = \frac{1}{(-4)^2} = \frac{1}{16}$$

(c) First apply the rule for a power of a power.

$$\left[\left(\frac{3}{4}\right)^{-1}\right]^2 = \left(\frac{3}{4}\right)^{-2} = \frac{1}{(\frac{3}{4})^2} = \frac{1}{\frac{9}{16}} = \frac{16}{9}$$

(d) Note that the two powers of 2 can be combined.

$$2^5 3^{-2} 2^{-3} = 2^5 2^{-3} 3^{-2} = 2^2 \cdot \frac{1}{3^2} = \frac{4}{9}$$

Write without negative exponents and simplify.

7. 5^{-2} 8. $(-5)^{-2}$ 9. -5^{-2} 10. 2^{-5}

11. $(-2)^{-5}$ 12. $\left(\dfrac{1}{5}\right)^{-2}$ 13. $\left(\dfrac{-2}{3}\right)^{-3}$ 14. $-\dfrac{2^{-3}}{3}$

15. $\dfrac{2^{-2}}{3^{-3}}$ 16. $\left[\left(\dfrac{2}{3}\right)^{-2}\right]^2$ 17. $\left[\left(\dfrac{3}{2}\right)^{-2}\right]^{-2}$ 18. $\dfrac{4^0 + 0^4}{4^{-1}}$

19. $\dfrac{2^{-2} - 4^{-3}}{(-2)^2 + (-4)^0}$ 20. $\dfrac{3^{-1} + 2^{-3}}{(-1)^3 + (-3)^2}$

21. $3^3 \cdot 2^{-3} \cdot 3^{-5}$ 22. $4^2 \cdot 4^{-4} \cdot 3^0$

EXAMPLE B (Products and quotients)   Use Rules 4 and 5 to simplify

(a) $(2x)^6$;   (b) $\left(\dfrac{2x}{3}\right)^4$;   (c) $(x^{-1}y^2)^{-3}$.

Solution.

(a) $(2x)^6 = 2^6 x^6 = 64x^6$

(b) $\left(\dfrac{2x}{3}\right)^4 = \dfrac{(2x)^4}{3^4} = \dfrac{2^4 x^4}{3^4} = \dfrac{16x^4}{81}$

(c) $(x^{-1}y^2)^{-3} = (x^{-1})^{-3}(y^2)^{-3} = x^3 y^{-6} = x^3 \cdot \dfrac{1}{y^6} = \dfrac{x^3}{y^6}$

Simplify, writing your answer without negative exponents.

23. $(3x)^4$
24. $\left(\dfrac{2}{y}\right)^5$
25. $(xy^2)^6$
26. $\left(\dfrac{y^2}{3z}\right)^4$

27. $\left(\dfrac{2x^2 y}{w^3}\right)^4$
28. $\left(\dfrac{\sqrt{2}x}{3}\right)^4$
29. $\left(\dfrac{3x^{-1}y^2}{z^2}\right)^3$
30. $\left(\dfrac{2x^{-2}y}{z^{-1}}\right)^2$

ⓘ 31. $\left(\dfrac{\sqrt{5}i}{x^{-2}}\right)^4$ ⓘ 32. $(i\sqrt{3}x^{-2})^6$
33. $(4y^3)^{-2}$
34. $(x^3 z^{-2})^{-1}$

35. $\left(\dfrac{5x^2}{ab^{-2}}\right)^{-1}$
36. $\left(\dfrac{3x^2 y^{-2}}{2x^{-1}y^4}\right)^{-3}$

EXAMPLE C (Simplifying complicated expressions)   Simplify

(a) $\dfrac{4ab^{-2}c^3}{a^{-3}b^3 c^{-1}}$;   (b) $\left[\dfrac{(2xz^{-2})^3(x^{-2}z)}{2xz^2}\right]^4$;   (c) $(a^{-1} + b^{-2})^{-1}$.

Solution.

(a) $\dfrac{4ab^{-2}c^3}{a^{-3}b^3 c^{-1}} = \dfrac{4a(1/b^2) \cdot c^3}{(1/a^3)b^3(1/c)} = \dfrac{4ac^3/b^2}{b^3/(a^3 c)} = \dfrac{4ac^3}{b^2} \cdot \dfrac{a^3 c}{b^3} = \dfrac{4a^4 c^4}{b^5}$

In simplifying expressions like the one above, a *factor can be moved from numerator to denominator or vice versa by changing the sign of its exponent.* That is important enough to remember. Let us do part (a) again using this fact.

$$\dfrac{4ab^{-2}c^3}{a^{-3}b^3 c^{-1}} = \dfrac{4aa^3 c^3 c}{b^2 b^3} = \dfrac{4a^4 c^4}{b^5}$$

(b) $\left[\dfrac{(2xz^{-2})^3(x^{-2}z)}{2xz^2}\right]^4 = \left[\dfrac{8x^3 z^{-6}x^{-2}z}{2xz^2}\right]^4$

$= \left[\dfrac{8xz^{-5}}{2xz^2}\right]^4$

$= \left[\dfrac{4}{z^2 z^5}\right]^4 = \dfrac{256}{z^{28}}$

(c) $(a^{-1} + b^{-2})^{-1} = \left(\dfrac{1}{a} + \dfrac{1}{b^2}\right)^{-1} = \left(\dfrac{b^2 + a}{ab^2}\right)^{-1} = \dfrac{ab^2}{b^2 + a}$

Note the difference between a product and a sum.

$$(a^{-1} \cdot b^{-2})^{-1} = a \cdot b^2$$

but

$$(a^{-1} + b^{-2})^{-1} \neq a + b^2$$

$(a^{-1} + b^{-1})^{-1} \neq a + b$

$(a^{-1} + b^{-1})^{-1} = \left(\dfrac{1}{a} + \dfrac{1}{b}\right)^{-1}$

$= \left(\dfrac{b + a}{ab}\right)^{-1} = \dfrac{ab}{b + a}$

Simplify, leaving your answer free of negative exponents.

37. $\dfrac{2x^{-3}y^2z}{x^3y^4z^{-2}}$

38. $\dfrac{3x^{-5}y^{-3}z^4}{9x^2yz^{-1}}$

39. $\left(\dfrac{-2xy}{z^2}\right)^{-1}(x^2y^{-3})^2$

40. $(4ab^2)^3\left(\dfrac{-a^3}{2b}\right)^2$

41. $\dfrac{ab^{-1}}{(ab)^{-1}} \cdot \dfrac{a^2b}{b^{-2}}$

42. $\dfrac{3(b^{-2}d)^4(2bd^3)^2}{(2b^2d^3)(b^{-1}d^2)^5}$

43. $\left[\dfrac{(3b^{-2}d)(2)(bd^3)^2}{12b^3d^{-1}}\right]^5$

44. $\left[\dfrac{(ab^2)^{-1}}{(ba^2)^{-2}}\right]^{-1}$

45. $(a^{-2} + a^{-3})^{-1}$

46. $a^{-2} + a^{-3}$

47. $\dfrac{x^{-1}}{y^{-1}} - \left(\dfrac{x}{y}\right)^{-1}$

48. $(x^{-1} - y^{-1})^{-1}$

Miscellaneous Problems

Simplify the expressions in Problems 49–59, leaving your answer free of negative exponents.

49. $(\sqrt{3}x^2y^{-2})^2$

50. $(\tfrac{1}{3}x^{-1}z^3)^3$

51. $\dfrac{(3x^{-1}y)^2}{3xy^{-2}}$

52. $\left(\dfrac{3x^{-1}y}{6xy^{-2}}\right)^2$

53. $[(2x^{-2})^3(3z^{-1}x)^2]^2$

54. $\left(\dfrac{5x^2z^{-3}}{2x^{-1}z^2)^3}\right)^{-1}$

55. $x^2 + x^{-2}$

56. $(x + x^{-1})^2$

57. $(x + x^{-1})^{-1}$

58. $\left(1 - \dfrac{x - 1}{x + 1}\right)^{-1}$

59. $[1 - (1 + x^{-1})^{-1}]^{-1}$

60. Express each of the following as a power of 5.

   (a) $(25)^{-2}$ (b) $\left(\dfrac{1}{5^3}\right)^{-2}$ (c) $(.2)(5^{-4})^{-2}$

61. Express each of the following as a power of 2.
   (a) $\tfrac{1}{32}$ (b) 1   (c) $\tfrac{1}{2} \cdot \tfrac{1}{16} \cdot \tfrac{1}{8} \cdot \tfrac{1}{4}$

62. Which is larger, 2^{1000} or $(1000)^{100}$?

63. G. P. Jetty has agreed to pay his private secretary according to the following plan: 1¢ the first day, 2¢ the second day, 4¢ the third day, and so on.
   (a) About how much will the secretary make on the 20th day? (Recall that 2^{10} is approximately 1000).
   (b) About how much will the secretary make on the 50th day?

64. Refer to Problem 63.
   (a) How much will the secretary make during the first 3 days? 4 days? 5 days?

(b) How do your answers in part (a) compare with the amounts the secretary made on the fourth day? Fifth day? Sixth day?

(c) Guess at a formula for the total amount the secretary will make in *n* days.

(d) About how much will the secretary make altogether during the first 30 days? 40 days?

65. Refer to Problem 64. If G. P. Jetty is worth $2 billion and his secretary starts work on January 1, about when will G. P. Jetty go broke?

66. Consider Franklin Figit's paper folding problem with which we began this section. If the pile of paper covers 1 square inch after 40 folds, approximately how much area did it cover at the beginning?

A new revolution is taking place in technology today. It both parallels and completes the Industrial Revolution that started a century ago. The first phase of the Industrial Revolution meant the mechanization, then the electrification of brawn. The new revolution means the mechanization and electrification of brains.

Harry M. Davis

2-2
Calculators
and Scientific
Notation

Most people can do arithmetic if they have to. But few do it with either enthusiasm or accuracy. Frankly, it is a rather dull subject. The spectacular sales of pocket electronic calculators demonstrate both our need to do arithmetic and our distaste for it.

Pocket calculators vary greatly in what they can do. The simplest perform only the four arithmetic operations and sell for under $10. The most sophisticated are programmable and can automatically perform dozens of operations in sequence. For this course, a standard scientific calculator is ideal. In addition to the four arithmetic operations, it will calculate values for the exponential and logarithmic functions and the trigonometric and inverse trigonometric functions.

Two kinds of logic are used in pocket calculators, **reverse Polish** logic and **algebraic** logic. The former avoids the use of parentheses, but is highly efficient once one learns its rules. Algebraic logic uses parentheses and mimics the procedures of ordinary algebra. The Texas Instruments Model 30 shown at the beginning of this section uses algebra logic and will serve to illustrate our discussion of calculators. You will need to study the instruction book for any other calculator to understand the minor modifications required to use that calculator.

Since calculators can display only a fixed number of digits (usually 8 or 10), we face an immediate problem. How shall we handle very large or very small numbers on a calculator? The answer depends on a notational device that was invented long before pocket calculators.

SCIENTIFIC NOTATION

It is in science that we are most likely to meet very large or very small numbers. For example, the speed of light is 29,979,000,000 centimeters per second. At the other extreme, the mass of the proton is .00000000000000000000000167 grams. These numbers are unwieldy primarily because of the large number of zeros, zeros which serve only to place the decimal point. The significant digits are 29979 in the first case and 167 in the second. Now note that

$$29{,}979{,}000{,}000 = 2.9979 \times 10^{10}$$

$$.00000000000000000000000167 = 1.67 \times 10^{-24}$$

Both of these numbers have been rewritten in scientific notation.

A positive number N is in **scientific notation** when it is written in the form

$$N = c \times 10^n$$

where n is an integer and c is a real number such that $1 \le c < 10$. To put N in scientific notation, place the decimal point after the first nonzero digit of N, the so-called standard position. Then count the number of places from there to the original position of the decimal point. This number, taken as positive if counted to the right and negative if counted to the left, is the exponent n to be used as the power of 10. Here is the process illustrated for the number 3,651,000.

$$3651000 = 3.651 \times 10^6$$

6 places

Calculations with large or small numbers are easily accomplished in scientific notation. Suppose we wish to calculate

$$p = \frac{(3{,}200{,}000{,}000)(.0000000284)}{.00000000128}$$

First, we write

$$3{,}200{,}000{,}000 = 3.2 \times 10^9$$

$$.0000000284 = 2.84 \times 10^{-8}$$

$$.00000000128 = 1.28 \times 10^{-9}$$

Then

$$p = \frac{(3.2 \times 10^9)(2.84 \times 10^{-8})}{1.28 \times 10^{-9}}$$

$$= \frac{(3.2)(2.84)}{1.28} \times 10^{9-8-(-9)}$$

$$= 7.1 \times 10^{10}$$

ENTERING DATA INTO A CALCULATOR

With scientific notation in hand, we can explain how to get a number into a typical calculator. To enter 238.75, simply press in order the keys 2 3 8 . 7 5.

If the negative of this number is desired, press the same keys and then press the change sign key $\boxed{+/-}$. Numbers larger than 10^8 or smaller than 10^{-8} must be entered in scientific notation. For example, 2.3875×10^{19} would be entered as follows.

$$2.3875 \;\boxed{EE}\; 19$$

The key $\boxed{EE}$ (which stands for *enter exponent*) controls the two places at the extreme right of the display. They are reserved for the exponent on 10. After pressing the indicated keys, the display will read

$$\boxed{2.3875 \qquad 19}$$

If you press

$$2.3875 \;\boxed{+/-}\; \boxed{EE}\; 19 \;\boxed{+/-}$$

the display will read

$$\boxed{-2.3875 \quad -19}$$

which stands for the number -2.3875×10^{-19}.

In making a complicated calculation, you may enter some numbers in standard notation and others in scientific notation. The calculator understands either form and makes the proper translations. If any of the entered data is in scientific notation, it will display the answer in this format. Also, if the answer is too large or too small for standard format, the calculator will automatically convert the result of a calculation to scientific notation.

DOING ARITHMETIC

The five keys $\boxed{+}$, $\boxed{-}$, $\boxed{\times}$, $\boxed{\div}$, and $\boxed{=}$ are the workhorses for arithmetic in any calculator using algebraic logic. To perform the calculation

$$175 + 34 - 18$$

simply press the keys indicated below.

$$175 \boxed{+} 34 \boxed{-} 18 \boxed{=}$$

The answer 191 will appear in the display.

Or consider (175)(14)/18. Press

$$175 \boxed{\times} 14 \boxed{\div} 18 \boxed{=}$$

and the calculator will display 136.11111.

An expression involving additions (or subtractions) and multiplications (or divisions) may be ambiguous. For example, $2 \times 3 + 4 \times 5$ could have several meanings depending on which operations are performed first.

(i) $\qquad\qquad 2 \times (3 + 4) \times 5 = 70$

(ii) $\qquad\qquad (2 \times 3) + (4 \times 5) = 26$

(iii) $\qquad\qquad [(2 \times 3) + 4] \times 5 = 50$

(iv) $\qquad\qquad 2 \times [3 + (4 \times 5)] = 46$

Parentheses are used in mathematics and in calculators to indicate the order in which operations are to be performed. To do calculation (i), press

$$2 \boxed{\times} \boxed{(} 3 \boxed{+} 4 \boxed{)} \boxed{\times} 5 \boxed{=}$$

Similarly to do calculation (iii), press

$$\boxed{(} \boxed{(} 2 \boxed{\times} 3 \boxed{)} \boxed{+} 4 \boxed{)} \boxed{\times} 5 \boxed{=}$$

Recall that, in arithmetic, we have an agreement that when no parentheses are used, multiplications and divisions are done before additions and subtractions. Thus

$$2 \times 3 + 4 \times 5$$

is interpreted as $(2 \times 3) + (4 \times 5)$. The same convention is used in most calculators. Pressing.

$$2 \boxed{\times} 3 \boxed{+} 4 \boxed{\times} 5 \boxed{=}$$

will yield the answer 26. Similarly, for calculation (iii), pressing

$$\boxed{(} 2 \boxed{\times} 3 \boxed{+} 4 \boxed{)} \boxed{\times} 5 \boxed{=}$$

will yield 50, since within the parentheses, the calculator will do the multiplication first. However, when in doubt use parentheses, since without them it is easy to make errors.

SPECIAL FUNCTIONS

Most scientific calculators have keys for finding powers and roots of a number. On our sample calculator, the $\boxed{y^x}$ key is used to raise a number y to the xth power. For example, to calculate $2.75^{-.34}$, press

$$2.75 \;\boxed{y^x}\; .34 \;\boxed{+/-}\; \boxed{=}$$

and the correct result, .70896841, will appear in the display.

Finding a root is the inverse of raising to a power. For example, taking a cube root is the inverse operation of cubing. Thus, to calculate $\sqrt[3]{17}$, press 17 $\boxed{\text{INV}}$ $\boxed{y^x}$ 3 $\boxed{=}$ and you will get 2.5712816. In using the $\boxed{y^x}$ key, the calculator insists that y be positive. However, x may be either positive or negative.

Square roots occur so often that there is a special key for them on some calculators. To calculate $\sqrt{17}$, simply press 17 $\boxed{\sqrt{}}$ and you will immediately get 4.1231056.

Our introduction to pocket calculators has been very brief. The problem set will give you practice in using your particular model. To clear up any difficulties and to learn other features of your calculator, consult your instruction book.

THE METRIC SYSTEM

Several of the examples of this section have involved metric units; it is time we said a word about the metric system of measurement. It has long been recognized by most countries that the metric system, with its emphasis on 10 and powers of 10, offers an attractive way to measure length, volume and weight. Only in the United States do we hang on to our hodgepodge of inches, feet, miles, pounds, and quarts. Even here, however, it appears that the metric system will gradually win acceptance.

The following table summarizes the metric system. It highlights the relationship to powers of 10 within the metric system and gives some of the conversion factors that relate the metric system to our English system. Using the table, you should be able to answer questions such as the following.

1. How many kilometers per hour are equivalent to 65 miles per hour?

TABLE 1 The Metric System

Length	Volume	Weight
kilometer $= 10^3$ meter	kiloliter $= 10^3$ liter	kilogram $= 10^3$ gram
hectometer $= 10^2$ meter	hectoliter $= 10^2$ liter	hectogram $= 10^2$ gram
dekameter $= 10$ meter	dekaliter $= 10$ liter	dekagram $= 10$ gram
meter $= 1$ meter	liter $= 1$ liter	gram $= 1$ gram
decimeter $= 10^{-1}$ meter	deciliter $= 10^{-1}$ liter	decigram $= 10^{-1}$ gram
centimeter $= 10^{-2}$ meter	centiliter $= 10^{-2}$ liter	centigram $= 10^{-2}$ gram
millimeter $= 10^{-3}$ meter	milliliter $= 10^{-3}$ liter	milligram $= 10^{-3}$ gram
1 kilometer $\approx .62$ miles	1 liter ≈ 1.057 quarts	1 kilogram ≈ 2.20 pounds
1 inch ≈ 2.54 centimeters	1 liter $= 10^3$ cubic centimeters	1 pound ≈ 453.6 grams

2. How many grams are equivalent to 200 pounds?

The answer to the two questions we posed are:

1. 65 miles per hour $\approx 65/.62$, or 105 kilograms per hour;
2. 200 pounds $\approx 200(453.6)$ grams $= 9.07 \times 10^4$ grams.

Problem Set 2-2

Write each of the following numbers in scientific notation.

1. 341,000,000
2. 25 billion
3. .0000000513
4. .00000000012
5. .0000000001245
6. 0000000000012578

Calculate, leaving your answers in scientific notation.

7. $(1.2 \times 10^5)(7 \times 10^{-9})$

8. $(2.4 \times 10^{-11})(1.2 \times 10^{16})$

9. $\dfrac{(0.000021)(240000)}{7000}$

10. $\dfrac{(36,000,000)(.000011)}{.0000033}$

11. $(54)(.00005)(2,000,000)^2$

12. $\dfrac{(3400)^2(400,000)^3}{(.017)^2}$

EXAMPLE A (Conversions between units)
(a) Convert 2.56×10^4 kilograms to centimeters.
(b) Convert 3.42×10^2 kilograms to ounces.
(c) Convert 43.8 cubic meters to liters.

Solution.

(a) 2.56×10^4 kilometers $= (2.56 \times 10^4)(10^3)$ meters

$$= (2.56 \times 10^4)(10^3)(10^2) \text{ centimeters}$$

$$= 2.56 \times 10^9 \text{ centimeters}$$

(b) 3.42×10^2 kilograms $\approx (3.42 \times 10^2)(2.20)$ pounds

$$= (3.42 \times 10^2)(2.20)(16) \text{ ounces}$$

$$\approx 1.20 \times 10^4 \text{ ounces}$$

(c) 43.8 cubic meters $= (43.8)(100)^3$ cubic centimeters

$$= \dfrac{(43.8)(100)^3}{10^3} \text{ liters}$$

$$= 4.38 \times 10^4 \text{ liters}$$

Express each of the following in scientific notation.

13. The number of centimeters in 413.2 meters.
14. The number of millimeters in 1.32×10^4 kilometers.

15. The number of kilometers in 4×10^{15} millimeters.
16. The number of meters in 1.92×10^8 centimeters.
17. The number of millimeters in one yard (36 inches).
18. The number of grams in 4.1×10^3 pounds.

EXAMPLE B (Hierarchy of operations in a calculator) Most scientific calculators (with algebraic logic) perform operations in the following order.

1. Unary operations, such as taking square roots.
2. Multiplications and divisions from left to right.
3. Additions and subtractions from left to right.

Parentheses are used just as in ordinary algebra. Pressing the $\boxed{=}$ key will cause all pending operations to be performed. Calculate

$$\frac{3.12 + (4.15)(5.79)}{5.13 - 3.76}$$

Solution. This can be done in more than one way, and calculators may vary. On the author's model, either of the following sequences of keys will give the correct result.

3.12 $\boxed{+}$ 4.15 $\boxed{\times}$ 5.79 $\boxed{=}$ $\boxed{\div}$ $\boxed{(}$ 5.13 $\boxed{-}$ 3.76 $\boxed{)}$ $\boxed{=}$

5.13 $\boxed{-}$ 3.76 $\boxed{=}$ $\boxed{1/x}$ $\boxed{\times}$ $\boxed{(}$ 3.12 $\boxed{+}$ 4.15 $\boxed{\times}$ 5.79 $\boxed{)}$ $\boxed{=}$

The result is 19.816423, or 19.82 rounded to two decimal places.

$\boxed{c}$ *Use your pocket calculator to perform the calculations in Problems 19–38. We suggest that you begin by making a mental estimate of the answer. For example, the answer to Problem 21 might be estimated as $(3 - 6)(14 \times 50) = -2100$. Similarly, the answer to Problem 35 might be estimated as $(1.5)(10)/(20) = .75$. This will help you catch errors caused by pressing the wrong keys or failing to use parentheses properly.*

19. $34.1 - 49.95 + 64.2$
20. $7.465 + 3.12 - .0156$
21. $(3.42 - 6.71)(14.3 \times 51.9)$
22. $(21.34 + 2.37)(74.13 - 26.3)$
23. $\dfrac{514 + 31.9}{52.6 - 50.8}$
24. $\dfrac{547.3 - 832.7}{.0567 + .0416}$
25. $\dfrac{(6.34 \times 10^7)(537.8)}{1.23 \times 10^{-5}}$
26. $\dfrac{(5.23 \times 10^{16})(.0012)}{1.34 \times 10^{11}}$
27. $\dfrac{6.34 \times 10^7}{.00152 + .00341}$
28. $\dfrac{3.134 \times 10^{-8}}{5.123 + 6.1457}$
29. $\dfrac{532 + 1.346}{34.91} (1.75 - 2.61)$
30. $\dfrac{39.95 - 42.34}{15.76 -- 16.71} (5.31 \times 10^4)$
31. $(1.214)^3$
32. $(3.617)^{-2}$
33. $\sqrt[3]{1.215}$
34. $\sqrt[3]{1.5789}$
35. $\dfrac{(1.34)(2.345)^3}{\sqrt{364}}$
36. $\dfrac{(14.72)^{12}(59.3)^{11}}{\sqrt{17.1}}$

37. $\dfrac{\sqrt{130} - \sqrt{5}}{15^6 - 4^8}$ 38. $\dfrac{\sqrt{143.2} + \sqrt{36.1}}{(234.1)^4 - (11.2)^2}$

EXAMPLE C (A speed of light problem)   How long will it take a light ray to reach the earth from the sun? Assume that light travels 2.9979×10^{10} centimeters per second, that 1 mile is equivalent to 1.609 kilometers, and that it is 9.30×10^7 miles from the sun to the earth.

Solution.   The speed of light in kilometers per second is 2.9979×10^5, which we shall round to 2.998×10^5. The time required is

$$\frac{(9.30 \times 10^7)(1.609)}{2.998 \times 10^5} = 4.9912 \times 10^2 \approx 499$$

seconds.

[c] 39. How long will it take a light ray to travel from the moon to the earth? (The distance to the moon is 2.39×10^5 miles.)

[c] 40. How long would it take a rocket traveling 4500 miles per hour to reach the sun from the earth?

[c] 41. A light year is the distance light travels in one year (365.24 days). Our nearest star is 4.300 light years away. How many meters is that?

[c] 42. How long would it take a rocket going 4500 miles per hour to get from the earth to our nearest star (see Problem 41)?

[c] 43. What is the area in square meters of a rectangular field 2.3 light years by 4.5 light years?

[c] 44. What is the volume in cubic meters of a cube 4.3 light years on a side?

Miscellaneous Problems

In Problems 45–48, perform the indicated calculations, writing your answer in scientific notation.

[c] 45. $\dfrac{(2.16 \times 10^4)^3(92{,}400)}{(41{,}200)^2}$ [c] 46. $\dfrac{(2967)(4.16 \times 10^4)}{(9.2 \times 10^{-2})^2}$

[c] 47. $\sqrt[3]{29.64} - \dfrac{1}{(0.387)^2} + \sqrt{51.54}$ [c] 48. $\dfrac{(4.26)^2 + (9.32)^3}{5.8 \times 10^3}$

[c] 49. Find the average of the following weights, all measured in pounds: 184.3, 212, 112.8, 274.3, 147.

[c] 50. Calculate
$[(18.6 - 16)^2 + (22.4 - 16)^2 + (49.0 - 16)^2 + (13.3 - 16)^2]/4.$

[c] 51. Find the area of a rectangle whose dimensions are 39.22 and 11.83 millimeters.

[c] 52. Calculate the volume of a box whose dimensions are 21.92, 18.43, and 12.02 centimeters.

[c] 53. How many centimeters are there in a mile?

[c] 54. Find the volume of a cone whose base radius is 23.2 centimeters and whose height is 42.8 centimeters. (The volume of a cone is given by $V = \pi r^2 h / 3$.)

9.2 in.

10.4 in.

14.3 in.

[c] 55. How many cubic feet of helium are there in a spherical balloon of radius 40.9 meters? (The volume of a sphere is given by $V = 4\pi r^3/3$.)

[c] 56. Find the area of the trapezoid shown in the margin.

[c] 57. If I walk 43.6 meters due east and then 97.6 meters due north, how far am I from my starting point?

[c] 58. Find the area of the ring shown in the margin if the outer circle has radius 13.25 centimeters and the inner circle has radius 6.42 centimeters.

[c] 59. Suppose that a clock strikes once at 1 o'clock, twice at 2 o'clock, and so on. How many times will it strike in a century? Assume that a century has 25 leap years.

[c] 60. Henry Clemens is 85 years old today and has seen 21 leap years come and go. Assuming that his heart has beat 74 times per minute over all his years, find the total number of beats during his 85 years.

One mole of any substance is an amount equal to its molecular weight in grams. Avogadro's number (6.02×10^{23}) is the number of molecules in a mole. The atomic weights of hydrogen, carbon, and oxygen are 1, 12, and 16, respectively. Thus the molecular weight of water (H_2O) is $2 \cdot 1 + 1 \cdot 16 = 18$.

[c] 61. How many molecules are there in 36 grams of water? (*Hint:* Thirty-six grams of water equal 2 moles, since the molecular weight of water is 18.)

[c] 62. How many molecules are there in 56.3 grams of water?

[c] 63. How many molecules are there in 51.2 grams of carbon dioxide (CO_2)? Note that the molecular weight of carbon dioxide is $1 \cdot 12 + 2 \cdot 16 = 44$.)

[c] 64. How many molecules are there in 36 grams of cholesterol ($C_{27}H_{46}O$)?

[c] 65. One mole of a gas occupies 22.414 liters under standard conditions. Find the volume occupied by 51.2 grams of the gas carbon dioxide under these conditions.

[c] 66. How much does one molecule of water weigh?

$$3x^2 + 11x + 9$$

$$\frac{3}{2}x - 5$$

$$\sqrt{3}x^5 + \pi x^3 + 2x^2 + \frac{5}{4}$$

$$-3x^4 + 2x^3 - 3x$$

$$5x^{12}$$

$$39$$

3.13

Poly-what?

polygamist   A person who has several spouses.
polyglot   A person who speaks several languages.
polygon   A plane figure which has several sides.
polymath   A person who knows much about many subjects.
polynomial   An expression which is the sum of several terms.
polyphony   A musical combination of several harmonizing parts.

2-3
Polynomials

The dictionary definition of polynomial given above is suggestive, but it is not nearly precise enough for mathematicians. The expression $2x^{-2} + \sqrt{x} + 1/x$ is a sum of several terms, but it is not a polynomial. On the other hand, $3x^4$ is just one term; yet it is a perfectly good polynomial. Fundamentally, a **real polynomial in x** is any expression that can be obtained from the real numbers and x using only the operations of addition, subtraction, and multiplication. For example, we can get $3 \cdot x \cdot x \cdot x \cdot x$ or $3x^4$ by multiplication. We can get $2x^3$ by the same process and then have

$$3x^4 + 2x^3$$

by addition. We could never get $2x^{-2} = 2/x^2$ or $\sqrt{x}$; the first involves a division and the second involves taking a root. Thus $2/x^2$ and $\sqrt{x}$ are not polynomials. Try to convince yourself that all of the expressions in the box above are polynomials.

There is another way to define a polynomial in fewer words. A **real polynomial in x** is any expression of the form

$$a_n x^n + a_{n-1} x^{n-1} + a_{n-2} x^{n-2} + \cdots + a_1 x + a_0$$

where the a's are real numbers and n is a nonnegative integer. We shall not have much to do with complex polynomials in x, but they are defined in exactly the same way, except that the a's are allowed to be complex numbers. In either case, we refer to the a's as **coefficients.** The **degree** of a polynomial is the largest exponent that occurs in the polynomial in a term with a nonzero coefficient. Here are several more examples.

1. $\frac{3}{2}x - 5$ is a first degree (or **linear**) polynomial in x.
2. $3y^2 - 2y + 16$ is a second degree (or **quadratic**) polynomial in y.
3. $\sqrt{3}t^5 - \pi t^2 - 17$ is a fifth degree polynomial in t.

highest

4. $5x^3$ is a third degree polynomial in x. It is also called a **monomial** since it has only one term.

5. 13 is a polynomial of degree zero. Does that seem strange to you? If it helps, think of it as $13x^0$. In general, any nonzero constant polynomial has degree zero. We do not define the degree of the zero polynomial.

The whole subject of polynomials would be pretty dull and almost useless if it stopped with a definition. Fortunately, polynomials—like numbers—can be manipulated. In fact, they behave something like the integers. Just as the sum, difference, and product of two integers are integers, so the sum, difference, and product of two polynomials are polynomials. Notice that we did not mention division; that subject is discussed in Section 2-6 and in Chapter 7.

ADDITION

Adding two polynomials is a snap. Treat x like a number and use the commutative, associative, and distributive properties freely. When you are all done, you will discover that you have just grouped like terms (that is, terms of the same degree) and added their coefficients. Here, for example, is how we add $x^3 + 2x^2 + 7x + 5$ and $x^2 - 3x - 4$.

$$(x^3 + 2x^2 + 7x + 5) + (x^2 - 3x - 4)$$
$$= x^3 + (2x^2 + x^2) + (7x - 3x)$$
$$+ (5 - 4) \qquad \text{(associative and}$$
$$\text{commutative properties)}$$
$$= x^3 + (2 + 1)x^2 + (7 - 3)x + 1 \quad \text{(distributive property)}$$
$$= x^3 + 3x^2 + 4x + 1$$

How important are the parentheses in this example? Actually, they are indispensable only in the third line, where we used the distributive property. Why did we use them in the first and second lines? Because they emphasize what is happening. In the first line, they show which polynomials are added; in the second line, they draw attention to the terms being grouped. To shed additional light, notice that

$$x^3 + 3x^2 + 4x + 1$$

is the correct answer not only for

$$(x^3 + 2x^2 + 7x + 5) + (x^2 - 3x - 4)$$

but also for

$$(x^3 + 2x^2) + (7x + 5 + x^2) + (-3x - 4)$$

and even for

$$(x^3) + (2x^2) + (7x) + (5) + (x^2) + (-3x) + (-4)$$

SUBTRACTION

How do we subtract two polynomials? We replace the subtracted polynomial by its negative and add. For example, we rewrite

$$(3x^2 - 5x + 2) - (5x^2 + 4x - 4)$$

as

$$(3x^2 - 5x + 2) + (-5x^2 - 4x + 4)$$

Then, after grouping like terms, we obtain

$$(3x^2 - 5x^2) + (-5x - 4x) + (2 + 4)$$
$$= (3 - 5)x^2 + (-5 - 4)x + (2 + 4)$$
$$= -2x^2 - 9x + 6$$

If you can go directly from the original problem to the answer, do so; we do not want to make simple things complicated. But be sure to note that

$$(3x^2 - 5x + 2) - (5x^2 + 4x - 4) \neq 3x^2 - 5x + 2 - 5x^2 + 4x - 4$$

The minus sign in front of $(5x^2 + 4x - 4)$ changes the sign of all three terms.

MULTIPLICATION

The distributive property is the basic tool in multiplication. Here is a simple example using $a(b + c) = ab + ac$.

$$(3x^2)(2x^3 + 7) = (3x^2)(2x^3) + (3x^2)(7)$$
$$= 6x^5 + 21x^2$$

Here is a more complicated example, which uses the distributive property $(a + b)c = ac + bc$ at the first step.

$$(3x - 4)(2x^3 - 7x + 8) = (3x)(2x^3 - 7x + 8) + (-4)(2x^3 - 7x + 8)$$
$$= (3x)(2x^3) + (3x)(-7x) + (3x)(8) + (-4)(2x^3)$$
$$+ (-4)(-7x) + (-4)(8)$$
$$= 6x^4 - 21x^2 + 24x - 8x^3 + 28x - 32$$
$$= 6x^4 - 8x^3 - 21x^2 + 52x - 32$$

Notice that each term of $3x - 4$ multiplies each term of $2x^3 - 7x + 8$.

If the process just illustrated seems unwieldy, you may find the format below helpful.

$$
\begin{array}{r}
2x^3 - 7x + 8 \\
3x - 4 \\
\hline
6x^4 \qquad\ -21x^2 + 24x \\
-\ 8x^3 \qquad\quad +\ 28x - 32 \\
\hline
6x^4 - 8x^3 - 21x^2 + 52x - 32
\end{array}
$$

When both polynomials are linear (that is, of the form $ax + b$), there is a handy shortcut. For example, just one look at $(x + 4)(x + 5)$ convinces us that the product has the form $x^2 + (\quad) + 20$. It is the middle term that may cause a little trouble. Think of it this way.

$$\overset{5x}{\underset{4x}{}} \qquad \overset{5x + 4x}{}$$

$$(x + 4)(x + 5) = x^2 + (\quad) + 20$$
$$= x^2 + 9x + 20$$

Similarly,

$$\overset{10x}{\underset{-3x}{}} \qquad \overset{10x - 3x}{}$$

$$(2x - 3)(x + 5) = 2x^2 + (\quad) - 15$$
$$= 2x^2 + 7x - 15$$

Finally,

$$\overset{-21x}{\underset{10x}{}} \qquad \overset{-21x + 10x}{}$$

$$(3x + 2)(5x - 7) = 15x^2 + (\quad) - 14$$
$$= 15x^2 - 11x - 14$$

Soon you should be able to find such simple products in your head.

THREE SPECIAL PRODUCTS

Some products occur so often that they deserve to be highlighted.

$$(x + 5)^2 = x^2 + 25$$
$$(x + 5)^2 = x^2 + 10x + 25$$

$$(x + a)(x - a) = x^2 - a^2$$
$$(x + a)^2 = x^2 + 2ax + a^2$$
$$(x - a)^2 = x^2 - 2ax + a^2$$

Here are some illustrations.

$$(x + 7)(x - 7) = x^2 - 7^2 = x^2 - 49$$
$$(x + 3)^2 = x^2 + 2 \cdot 3x + 3^2 = x^2 + 6x + 9$$
$$(x - 4)^2 = x^2 - 2 \cdot 4x + 4^2 = x^2 - 8x + 16$$

POLYNOMIALS IN SEVERAL VARIABLES

In the box above, we assumed that you would think of x as a variable and a as a constant. However, that is not necessary. If we consider both x and a to be variables, then expressions like $x^2 - a^2$ and $x^2 + 2ax + a^2$ are polynomials in two variables. Other examples are

$$x^2y + 3xy + y \qquad u^3 + 3u^2v + 3uv^2 + v^3$$

Polynomials in several variables present little in the way of new ideas; we defer further consideration of them to Example C.

Problem Set 2-3

Decide whether the given expression is a polynomial. If it is, give its degree.

1. $3x^2 - x + 2$
2. $4x^5 - x$
3. $\pi s^5 - \sqrt{2}$
4. $3\sqrt{2}t$
5. $16\sqrt{2}$
6. $511/\sqrt{2}$
7. $3t^2 + \sqrt{t} + 1$
8. $t^2 + 3t + 1/t$
9. $3t^{-2} + 2t^{-1} + 5$
10. $5 + 4t + 6t^{10}$

Perform the indicated operations and simplify. Write your answer as a polynomial in descending powers of the variable.

11. $(2x - 7) + (-4x + 8)$
12. $(\frac{3}{2}x - \frac{1}{4}) + \frac{5}{6}x$
13. $(2x^2 - 5x + 6) + (2x^2 + 5x - 6)$
14. $(\sqrt{3}t + 5) + (6 - 4 - 2\sqrt{3}t)$
15. $(5 - 11x^2 + 4x) + (x - 4 + 9x^2)$
16. $(x^2 - 5x + 4) + (3x^2 + 8x - 7)$
17. $(2x - 7) - (-4x + 8)$
18. $(\frac{3}{2}x - \frac{1}{4}) - \frac{5}{6}x$
19. $(2x^2 - 5x + 6) - (2x^2 + 5x - 6)$
20. $y^3 - 4y + 6 - (3y^2 + 6y - 3)$
21. $5x(7x - 11) + 19$
22. $-x^2(7x^3 - 5x + 1)$
23. $(t + 5)(t + 11)$
24. $(t - 5)(t + 13)$
25. $(x + 9)(x - 10)$
26. $(x - 13)(x - 7)$
27. $(2t - 1)(t + 7)$
28. $(3t - 5)(4t - 2)$
29. $(4 + y)(y - 2)$
30. $1 + y(y - 2)$
31. $(3.41x - 2.53)(2.34x + 1.77)$ [c]
32. $(.66x + .87)(.41x - 3.12)$ [c]

EXAMPLE A (**More on special products**)   Find each of the following products.
(a) $(2t + 9)(2t - 9)$
(b) $(2x^2 - 3x)^2$
(c) $[(x^2 + 2) + x][(x^2 + 2) - x]$

Solution.   We appeal to the boxed results on page 62.
(a) $(2t + 9)(2t - 9) = (2t)^2 - 9^2 = 4t^2 - 81$
(b) $(2x^2 - 3x)^2 = (2x^2)^2 - 2(2x^2)(3x) + (3x)^2 = 4x^4 - 12x^3 + 9x^2$
(c) We apply the first formula in the box with $x^2 + 2$ and x playing the roles of x and a, respectively.

$$[(x^2 + 2) + x][(x^2 + 2) - x] = (x^2 + 2)^2 - x^2$$
$$= x^4 + 4x^2 + 4 - x^2$$
$$= x^4 + 3x^2 + 4$$

Use the special product formulas to perform the following multiplications. Write your answer as a polynomial in descending powers of the variable.

33. $(x + 10)^2$

34. $(y + 12)^2$

35. $(x + 8)(x - 8)$

36. $(t - 5)(t + 5)$

37. $(2t - 5)^2$

38. $(3s + 11)^2$

39. $(2x^4 + 5x)(2x^4 - 5x)$

40. $(u^3 + 2u^2)(u^3 - 2u^2)$

41. $[(t + 2) + t^3]^2$

42. $[(1 - x) + x^2]^2$

43. $[(t + 2) + t^3][(t + 2) - t^3]$

44. $[(1 - x) + x^2][(1 - x) - x^2]$

ⓒ 45. $(2.3x - 1.4)^2$

ⓒ 46. $(2.43x - 1.79)(2.43x + 1.79)$

EXAMPLE B (Expanding cubes)   It is easy to show that

$$(x + a)^3 = x^3 + 3ax^2 + 3a^2x + a^3$$
$$(x - a)^3 = x^3 - 3ax^2 + 3a^2x - a^3$$

Use these results to find (a) $(x - 5)^3$;   (b) $(2x^2 + 3x)^3$.

Solution.

(a) $(x - 5)^3 = x^3 - 3 \cdot 5x^2 + 3 \cdot 5^2x - 5^3$
$$= x^3 - 15x^2 + 75x - 125$$

(b) $(2x^2 + 3x)^3 = (2x^2)^3 + 3(2x^2)^2(3x) + 3(2x^2)(3x)^2 + (3x)^3$
$$= 8x^6 + 36x^5 + 54x^4 + 27x^3$$

Expand the following cubes.

47. $(x + 2)^3$

48. $(x - 4)^3$

49. $(2t - 3)^2$

50. $(3u + 1)^3$

51. $(2t + t^2)^3$

52. $(4 - 3t^2)^3$

53. $[(2t + 1) + t^2]^3$

54. $[u^2 - (u + 1)]^3$

EXAMPLE C (Multiplying polynomials in several variables)   Find the following products.

(a) $(3x + 2y)(x - 5y)$;   (b) $(2xy - 3z)^2$

Solution.

(a) $(3x + 2y)(x - 5y) = 3x^2 - 13xy - 10y^2$

(b) $(2xy - 3z)^2 = (2xy)^2 - 2(2xy)(3z) + (3z)^2$
$$= 4x^2y^2 - 12xyz + 9z^2$$

Perform the indicated operations and simplify.

55. $(x - 3y)^2$

56. $(2x + y)^2$

57. $(3x - 2y)(3x + 2y)$

58. $(u^2 - 2v)(u^2 + 2v)$

59. $(3x - y)(4x + 5y)$ 60. $(2x + y)(6x + 5y)$

61. $(2x^2y + z)(x^2y - z)$ 62. $(3x + 5y^2z)(x - y^2z)$

63. $(t + 1 + s)(t + 1 - s)$ 64. $(3u + 2v + 1)^2$

65. $(2t - 3s)^3$ 66. $(u + 4v)^3$

Miscellaneous Problems

Perform the indicated operations and simplify. Write your answer as a polynomial in descending powers of the variable.

67. $2x^2 - 4x + 5 - 3x(x^2 + 7x - 2)$ 68. $(2y - 4)(y + 3) + (y^2 - 9)$

69. $(y + 3)(2y - 5)$ 70. $(2s - 1)(3s + 2)$

71. $(2y + 5)^2$ 72. $(3x - 2)^2$

73. $(2z - 7)(2z + 7)$ 74. $(5x + 4)(5x - 4)$

75. $(x^2 + 4)(x^2 - 3)$ 76. $(y^3 + 2)(y^3 - 5)$

77. $(2x^2 + x - 5)(3x + 2)$ 78. $(3x^2 - 4x + 1)(2x^2 - 5)$

79. $(x + \sqrt{2})(x - \sqrt{2})$ □ 80. $(x + 2i)(x - 2i)$

81. $(x - 2\sqrt{7})^2$ 82. $(\sqrt{2}x + 5)^2$

83. $(2x + 1)^2 - (2x - 1)^2$ 84. $4(x + 2)^2 - (x + 4)^2$

85. $(2x + 1)^3$ 86. $(3y + 2)^3$

ⓒ 87. $(2.39x + 4.12)^2 - 4.13x$ ⓒ 88. $(2.39x - 4.12)(2.39x - 4.12)$
$\qquad\qquad\qquad\qquad\qquad\qquad\qquad\qquad + (1.12x)^2$

The remaining problems involve polynomials in several variables. In each case, do the indicated operations and simplify.

89. $(5x - y)(5x + y)$ 90. $(2x + 3y)(2x - 3y)$

91. $(5x - y)^2 + y^2$ 92. $(2x + 3y)^2 + 3x^2$

93. $(2x - 3y)(4x + y)$ 94. $(2x + 4y)(x - y)$

95. $(x + y)(x^2 - xy + y^2)$ 96. $(x - y)(x^2 + xy + y^2)$

97. $(x + 2y)^3$ 98. $(x - 2y)^3$

99. $(x^2 + y^2)(x^2 - y^2)$ 100. $(x^2 + 2xy)(x^2 - 2xy)$

101. $(x^2 + 2y^2)(x^2 - 3y^2)$ 102. $(2x^2 - 3y^2)(2x^2 + 4y^2)$

103. $(3xy + z)(2xy - 3z)$ 104. $(2xy + 3uv)(2xy - 4uv)$

105. $(x^2 + xy + y^2)(x^2 + 2xy - y^2)$ 106. $(x^2 - 3xy + y^2)(x^2 + y^2)$

Polynomial to be Factored	Johnny's Answer	Teacher's Comments
1. $x^6 + 2x^2$	$x^2(x^3 + 2)$	Wrong. Have you forgotten that $x^2x^3 = x^5$? Right answer: $x^2(x^4 + 2)$
2. $x^2 + 5x + 6$	$(x + 6)(x + 1)$	Wrong. You didn't check the middle term. Right answer: $(x + 2)(x + 3)$
3. $x^2 - 4y^2$	$(x + 2y)(x - 2y)$	Right.
4. $x^2 + 4y^2$	$(x + 2y)^2$	Wrong. $(x + 2y)^2 = x^2 + 4xy + 4y^2$ Right answer: $x^2 + 4y^2$ doesn't factor using real coefficients.
5. $x^2y^2 + 6xy + 9$	Impossible	Wrong. x^2y^2 and 9 are squares. You should have suspected a perfect square. Right answer: $(xy + 3)^2$

2-4 Factoring Polynomials

To factor 90 means to write it as a product of smaller numbers; to factor it completely means to write it as a product of primes, that is, numbers that cannot be factored further. Thus we have factored 90 when we write $90 = 9 \cdot 10$, but it is not factored completely until we write

$$90 = 2 \cdot 3 \cdot 3 \cdot 5$$

Similarly to **factor** a polynomial means to write it as a product of simpler polynomials; to **factor** a polynomial **completely** is to write it as a product of polynomials that cannot be factored further. Thus when we write

$$x^3 - 9x = x(x^2 - 9)$$

we have factored $x^3 - 9x$, but not until we write

$$x^3 - 9x = x(x + 3)(x - 3)$$

have we factored $x^3 - 9x$ completely.

Now why can't Johnny factor? He can't factor because he can't multiply. If he doesn't know that

$$(x + 2)(x + 3) = x^2 + 5x + 6$$

he certainly is not going to know how to factor $x^2 + 5x + 6$. That is why we urge you to memorize the special product formulas on page 67. Of course, a product formula is also a factoring formula when read from right to left.

To urge memorization may be a bit old-fashioned, but we suggest that a fact, once memorized, becomes a permanent friend. It is best to memorize in words. For example, read formula 3 as follows: the square of a sum of two terms is the first squared plus twice their product plus the second squared.

$$\boxed{\begin{array}{l}
\text{PRODUCT FORMULAS} \longrightarrow \\
\longleftarrow \text{\quad\quad FACTORING FORMULAS} \\[4pt]
\text{1. } a(x + y + z) = ax + ay + az \\
\text{2. } (x + a)(x + b) = x^2 + (a + b)x + ab \\
\text{3. } (x + y)^2 = x^2 + 2xy + y^2 \\
\text{4. } (x - y)^2 = x^2 - 2xy + y^2 \\
\text{5. } (x + y)(x - y) = x^2 - y^2 \\
\text{6. } (x + y)^3 = x^3 + 3x^2y + 3xy^2 + y^3 \\
\text{7. } (x - y)^3 = x^3 - 3x^2y + 3xy^2 - y^3 \\
\text{8. } (x + y)(x^2 - xy + y^2) = x^3 + y^3 \\
\text{9. } (x - y)(x^2 + xy + y^2) = x^3 - y^3
\end{array}}$$

TAKING OUT A COMMON FACTOR

This, the simplest factoring procedure, is based on formula 1 above. You should always try this process first. Take Johnny's first problem as an example. Both terms of $x^6 + 2x^2$ have x^2 as a factor, so we take it out.

$$x^6 + 2x^2 = x^2(x^4 + 2)$$

Always factor out as much as you can. Taking 2 out of $4xy^2 - 6x^3y^4 + 8x^4y^2$ is not nearly enough, though it is a common factor; taking out $2xy$ is not enough either. You should take out $2xy^2$. Then

$$4xy^2 - 6x^3y^4 + 8x^4y^2 = 2xy^2(2 - 3x^2y^2 + 4x^3)$$

FACTORING BY TRIAL AND ERROR

In factoring, as in life, success often results from trying and trying again. What does not work is systematically eliminated; eventually, effort is rewarded. Let us see how this process works on $x^2 - 5x - 14$. We need to find numbers a and b such that

$$x^2 - 5x - 14 = (x + a)(x + b)$$

Since ab must equal -14, two possibilities immediately occur to us: $a = 7$ and $b = -2$ or $a = -7$ and $b = 2$. Try them both to see if one works.

$$(x + 7)(x - 2) = x^2 + 5x - 14$$

$$(x - 7)(x + 2) = x^2 - 5x - 14 \qquad \text{Success!}$$

The brackets help us calculate the middle term, the crucial step in this kind of factoring.

Here is a tougher factoring problem: Factor $2x^2 + 13x - 15$. It is a safe bet that if $2x^2 + 13x - 15$ factors at all, then

$$2x^2 + 13x - 15 = (2x + a)(x + b)$$

Since $ab = -15$, we are likely to try combinations of 3 and 5 first.

$$(2x + 5)(x - 3) = 2x^2 - x - 15$$

$$(2x - 5)(x + 3) = 2x^2 + x - 15$$

$$(2x + 3)(x - 5) = 2x^2 - 7x - 15$$

$$(2x - 3)(x + 5) = 2x^2 + 7x - 15$$

Discouraging, isn't it? But that is a poor reason to give up. Maybe we have missed some possibilities. We have, since combinations of 15 and 1 might work.

$$(2x - 15)(x + 1) = 2x^2 - 13x - 15$$

$$(2x + 15)(x - 1) = 2x^2 + 13x - 15 \qquad \text{Success!}$$

When you have had a lot of practice, you will be able to speed up the process. You will simply write

$$2x^2 + 13x - 15 = (2x + ?)(x + ?)$$

and mentally try the various possibilities until you find the right one. Of course, it may happen, as in the case of $2x^2 - 4x + 5$, that you cannot find a factorization.

PERFECT SQUARES

Certain second degree (quadratic) polynomials are a breeze to factor.

$$x^2 + 10x + 25 = (x + 5)(x + 5) = (x + 5)^2$$

$$x^2 - 12x + 36 = (x - 6)^2$$

$$4x^2 + 12x + 9 = (2x + 3)^2$$

These are modeled after the special product formulas 3 and 4, which we now write with a and b replacing x and y.

$$a^2 + 2ab + b^2 = (a + b)^2$$
$$a^2 - 2ab + b^2 = (a - b)^2$$

We look for first and last terms that are squares, say of a and b. Then we ask if the middle term is twice their product.

But we need to be very flexible; a and b might be quite complicated. Consider $x^4 + 2x^2y^3 + y^6$. The first term is the square of x^2 and the last is the square of y^3; the middle term is twice their product.

$$x^4 + 2x^2y^3 + y^6 = (x^2)^2 + 2(x^2)(y^3) + (y^3)^2$$
$$= (x^2 + y^3)^2$$

Similarly,

$$y^2z^2 - 6ayz + 9a^2 = (yz - 3a)^2.$$

However,

$$a^4b^2 + 6a^2bc + 4c^2 \neq (a^2b + 2c)^2$$

since the middle term does not check.

DIFFERENCE OF SQUARES

Do you see a common feature in the following polynomials?

$$x^2 - 16 \qquad y^2 - 100 \qquad 4y^2 - 9b^2$$

Each is the difference of two squares. From one of our special product formulas (formula 5), we know that

$$\boxed{a^2 - b^2 = (a + b)(a - b)}$$

Thus

$$x^2 - 16 = (x + 4)(x - 4)$$
$$y^2 - 100 = (y + 10)(y - 10)$$
$$4y^2 - 9b^2 = (2y + 3b)(2y - 3b)$$

SUM AND DIFFERENCE OF CUBES

Now we are ready for some high-class factoring. Consider $8x^3 + 27$ and $x^3z^3 - 1000$. The first is a sum of cubes and the second is a difference of cubes. The secrets to success are the two special product formulas for cubes, restated here.

$$\boxed{\begin{aligned} a^3 + b^3 &= (a + b)(a^2 - ab + b^2) \\ a^3 - b^3 &= (a - b)(a^2 + ab + b^2) \end{aligned}}$$

To factor $8x^3 + 27$, replace a by $2x$ and b by 3 in the first formula.

$$8x^3 + 27 = (2x)^3 + 3^3 = (2x + 3)[(2x)^2 - (2x)(3) + 3^2]$$
$$= (2x + 3)(4x^2 - 6x + 9)$$

Similarly, to factor $x^3z^3 - 1000$, let $a = xz$ and $b = 10$ in the second formula.

$$x^3z^3 - 1000 = (xz)^3 - 10^3 = (xz - 10)(x^2z^2 + 10xz + 100)$$

Someone is sure to make a terrible mistake and write

$$x^3 + y^3 = (x + y)^3 \qquad \text{Wrong!!}$$

Remember that

$$(x + y)^3 = x^3 + 3x^2y + 3xy^2 + y^3$$

The boxed caution panel:

$x^3 - 8 = (x - 2)(x^2 + 4)$
$x^3 - 8 = (x - 2)(x^2 - 4x + 4)$
$x^3 - 8 = (x - 2)(x^2 + 2x + 4)$

TO FACTOR OR NOT TO FACTOR

Which of the following can be factored?

(i) $\qquad x^2 - 4$

(ii) $\qquad x^2 - 6$

(iii) $\qquad x^2 + 16$

Did you say only the first one? You are correct if we insist on integer coefficients, or as we say, if we **factor over the integers.** But if we factor over the real numbers (that is, insist only that the coefficients be real), then the second polynomial can be factored.

$$x^2 - 6 = (x + \sqrt{6})(x - \sqrt{6})$$

If we factor over the complex numbers, even the third polynomial can be factored.

$$x^2 + 16 = (x + 4i)(x - 4i)$$

For this reason, we should always spell out what kind of coefficients we permit in the answer. We give specific directions in the following problem set. Incidentally, before trying the problems, review the opening panel of this section. It should make good sense now.

Problem Set 2-4

Factor completely over the integers (that is, allow only integer coefficients in your answers).

1. $x^2 + 5x$
2. $y^3 + 4y^2$
3. $x^2 + 5x - 6$
4. $x^2 + 5x + 4$
5. $y^4 - 6y^3$
6. $t^4 + t^2$
7. $y^2 + 4y - 12$
8. $z^2 - 3z - 40$
9. $y^2 + 8y + 16$
10. $9x^2 + 24x + 16$
11. $4x^2 - 12xy + 9y^2$
12. $9x^2 - 6x + 1$
13. $y^2 - 64$
14. $x^2 - 4y^2$
15. $1 - 25b^2$

16. $9x^2 - 64y^2$

17. $4z^2 - 4z - 3$

18. $7x^2 - 19x - 6$

19. $20x^2 + 3xy - 2y^2$

20. $4x^2 + 13xy - 12y^2$

21. $x^3 + 27$

22. $y^3 - 27$

23. $a^3 - 8b^3$

24. $8a^3 - 27b^3$

25. $x^3 - x^3y^3$

26. $x^6 + x^3y^3$

27. $x^2 - 3$

28. $y^2 - 5$

29. $3x^2 - 4$

Factor completely over the real numbers.

30. $x^2 - 3$

31. $y^2 - 5$

32. $3x^2 - 4$

33. $5z^2 - 4$

34. $t^4 - t^2$

35. $t^4 - 2t^2$

36. $x^2 + 2\sqrt{2}x + 2$

37. $y^2 - 2\sqrt{3}y + 3$

38. $x^2 + 4y^2$

39. $x^2 + 9$

40. $4x^2 + 1$

☐ *Factor completely over the complex numbers.*

41. $x^2 + 9$

42. $4x^2 + 1$

EXAMPLE A (Factoring by substitution)   Factor completely over the integers.

(a) $3x^4 + 10x^2 - 8$

(b) $(x + 2y)^2 - 3(x + 2y) - 10$

Solution.

(a) Replace x^2 by u (or some other favorite letter of yours). Then

$$3x^4 + 10x^2 - 8 = 3u^2 + 10u - 8$$

But we know how to factor the latter:

$$3u^2 + 10u - 8 = (3u - 2)(u + 4)$$

Thus, when we go back to x, we get

$$3x^4 + 10x^2 - 8 = (3x^2 - 2)(x^2 + 4)$$

Neither of these quadratic polynomials factors further (using integer coefficients), so we are done.

(b) Here we could let $u = x + 2y$ and then factor the resulting quadratic polynomial. But this time, let us do that step mentally and write

$$(x + 2y)^2 - 3(x + 2y) - 10 = [(x + 2y) + ?][(x + 2y) - ?]$$
$$= (x + 2y + 2)(x + 2y - 5)$$

Factor completely over the integers. If you can factor without actually making a substitution, that is fine.

43. $x^6 + 9x^3 + 14$

44. $x^4 - x^2 - 6$

45. $4x^4 - 37x^2 + 9$

46. $6y^4 + 13y^2 - 5$

47. $(x + 4y)^2 + 6(x + 4y) + 9$

48. $(m - n)^2 + 5(m - n) + 4$

49. $x^4 - x^2y^2 - 6y^4$

50. $x^4y^4 + 5x^4y^2 + 6x^4$

EXAMPLE B (Factoring in stages)   Factor $x^6 - y^6$ over the integers.

Solution.   First think of this expression as a difference of squares. Then factor again.

$$x^6 - y^6 = (x^3 - y^3)(x^3 + y^3)$$
$$= (x - y)(x^2 + xy + y^2)(x + y)(x^2 - xy + y^2)$$

Factor completely over the integers.

51. $x^6 - 64$ 52. $x^4 - y^4$ 53. $x^8 - x^4y^4$
54. $x^9 - 64x^3$ 55. $x^6 + y^6$ (sum of cubes)   56. $a^6 + 64$

EXAMPLE C (Factoring by grouping)   Factor.
   (a) $am - an - bm + bn$
   (b) $a^2 - 4ab + 4b^2 - c^2$

Solution.   To factor expressions involving more than three terms will usually require grouping of some of the terms together.
   (a) $am - an - bm + bn = (am - an) - (bm - bn)$
$$= a(m - n) - b(m - n)$$
$$= (a - b)(m - n)$$

   Note that $m - n$ was a common factor of both terms. Only after removing that factor did we have a factored form.
   (b) $a^2 - 4ab + 4b^2 - c^2 = (a^2 - 4ab + 4b^2) - c^2$
$$= (a - 2b)^2 - c^2 \quad \text{(difference of squares)}$$
$$= [(a - 2b) + c][(a - 2b) - c]$$
$$= (a - 2b + c)(a - 2b - c)$$

Factor completely over the integers.

57. $x^3 - 4x^2 + x - 4$ 58. $y^3 + 3y^2 - 2y - 6$
59. $4x^2 - 4x + 1 - y^2$ 60. $9a^2 - 4b^2 - 12b - 9$
61. $3x + 3y - x^2 - xy$ 62. $y^2 - 3y + xy - 3x$
63. $x^2 + 6xy + 9y^2 + 2x + 6y$ 64. $y^2 + 4xy + 4x^2 - 3y - 6x$
65. $x^2 + 2xy + y^2 + 3x + 3y + 2$
66. $a^2 - 2ab + b^2 - c^2 + 4cd - 4d^2$

EXAMPLE D (Factoring by adding and subtracting the same thing)   Factor $x^4 + 4$ over the integers.

Solution.   Most people would bet money that this cannot be factored. We have to admit that it is tricky. But see what happens when we add and subtract $4x^2$.

$$x^4 + 4 = x^4 + 4x^2 + 4 - 4x^2$$
$$= (x^4 + 4x^2 + 4) - 4x^2$$
$$= (x^2 + 2)^2 - (2x)^2 \qquad \text{(difference of squares)}$$
$$= (x^2 + 2 + 2x)(x^2 + 2 - 2x)$$
$$= (x^2 + 2x + 2)(x^2 - 2x + 2)$$

Factor completely over the integers.

67. $x^4 + 64$ 68. $y^8 + 4$ 69. $x^4 + x^2 + 1$ 70. $x^8 + 3x^4 + 4$

Miscellaneous Problems

Factor completely over the integers.

71. $5x^2 - 10x$

72. $3x^2 - 18x + 27$

73. $4x^2 + 4x + 1$

74. $9 - 4m^2$

75. $4y^2 - 1$

76. $6x^2 - 5x + 1$

77. $2y^2 - 7y + 5$

78. $6x^2 - 5x - 6$

79. $4x^4 - 16$

80. $4x^4 - 4x$

81. $a^2 + 2ab + b^2 - 25$

82. $b^2 + 4bc + 4c^2 + 3b + 6c$

83. $a^2 + 2ab + b^2 - 2a - 2b$

84. $b^2 + 4bc + 4c^2 - 1$

85. $-2xy^3 + 54x$

86. $x^3y^2 + x^2y^3 + 2xy^4$

87. $x^4 - y^2 + 4y - 4$

88. $a^6 - b^6$

89. $(a + 2b)^2 - (a + 2b) - 6$

90. $x^2 + 3x + 4$

91. $x^2 + x + 1$

92. $x^4 + 3x^2 + 4$

93. $x^4 + x^2 + 1$

94. $4b^4 - 37b^2 + 9$

95. $(x + 3)^2(x - 7)^3 + (x + 3)(x - 7)^2$

96. $(x - 4)(x + 5)^4 - (x - 4)^2(x + 5)^3$

c *Factor completely over the real numbers.*

97. $x^2 - 3.124$ 98. $x^3 - 5.13$ 99. $x^3 + 4.91$ 100. $2x^2 - 3.22$

i *Factor completely over the complex numbers.*

101. $4x^2 + 9$

102. $9z^2 + 1$

103. $x^3 + 8i$ (*Hint:* $8i = -(2i)^3$)

104. $27x^3 - i$

A Riddle for Thomas Jefferson

Riddle: What do the following have in common?

(i) The Declaration of Independence

(ii) $\dfrac{x^2y^5 + 13xy^7 + 25x^4y^2 + 17}{3xy^3 + 105y^2 + 14x - 2x^5y^7}$

Answer: They are both rational expressions.

"I'd say the second expression was a mite more rational!"

Thomas Jefferson

2-5 Rational Expressions

We admit that the riddle above is a bit silly, but it allowed us to insert Jefferson's picture and to make a point that is not well known. Thomas Jefferson was trained in mathematics and wrote that he had been profoundly influenced by his college mathematics teacher.

Recall that the quotient (ratio) of two integers is a rational number. A quotient (ratio) of two polynomials is called a **rational expression.** Here are some examples involving one variable.

$$\frac{x}{x + 1} \qquad \frac{3x^2 + 6}{x^2 + 2x} \qquad \frac{x^{15} + 3x}{6x^2 + 11}$$

The example in the riddle is a rational expression in two variables.

We add, subtract, multiply, and divide rational expressions by the same rules we used for rational numbers (see Section 1-2). The result is always a rational expression. That should not surprise us, since we had the same kind of experience with rational numbers.

Our immediate task is to define the reduced form for a rational expression. This, too, generalizes a notion we had for rational numbers.

REDUCED FORM

A rational expression is in **reduced form** if its numerator and denominator have no (nontrivial) common factor. For example, $x/(x + 2)$ is in reduced form, but $x^2/(x^2 + 2x)$ is not, since

$$\frac{x^2}{x^2 + 2x} = \frac{\not{x} \cdot x}{\not{x}(x + 2)} = \frac{x}{x + 2}$$

To reduce a rational expression, we factor numerator and denominator and divide out, or cancel, common factors. Here are two examples.

$$\frac{x^3 + 3x^2}{x^3} \cancel{=} 1 + 3x^2$$

$$\frac{x^3 + 3x^2}{x^3} = \frac{x^2(x + 3)}{x^3}$$

$$= \frac{x + 3}{x}$$

$$\frac{x^2 + 7x + 10}{x^2 - 25} = \frac{(x + 5)(x + 2)}{(x + 5)(x - 5)} = \frac{x + 2}{x - 5}$$

$$\frac{2x^2 + 5xy - 3y^2}{2x^2 + xy - y^2} = \frac{(2x - y)(x + 3y)}{(2x - y)(x + y)} = \frac{x + 3y}{x + y}$$

ADDITION AND SUBTRACTION

We add (or subtract) rational expressions by rewriting them so that they have the same denominator and then adding (or subtracting) the new numerators. Suppose we want to add

$$\frac{3}{x} + \frac{2}{x + 1}$$

The appropriate common denominator is $x(x + 1)$. Remember that we can multiply numerator and denominator of a fraction by the same thing. Accordingly,

$$\frac{3}{x} + \frac{2}{x + 1} = \frac{3(x + 1)}{x(x + 1)} + \frac{x \cdot 2}{x(x + 1)} = \frac{3x + 3 + 2x}{x(x + 1)} = \frac{5x + 3}{x(x + 1)}$$

The same procedure is used in subtraction.

$$\frac{2}{x + 1} - \frac{3}{x} = \frac{x \cdot 2}{x(x + 1)} - \frac{3(x + 1)}{x(x + 1)} = \frac{2x - (3x + 3)}{x(x + 1)}$$

$$= \frac{2x - 3x - 3}{x(x + 1)} = \frac{-x - 3}{x(x + 1)}$$

Here is a more complicated example. Study each step carefully.

Begin by factoring

$(x - 1)^2 (x + 1)$ is the lowest common denominator

Forgetting the parentheses around $2x^2 + 3x + 1$ would be a serious blunder

No cancellation is possible since $x^2 - 6x - 1$ doesn't factor over the integers

$$\frac{3x}{x^2 - 1} - \frac{2x + 1}{x^2 - 2x + 1} = \frac{3x}{(x - 1)(x + 1)} - \frac{2x + 1}{(x - 1)^2}$$

$$= \frac{3x(x - 1)}{(x - 1)^2(x + 1)} - \frac{(2x + 1)(x + 1)}{(x - 1)^2(x + 1)}$$

$$= \frac{3x^2 - 3x - (2x^2 + 3x + 1)}{(x - 1)^2(x + 1)}$$

$$= \frac{3x^2 - 3x - 2x^2 - 3x - 1}{(x - 1)^2(x + 1)}$$

$$= \frac{x^2 - 6x - 1}{(x - 1)^2(x + 1)}$$

MULTIPLICATION

We multiply rational expressions in the same manner as we do rational numbers; that is, we multiply numerators and multiply denominators. For example,

$$\frac{3}{x+5} \cdot \frac{x+2}{x-4} = \frac{3(x+2)}{(x+5)(x-4)} = \frac{3x+6}{x^2+x-20}$$

Sometimes we need to reduce the product, if we want the simplest possible answer. Here is an illustration.

$$\frac{2x-3}{x+5} \cdot \frac{x^2-25}{6xy-9y} = \frac{(2x-3)(x^2-25)}{(x+5)(6xy-9y)}$$

$$= \frac{(2x-3)(x-5)(x+5)}{(x+5)(3y)(2x-3)}$$

$$= \frac{x-5}{3y}$$

This example shows that it is a good idea to do as much factoring as possible at the outset. That is what set up the cancellation.

DIVISION

There are no real surprises with division, as we simply invert the divisor and multiply. Here is a nontrivial example.

$$\frac{x^2-5x+4}{2x+6} \div \frac{2x^2-x-1}{x^2+5x+6} = \frac{x^2-5x+4}{2x+6} \cdot \frac{x^2+5x+6}{2x^2-x-1}$$

$$= \frac{(x^2-5x+4)(x^2+5x+6)}{(2x+6)(2x^2-x-1)}$$

$$= \frac{(x-4)(x-1)(x+2)(x+3)}{2(x+3)(x-1)(2x+1)}$$

$$= \frac{(x-4)(x+2)}{2(2x+1)}$$

Problem Set 2-5

Reduce each of the following.

1. $\dfrac{x+6}{x^2-36}$

2. $\dfrac{x^2-1}{4x-4}$

3. $\dfrac{y^2+y}{5y+5}$

4. $\dfrac{x^2-7x+6}{x^2-4x-12}$

5. $\dfrac{(x+2)^3}{x^2-4}$

6. $\dfrac{x^3+a^3}{(x+a)^2}$

7. $\dfrac{zx^2+4xyz+4y^2z}{x^2+3xy+2y^2}$

8. $\dfrac{x^3-27}{3x^2+9x+27}$

Perform the indicated operations and simplify.

9. $\dfrac{5}{x-2} + \dfrac{4}{x+2}$

10. $\dfrac{5}{x-2} - \dfrac{4}{x+2}$

11. $\dfrac{5x}{x^2-4} + \dfrac{3}{x+2}$

12. $\dfrac{3}{x} - \dfrac{2}{x+3} + \dfrac{1}{x^2+3x}$

13. $\dfrac{2}{xy} + \dfrac{3}{xy^2} - \dfrac{1}{x^2y^2}$

14. $\dfrac{x+y}{xy^3} - \dfrac{x-y}{y^4}$

15. $\dfrac{x+1}{x^2-4x+4} + \dfrac{4}{x^2+3x-10}$

16. $\dfrac{x^2}{x^2-x+1} - \dfrac{x+1}{x}$

EXAMPLE A (The three signs of a fraction)   Simplify

$$\frac{x}{3x-6} - \frac{2}{2-x}$$

Solution.

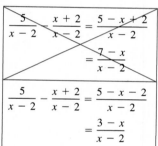

$$\frac{x}{3x-6} - \frac{2}{2-x} = \frac{x}{3(x-2)} - \frac{2}{2-x}$$

Now we make a crucial observation. Notice that

$$-(2-x) = -2 + x = x - 2$$

That is, $2 - x$ and $x - 2$ are negatives of each other. Thus the expression above may be rewritten as

$$\frac{x}{3(x-2)} - \frac{2}{2-x} = \frac{x}{3(x-2)} - \frac{2}{-(x-2)}$$

$$= \frac{x}{3(x-2)} + \frac{2}{x-2}$$

$$= \frac{x}{3(x-2)} + \frac{6}{3(x-2)}$$

$$= \frac{x+6}{3(x-2)}$$

When we replaced $-\dfrac{2}{-(x-2)}$ by $\dfrac{2}{x-2}$, we used the fact that

$$-\frac{a}{-b} = \frac{a}{b}$$

Keep in mind that a fraction has three sign positions: numerator, denominator, and total fraction. You may change any two of them without changing the value of the fraction. Thus

$$\frac{a}{b} = -\frac{a}{-b} = -\frac{-a}{b} = \frac{-a}{-b}$$

Simplify.

17. $\dfrac{4}{2x-1} + \dfrac{x}{1-2x}$

18. $\dfrac{x}{6x-2} - \dfrac{3}{1-3x}$

19. $\dfrac{2}{6y-2} + \dfrac{y}{9y^2-1} - \dfrac{2y+1}{1-3y}$

20. $\dfrac{x}{4x^2-1} + \dfrac{2}{4x-2} - \dfrac{3x+1}{1-2x}$

21. $\dfrac{m^2}{m^2-2m+1} - \dfrac{1}{3-3m}$

22. $\dfrac{2x}{x^2-y^2} + \dfrac{1}{x+y} + \dfrac{1}{y-x}$

In Problems 23–30, multiply and express in simplest form, as illustrated in the text.

23. $\dfrac{5}{2x-1} \cdot \dfrac{x}{x+1}$

24. $\dfrac{3}{x^2-2x} \cdot \dfrac{x-2}{x}$

25. $\dfrac{x+2}{x^2-9} \cdot \dfrac{x+3}{x^2-4}$

26. $\left(1 + \dfrac{1}{x+2}\right)\left(\dfrac{4}{3x+9}\right)$

27. $x^2 y^4 \left(\dfrac{x}{y^2} - \dfrac{y}{x^2}\right)$

28. $\dfrac{5x^2}{x^3+y^3}\left(\dfrac{1}{xy^2} - \dfrac{1}{x^2 y^2} + \dfrac{1}{x^3 y}\right)$

29. $\left(\dfrac{x^2+5x}{x^2-16}\right)\left(\dfrac{x^2-2x-24}{x^2-x-30}\right)$

30. $\left(\dfrac{x^3-125}{2x^3-10x^2}\right)\left(\dfrac{7x}{x^3+5x^2+25x}\right)$

Express the quotients in Problems 31–37 in simplest form, as illustrated in the text.

31. $\dfrac{\dfrac{5}{2x-1}}{\dfrac{x}{x+1}}$

32. $\dfrac{\dfrac{5}{2x-1}}{\dfrac{x}{4x^2-1}}$

33. $\dfrac{\dfrac{x+2}{x^2-4}}{x}$

34. $\dfrac{\dfrac{x+2}{x^2-3x}}{\dfrac{x^2-4}{x}}$

35. $\dfrac{\dfrac{x^2+a^3}{x^3-a^3}}{\dfrac{x+2a}{(x-a)^2}}$

36. $\dfrac{1 + \dfrac{2}{b}}{1 - \dfrac{4}{b^2}}$

37. $\dfrac{\dfrac{y^2+y-2}{y^2+4y}}{\dfrac{2y^2-8}{y^2+2y-8}}$

EXAMPLE B (A quotient arising in calculus)   Simplify

$$\dfrac{\dfrac{2}{x+h} - \dfrac{2}{x}}{h}$$

Solution.   This expression may look artificial, but it is one you are apt to find in calculus. It represents the average rate of change in $2/x$ as x changes to $x + h$. We begin by doing the subtraction in the numerator.

$$\dfrac{\dfrac{2}{x+h} - \dfrac{2}{x}}{h} = \dfrac{\dfrac{2x - 2(x+h)}{(x+h)x}}{h} = \dfrac{\dfrac{2x - 2x - 2h}{(x+h)x}}{\dfrac{h}{1}}$$

$$= \dfrac{-2h}{(x+h)x} \cdot \dfrac{1}{h} = \dfrac{-2}{(x+h)x}$$

Simplify each of the following.

38. $\dfrac{\dfrac{4}{x+h} - \dfrac{4}{x}}{h}$

39. $\dfrac{\dfrac{1}{2x+2h+3} - \dfrac{1}{2x+3}}{h}$

40. $\dfrac{\dfrac{x+h}{x+h+4} - \dfrac{x}{x+4}}{h}$

41. $\dfrac{\dfrac{1}{(x+h)^2} - \dfrac{1}{x^2}}{h}$

EXAMPLE C (Four-story fractions)   Simplify

$$\dfrac{\dfrac{x}{x-4} - \dfrac{3}{x+3}}{\dfrac{1}{x} + \dfrac{2}{x-4}}$$

Solution.   *Method 1*   (Simplify the numerator and denominator separately and then divide.)

$$\dfrac{\dfrac{x}{x-4} - \dfrac{3}{x+3}}{\dfrac{1}{x} + \dfrac{2}{x-4}} = \dfrac{\dfrac{x(x+3) - 3(x-4)}{(x-4)(x+3)}}{\dfrac{x-4+2x}{x(x-4)}}$$

$$= \dfrac{\dfrac{x^2 + 3x - 3x + 12}{(x-4)(x+3)}}{\dfrac{3x-4}{x(x-4)}}$$

$$= \dfrac{x^2 + 12}{(\cancel{x-4})(x+3)} \cdot \dfrac{x(\cancel{x-4})}{3x-4}$$

$$= \dfrac{x(x^2 + 12)}{(x+3)(3x-4)}$$

Method 2   (Multiply the fractions in the numerator and denominator by a common denominator, in this case, $(x-4)(x+3)x$.)

$$\dfrac{\dfrac{x}{x-4} - \dfrac{3}{x+3}}{\dfrac{1}{x} + \dfrac{2}{x-4}} = \dfrac{\left(\dfrac{x}{x-4} - \dfrac{3}{x+3}\right)(x-4)(x+3)x}{\left(\dfrac{1}{x} + \dfrac{2}{x-4}\right)(x-4)(x+3)x}$$

$$= \dfrac{x^2(x+3) - 3x(x-4)}{(x-4)(x+3) + 2x(x+3)}$$

$$= \dfrac{x^3 + 3x^2 - 3x^2 + 12x}{(x+3)(x-4+2x)}$$

$$= \dfrac{x(x^2 + 12)}{(x+3)(3x-4)}$$

Simplify, using either of the above methods.

42. $\dfrac{\dfrac{1}{x+2} - \dfrac{3}{x^2-4}}{\dfrac{3}{x-2}}$

43. $\dfrac{\dfrac{y}{y+4} - \dfrac{2}{y^2+5y+4}}{\dfrac{4}{y+1} + \dfrac{3}{y+4}}$

44. $\dfrac{\dfrac{1}{x} - \dfrac{1}{x-2} + \dfrac{3}{x^2-2x}}{\dfrac{x}{x-2} + \dfrac{3}{x}}$

45. $\dfrac{\dfrac{a^2}{b^2} - \dfrac{b^2}{a^2}}{\dfrac{a}{b} - \dfrac{b}{a}}$

46. $\dfrac{n - \dfrac{n^2}{n-m}}{1 + \dfrac{m^2}{n^2-m^2}}$

47. $\dfrac{\dfrac{x^2}{x-y} - x}{\dfrac{y^2}{x-y} + y}$

48. $1 - \dfrac{x - (1/x)}{1 - (1/x)}$

49. $\dfrac{y - \dfrac{1}{1+(1/y)}}{y + \dfrac{1}{y-(1/y)}}$

Miscellaneous Problems

In each of the following, perform the indicated operations and simplify.

50. $\dfrac{1-2x}{x-4} + \dfrac{x+2}{3x-12}$

51. $x + y + \dfrac{x^2}{x-y}$

52. $\dfrac{x+y}{z} - \left(\dfrac{x}{z} + \dfrac{y}{z}\right)$

53. $\dfrac{2x+1}{x^2+4x-60} - \dfrac{3x}{2x-12}$

54. $\dfrac{\dfrac{25-9u^2}{2u+3}}{5u+3u^2}$

55. $\dfrac{\dfrac{x^3-y^3}{2x+3y}}{\dfrac{x-y}{4x^2-9y^2}}$

56. $\dfrac{x^2-4}{x^2-5x} \cdot \dfrac{x^2-10x+25}{x^2+3x-10}$

57. $\dfrac{3x}{4x^2-9} - \dfrac{5}{3-2x}$

58. $\dfrac{1}{x}\left(x + \dfrac{1}{x}\right)^{-1}$

59. $\left(\dfrac{x-2}{x}\right)^2 \div (3x-6)$

CHAPTER SUMMARY

An **exponent** is a numerical superscript placed on a number to indicate that a certain operation is to be performed on that number. In particular

$$b^4 = b \cdot b \cdot b \cdot b \qquad b^0 = 1 \qquad b^{-3} = \dfrac{1}{b^3} = \dfrac{1}{b \cdot b \cdot b}$$

Exponents mix together according to five laws called the **rules of exponents.**

A number is in **scientific notation** when it appears in the form $c \times 10^n$, where $1 \leq |c| < 10$ and n is an integer. Very small and very large numbers are commonly written this way. Pocket calculators often use scientific notation in their displays. These electronic marvels are designed to take the drudgery out of arithmetic calculations and are a valuable tool in an algebra-trigonometry course.

An expression of the form

$$a_n x^n + a_{n-1} x^{n-1} + \cdots + a_1 x + a_0$$

is called a **polynomial** in x. The exponent n is its **degree** (provided $a_n \neq 0$); the a's are its **coefficients.** Polynomials can be added, subtracted, and multiplied, the result in each case being another polynomial.

To **factor** a polynomial is to write it as a product of simpler polynomials; to **factor over the integers** is to write a polynomial as a product of polynomials with integer coefficients. Here are five examples.

$$x^2 - 2ax = x(x - 2a)$$ Common factor
$$4x^2 - 25 = (2x + 5)(2x - 5)$$ Difference of squares
$$6x^2 + x - 15 = (2x - 3)(3x + 5)$$ Trial and error
$$x^2 + 14x + 49 = (x + 7)^2$$ Perfect square
$$x^3 + 1000 = (x + 10)(x^2 - 10x + 100)$$ Sum of cubes

A quotient (ratio) of two polynomials is called a **rational expression.** The expression is in **reduced form** if its numerator and denominator have no nontrivial common factors. We add, subtract, multiply, and divide rational expressions in much the same way as we do rational numbers.

CHAPTER REVIEW PROBLEM SET

Simplify, leaving your answer free of negative exponents.

1. $\left(\dfrac{3}{4}\right)^{-2}$

2. $\left(\dfrac{5}{6}\right)^2 \left(\dfrac{5}{6}\right)^{-4}$

3. $\left(\dfrac{5}{6} + \dfrac{1}{3}\right)^{-2}$

4. $\dfrac{2x^{-2}y^2}{4xy^{-3}}$

5. $(3x^{-2}y^3)^{-3}$

6. $\dfrac{(a^{-2}b)^2(2ab^{-3})^{-1}}{(ab^{-2})^3}$

Express in scientific notation.

7. $1,382,000$

8. $(3.1)10^4(2.2)10^{-7}$

9. $\dfrac{(6.5)10^4}{(1.3)10^{-3}}$

Decide which of the following are polynomials. Give the degree of each polynomial.

10. $\frac{3}{2}x^2 - \sqrt{2}x + \pi$

11. $10x^3 + 5\sqrt{x} + 6$

12. $4t^{-2} + 5t^{-1} + 6$

13. $\dfrac{4}{x^3 + 11}$

Perform the indicated operations and simplify.

14. $(3x - 5) + (2x^2 - 2x + 11)$

15. $5 - x^3 - (x^2 - 2x^3 + x - 2)$

16. $(2x - 3)(x + 5)$

17. $(2x - 1)^2 - 4x^2$

18. $(z^3 + 4)(z^3 - 4)$

19. $(2x^2 - 3w)(x^2 + 4w)$

20. $(x + 2a)(x^2 - 2ax + 4a^2)$

21. $(2y - 3)(3y^2 - 5y + 6)$

22. $(3t^2 - t + 1)^2$

23. $(a + bcd)(a - bcd)$

In Problems 24–33, factor completely over the integers.

24. $2x^4 - x^3 + 11x^2$

25. $y^2 - 7y + 12$

26. $6z^2 + z - 1$

27. $49a^2 - 25$

28. $9c^2 - 24cd + 16d^2$

29. $a^3 - 27$

30. $8a^3b^3 + 1$

31. $x^8 - x^4y^4$

32. $x^2 + 2xy + y^2 - z^4$

33. $4c^2 - d^2 - 6c - 3d$

34. Factor $9x^2 - 11$ over the real numbers.

$\boxed{\text{i}}$ 35. Factor $x^2 + 16$ over the complex numbers.

Reduce each of the following.

36. $\dfrac{x^3 - 8}{2x - 4}$

37. $\dfrac{2x - 2x^2}{x^3 - 2x^2 + x}$

Perform the indicated operations and simplify.

38. $\dfrac{18}{x^2 + 3x} - \dfrac{4}{x} + \dfrac{6}{x + 3}$

39. $\dfrac{x^2 + x - 6}{x^2 - 1} \cdot \dfrac{x^2 + x - 2}{x^2 + 5x + 6}$

40. $\dfrac{\dfrac{x}{x - 3} - \dfrac{2}{x^2 - 4x + 3}}{\dfrac{5}{x - 1} + \dfrac{5}{x - 3}}$

3

EQUATIONS AND INEQUALITIES

As the sun eclipses the stars by its brilliancy, so the man of knowledge will eclipse the fame of others in the assemblies of the people if he proposes algebraic problems, and still more if he solves them.

Brahmagupta

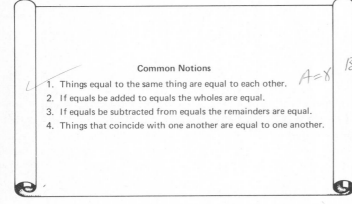

Common Notions

1. Things equal to the same thing are equal to each other.
2. If equals be added to equals the wholes are equal.
3. If equals be subtracted from equals the remainders are equal.
4. Things that coincide with one another are equal to one another.

$A = X$ $B = X$ then $A = B$

Euclid's *Elements,* composed about 300 B.C., has been reprinted in several hundred editions. It is, next to the Bible, the most influential book ever written. The common notions at the left are four of the ten basic axioms with which Euclid begins his book.

3-1
Equations

It was part of Euclid's genius to recognize that our usage of the word *equals* is fundamental to all that we do in mathematics. But to describe that usage may not be as simple as Euclid thought. When we write

$$.25 + \frac{3}{4} + \frac{1}{3} - (.3333\ldots) = 1$$

we certainly do not mean that the symbol on the left coincides with the one on the right. Instead, we mean that both symbols, the complicated one and the simple one, stand for (or name) the same number. That is the basic meaning of *equals* as used in this book.

But having said that, we must make another distinction. When we write

$$x^2 - 25 = (x - 5)(x + 5)$$

and

$$x^2 - 25 = 0$$

we have two quite different things in mind. In the first case, we are making an assertion. We claim that no matter what number x represents, the expressions on the left and right of the equality stand for the same number. This certainly cannot be our meaning in the second case. There we are asking a question: What numbers can x symbolize so that both sides of the equality $x^2 - 25 = 0$ stand for the same number?

An equality that is true for all values of the variable is called an **identity.** One that is true only for some values is called a conditional **equation.** And here are the corresponding jobs for us to do. We **prove** identities, but we **solve** (or find the solutions of) equations. Both jobs will be very important in this book; however, it is the second that interests us most right now.

SOLVING EQUATIONS

Sometimes we can solve an equation by inspection. It takes no mathematical apparatus and little imagination to see that

$$x + 4 = 6$$

has $x = 2$ as a solution. On the other hand, to solve

$$2x^2 + 8x = 8x + 18$$

is quite a different matter. For this kind of equation, we need some machinery. Our general strategy is to modify an equation one step at a time until it is in a form where the solution is obvious. Of course, we must be careful that the modifications we make do not change the solutions. Here, too, Euclid pointed the way.

RULES FOR MODIFYING EQUATIONS

1. Adding the same quantity to (or subtracting the same quantity from) both sides of an equation does not change its solutions.
2. Multiplying (or dividing) both sides of an equation by the same nonzero quantity does not change its solutions.

Consider $2x^2 + 8x = 8x + 18$ again. One way to solve this equation is to use the following steps.

Given equation:	$2x^2 + 8x = 8x + 18$
Subtract $8x$:	$2x^2 = 18$
Divide by 2:	$x^2 = 9$
Take square roots:	$x = 3$ or -3

Thus the solutions of $2x^2 + 8x = 8x + 18$ are 3 and -3. In Section 3-4, we will solve equations like this one by other methods, but the rules stated above will continue to play a fundamental role.

1st power

LINEAR EQUATIONS

The simplest kind of equation to solve is one in which the *variable* (also called the *unknown*) occurs only to the first power. Consider

$$12x - 9 = 5x + 5$$

Our procedure is to use the rules for modifying equations to bring all the terms in x to one side and the constant terms to the other and then to divide by the coefficient of x. The result is that we have x all alone on one side of the equation and a number (the solution) on the other.

Given equation: $12x - 9 = 5x + 5$
Add 9: $12x = 5x + 14$
Subtract $5x$: $7x = 14$
Divide by 7: $x = 2$

It is always a good idea to check your answer. In the original equation, replace x by the value that you found, to see if a true statement results.

$$12(2) - 9 \stackrel{?}{=} 5(2) + 5$$

$$15 = 15$$

An equation of the form $ax + b = 0 \ (a \neq 0)$ is called a **linear equation.** It has one solution, $x = -b/a$. Many equations not initially in this form can be transformed to it using the rules we have learned.

EQUATIONS THAT CAN BE CHANGED TO LINEAR FORM

Consider

$$\frac{2}{x + 1} = \frac{3}{2x - 2}$$

If we agree to exclude $x = -1$ and $x = 1$ from consideration, then $(x + 1)(2x - 2)$ is not zero, and we may multiply both sides by that expression. We get

$$\frac{2}{x + 1}(x + 1)(2x - 2) = \frac{3}{2x - 2}(x + 1)(2x - 2)$$

$$2(2x - 2) = 3(x + 1)$$

$$4x - 4 = 3x + 3$$

$$x = 7$$

As usual, we check our solution in the original equation.

$$\frac{2}{7 + 1} \stackrel{?}{=} \frac{3}{14 - 2}$$

$$\frac{2}{8} = \frac{3}{12}$$

So $x = 7$ is a solution.

The importance of checking is illustrated by our next example.

$$\frac{3x}{x - 3} = 1 + \frac{9}{x - 3}$$

To solve, we multiply both sides by $x - 3$ and then simplify.

$$\frac{3x}{x - 3}(x - 3) = \left(1 + \frac{9}{x - 3}\right)(x - 3)$$

$$3x = x - 3 + 9$$

$$2x = 6$$

$$x = 3$$

When we check in the original equation, we get

$$\frac{3 \cdot 3}{3 - 3} \stackrel{?}{=} 1 + \frac{9}{3 - 3}$$

This is nonsense, since it involves division by zero. What went wrong? If $x = 3$, then in our very first step we actually multiplied both sides of the equation by zero, a forbidden operation. Thus the given equation has no solution.

The strategy of multiplying both sides by $x - 3$ in this example was appropriate, even though it initially led us to an incorrect answer. We did not worry, because we knew that in the end we were going to run a check. We should always check answers, especially in any situation in which we have multiplied by an expression involving the unknown. Such a multiplication may introduce an *extraneous* solution (but never results in the loss of a solution).

Here is another example of a similar type of equation.

$$\frac{x + 4}{(x + 1)(x - 2)} - \frac{3}{x + 1} - \frac{2}{x - 2} = \frac{-8}{(x + 1)(x - 2)}$$

We have no choice but to multiply both sides of the equation by $(x + 1)$ $(x - 2)$. This gives

$$x + 4 - 3(x - 2) - 2(x + 1) = -8$$

$$x + 4 - 3x + 6 - 2x - 2 = -8$$

$$-4x + 8 = -8$$

$$-4x = -16$$

$$x = 4$$

At this point, $x = 4$ is an apparent solution; however, we are not sure until we check it in the original equation.

$$\frac{4+4}{(4+1)(4-2)} - \frac{3}{4+1} - \frac{2}{4-2} \stackrel{?}{=} \frac{-8}{(4+1)(4-2)}$$

$$\frac{8}{5\cdot 2} - \frac{3}{5} - \frac{2}{2} \stackrel{?}{=} \frac{-8}{5\cdot 2}$$

$$\frac{8}{10} - \frac{6}{10} - \frac{10}{10} = \frac{-8}{10}$$

It works, so $x = 4$ is a solution.

Problem Set 3-1

Determine which of the following are identities and which are conditional equations.

1. $2(x + 4) = 8$
2. $2(x + 4) = 2x + 8$
3. $3(2x - \frac{2}{3}) = 6x - 2$
4. $2x - 4 - \frac{2}{3}x = \frac{4}{3}x - 4$
5. $\frac{2}{3}x + 4 = \frac{1}{2}x - 1$
6. $3(x - 2) = 2(x - 3) + x$
7. $(x + 2)^2 = x^2 + 4$
8. $x(x + 2) = x^2 + 2x$
9. $x^2 - 9 = (x + 3)(x - 3)$
10. $x^2 - 5x + 6 = (x - 1)(x - 6)$

Solve each of the following equations.

11. $4x - 3 = 3x - 1$
12. $2x + 5 = 5x + 14$
13. $2t + \frac{1}{2} = 4t - \frac{7}{2} + 8t$
14. $y + \frac{1}{3} = 2y - \frac{2}{3} - 6y$
15. $3(x - 2) = 5(x - 3)$
16. $4(x + 1) = 2(x - 3)$
17. $\sqrt{3}z + 4 = -\sqrt{3}z + 8$
18. $\sqrt{2}x + 1 = x + \sqrt{2}$

[c] 19. $3.23x - 6.15 = 1.41x + 7.63$

(First obtain $x = \dfrac{7.63 + 6.15}{3.23 - 1.41}$ and then use the calculator.)

[c] 20. $42.1x + 11.9 = 1.03x - 4.32$

[c] 21. $(6.13 \times 10^{-8})x + (5.34 \times 10^{-6}) = 0$

[c] 22. $(5.11 \times 10^{11})x - (6.12 \times 10^{12}) = 0$

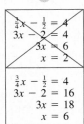

EXAMPLE A (Equations involving fractions)   Solve $\frac{2}{3}x - \frac{3}{4} = \frac{7}{6}x + \frac{1}{2}$.

Solution.   When an equation is cluttered up with many fractions, the best first step may be to get rid of them. To do this, multiply both sides by the lowest common denominator (in this case, 12). Then proceed as usual.

$$12\left(\frac{2}{3}x - \frac{3}{4}\right) = 12\left(\frac{7}{6}x + \frac{1}{2}\right)$$

$$12\left(\frac{2}{3}x\right) - 12\left(\frac{3}{4}\right) = 12\left(\frac{7}{6}x\right) + 12\left(\frac{1}{2}\right) \qquad \text{(distributive property)}$$

$$8x - 9 = 14x + 6$$

$$8x = 14x + 15 \qquad \text{(add 9)}$$

$$-6x = 15 \qquad \text{(subtract } 14x)$$

$$x = \frac{15}{-6} = -\frac{5}{2} \qquad \text{(divide by } -6)$$

The solution should now be checked in the original equation. We leave the check to the student.

Solve by first clearing the fractions.

23. $\frac{2}{3}x + 4 = \frac{1}{2}x$ 24. $\frac{2}{3}x - 4 = \frac{1}{2}x + 4$

25. $\frac{9}{10}x + \frac{5}{8} = \frac{1}{5}x + \frac{9}{20}$ 26. $\frac{1}{3}x + \frac{1}{4} = \frac{1}{5}x + \frac{1}{6}$

27. $\frac{3}{4}(x - 2) = \frac{9}{5}$ 28. $\frac{x}{8} = \frac{2}{3}(2 - x)$

The following equations are nonlinear equations that become linear when cleared of fractions. Solve each equation and check your solutions as some might be extraneous (see page 87).

29. $\frac{5}{x + 2} = \frac{2}{x - 1}$ 30. $\frac{10}{2x - 1} = \frac{14}{x + 4}$

31. $\frac{2}{x - 3} + \frac{3}{x - 7} = \frac{7}{(x - 3)(x - 7)}$ 32. $\frac{2}{x - 1} + \frac{3}{x + 1} = \frac{19}{x^2 - 1}$

33. $\frac{x}{x - 2} = 2 + \frac{2}{x - 2}$ 34. $\frac{x}{2x - 4} - \frac{2}{3} = \frac{7 - 2x}{3x - 6}$

Sometimes an equation that appears to be quadratic is actually equivalent to a linear equation. For example, if we subtract x^2 from both sides of the equation $x^2 + 3x = x^2 + 5$, we see that it is equivalent to $3x = 5$. Use this idea to solve each of the following.

35. $x^2 + 4x = x^2 - 3$ 36. $x^2 - 2x = x^2 + 3x + 20$

37. $(x - 4)(x + 5) = (x + 2)(x + 3)$ 38. $(2x - 1)(2x + 3) = 4x^2 + 6$

Sometimes an equation involving a radical becomes linear when both sides are raised to the same (appropriate) power. Solve each of the following equations. Check your answers, since raising to powers may introduce extraneous solutions.

39. $\sqrt{5 - 2x} = 5$ 40. $\sqrt{3x + 7} = 4$

41. $\sqrt[3]{1 - 3x} = 4$ 42. $\sqrt[3]{4x - 3} = -3$

EXAMPLE B (Solving for one variable in terms of others) Solve $I = nE/(R + nr)$ for n.

Solution. This is a typical problem in science in which an equality relates several variables and we want to solve for one of them in terms of the others. To do this, we proceed as if the other variables were simply numbers, which, after all, is what every variable represents.

$$I = \frac{nE}{R + nr} \qquad \text{(original equality)}$$

$$(R + nr)I = nE \qquad \text{(multiply by } R + nr)$$

$$RI + nrI = nE \qquad \text{(distributive property)}$$

$$nrI - nE = -RI \qquad \text{(subtract } nE \text{ and } RI)$$

$$n(rI - E) = -RI \qquad \text{(factor)}$$

$$n = \frac{-RI}{rI - E} \qquad \text{(divide by } rI - E)$$

Solve for the indicated variable in terms of the remaining variables.

43. $A = P + Prt$ for P

44. $R = \dfrac{E}{L - 5}$ for L

45. $I = \dfrac{nE}{R + nr}$ for r

46. $mv = Ft + mv_0$ for m

47. $A = 2\pi r^2 + 2\pi rh$ for h

48. $F = \frac{9}{5}C + 32$ for C

49. $R = \dfrac{R_1 R_2}{R_1 + R_2}$ for R_1

50. $\dfrac{1}{R} = \dfrac{1}{R_1} + \dfrac{1}{R_2} + \dfrac{1}{R_3}$ for R_2

Miscellaneous Problems

Solve the equations in Problems 51–62.

51. $5x - 11 = 2x + 13$

52. $-2x + 29 = 3x + 74$

53. $\frac{2}{5}y - 4 = \frac{1}{3}y$

54. $\frac{5}{7}a = \frac{2}{3}a - \frac{4}{3}$

[c] 55. $0.081x - 0.123 = 0.152x$

[c] 56. $42.34(x - 1.35) = 432.78$

57. $\dfrac{2}{x - 3} = \dfrac{5}{x + 2}$

58. $\dfrac{4}{2x + 1} = \dfrac{-3}{x - 4}$

59. $\dfrac{2x}{x - 3} = \dfrac{2x + 1}{x + 2}$

60. $\dfrac{2}{(x + 2)(x - 1)} - \dfrac{1}{x + 2}$
$$= \dfrac{3}{x - 1}$$

61. $\dfrac{x}{x + 3} = -1 - \dfrac{3}{x + 3}$

62. $\dfrac{x^2}{x - 2} = \dfrac{4}{x - 2}$

63. The Celsius and Fahrenheit temperature scales are related by the formula $C = \frac{5}{9}(F - 32)$.
    (a) Solve for F in terms of C.
    (b) What reading on the Fahrenheit scale corresponds to 35 degrees Celsius?
    (c) How warm is it (in Fahrenheit degrees) when the Celsius reading equals the Fahrenheit reading?
    (d) How warm is it (in degrees Fahrenheit) when the Fahrenheit reading is 50 degrees higher than the Celsius reading?
    (e) How warm is it (in Celsius degrees) when the Celsius reading is one-third of the Fahrenheit reading?

64. If a principal P is invested at the simple interest rate r for t years, then the accumulated amount after t years is given by $A = P + Prt$. For example,

$2000 invested at 8 percent simple interest will accumulate to $2000 + 2000(.08)(10)$, or $3600, in 10 years.

(a) If $2000 is invested at 9 percent simple interest, to what sum will it accumulate in 5 years?

(b) How long will it take $2000 to grow to $5000 if invested at 9 percent simple interest?

(c) A principal of $2000 grew to $4000 in 8 years at a certain interest rate r. Find r.

65. A stone is thrown vertically upward with a velocity of 80 feet per second. After t seconds, its velocity is given by $v = 80 - 32t$.

(a) When is the velocity zero?

(b) When is the velocity -40 feet per second (a negative velocity means that the stone is falling)?

66. The book value (in dollars) of a bulldozer t years after purchase is given by

$$B = 150{,}000 - 21{,}000t$$

(a) When will the book value be half the original value?

(b) When will the book value reach the scrap value of $31,000?

René Descartes
1596–1650

Rules for the Direction of the Mind

1. Reduce any kind of problem to a mathematical problem.
2. Reduce any kind of mathematical problem to a problem of algebra.
3. Reduce any problem of algebra to the solution of a single equation.

3-2
Applications Using One Unknown

Besides being a mathematician, Descartes was a first-rate philosopher. His name appears prominently in every philosophy text. We admit that his rules for the direction of the mind are overstated. Not every problem in life can be solved this way. But they do suggest a style of thinking that has been very fruitful, especially in the sciences. We intend to exploit it.

A TYPICAL WORD PROBLEM

Sometimes a little story can illustrate some big ideas.

John plans to take his wife Helen out to dinner. Concerned about their financial situation, Helen asks him point-blank, "How much money do you have?" Never one to give a simple answer when a complicated one will do, John replies, "If I had $12 more than I have and then doubled that amount, I'd be $60 richer than I am." Helen's response is best left unrecorded.

The problem, of course, is to find out exactly how much money John has. Our task is to take a complicated word description, translate it into mathematical symbols, and then let the machinery of algebra grind out the answer. First we introduce a symbol x. It usually will stand for the principal unknown in the problem. But we need to be very precise. It is not enough to let x be John's money, or even to let x be the amount of John's money, though that is better. The symbol x must represent a number. What we should say is

Let x be the number of dollars John has.

Our story puts restrictions on x; x must satisfy a specified condition. That condition must be translated into an equation. We shall do it by bits and pieces.

WORD PHRASE	*ALGEBRAIC TRANSLATION*
how much John has	x
$12 more than he has	$x + 12$
double that amount	$2(x + 12)$
$60 richer than he is	$x + 60$

Read John's answer again. It says that the expressions $2(x + 12)$ and $x + 60$ are equal. Thus

$$2(x + 12) = x + 60$$

$$2x + 24 = x + 60$$

$$x = 36$$

John has $36, enough for a pretty good dinner for two—even at today's prices.

A DISTANCE-RATE PROBLEM

Problems involving rates and distances occur very frequently in physics. Usually their solution involves use of the formula $D = RT$, which stands for "distance equals rate multiplied by time."

At 2:00 P.M., Slowpoke left Kansas City traveling due east at 45 miles per hour. An hour later, Speedy started after him going 60 miles per hour. When will Speedy catch up with Slowpoke?

Most of us can grasp the essential features in a picture more readily than in a mass of words. That is why all good mathematicians make sketches that summarize what is given.

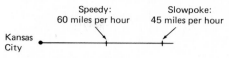

Next assign the unknown. Be precise.

Poor: Let t be time.

Better: Let t be the time when Speedy catches up.

Good: Let t be the number of hours after 2:00 P.M. when Speedy catches up with Slowpoke.

Notice two things:

1. Slowpoke drove t hours: Speedy, starting an hour later, drove only $t - 1$ hours.
2. Both drove the same distance.

Now use the formula $D = RT$ to conclude that Slowpoke drove a distance of $45t$ miles and Speedy drove a distance of $60(t - 1)$ miles. By statement 2, these are equal.

$$45t = 60(t - 1)$$

$$45t = 60t - 60$$

$$60 = 15t$$

$$4 = t$$

Speedy will catch up with Slowpoke 4 hours after 2:00 P.M., or at 6:00 P.M.

A MIXTURE PROBLEM

Here is a problem from chemistry.

How many liters of a 60 percent solution of nitric acid should be added to 10 liters of a 30 percent solution to obtain a 50 percent solution?

We are certain that a picture will help most students with this problem.

We have indicated on the picture what x represents, but let us be specific.

Let x be the number of liters of 60 percent solution to be added.

Now we make a crucial observation, one that will seem obvious once you think about it.

$$\begin{pmatrix} \text{The amount of} \\ \text{pure acid we} \\ \text{start with} \end{pmatrix} + \begin{pmatrix} \text{The amount of} \\ \text{pure acid we} \\ \text{add} \end{pmatrix} = \begin{pmatrix} \text{The amount of} \\ \text{pure acid we} \\ \text{end with} \end{pmatrix}$$

In symbols, this becomes

$$(.30)(10) + (.60)x = (.50)(10 + x)$$

The big job has been accomplished; we have the equation. After multiplying both sides by 10 to clear the equation of decimal fractions, we can easily solve for x.

$$(3)(10) + 6x = 5(10 + x)$$
$$30 + 6x = 50 + 5x$$
$$x = 20$$

We should add 20 liters of 60 percent solution.

SUMMARY

One hears students say that they have a mental block when it comes to "word problems." Yet most of the problems of the real world are initially stated in words. We want to destroy those mental blocks and give you one of the most satisfying experiences in mathematics. We believe you can learn to do word problems. Here are a few simple suggestions.

1. Read the problem very carefully so that you know exactly what it says and what it asks.
2. Draw a picture or make a chart that summarizes the given information.
3. Identify the unknown quantity and assign a letter to it. Be sure it represents a number (for example, of dollars, of miles, or of liters).
4. Note the condition or restriction that the problem puts on the unknown. Translate it into an equation.
5. Solve the equation. Your result is a specific number. The unknown has become known.
6. State your conclusion in words (for example, Speedy will catch up to Slowpoke at 4 hours after 2:00 P.M., or at 6:00 P.M.).
7. Note whether your answer is reasonable. If, for example, you found that Speedy would not catch up with Slowpoke until 4000 hours after 2:00 P.M., you should suspect that you have made a mistake.

Problem Set 3-2

12 inches

1. The sum of 15 and twice a certain number is 33. Find that number.

2. Tom says to Jerry: I am thinking of a number. When I subtract 5 from that number and then multiply the result by 3, I get 42. What was my number? Jerry figured it out. Can you?

3. Find the number for which twice the number is 12 less than 3 times the number.

4. The result of adding 28 to 4 times a certain number is the same as subtracting 5 from 7 times that number. Find the number.

5. The sum of three consecutive positive integers is 72. Find the smallest one.

6. The perimeter (distance around) of the rectangle shown in the margin is 31 inches. Find the width x.

7. A wire 130 centimeters long is bent into the shape of a rectangle which is 3 centimeters longer than it is wide. Find the width of the rectangle.

8. A rancher wants to put 2850 pounds of feed into two empty bins. If she wants the larger bin to contain 750 pounds more of the feed than the smaller one, how much must she put into each?

9. Mary scored 61 on her first math test. What must she score on a second test to bring her average up to 75?

10. Henry has scores of 61, 73, and 82 on his first three tests. How well must he do on the fourth and final test to wind up with an average of 75 for the course?

11. A change box contains 21 dimes. How many quarters must be put in to bring the total amount of change to $3.85?

12. A change box contains $3.00 in dimes and nothing else. A certain number of dimes are taken out and replaced by an equal number of quarters, with the result that the box now contains $4.20. How many dimes are taken out?

13. A woman has $4.45 in dimes and quarters in her purse. If there are 25 coins in all, how many dimes are there?

14. Young Amy has saved up to $8.05 in nickels and quarters. She has 29 more nickels than quarters. How many quarters does she have?

15. Read the example (in this section) about Slowpoke and Speedy again. When will Speedy be 100 miles ahead of Slowpoke?

16. Two long-distance runners start out from the same spot on an oval track which is $\frac{1}{2}$ mile around. If one runs at 6 miles per hour and the other at 7 miles an hour, when will the faster runner be one lap ahead of the slower runner?

17. The City of Harmony is 455 miles from the city of Dissension. At 12:00 noon Paul Haymaker leaves Harmony traveling at 60 miles per hour toward Dissension. Simultaneously, Nick Ploughman starts from Dissension heading toward Harmony, managing only 45 miles per hour. When will they meet?

18. Suppose in Problem 17, Mr. Ploughman starts at 3:00 P.M. At what time will the two drivers meet?

19. Luella can row 1 mile upstream in the same amount of time that it takes her to row two miles downstream. If the rate of the current is 3 miles per hour, how fast can she row in still water?

20. An airplane flew with the wind for 1 hour and returned the same distance against the wind in $1\frac{1}{2}$ hours. If the speed of the plane in still air is 300 miles per hour, find the speed of the wind.

21. A father is three times as old as his son, but 15 years from now he will be only twice as old as his son. How old is his son now?

22. Jim Warmath was in charge of ticket sales at a football game. The price for general admission was $3.50, while reserved seat tickets sold for $5.00. He lost track of the ticket count, but he knew that 110 more general admission tickets had been sold than reserve seat tickets and the total gate receipts were $980. See if you can find out how many general admission tickets were sold.

23. How many cubic centimeters of a 40 percent solution of hydrochloric acid should be added to 2000 cubic centimeters of a 20 percent solution to obtain a 35 percent solution?

24. In Problem 23, how much of the 40 percent solution would have to be added in order to have a 39 percent solution?

25. A tank contains 1000 liters of 30 percent brine solution. Boiling off water from the solution will increase the percentage of salt. How much water should be boiled off to achieve a 35 percent solution?

26. Sheila Carlson invested $10,000 in a savings and loan association, some at 7 percent (simple interest) per year and the rest at $8\frac{1}{2}$ percent. How much did she invest at 7 percent if the total amount of interest for one year was $796?

27. At the Style King shop, a man's suit was marked down 15 percent and sold at $123.25. What was the original price?

28. Mr. Titus Canby bickered with a furrier over the price of a fur coat he intended to buy for his wife. The furrier offered to reduce the price by 10 percent. Titus was still not satisfied; he said he would buy the coat if the furrier would come down an additional $200 on the price. The furrier agreed and sold the coat for $1960. What was the original price?

29. The Conkwrights plan to put in a concrete drive from the street to the garage. The drive is 36 feet long and they plan to make it 4 inches thick. Since there is a delivery charge for less than 4 cubic yards of ready mixed concrete, the Conkwrights have decided to make the drive just wide enough to use 4 cubic yards. How wide should they make it?

30. Tom can do a certain job in 3 days, Dick can do it in 4 days, and Harry can do it in 5 days. How long will it take them working together? (*Hint:* In one day Tom can do $\frac{1}{3}$ of the job; in x days, he can do $x/3$ of the job.)

31. It takes Jack 5 days to hoe his vegetable garden. Jack and Jill together can do it in 3 days. How long would it take Jill to hoe the garden by herself?

EXAMPLE (Balancing weight problems) Susan, who weighs 80 pounds, wants to ride on a seesaw with her father, who weighs 200 pounds. The plank is 20 feet long and the fulcrum is at the center. If Susan sits at the very end of her side, how far from the fulcrum should her father sit to achieve balance?

Solution. Let x be the number of feet from the fulcrum to the point where Susan's dad sits. A law of physics demands that *the weight times the distance from the fulcrum* must be the same for both sides in order to have balance. For Susan, weight times distance is $80 \cdot 10$; for her dad, it is $200x$. This gives us the equation

$$200x = 800$$

from which we get $x = 4$. Susan's father should sit 4 feet from the fulcrum.

32. Where should Susan's father sit if Susan moves 2 feet closer to the fulcrum?

33. If Susan sits at one end with Roscoe, a 12-pound puppy, in her arms, where should her father sit?

Find x in each of the following. Assume in each case that the plank is 20 feet long, that the fulcrum is at the center, and that the weights balance.

34.

35.

36.

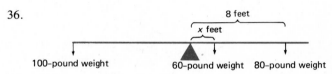

(*Hint:* On the right side, you add the two products.)

37.

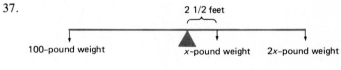

38.

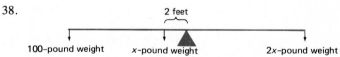

The Farmer's Riddle

I have a collection of hens and rabbits. These animals have 50 heads and 140 feet. How many hens and how many rabbits do I have?

3-3
Two Equations
in Two Unknowns

No one has thought more deeply or written more wisely about problem solving than George Polya. In his book, *Mathematical Discovery* (Volume 1, Wiley, 1962), Polya uses the farmer's riddle as the starting point for a brilliant essay on setting up equations to solve problems. He suggests three different approaches that we might take to untangle the riddle.

TRIAL AND ERROR

Hens	Rabbits	Feet
50	0	100
0	50	200
25	25	150
28	22	144
30	20	140

There are 50 animals altogether. They cannot all be hens as that would give only 100 feet. Nor can all be rabbits; that would give 200 feet. Surely the right answer is somewhere between these extremes. Let us try 25 of each. That gives 50 hen-feet and 100 rabbit-feet, or a total of 150, which is too many. We need more hens and fewer rabbits. Well, try 28 hens and 22 rabbits. It does not work. Try 30 hens and 20 rabbits. There it is! That gives 60 feet plus 80 feet, just what we wanted.

BRIGHT IDEA

Let us use a little imagination. Imagine that we catch the hens and rabbits engaged in a weird new game. The hens are all standing on one foot and the rabbits are hopping around on their two hind feet. In this remarkable situation, only half of the feet, that is 70 feet, are in use. We can think of 70 as counting each hen once and each rabbit twice. If we subtract the total number of animals, namely 50, we will have the number of rabbits. There it is! There have to be 70 − 50 = 20 rabbits; and that leaves 30 hens.

ALGEBRA

Trial and error is time consuming and inefficient, especially in problems with many possibilities. And we cannot expect a brilliant idea to come along for every problem. We need a systematic method that depends neither on guess work nor on sudden visions. Algebra provides such a method. To use it, we must translate the problem into algebraic symbols and set up equations.

ENGLISH	*ALGEBRAIC SYMBOLS*
The farmer has a certain number of hens	x
and a certain number of rabbits.	y
These animals have 50 heads	$x + y = 50$
and 140 feet.	$2x + 4y = 140$

Now we have two unknowns, x and y, but we also have two equations relating them. We want to find the values for x and y that satisfy both equations at the same time. There are two standard methods.

METHOD OF ADDITION OR SUBTRACTION

We learned two rules for modifying equations in Section 3-1. Here is another rule, especially useful in solving a **system of equations,** that is, a set of several equations in several unknowns.

> RULE 3
>
> You may add one equation to another (or subtract one equation from another) without changing the simultaneous solutions of a system of equations.

Here is how this rule is used to solve the farmer's riddle.

Given equations:
$$\begin{cases} x + y = 50 \\ 2x + 4y = 140 \end{cases}$$

Multiply the first equation by (-2): $\quad -2x - 2y = -100$
Write the second equation: $\quad \underline{2x + 4y = 140}$

Add the two equations: $\quad 2y = 40$

Multiply by $\frac{1}{2}$: $\quad \boxed{y = 20}$

Substitute $y = 20$ into one of the original equations (in this case, we shall use the first one): $\quad x + 20 = 50$

Add -20: $\quad \boxed{x = 30}$

The key idea is this: Multiplying the first equation by -2 makes the coefficients of x in the two equations negatives of each other. Addition of the two equations eliminates x, leaving one equation in the single unknown y. The resulting equation can be solved by methods learned earlier (see Section 3-1).

METHOD OF SUBSTITUTION

Consider the same pair of equations again.

$$\begin{cases} x + y = 50 \\ 2x + 4y = 140 \end{cases}$$

We may solve the first equation for y in terms of x and substitute the result in the second equation

$$y = 50 - x$$

$$2x + 4(50 - x) = 140$$

$$2x + 200 - 4x = 140$$

$$-2x = -60$$

$$\boxed{x = 30}$$

Then substitute the value obtained for x in the expression for y.

$$\boxed{y = 50 - 30 = 20}$$

Naturally, our results agree with those obtained earlier.

Whichever method we use, it is a good idea to check our answer against the original problem. Thirty chickens and 20 rabbits do have a total of 50 heads and they do have $(30)(2) + (20)(4) = 140$ feet in all.

A DISTANCE-RATE PROBLEM

An airplane, flying with the help of a strong tail wind, covered 1200 miles in 2 hours. However, the return trip flying into the wind took $2\frac{1}{2}$ hours. How fast would the plane have flown in still air and what was the speed of the wind, assuming both rates to be constant?

Here is the solution. Let

$$x = \text{speed of the plane in still air in miles per hour}$$

$$y = \text{speed of the wind in miles per hour}$$

Then

$$x + y = \text{speed of the plane with the wind}$$

$$x - y = \text{speed of the plane against the wind}$$

Next, we recall the familiar formula $D = RT$, or distance equals rate multiplied by time. Applying it in the form $TR = D$ to the two trips yields

$$2(x + y) = 1200 \qquad \text{(with wind)}$$

$$\tfrac{5}{2}(x - y) = 1200 \qquad \text{(against wind)}$$

or equivalently

$$2x + 2y = 1200$$

$$5x - 5y = 2400$$

To eliminate y, we multiply the first equation by 5 and the second by 2, and then add the two equations.

$$10x + 10y = 6000$$

$$\underline{10x - 10y = 4800}$$

$$20x = 10{,}800$$

$$x = 540$$

Finally, substituting $x = 540$ in the first of the original equations gives

$$2 \cdot 540 + 2y = 1200$$

$$1080 + 2y = 1200$$

$$2y = 120$$

$$y = 60$$

We conclude that the plane's speed in still air was 540 miles per hour and that the wind speed was 60 miles per hour.

A check against the original statement of the problem shows that our answers are correct. With the wind, the plane will travel at 600 miles per hour and will cover 1200 miles in 2 hours. Against the wind, the plane will fly at 480 miles per hour and take $2\frac{1}{2}$ hours to cover 1200 miles.

Problem Set 3-3

In each of the following, find the values for the two unknowns that satisfy both equations. Use whichever method you prefer.

1. $2x + 3y = 13$
   $y = 13$

2. $2x - 3y = 7$
   $x = -4$

3. $2u - 5v = 23$
   $2u = 3$

4. $5s + 6t = 2$
   $3t = -4$

5. $7x + 2y = -1$
   $y = 4x + 7$

6. $7x + 2y = -1$
   $x = -5y + 14$

7. $y = -2x + 11$
   $y = 3x - 9$

8. $x = 5y$
   $x = -3y - 24$

9. $x - y = 14$
   $x + y = -2$

10. $2x - 3y = 8$
    $4x + 3y = 16$

11. $2s - 3t = -10$
    $5s + 6t = 29$

12. $2w - 3z = -23$
    $8w + 2z = -22$

13. $5x - 4y = 19$
    $7x + 3y = 18$

14. $4x - 2y = 16$
    $6x + 5y = 24$

(*Hint:* In Problem 13, multiply top equation by 3 and the bottom one by 4; then add.)

15. $7x - 4y = 0$
$2x + 7y = 57$

16. $2a + 3b = 0$
$3a - 2b = \frac{13}{2}$

17. $\frac{2}{3}x + y = 4$
$x + 2y = 5$

18. $\frac{3}{4}x - \frac{1}{2}y = 12$
$x + y = -8$

19. $.125x - .2y = 3$
$.75x + .3y = 10.5$

20. $.13x - .24y = 1$
$2.6x + \quad 4y = -2.4$

21. $\dfrac{4}{x} + \dfrac{3}{y} = 17$

$\dfrac{1}{x} - \dfrac{3}{y} = -7$

22. $\dfrac{4}{x} - \dfrac{2}{y} = 12$

$\dfrac{5}{x} + \dfrac{1}{y} = 8$

(*Hint:* Let $u = 1/x$ and $v = 1/y$. Solve for u and v and then find x and y.)

23. $\dfrac{2}{\sqrt{x}} - \dfrac{1}{\sqrt{y}} = \dfrac{2}{3}$

$\dfrac{1}{\sqrt{x}} + \dfrac{2}{\sqrt{y}} = \dfrac{7}{6}$

24. $\dfrac{1}{x - 2} - \dfrac{2}{y + 3} = 3$

$\dfrac{5}{x - 2} + \dfrac{2}{y + 3} = 3$

EXAMPLE (Three unknowns) Find values for x, y, and z that satisfy all three of the following equations.

$$2x + y - 3z = -9$$

$$x - 2y + 4z = 17$$

$$3x - y - z = 2$$

Solution. The idea is to eliminate one of the unknowns from two different pairs of equations. This gives us just two equations in two unknowns, which we solve as before. Let us eliminate y from the first two equations. To do this, we multiply the first equation by 2 and add to the second.

$$4x + 2y - 6z = -18$$
$$\underline{x - 2y + 4z = 17}$$
$$5x \quad\quad - 2z = -1$$

Next, we eliminate y from the first and third equations by simply adding them.

$$2x + y - 3z = -9$$
$$\underline{3x - y - \quad z = 2}$$
$$5x \quad\quad - 4z = -7$$

Our problem thus reduces to solving the following system of two equations in two unknowns.

$$5x - 2z = -1$$

$$5x - 4z = -7$$

We leave this for you to do. You should get $z = 3$ and $x = 1$. If you substitute these values for x and z in any of the original equations, you will find that $y = -2$. Thus

$$x = 1 \qquad y = -2 \qquad z = 3$$

Solve each of these systems for x, y, and z.

25. $4x - y + 2z = 2$
    $-3x + y - 4z = -1$
    $x \qquad + 5z = 1$

26. $x + 4y - 8z = -10$
    $3x - y + 5z = 12$
    $-4x + 2y + z = -9$

27. $2x + 3y + 4z = -6$
    $-x + 4y - 6z = 6$
    $3x - 2y + 2z = 2$

28. $3x \qquad + z = 0$
    $3x + 2y + z = 4$
    $9x + 5y + 10z = 3$

Miscellaneous Problems

29. Find two numbers whose sum is 18 and whose difference is 4.

30. One number is three times as large as another and their sum is 14. Find the two numbers.

31. A man sold two lots at a total price of $13,000. If he received $1400 more for one lot than for the other, what was the selling price of each lot?

32. Ian Stockton's estate is valued at $5000 more than three times as much as his wife's estate. The combined value of their estates is $185,000. Find the value of each estate.

33. Mrs. Goldthorpe invests $1\frac{1}{2}$ times as much in savings certificates as in government bonds. If her total investment amounts to $47,500, how much does she invest in each?

34. A ticket manager sold 700 more general admission tickets than reserved seat tickets. How many of each kind did he sell if he sold 2300 tickets in all?

35. The attendance at a professional football game was 45,000 and the total gate receipts were $385,000. If each person bought either an $8 ticket or a $10 ticket, how many tickets of each kind were sold?

36. A change box contains $30.70 in quarters and dimes. If there are 190 coins in all, how many quarters and how many dimes are there?

37. Johnny boasts that he has 8 more than 4 times as many nickels as he has dimes. When his mother asks him what it all amounts to, his reply is $1.30. How many nickels and how many dimes does he have?

38. A woman needs to borrow $80,000 for a business venture. She is able to get an 11 percent loan at a bank and a 13 percent loan at a savings and loan association. In each case, the interest is payable after one year. How much does she borrow from each institution if the total interest due after one year is $9300?

39. A tourist takes a trip of 700 miles, driving at an average speed of 30 miles per hour before lunch and at an average speed of 50 miles per hour after lunch. His time on the road after lunch is one hour more than twice the time on the road before lunch. How many hours does he drive before stopping for lunch? How long is he on the road after lunch?

40. James Carson rows 14 miles downstream on the Mississippi River and then

rows back. If it takes him 2 hours going downstream and 7 hours coming back, what is the rate of the current and how fast can he row in still water?

41. A grocer has some coffee worth $1.69 per pound and some worth only $1.26 per pound. How much of each kind of coffee should she mix together to get 100 pounds worth $1.53 per pound?

42. A solution that is 40 percent alcohol is to be mixed with one that is 90 percent alcohol to obtain 100 liters of 60 percent alcohol solution. How many liters of each will be used?

43. Susan Sharp paid $4800 for some dresses and coats. She paid $40 for each dress and $100 for each coat. She sold the dresses at 20 percent profit and the coats at 50 percent profit. Her total profit was $1800. How many coats and how many dresses did she buy?

44. Workers in a certain factory are classified into two groups, depending upon their skills for their jobs. Group 1 workers are paid $8.00 per hour and group 2 workers are paid $5.00 per hour. In negotiations for a new contract, the union demands that workers in the second group have their hourly wages brought up to $\frac{2}{3}$ of those for the workers in the first group. The company has 55 employees in group 1 and 40 in group 2, all of whom work a 40-hour week. If the company is prepared to increase its weekly payroll by $5760, what hourly wages should be proposed for each class of workers?

Old Stuff

By 2000 B.C. Babylonian arithmetic had evolved into a well-developed rhetorical algebra. Not only were quadratic equations solved, both by the equivalent of substituting in a general formula and by completing the square, but some cubic (third degree) and biquadratic (fourth degree) equations were discussed.

Howard Eves

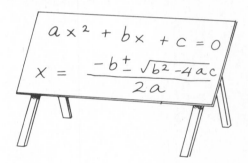

3-4
Quadratic Equations

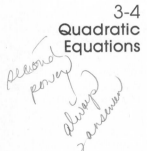

A linear (first degree) equation may be put in the form $ax + b = 0$. We have seen that such an equation has exactly one solution, $x = -b/a$. That is simple and straightforward; no one is likely to stumble over it. But even the ancient Babylonians knew that equation solving goes far beyond this simple case. In fact, a good part of mathematical history revolves around attempts to solve more and more complicated equations.

The next case to consider is the second degree, or **quadratic**, equation—that is, an equation of the form

$$ax^2 + bx + c = 0 \qquad (a \neq 0)$$

Here are some examples.

$$\text{(i)} \qquad x^2 - 4 = 0$$

$$\text{(ii)} \qquad x^2 - x - 6 = 0$$

$$\text{(iii)} \qquad 8x^2 - 2x = 1$$

$$\text{(iv)} \qquad x^2 = 6x - 2$$

While equations (iii) and (iv) do not quite fit the pattern, we accept them because they readily transform to equations in standard form.

$$\text{(iii)} \qquad 8x^2 - 2x - 1 = 0$$

$$\text{(iv)} \qquad x^2 - 6x + 2 = 0$$

SOLUTION BY FACTORING

All of us remember that 0 times any number is 0. Just as important but sometimes forgotten is the fact that if the product of two numbers is 0, then one or both of the factors must be 0.

If $u = 0$ or $v = 0$, then $u \cdot v = 0$
If $u \cdot v = 0$, then either $u = 0$, or $v = 0$, or both.

This fact allows us to solve any quadratic equation which has 0 on one side provided we can factor its other side. Simply factor, set each factor equal to 0, and solve the resulting linear equations. We illustrate.

$$\text{(i)} \qquad x^2 - 4 = 0$$
$$(x - 2)(x + 2) = 0$$
$$x - 2 = 0 \qquad x + 2 = 0$$
$$x = 2 \qquad\qquad x = -2$$

$$\text{(ii)} \qquad x^2 - x - 6 = 0$$
$$(x - 3)(x + 2) = 0$$
$$x - 3 = 0 \qquad x + 2 = 0$$
$$x = 3 \qquad\qquad x = -2$$

$$\text{(iii)} \qquad 8x^2 - 2x - 1 = 0$$
$$(4x + 1)(2x - 1) = 0$$
$$4x + 1 = 0 \qquad 2x - 1 = 0$$
$$x = -\frac{1}{4} \qquad\qquad x = \frac{1}{2}$$

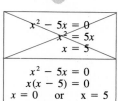

Equation (iv) remains unsolved; we do not know how to factor its left side. For this equation, we need a more powerful method. First, however, we need a brief discussion of square roots.

SQUARE ROOTS

The number 9 has two square roots, 3 and -3. In fact, every positive number has two square roots, one positive and the other negative. If a is positive, its **positive square root** is denoted by $\sqrt{a}$. Thus $\sqrt{9} = 3$. Do not write $\sqrt{9} = -3$ or $\sqrt{9} = \pm 3$; both are wrong. But you can say that the two square roots of 9 are $\pm\sqrt{9}$ (or ± 3) and that the two square roots of 7 are $\pm\sqrt{7}$.

Here are two important properties of square roots, valid for any positive numbers a and b.

$$\sqrt{ab} = \sqrt{a}\sqrt{b}$$

$$\sqrt{\frac{a}{b}} = \frac{\sqrt{a}}{\sqrt{b}}$$

For example,

$$\sqrt{4 \cdot 16} = \sqrt{4}\sqrt{16} = 2 \cdot 4 = 8$$

$$\sqrt{28} = \sqrt{4 \cdot 7} = \sqrt{4}\sqrt{7} = 2\sqrt{7}$$

$$\sqrt{\frac{4}{9}} = \frac{\sqrt{4}}{\sqrt{9}} = \frac{2}{3}$$

The square roots of a negative number are imaginary. For example, the two square roots of -9 are $3i$ and $-3i$, since

$$(3i)^2 = 3^2 i^2 = 9(-1) = -9$$

$$(-3i)^2 = (-3)^2 i^2 = 9(-1) = -9$$

In fact, if a is positive, the two square roots of $-a$ are $\pm\sqrt{a}\,i$. And in this case, the symbol $\sqrt{-a}$ will denote $\sqrt{a}\,i$. Thus $\sqrt{-7} = \sqrt{7}\,i$.

COMPLETING THE SQUARE

Consider equation (iv) again.

$$x^2 - 6x + 2 = 0$$

We may write it as

$$x^2 - 6x = -2$$

Now add 9 to both sides, making the left side a perfect square, and factor.

$$x^2 - 6x + 9 = -2 + 9$$

$$(x - 3)^2 = 7$$

This means that $x - 3$ must be one of the two square roots of 7—that is,

$$x - 3 = \pm\sqrt{7}$$

Hence

$$x = 3 + \sqrt{7} \quad \text{or} \quad x = 3 - \sqrt{7}$$

You may ask how we knew that we should add 9. Any expression of the form $x^2 + px$ becomes a perfect square when $(p/2)^2$ is added, since

$$x^2 + px + \left(\frac{p}{2}\right)^2 = \left(x + \frac{p}{2}\right)^2$$

For example, $x^2 + 10x$ becomes a perfect square when we add $(10/2)^2$ or 25.

$$x^2 + 10x + 25 = (x + 5)^2$$

The rule for completing the square (namely, add $(p/2)^2$) works only when the coefficient of x^2 is 1. However, that fact causes no difficulty for quadratic equations. If the leading coefficient is not 1, we simply divide both sides by this coefficient and then complete the square. We illustrate with the equation

$$2x^2 - x - 3 = 0$$

We divide both sides by 2 and proceed as before.

$$x^2 - \frac{1}{2}x - \frac{3}{2} = 0$$

$$x^2 - \frac{1}{2}x = \frac{3}{2}$$

$$x^2 - \frac{1}{2}x + \left(\frac{1}{4}\right)^2 = \frac{3}{2} + \left(\frac{1}{4}\right)^2$$

$$\left(x - \frac{1}{4}\right)^2 = \frac{25}{16}$$

$$x - \frac{1}{4} = \pm\frac{5}{4}$$

$$x = \frac{1}{4} + \frac{5}{4} = \frac{3}{2} \quad \text{or} \quad x = \frac{1}{4} - \frac{5}{4} = -1$$

THE QUADRATIC FORMULA

The method of completing the square works on any quadratic equation. But there is a way of doing this process once and for all. Consider the general quadratic equation

$$ax^2 + bx + c = 0$$

with real coefficients $a \neq 0$, b, and c. First add $-c$ to both sides and then divide by a to obtain

$$x^2 + \frac{b}{a}x = -\frac{c}{a}$$

Next complete the square by adding $(b/2a)^2$ to both sides, and then simplify.

$$x^2 + \frac{b}{a}x + \left(\frac{b}{2a}\right)^2 = -\frac{c}{a} + \left(\frac{b}{2a}\right)^2$$

$$\left(x + \frac{b}{2a}\right)^2 = -\frac{c}{a} + \frac{b^2}{4a^2}$$

$$\left(x + \frac{b}{2a}\right)^2 = \frac{b^2 - 4ac}{4a^2}$$

Finally take the square root of both sides.

$$x + \frac{b}{2a} = \pm\frac{\sqrt{b^2 - 4ac}}{2a}$$

or

$$x = \frac{-b}{2a} \pm \frac{\sqrt{b^2 - 4ac}}{2a}$$

We call this result the **quadratic formula** and normally write it as follows.

$$x = \frac{-b \pm \sqrt{b^2 - 4ac}}{2a}$$

Let us see how it works on example (iv).

$$x^2 - 6x + 2 = 0$$

Here $a = 1$, $b = -6$, and $c = 2$. Thus

$$x = \frac{-(-6) \pm \sqrt{36 - 4 \cdot 2}}{2}$$

$$= \frac{6 \pm \sqrt{28}}{2}$$

$$= \frac{6 \pm \sqrt{4 \cdot 7}}{2}$$

$$= \frac{6 \pm 2\sqrt{7}}{2}$$

$$= \frac{2(3 \pm \sqrt{7})}{2}$$

$$= 3 \pm \sqrt{7}$$

As a second example, consider $2x^2 - 4x + \frac{25}{8} = 0$. Here $a = 2$, $b = -4$, and $c = \frac{25}{8}$. The quadratic formula gives

$$x = \frac{4 \pm \sqrt{16 - 25}}{4} = \frac{4 \pm \sqrt{-9}}{4} = \frac{4 \pm 3i}{4}$$

The expression $b^2 - 4ac$ that appears under the square root sign in the quadratic formula is called the **discriminant.** It determines the character of the solutions.

1. If $b^2 - 4ac > 0$, there are two real solutions.
2. If $b^2 - 4ac = 0$, there is one real solution.
3. If $b^2 - 4ac < 0$, there are two nonreal solutions.

Problem Set 3-4

EXAMPLE A (Simplifying square roots)   Simplify

(a) $\sqrt{54}$;   (b) $\dfrac{2 + \sqrt{48}}{4}$;   (c) $\dfrac{\sqrt{6}}{\sqrt{150}}$

Solution.

(a) $\sqrt{54} = \sqrt{9 \cdot 6} = \sqrt{9}\sqrt{6} = 3\sqrt{6}$

Here we factored out the largest square in 54, namely 9, and then used the first property of square roots.

(b) $\dfrac{2 + \sqrt{48}}{4} = \dfrac{2 + \sqrt{16 \cdot 3}}{4} = \dfrac{2 + 4\sqrt{3}}{4} = \dfrac{2(1 + 2\sqrt{3})}{4}$

$$= \dfrac{1 + 2\sqrt{3}}{2}$$

If you are tempted to continue as follows,

$$\frac{1 + \cancel{2}\sqrt{3}}{\cancel{2}} = 1 + \sqrt{3}$$

resist the temptation. That cancellation is wrong, because 2 is not a factor of the entire numerator but only of $2\sqrt{3}$.

(c) $\dfrac{\sqrt{6}}{\sqrt{150}} = \sqrt{\dfrac{6}{150}} = \sqrt{\dfrac{1}{25}} = \dfrac{1}{5}$

Sometimes, as in this case, it is best to write a quotient of two square roots as a single square root and then simplify.

Simplify each of the following.

1. $\sqrt{50}$

2. $\sqrt{300}$

3. $\sqrt{\frac{1}{4}}$

4. $\sqrt{\dfrac{3}{27}}$

5. $\dfrac{\sqrt{45}}{\sqrt{20}}$

6. $\sqrt{.04}$

7. $\sqrt{11^2 \cdot 4}$

8. $\dfrac{\sqrt{2^3 \cdot 5}}{\sqrt{2 \cdot 5^3}}$

9. $\dfrac{5 + \sqrt{72}}{5}$

10. $\dfrac{4 - \sqrt{12}}{2}$

11. $\dfrac{18 + \sqrt{-9}}{6}$

12. $\dfrac{3 + \sqrt{-8}}{3}$

EXAMPLE B (Quadratics that are already perfect squares) Solve (a) $x^2 = 9$; (b) $(x + 3)^2 = 17$; (c) $(2y - 5)^2 = 16$.

Solution. The easiest way to solve these equations is to take square roots of both sides.

(a) $x = 3$ or $x = -3$

(b) $x + 3 = \pm\sqrt{17}$

$\quad x = -3 + \sqrt{17}$ or $x = -3 - \sqrt{17}$

(c) $2y - 5 = \pm 4$

$\quad 2y = 5 + 4$ or $2y = 5 - 4$

$\quad y = \frac{9}{2}$ or $y = \frac{1}{2}$

Solve by the method above.

13. $x^2 = 25$
14. $x^2 = 14$
15. $(x - 3)^2 = 16$
16. $(x + 4)^2 = 49$
17. $(2x + 5)^2 = 100$
18. $(3y - \frac{1}{3})^2 = 25$
19. $m^2 = -9$
20. $(m - 6)^2 = -36$

Solve by factoring.

21. $x^2 = 3x$
22. $2x^2 - 5x = 0$
23. $x^2 - 9 = 0$
24. $x^2 - \frac{9}{4} = 0$
25. $m^2 - .0144 = 0$
26. $x^2 - x - 2 = 0$
27. $x^2 - 3x - 10 = 0$
28. $x^2 + 13x + 22 = 0$
29. $3x^2 + 5x - 2 = 0$
30. $3x^2 + x - 2 = 0$
31. $6x^2 - 13x - 28 = 0$
32. $10x^2 + 19x - 15 = 0$

Solve by completing the square.

33. $x^2 + 8x = 9$
34. $x^2 - 12x = 45$
35. $z^2 - z = \frac{3}{4}$
36. $x^2 + 5x = 2\frac{3}{4}$
37. $x^2 + 4x = -9$
38. $x^2 - 14x = -65$

Solve by using the quadratic formula.

39. $x^2 + 8x + 12 = 0$
40. $x^2 - 2x - 15 = 0$
41. $x^2 + 5x + 3 = 0$
42. $z^2 - 3z - 8 = 0$
43. $3x^2 - 6x - 11 = 0$
44. $4t^2 - t - 3 = 0$
45. $x^2 + 5x + 5 = 0$
46. $y^2 + 8y + 10 = 0$
47. $2z^2 - 6z + 11 = 0$
48. $x^2 + x + 1 = 0$

Solve using the quadratic formula. Write your answers rounded to four decimal places.

49. $2x^2 - \pi x - 1 = 0$ 50. $3x^2 - \sqrt{2}x - 3\pi = 0$

51. $x^2 + .8235x - 1.3728 = 0$ 52. $5x^2 - \sqrt{3}x - 4.3213 = 0$

EXAMPLE C (Solving for one variable in terms of the other)   Solve for y in terms of x.

(a) $y^2 - 2xy + x^2 - 2 = 0$ (b) $(y - 3x)^2 + 3(y - 3x) - 4 = 0$

Solution.

(a) We use the quadratic formula with $a = 1$, $b = -2x$, and $c = x^2 - 2$.

$$y = \frac{2x \pm \sqrt{4x^2 - 4(x^2 - 2)}}{2} = \frac{2x \pm \sqrt{8}}{2} = \frac{2x \pm 2\sqrt{2}}{2}$$

$$= x \pm \sqrt{2}$$

So $y = x + \sqrt{2}$ or $y = x - \sqrt{2}$.

(b) If we substitute z for $y - 3x$, the equation becomes

$$z^2 + 3z - 4 = 0$$

We solve this equation for z.

$$(z + 4)(z - 1) = 0$$
$$z = -4 \quad \text{or} \quad z = 1$$

Thus

$$y - 3x = -4 \quad \text{or} \quad y - 3x = 1$$
$$y = 3x - 4 \quad \text{or} \quad y = 3x + 1$$

Solve for y in terms of x.

53. $(y - 2)^2 = 4x^2$ 54. $(y + 3x)^2 = 9$

55. $(y + 3x)^2 = 9x^2$ 56. $4y^2 + 4xy - 5 + x^2 = 0$

57. $(y + 2x)^2 - 8(y + 2x)$ 58. $(x - 2y + 3)^2 - 3(x - 2y + 3)$
      $+ 15 = 0$ $+ 2 = 0$

Miscellaneous Problems

Solve equations 59–70 using any method you like.

59. $x^2 = 144$ 60. $3x^2 = 75$

61. $(x - 2)(x + \frac{3}{2}) = 0$ 62. $(2y + 1)^2 = \frac{9}{4}$

63. $2x^2 + 5x - 3 = 0$ 64. $x^2 + 12x = -40$

65. $4m^2 + 5m = 0$ 66. $4m^2 + 5m + 1 = 0$

67. $x^2 + 2x - 5 = 0$ 68. $2x^2 + x = x^2 - 2x + 3$

[i] 69. $x^2 - 4x + 5 = 0$ [i] 70. $2x^2 - 7x + 10 = 0$

71. A rectangle has a perimeter of 40 feet and an area of 91 square feet. Find the dimensions of the rectangle.

72. The sum of the squares of three consecutive positive integers is 365. Find the smallest of the three integers.

[c] 73. A garden in the form of a square is surrounded by a walk 2 feet wide. What are the dimensions of the garden if the combined area of the garden and the walk is 800 square meters?

[c] 74. Find the radius of a circle whose area is twice the area of a square which is 11 inches on a side.

75. A square piece of cardboard is used to construct a tray by cutting out one-inch squares from the four corners and then turning up the four flaps. Find the size of the original square if the resulting tray has a volume of 128 cubic inches.

76. One leg of a right triangle is 2 inches longer than the other leg. If the hypotenuse is $\frac{5}{2}\sqrt{10}$ inches long, how long are the legs?

77. A rock is thrown vertically upward at an initial speed of 128 feet per second. Assume that its distance from the ground t seconds later is $(-16t^2 + 128t)$ feet. (This is approximately correct if we neglect air resistance.)
    (a) At what time will the rock strike the ground?
    (b) When will the rock be 192 feet from the ground?
    (c) When will the rock be 400 feet from the ground?

78. The distance from St. Charles to St. Agnes is 360 miles. Jones and Smith, driving in separate cars, start out from St. Charles for St. Agnes at the same time. Jones drives 5 miles per hour faster than Smith and arrives at St. Agnes 1 hour earlier. How fast is each one driving?

In each of the following, find the values for the two unknowns that satisfy both equations.

79. $x^2 + y^2 = 25$
    $3x + y = 13$

80. $xy = 20$
    $3 = 2x - y$

(*Hint:* Solve the linear equation for y in terms of x and substitute this value in the other equation.)

81. $2x^2 - xy + y^2 = 11$
    $x - 2y = 4$

82. $x^2 + 4y^2 = 8$
    $3x^2 + 2y^2 = 14$

	Equation	Inequality
	$-3x + 7 = 2$	$-3x + 7 < 2$
	$-3x = -5$	$-3x < -5$
	$x = \dfrac{5}{3}$	$x > \dfrac{5}{3}$

3-5
Inequalities

Solving an inequality is very much like solving an equation, as the example above demonstrates. However, there are dangers in proceeding too mechanically. It will be important to think at every step.

Recall the distinction we made between identities and equations in Section 3-1. A similar distinction applies to inequalities. An inequality which is true for all values of the variables is called an **unconditional inequality.** Examples are

$$(x - 3)^2 + 1 > 0$$

and

$$|x| \le |x| + |y|$$

Most inequalities (for example, $-3x + 7 < 2$) are true only for some values of the variables; we call them **conditional inequalities.** Our primary task in this section is to solve conditional inequalities, that is, to find all those numbers which make a conditional inequality true.

LINEAR INEQUALITIES

To solve the linear inequality $Ax + B < C$, we try to rewrite it in successive steps until the variable x stands by itself on one side of the inequality (see opening display). This depends primarily on the properties stated in Section 1-5 and repeated here.

> PROPERTIES OF INEQUALITIES
>
> 1. (Transitivity). If $a < b$ and $b < c$, then $a < c$.
> 2. (Addition). If $a < b$, then $a + c < b + c$.
> 3. (Multiplication). If $a < b$ and $c > 0$, then $a \cdot c < b \cdot c$.
>   If $a < b$ and $c < 0$, then $a \cdot c > b \cdot c$.

We illustrate the use of these properties, applied to $\le$ rather than $<$, by solving the following inequality.

$$-2x + 6 \leq 18 + 4x$$

Add $-4x$: $\qquad -6x + 6 \leq 18$

Add -6: $\qquad\qquad -6x \leq 12$

Multiply by $-\frac{1}{6}$: $\qquad\qquad x \geq -2$

By rights, we should check this solution. All we know so far is that any value of x that satisfies the original inequality satisfies $x \geq -2$. Can we go in the opposite direction? Yes, because every step is reversible. For example, starting with $x \geq -2$, we can multiply by -6 to get $-6x \leq 12$. In practice, we do not actually carry out this check as we recognize that Property 2 can be restated:

$$\boxed{a < b \text{ is equivalent to } a + c < b + c.}$$

There are similar restatements of Property 3.

QUADRATIC INEQUALITIES

To solve

$$x^2 - 2x - 3 > 0$$

we first factor, obtaining

$$(x + 1)(x - 3) > 0$$

Next we ask ourselves when the product of two numbers is positive. There are two cases; either both factors are negative or both factors are positive.

Case 1 (Both negative) We want to know when both factors are negative, that is, we seek to solve $x + 1 < 0$ and $x - 3 < 0$ simultaneously. The first gives $x < -1$ and the second gives $x < 3$. Together they give $x < -1$.

Case 2 (Both positive) Both factors are positive when $x + 1 > 0$ and $x - 3 > 0$, that is, when $x > -1$ and $x > 3$. These give $x > 3$.

The solution set for the original inequality is the union of the solution sets for the two cases. In set notation, it may be written either as

$$\{x : x < -1 \text{ or } x > 3\}$$

or as

$$\{x : x < -1\} \cup \{x : x > 3\}$$

The chart at the top of the next page summarizes what we have learned.

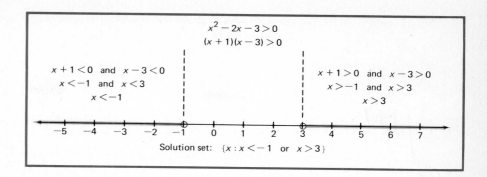

SPLIT-POINT METHOD

The preceding example could have been approached in a slightly different way using the notion of split-points. The solutions of $(x + 1)(x - 3) = 0$, which are -1 and 3, serve as split-points that divide the real line into the three intervals: $x < -1$, $-1 < x < 3$, and $3 < x$. Since $(x + 1)(x - 3)$ can change sign only at a split-point, it must be of one sign (that is, be either always positive or always negative) on each of these intervals. To determine which of them make up the solution set of the inequality $(x + 1)(x - 3) > 0$, all we need to do is pick a single (arbitrary) point from each interval and test it for inclusion in the solution set. If it passes the test, the entire interval from which it was drawn belongs to the solution set.

To show how this method works, let us consider the third degree inequality

$$(x + 2)(x - 1)(x - 4) < 0$$

The solutions of the corresponding equation

$$(x + 2)(x - 1)(x - 4) = 0$$

are -2, 1, and 4. They break the real line into the four intervals $x < -2$, $-2 < x < 1$, $1 < x < 4$, and $4 < x$. Suppose we pick -3 as the test point for the interval $x < -2$. We see that -3 makes each of the three factors $x + 2$, $x - 1$, and $x - 4$ negative, and so it makes $(x + 2)(x - 1)(x - 4)$ negative. You should pick test points from each of the other three intervals to verify the results shown below.

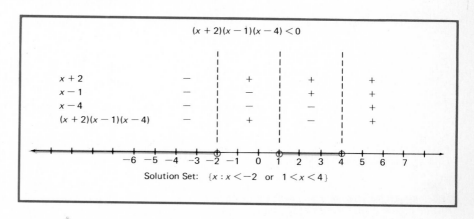

INEQUALITIES WITH ABSOLUTE VALUES

One basic fact to remember is that

$$|x| < a$$

means the same thing as the double inequality

$$-a < x < a$$

A second fact is that

$$|x| > a, \qquad a > 0$$

means

$$x < -a \quad \text{or} \quad x > a$$

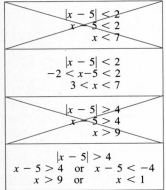

To solve the inequality $|3x - 2| < 4$, proceed as follows.

Given inequality:	$|3x - 2| < 4$
Remove absolute value:	$-4 < 3x - 2 < 4$
Add 2:	$-2 < 3x \qquad < 6$
Multiply by $\frac{1}{3}$:	$-\frac{2}{3} < \quad x \qquad < 2$

Here, as in the preceding examples, we could use the split-point method. The solutions of the equation $|3x - 2| = 4$, which are $-\frac{2}{3}$ and 2, would be the split-points. The solution would again be $-\frac{2}{3} < x < 2$.

AN APPLICATION

A student wishes to get a grade of B in Mathematics 16. On the first four tests, he got 82 percent, 63 percent, 78 percent, and 90 percent, respectively. A grade of B requires an average between 80 percent and 90 percent, inclusive. What grade on the fifth test would qualify this student for a B?

Let x represent the grade (in percent) on the fifth test. The inequality to be satisfied is

$$80 \le \frac{82 + 63 + 78 + 90 + x}{5} \le 90$$

This can be rewritten successively as

$$80 \le \frac{313 + x}{5} \le 90$$

$$400 \le 313 + x \le 450$$

$$87 \le \qquad x \le 137$$

A score greater than 100 is impossible, so the actual solution to this problem is $87 \le x \le 100$.

Problem Set 3-5

Which of the following inequalities are unconditional and which are conditional?

1. $x \geq 0$
2. $x^2 \geq 0$
3. $x^2 + 1 > 0$
4. $x^2 > 1$
5. $x - 2 < -5$
6. $2x + 3 > -1$
7. $x(x + 4) \leq 0$
8. $(x - 1)(x + 2) > 0$
9. $(x + 1)^2 > x^2$
10. $(x - 2)^2 \leq x^2$
11. $(x + 1)^2 > x^2 + 2x$
12. $(x - 2)^2 > x(x - 4)$

Solve each of the following inequalities and show the solution set on the real number line.

13. $3x + 7 < x - 5$
14. $-2x + 11 > x - 4$
15. $\frac{2}{3}x + 1 > \frac{1}{2}x - 3$
16. $3x - \frac{1}{2} < \frac{1}{2}x + 4$
   (*Hint:* First get rid of the fractions.)
17. $\frac{3}{4}x - \frac{1}{2} < \frac{1}{6}x + 2$
18. $\frac{2}{7}x + \frac{1}{3} \leq -\frac{2}{3}x + \frac{15}{14}$
19. $(x - 2)(x + 5) \leq 0$
20. $(x + 1)(x + 4) \geq 0$
21. $(2x - 1)(x + 3) > 0$
22. $(3x + 2)(x - 2) < 0$
23. $x^2 - 5x + 4 \geq 0$
24. $x^2 + 4x + 3 \leq 0$
   (*Hint:* Factor the left side.)
25. $2x^2 - 7x + 3 < 0$
26. $3x^2 - 5x - 2 > 0$
27. $|2x + 3| < 2$
28. $|2x - 4| \leq 3$
29. $|-2x - 1| \leq 1$
30. $|-2x + 3| > 2$
31. $(x + 4)x(x - 3) \geq 0$
32. $(x + 3)x(x - 3) \geq 0$
33. $(x - 2)^2(x - 5) < 0$
34. $(x + 1)^2(x - 1) > 0$

EXAMPLE A (Inequalities involving quotients) Solve the following inequality.

$$\frac{3}{x - 2} > \frac{2}{x}$$

Solution. We rewrite the inequality as follows:

$$\frac{3}{x - 2} - \frac{2}{x} > 0 \qquad \text{(add } -2/x \text{ to both sides)}$$

$$\frac{3x - 2(x - 2)}{(x - 2)x} > 0 \qquad \text{(combine fractions)}$$

$$\frac{x + 4}{(x - 2)x} > 0 \qquad \text{(simplify numerator)}$$

The factors $x + 4$, x, and $x - 2$ in the numerator and denominator determine the three split-points $-4, 0$, and 2. The chart on the next page shows the signs of $x + 4$, x, $x - 2$, and $(x + 4)/(x - 2)x$ on each of the intervals determined by the split-points, as well as the solution set.

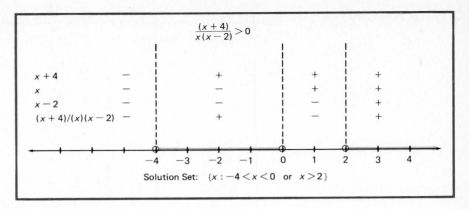

$$\frac{(x+4)}{x(x-2)} > 0$$

$x+4$	$-$	$+$	$+$	$+$
x	$-$	$-$	$+$	$+$
$x-2$	$-$	$-$	$-$	$+$
$(x+4)/(x)(x-2)$	$-$	$+$	$-$	$+$

Solution Set: $\{x : -4 < x < 0 \text{ or } x > 2\}$

Solve each of the following inequalities.

35. $\dfrac{x-5}{x+2} \le 0$ 36. $\dfrac{x+3}{x-2} > 0$ 37. $\dfrac{x(x+2)}{x-5} > 0$

38. $\dfrac{x-1}{(x-3)(x+3)} \ge 0$ 39. $\dfrac{5}{x-3} > \dfrac{4}{x-2}$ 40. $\dfrac{-3}{x+1} < \dfrac{2}{x-4}$

EXAMPLE B (Rewriting with absolute values)   Write $-2 < x < 8$ as an inequality involving absolute values.

Solution.   Look at this interval on the number line. It is 10 units long and its midpoint is at 3. A number x is in this interval provided that it is within a radius of 5 of this midpoint, that is, if

$$|x - 3| < 5$$

We can check that $|x - 3| < 5$ is equivalent to the original inequality by writing it as

$$-5 < x - 3 < 5$$

and then adding 3 to each member.

Write each of the following as an inequality involving absolute values.

41. $0 < x < 6$ 42. $0 < x < 12$ 43. $-1 \le x \le 7$

44. $-3 \le x \le 7$ 45. $2 < x < 11$ 46. $-10 < x < -3$

EXAMPLE C (Quadratic inequalities that cannot be factored by inspection)   Solve the inequality

$$x^2 - 5x + 3 \ge 0$$

Solution. Even though $x^2 - 5x + 3$ cannot be factored by inspection, we can solve the quadratic equation $x^2 - 5x + 3 = 0$ by use of the quadratic formula. We obtain the two solutions $(5 - \sqrt{13})/2 \approx .7$ and $(5 + \sqrt{13})/2 \approx 4.3$. These two numbers split the real line into three intervals from which the numbers 0, 1, and 5 could be picked as test points. Notice that $x = 0$ makes $x^2 - 5x + 3$ positive, $x = 1$ makes it negative, and $x = 5$ makes it positive. This gives us the solution set

$$\left\{ x : x \le \frac{5 - \sqrt{13}}{2} \text{ or } x \ge \frac{5 + \sqrt{13}}{2} \right\}$$

which can be pictured as follows.

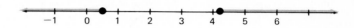

The split-points are included in the solution set because our inequality $x^2 - 5x + 3 \ge 0$ includes equality.

Solve each of the following inequalities and display the solution set on the number line.

47. $x^2 - 7 < 0$ 48. $x^2 - 12 > 0$

49. $x^2 - 4x + 2 \ge 0$ 50. $x^2 - 4x - 2 \le 0$

[c] 51. $x^2 + 6.32x + 3.49 > 0$ [c] 52. $x^2 + 4.23x - 2.79 < 0$

EXAMPLE D (Finding least values)   Find the least value that $x^2 - 4x + 9$ can take on.

Solution.   We use the method of completing squares to write $x^2 - 4x + 9$ as the sum of a perfect square and a constant.

$$x^2 - 4x + 9 = (x^2 - 4x + 4) + 5 = (x - 2)^2 + 5$$

Since the smallest value $(x - 2)^2$ can take on is zero, the smallest value $(x - 2)^2 + 5$ can assume is 5.

Find the least value each of the following expressions can take on.

53. $x^2 + 8x + 20$ 54. $x^2 + 10x + 40$

55. $x^2 - 2x + 101$ 56. $x^2 - 4x + 104$

Miscellaneous Problems

Solve each inequality and show its solution set on the number line.

57. $\frac{1}{2}x - \frac{3}{4} < \frac{2}{3}x$ 58. $2(x - \frac{3}{5}) \ge 3(x + \frac{1}{3})$

59. $(x + 4)(x - 2) > 0$ 60. $x^2 - 4x - 5 < 0$

61. $|3x - 2| \le 3$ 62. $|2x + 3| > 2$

63. $\dfrac{5}{x} \le \dfrac{2}{x - 3}$ 64. $(x - 1)(x + 2)^2(x - 5) \ge 0$

In Problems 65–68, find the set of values of k for which the solutions of the quadratic equation are real. Remember that the discriminant ($b^2 - 4ac$) needs to be greater than or equal to zero.

65. $x^2 + 3x + k = 0$ 66. $x^2 - kx + 10 = 0$

67. $x^2 + kx - 2 = 0$ 68. $x^2 + kx + k = 0$

69. Tom scored 73, 82, 69, and 94 points on four 100-point tests in Mathematics 30. Suppose that a grade of *B* requires an average between 75 percent and 85 percent.
   (a) What score on a 100-point final exam would qualify Tom for a *B*?
   (b) What score on a 200-point final would qualify him for a *B*?

70. A certain department of a company has a staff of six persons whose average annual salary is $17,500. An additional person will be hired. What salary can be offered if the average salary for the department cannot exceed $22,000?

71. A stone thrown vertically upward from the roof of a building has a height of ($-16t^2 + 64t + 80$) feet above the ground after *t* seconds.
   (a) What is the greatest height reached (see Example D of this section)?
   (b) During what time period is the stone higher than 80 feet?
   (c) At what time does the stone strike the ground?
   [c] (d) During what time interval on its way down is the stone less than 40 feet above the ground?

72. Podunk University hires two professors, Smith and Jones. Smith's starting salary is $18,500, while Jones starts at $21,000; however, Smith will get an annual increment of $800 while Jones will receive only $550. How soon will Smith's salary be higher than that of Jones?

73. A company has two garbage trucks to cover the weekly route of 800 miles. Truck A gets 5 miles per gallon of gas, while truck B gets 10 miles per gallon. If gasoline is rationed and the company is allowed only 90 gallons per week, what is the largest number of miles the company will dare drive truck A?

74. Company A will rent a car for $15 per day plus 10¢ per mile, while company B will rent a car for $10 a day plus 12¢ per mile. I need a car for 10 days. At least how many miles must I drive for me to consider renting from company A?

Solve the following inequalities.

75. $|x - 4| \le |x - 2|$ 76. $|x + 3| \ge |x - 3|$

77. $|x| < x + 4$ 78. $|x| + |x - 1| > 3$

Johnny's Dilemma

Although he had been warned against it a thousand times, Johnny still walked across the railroad bridge when he was in a hurry—which was most of the time. It was the day of the biggest football game of the season and Johnny was late. He was already one fourth of the way across the bridge when he heard something. Turning to look, he saw the train coming; it was exactly one bridge–length away from him. Which way should he run?

3-6
More Applications (Optional)

We would not suggest that the problem above is a typical application of algebra. Johnny would be well advised to forget about algebra, make an instant decision, and hope for the best. Nevertheless, Johnny's problem is intriguing. Moreover, it can serve to reemphasize the important principles of real life problem solving that we mentioned briefly at the end of Section 3-2.

First, **clarify the question that is asked**. "Which way should he run?" must mean "Which way should Johnny run to have the best chance of surviving?" That is still too vaguely stated for mathematical analysis. We think the question really means, "Which of the two directions allows Johnny to run at the slowest rate and still avoid the train?" Most questions that come to us from the real world are loosely stated. Our first job is always to pin down precisely (at least in our own minds) what the real question is. It is foolish to try to answer a question that we do not understand. That should be obvious, but it is often overlooked.

Johnny's problem appears to be difficult for another reason; there does not seem to be enough information. We do not know how long the bridge is, we do not know how fast the train is traveling, and we do not know how fast Johnny can run. We could easily despair of making any progress on the problem!

This leads us to state our second principle. **Organize the information you have.** The best way to do this is to draw a diagram or picture that somehow captures the essence of the problem. It should be an abstract or idealized picture. We shall represent people and trains by points and train tracks and bridges by line segments. These are only approximations to the real situation, but they are necessary if progress is to be made. After the picture is drawn, **label the key quantities of the problem with letters and write down what they represent.** One way to represent Johnny's situation is at the top of the next page.

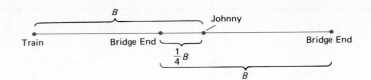

B = length of bridge in feet
S = speed of train in feet per second
Y = speed in feet per second that Johnny must run toward train to escape
Z = speed in feet per second that Johnny must run away from train to escape

Our next principle is this. **Write down the algebraic relationships that exist between the symbols you have introduced.** In Johnny's case, we need to remember the formula $D = RT$ (or $T = D/R$), which relates distance, rate, and time. Now, if Johnny runs toward the train just fast enough to escape, then his time to run the distance $B/4$ must equal the time for the train to go $3B/4$. That gives

(i) $$\frac{B/4}{Y} = \frac{3B/4}{S}$$

On the other hand, if Johnny runs away from the train, he will have to cover $3B/4$ in the same time that the train goes $7B/4$. Thus,

(ii) $$\frac{3B/4}{Z} = \frac{7B/4}{S}$$

Now we have done what is often the hardest part of the problem, translating it into algebraic equations. Our job now is to **manipulate the equations until a solution appears.** In the example we are considering, we must solve the equations for Y and Z in terms of S. When we do this, we get

(i) $$Y = \frac{1}{3}S$$

(ii) $$Z = \frac{3}{7}S$$

There remains a crucial step: **Interpret the result in the language of the original problem.** In Johnny's case, this step is easy. If he runs toward the train, he will need to run at $\frac{1}{3}$ the speed of the train to escape. If he runs away from the train, he must run at $\frac{3}{7}$ of the speed of the train. Clearly, he will have a better chance of making it if he runs toward the train.

Perhaps by now you need to be reassured. We do not expect you to go through a long-winded analysis like ours for every problem you meet. But we do think that you will have to apply the principles we have stated if you are to be successful at problem solving. The problem set that follows is designed to let you apply these principles to problems from several areas of applied mathematics.

Problem Set 3-6

The problems below are arranged according to type. Only the first set (rate-time problems) relates directly to the example of this section. However, all of them should require some use of the principles we have enunciated. Do not expect to solve all of the problems. But do accept them as challenges worthy of genuine effort.

Rate-Time Problems

1. Sound travels at 1100 feet per second in air and at 5000 feet per second in water. An explosion on a distant ship was recorded by the underwater recording device at a monitoring station. Thirteen seconds later the operator heard the explosion. How far was the ship from the station?

2. The primary wave and secondary wave of an earthquake travel away from the epicenter at rates of 8 and 4.8 kilometers per second, respectively. If the primary wave arrives at a seismic station 15 seconds ahead of the secondary wave, how far is the station from the epicenter?

3. Ricardo often flies from Clear Lake to Sun City and returns on the same day. On a windless day, he can average 120 miles per hour for the round trip, while on a windy day, he averaged 140 miles per hour one way and 100 miles per hour the other. It took him 15 minutes longer on the windy day. How far apart are Clear Lake and Sun City?

4. A classic puzzle problem goes like this: If a column of men 3 miles long is marching at 5 miles per hour, how long will it take a courier on a horse traveling at 25 miles per hour to deliver a message from the end of the column to the front and then return?

5. How long after 4:00 P.M. will the minute hand on a clock overtake the hour hand?

6. At what time between 4:00 and 5:00 do the hands of a clock form a straight line?

7. An old machine is able to do a certain job in 8 hours. Recently a new machine was installed. Working together, the two machines did the same job in 3 hours. How long would it take the new machine to do the job by itself?

8. Center City uses fire trucks to fill the village swimming pool. It takes one truck 3 hours to do the job and another 2 hours. How long would it take them working together?

9. An airplane takes off from a carrier at sea and flies west for 2 hours at 600 miles per hour. It then returns at 500 miles per hour. In the meantime, the carrier has traveled west at 30 miles per hours. When will the two meet?

10. A passenger train 480 feet long traveling at 75 miles per hour meets a freight train 1856 feet long traveling on parallel tracks at 45 miles per hour. How long does it take the trains to pass each other? (*Note:* Sixty miles per hour is equivalent to 88 feet per second.)

11. A medical company makes two types of heart valves, standard and deluxe. It takes 5 minutes on the lathe and 10 minutes on the drill press to make a standard valve, but 9 minutes on the lathe and 15 minutes on the drill press for a deluxe valve. On a certain day the lathe will be available for 4 hours and the drill press for 7 hours. How many valves of each kind should the company make that day if both machines are to be fully utilized?

12. Two boats travel at right angles to each other after leaving the dock at 1:00 P.M. At 3:00 P.M., they are 16 miles apart. If the first boat travels 6 miles per hour faster than the second, what are their rates?

13. A car is traveling at an unknown rate. If it traveled 15 miles per hour faster, it would take 90 minutes less to go 450 miles. How fast is the car going?

14. A boy walked along a level road for awhile, then up a hill. At the top of the hill he turned around and walked back to his starting point. He walked 4 miles per hour on level ground, 3 miles per hour uphill, and 6 miles per hour downhill, with the total trip taking 5 hours. How far did he walk altogether?

15. Jack and Jill live at opposite ends of the same street. Jack wanted to deliver a box at Jill's house and Jill wanted to leave some flowers at Jack's house. They started at the same moment, each walking at a constant speed. They met the first time 300 meters from Jack's house. On their return trip, they met 400 meters from Jill's house. How long is the street? (Assume that neither loitered at the other house nor when they met.)

16. Two cars are traveling toward each other on a straight road at the same constant speed. A plane flying at 350 miles per hour passes over the second car 2 hours after passing over the first car. The plane continues to fly in the same direction and is 2400 miles from the cars when they pass. Find the speed of the cars.

Science Problems

17. The illumination I in foot-candles on a surface d feet from a light source of c candlepower is given by the formula $I = c/d^2$. How far should an 80-candlepower light be placed from a surface to give the same illumination as a 20-candlepower light at 10 feet?

18. By experiment, it has been found that a car traveling v miles per hour will require d feet to stop, where $d = 0.044v^2 + 1.1v$. Find the velocity of a car if it took 176 feet to stop.

19. A bridge 200 feet long was built in the winter with no provision for expansion. In the summer, the supporting beams expanded in length by 8 inches, forcing the center of the bridge to drop. Assuming, for simplicity, that the bridge took the shape of a V, how far did the center drop? First guess at the answer and then work it out.

20. The distance s (in feet) traveled by an object in t seconds when it has an initial velocity v_0 and a constant acceleration a is given by $s = v_0t + \frac{1}{2}at^2$. An object was observed to travel 32 feet in 4 seconds and 72 feet in 6 seconds. Find the initial velocity and the acceleration.

21. A chemist has 5 kiloliters of 20 percent sulfuric acid solution. She wishes to increase its strength to 30 percent by draining off some and replacing it with 80 percent soluton. How much should she drain off?

22. How many liters each of a 35 percent alcohol solution and a 95 percent alcohol solution must be mixed to obtain 12 liters of an 80 percent alcohol solution?

23. One atom of carbon combines with 2 atoms of oxygen to form one molecule of carbon dioxide. The atomic weights of carbon and oxygen are 12.0 and 16.0, respectively. How many milligrams of oxygen are required to produce 4.52 milligrams of carbon dioxide?

24. Four atoms of iron (atomic weight 55.85) combine with 6 atoms of oxygen (atomic weight 16.00) to form 2 molecules of rust. How many grams of iron would there be in 79.85 grams of rust?

c 25. A sample weighing .5000 grams contained only sodium chloride and sodium bromide. The chlorine and bromine from this sample were precipitated together as silver chloride and silver bromide. This precipitate weighed 1.100 grams. Sodium chloride is 60.6 percent chlorine, sodium bromide is 77.6 percent bromine, silver chloride is 24.7 percent chlorine, and silver bromide is 42.5 percent bromine. Calculate the weights of sodium chloride and sodium bromide in the sample.

Business Problems

26. Sarah Tyler bought stock in the ABC Company on Monday. The stock went up 10 percent on Tuesday and then dropped 10 percent on Wednesday. If she sold the stock on Wednesday for $1000, what did she pay on Monday?

27. If Jane Witherspoon has $4182 in her savings account and wants to buy stock in the ABC Company at $59 a share, how many shares can she buy and still maintain a balance of at least $2000 in her account?

28. Six men plan to take a charter flight to Bear Lake in Canada for a fishing trip, sharing the cost equally. They discover that if they took three more men along, each share of the original six would be reduced by $150. What is the total cost of the charter?

29. Susan has a job at Jenny's Nut Shop. Jenny asks Susan to prepare 25 pounds of mixed nuts worth $1.74 per pound by using walnuts valued at $1.30 per pound and cashews valued at $2.30 per pound. How many pounds of each kind should Susan use?

30. Alec Brown plans to sell toy gizmos at the state fair. He can buy them at 40¢ apiece and will sell them for 65¢ each. It will cost him $200 to rent a booth for the 7-day fair. How many gizmos must he sell to just break even?

31. The cost of manufacturing a product is the sum of fixed plant costs (real estate taxes, utilities, and so on) and variable costs (labor, raw materials, and so on) that depend on the number of units produced. The profit P that the company makes in a year is given by

$$P = TR - (FC + VC)$$

where TR is the total revenue (total sales), FC is the total fixed cost, and VC is the total variable cost. A company that makes one product has $32,000 in total fixed costs. If the variable cost of producing one unit is $4 and if units can be sold at $6 each, find out how many units must be produced to give a profit of $15,000.

32. The ABC company has total fixed costs of $100,000 and total variable costs equal to 80 percent of total sales. What must the total sales be to yield a profit of $40,000? (See Problem 31.)

33. Do Problem 32 assuming that the company pays 30 percent income taxes on its profit and wants a profit of $40,000 after taxes.

34. The XYZ company has total fixed yearly costs of $120,000 and last year had total variable costs of $350,000 while selling 200,000 gizmos at $2.50 each. Competition will force the manager to reduce the sales price to $2.00 each

next year. If total fixed costs and variable costs per unit are expected to remain the same, how many gizmos will the company have to sell to have the same profit as last year?

35. Podunk University wishes to maintain a student-faculty ratio of 1 faculty member for every 15 undergraduates and 1 faculty member for each 6 graduate students. It costs the university $600 for each undergraduate student and $900 for each graduate student over and above what is received in tuition. If the university expects $2,181,600 in gifts (beyond tuition) next year and will have 300 faculty members, how many undergraduate and how many graduate students should it admit?

36. A department store purchased a number of smoke detectors at a total cost of $2000. In unpacking them, the stock boys damaged 8 of them so badly that they could not be sold. The remaining detectors were sold at a profit of $25 each, and a total profit of $400 was realized when all of them were sold. How many smoke detectors were originally purchased?

Geometry Problems

37. A flag that is 6 feet by 8 feet has a blue cross of uniform width on a white background. The cross extends to all 4 edges of the flag. The area of the cross and the background are equal. Find the width of the cross.

38. If a right triangle has hypotenuse 2 units longer than one leg and 4 units longer than the other, find the dimensions of the triangle.

39. The area and perimeter of a right triangle are both 30. Find its dimensions.

40. Assume that the earth is a sphere of radius 4000 miles. How far is the horizon from an airplane 5 miles high?

41. Three mutually tangent circles have centers, A, B, and C and radii a, b, and c, respectively. The lengths of the segments AB, BC, and CA are 13, 15, and 18, respectively. Find the lengths of the radii. Assume that each circle is outside of the other two.

42. A rectangle is inscribed in a circle of radius 5. Find the dimensions of the rectangle if its area is 40.

c 43. A ladder is standing against a house with its lower end 10 feet from the house. When the lower end is pulled 2 feet farther from the house, the upper end slides 3 feet down the house. How long is the ladder?

c 44. A trapezoid (a quadrilateral with two sides parallel) is inscribed in a square 12 inches on a side. One of the parallel sides is the diagonal of the square. If the trapezoid has area 24 square inches, how far apart are its parallel sides?

45. A 40-inch length of wire is cut in two. One of the pieces is bent to form a square and the other is bent to form a rectangle three times as long as wide. If the combined area of the square and the rectangle is $55\frac{3}{4}$ square inches, where was the wire cut?

CHAPTER SUMMARY

The equalities $(x + 1)^2 = x^2 + 2x + 1$ and $x^2 = 4$ are quite different in character. The first, called an **identity**, is true for all values of x. The second, called a **conditional equation**, is true only for certain values of x, in fact, only for $x = 2$ and $x = -2$. To

solve an equation is to find those values of the unknown which make the equality true; it is one of the major tasks of mathematics.

The equation $ax + b = 0\ (a \neq 0)$ is called a **linear equation** and has exactly one solution, $x = -b/a$. Similarly, $ax^2 + bx + c = 0\ (a \neq 0)$ is a **quadratic equation** and usually has two solutions. Sometimes they can be found by **factoring** the left side and setting both factors equal to zero. Another method that always works is **completing the square,** but the best general method is simply substituting in the **quadratic formula.**

$$x = \frac{-b \pm \sqrt{b^2 - 4ac}}{2a}$$

Here $b^2 - 4ac$, called the **discriminant**, plays a critical role. The equation has two real solutions, one real solution, or two nonreal solutions according as the discriminant is positive, zero, or negative.

Equations arise naturally in the study of word problems. Such problems may lead to one equation in one unknown but often lead to a **system** of several equations in several unknowns. In the latter case, our task is to find the values of the unknowns that satisfy all the equations of the system simultaneously.

Inequalities look like equations with the equal sign replaced by $<, \le, >,$ or $\ge$. The methods for solving **conditional inequalities** are very similar to those for conditional equations. One difference is that the direction of an inequality sign is reversed upon multiplication or division by a negative number. Another is that the set of solutions normally consists of one or more **intervals** of numbers, rather than a finite set. For example, the inequality $3x - 2 < 5$ has the solution set $\{x: x < \frac{7}{3}\}$.

CHAPTER REVIEW PROBLEM SET

1. Which are identities and which are conditional equations?
   (a) $3(x - 2) = 3x - 6$ (b) $3x - 2 = x - 6$
   (c) $(x + 2)^2 = x^2 + 4$ (d) $x^2 + 5x + 6 = (x + 3)(x + 2)$

2. Solve the following equations.

   (a) $3\left(x + \dfrac{1}{2}\right) = x - \dfrac{1}{3}$ (b) $\dfrac{6}{x - 5} = \dfrac{21}{x}$

   (c) $(x - 1)(2x + 1) = (2x - 3)(x + 2)$ (d) $\dfrac{x}{2x + 2} - 1 = \dfrac{8 - 3x}{6x + 6}$

3. In $s = \frac{1}{2}at^2 + v_0 t$, solve for v_0 in terms of the other variables.

4. Recall that the Fahrenheit and Celsius temperature scales are related by the equation $F = \frac{9}{5}C + 32$.
   (a) How cold is it in degrees Celsius when the Celsius reading is twice the Fahrenheit reading?
   (b) For what Fahrenheit temperatures is the Fahrenheit reading higher than the Celsius reading?

5. Solve the systems below.

   (a) $2x - 3y = 7$ (b) $\dfrac{1}{3}x - \dfrac{5}{6}y = 2$

   $\quad\; x + 4y = -2$ $\quad\; \dfrac{1}{2}x + y = -\dfrac{3}{2}$

6. Simplify.

   (a) $\dfrac{6 + \sqrt{18}}{12}$ (b) $\dfrac{10 - \sqrt{300}}{2}$ (c) $\dfrac{-2 + \sqrt{8}}{6}$

Solve the quadratic equations in Problems 7–16 by any method you choose.

7. $x^2 = 49$ 8. $(x - 2)(x + 5) = 0$

9. $(x + 3)^2 = 0$ 10. $(x - 2)^2 = 25$

11. $(2x + 1)^2 = 81$ 12. $x^2 - 9x + 20 = 0$

13. $x^2 = 4x$ 14. $x^2 + 2x - 4 = 0$

15. $3x^2 + x - 1 = 0$ 16. $x^2 + mx + 2n = 0$

Solve for y in terms of x.

17. $(y - 2x)^2 = 4$ 18. $(2y + 3x)^2 - 4(2y + 3x) + 3 = 0$

Solve the inequalities in Problems 19–22.

19. $-5x + 3 \geq 2x - 9$ 20. $x^2 + 5x - 6 \geq 0$

21. $x^2 + x - 3 < 0$ 22. $\dfrac{x - 4}{x + 1} > 0$

23. Jill Garcia has $10,000 to invest. How much should she put in the bank at 6 percent interest and how much in the credit union at 8 percent interest if she hopes to have $730 interest at the end of one year?

24. A fast train left Chicago at 6:00 A.M. traveling at 60 miles per hour. At 8:00 A.M., a slower train left St. Paul traveling at a constant rate. If these two cities are 450 miles apart and the two trains crashed head on at 11:00 A.M., what was the rate of the slower train?

25. John Appleseed rowed upstream for a distance of 4 miles in 2 hours. If he had rowed twice as hard and the current had been half as strong, he could have done it in $\frac{4}{7}$ hour. What was the rate of the current?

4

COORDINATES AND CURVES

And so Fermat and Descartes turned to the application of algebra to the study of geometry. The subject they created is called coordinate, or analytic, geometry; its central idea is the association of algebraic equations with curves and surfaces. This creation ranks as one of the richest and most fruitful veins of thought ever struck in mathematics.

Morris Kline

Pierre de Fermat
1601–1665

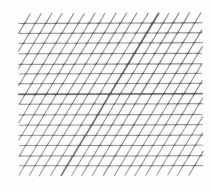

Why are Fermat and Descartes star-
ing at each other so intently?
Because both have axes to grind.

René Descartes
1596–1650

4-1
The Cartesian
Coordinate System

Two Frenchmen deserve credit for the idea of a coordinate system. Pierre de Fermat was a lawyer who made mathematics his hobby. In 1629, he wrote a paper which makes explicit use of coordinates to describe points and curves. René Descartes was a philosopher who thought mathematics could unlock the secrets of the universe. He published *La Géométrie* in 1637. It is a famous book and though it does emphasize the role of algebra in solving geometric problems, one finds only a hint of coordinates there. By virtue of having the idea first and more explicitly, Fermat ought to get the major credit. History can be a fickle friend; coordinates are known as Cartesian coordinates, named after René Descartes.

No matter who gets the credit, it was an idea whose time had come. It made possible the invention of calculus, one of the greatest inventions of the human mind. That invention was to come in 1665 at the hands of a 23-year-old genius named Isaac Newton. You will probably study calculus later on. There you will use the ideas of this chapter over and over.

REVIEW OF THE REAL LINE

Recall the real line, which was introduced in Section 1-3.

Every point on this line can be given a label, a real number, which specifies exactly where the point is. We call this label the **coordinate** of the point.

Consider now two points A and B with coordinates a and b, respectively. We will need a formula for the distance between A and B in terms of the coordinates a and b. The formula is

$$d(A, B) = |b - a|$$

and it is correct whether A is to the right or to the left of B. Note the two examples in the margin. In the first case,

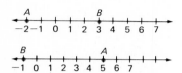

$$d(A, B) = |3 - (-2)| = |5| = 5$$

In the second,

$$d(A, B) = |-1 - 5| = |-6| = 6$$

CARTESIAN COORDINATES

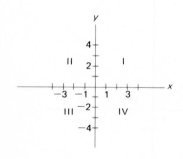

In the plane, produce two copies of the real line, one horizontal and the other vertical, so that they intersect at the zero points of the two lines. The two lines are called **coordinate axes;** their intersection is labeled with O and is called the **origin.** By convention, the horizontal line is called the **x-axis** and the vertical line is called the **y-axis.** The positive half of the x-axis is to the right; the positive half of the y-axis is upward. The coordinate axes divide the plane into four regions called **quadrants,** labeled I, II, III, and IV, as shown in the margin.

Each point P in the plane can now be assigned a pair of numbers called its **Cartesian coordinates.** If vertical and horizontal lines through P intersect the x- and y-axes at a and b, respectively, then P has coordinates (a, b). We call (a, b) an **ordered pair** of numbers because it makes a difference which number is first. The first number a is the **x-coordinate** (or abscissa); the second number b is the **y-coordinate** (or ordinate).

Conversely, take any ordered pair (a, b) of real numbers. The vertical line through a on the x-axis and the horizontal line through b on the y-axis meet in a point P whose coordinates are (a, b).

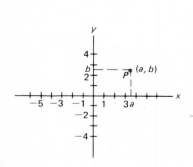

Think of it this way: The coordinates of a point are the address of that point. If you have found a house (or a point), you can read its address. Conversely, if you know the address of a house (or a point), you can always locate it. In the diagram at the top of page 132, we have shown the coordinates (addresses) of several points.

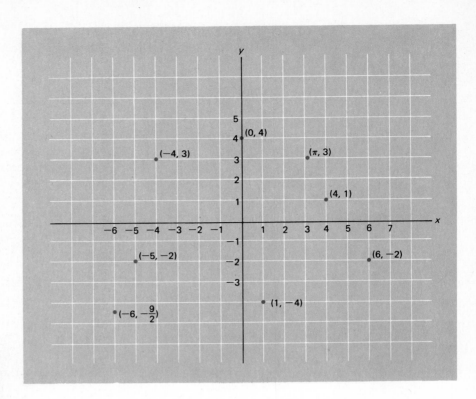

THE DISTANCE FORMULA

Consider the points P_1 and P_2 with coordinates $(-1, -2)$ and $(3, 1)$, respectively. The segment joining P_1 and P_2 is the hypotenuse of a right triangle with right angle at $P_3(3, -2)$ (see the diagram in the margin). We easily calculate the lengths of the two legs.

$$d(P_1, P_3) = |3 - (-1)| = 4 \qquad d(P_2, P_3) = |1 - (-2)| = 3$$

By the Pythagorean theorem (see Section 1-3),

$$d(P_1, P_2) = \sqrt{4^2 + 3^2} = \sqrt{25} = 5$$

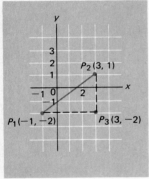

Next consider two arbitrary points $P_1, (x_1, y_1)$ and $P_2(x_2, y_2)$ that are not on the same horizontal or vertical line. They determine a right triangle with legs of length $|x_2 - x_1|$ and $|y_2 - y_1|$. By the Pythagorean theorem,

$$\boxed{d(P_1, P_2) = \sqrt{(x_2 - x_1)^2 + (y_2 - y_1)^2}}$$

This formula is known as the **distance formula.** You should check that it is valid even if P_1 and P_2 lie on the same vertical or horizontal line.

THE MIDPOINT FORMULA

Consider two points $A(x_1, y_1)$ and $B(x_2, y_2)$ in the plane and let $P(x, y)$ be the midpoint of the segment joining them. Drop perpendiculars from $A, P,$ and B to the x-axis as shown in the diagram. Then x is midway between x_1 and x_2, so

$$x - x_1 = x_2 - x$$

$$2x = x_1 + x_2$$

$$x = \frac{x_1 + x_2}{2}$$

Similar reasoning applies to y. The result, called the **midpoint formula,** says that the coordinates (x, y) of the midpoint P are given by

$$x = \frac{x_1 + x_2}{2} \qquad y = \frac{y_1 + y_2}{2}$$

For example, the midpoint of the segment joining $A(-3, -2)$ and $B(5, 9)$ has coordinates.

$$\left(\frac{-3 + 5}{2}, \frac{-2 + 9}{2} \right) = \left(1, \frac{7}{2} \right)$$

Problem Set 4-1

Find $d(A, B)$ where A and B are points on the number line having the given coordinates a and b.

1. $a = -3, b = 2$
2. $a = 5, b = -6$
3. $a = \frac{11}{4}, b = -\frac{5}{4}$
4. $a = \frac{31}{8}, b = \frac{13}{8}$
5. $a = 3.26, b = 4.96$
6. $a = -1.45, b = -5.65$
7. $a = 2 - \pi, b = \pi - 3$
8. $a = 4 + 2\sqrt{3}, b = -6 + \sqrt{3}$

Again, A and B are points on the number line. Given a and $d(A, B)$, what are the two possible values of b?

9. $a = 5, d(A, B) = 2$
10. $a = 9, d(A, B) = 5$
11. $a = -2, d(A, B) = 4$
12. $a = 3, d(A, B) = 7$
13. $a = \frac{5}{2}, d(A, B) = \frac{3}{4}$
14. $a = 1.8, d(A, B) = 2.4$

Plot $A, B, C,$ and D on a coordinate system. Name the quadrilateral ABCD—that is, is it a square, a rectangle, or what?

15. $A(4, 3), B(4, -3), C(-4, 3), D(-4, -3)$
16. $A(-1, 6), B(0, 5), C(3, 2), D(5, 0)$
17. $A(1, 3), B(2, 6), C(4, 7), D(3, 4)$
18. $A(0, 2), B(3, 0), C(2, -1), D(-1, 1)$

In Problems 19–26, find $d(P_1, P_2)$ where P_1 and P_2 have the given coordinates. Also find the coordinates of the midpoint of the segment P_1P_2.

19. $(2, -1)$, $(5, 3)$ 20. $(2, 1)$, $(7, 13)$
21. $(4, 2)$, $(2, 4)$ 22. $(-1, 5)$, $(6, 7)$
23. $(\sqrt{3}, 0)$, $(0, \sqrt{6})$ 24. $(\sqrt{2}, 0)$, $(0, -\sqrt{7})$
[c] 25. $(1.234, -5.132)$, $(6.714, 8.341)$
[c] 26. $(-42.1, 16.3)$, $(12.2, -5.3)$
27. The points A, B, C, and D of Problem 17 form a parallelogram.
   (a) Find $d(A, B)$, $d(B, C)$, $d(C, D)$, and $d(D, A)$.
   (b) Find the coordinates of the midpoints of the diagonals AC and BD of the parallelogram $ABCD$.
   (c) Your answers to parts (a) and (b) agree with what facts about a parallelogram?
28. The points $(3, -1)$ and $(3, 3)$ are two vertices of a square. Give two pairs of other possible vertices. Can you give a third pair?
29. Let $ABCD$ be a rectangle whose sides are parallel to the coordinate axes. Find the coordinates of B and D if the coordinates of A and C are as given.
   (a) $(-2, 0)$ and $(4, 3)$ (Draw a picture.)
   (b) $(2, -1)$ and $(8, 7)$
30. Show that the triangle whose vertices are $(5, 3)$, $(-2, 4)$ and $(10, 8)$ is isosceles.
31. Use the distance formula to show that the triangle whose vertices are $(2, -4)$, $(4, 0)$, and $(8, -2)$ is a right triangle.
32. (a) Find the point on the y-axis that is equidistant from the points $(3, 1)$ and $(6, 4)$. (*Hint:* Let the unknown point be $(0, y)$.)
   (b) Find the point on the x-axis that is equidistant from $(3, 1)$ and $(6, 4)$.
33. Use the distance formula to show that the three points are on a line.
   (a) $(0, 0)$, $(3, 4)$, $(-6, -8)$
   (b) $(-4, 1)$, $(-1, 5)$, $(5, 13)$

EXAMPLE (Point-of-division formula)   Let $A(x_1, y_1)$ and $B(x_2, y_2)$ be the endpoints of a line segment and let t be a number between 0 and 1. The point $P(x, y)$ on this segment satisfying

$$d(A, P) = t \cdot d(A, B)$$

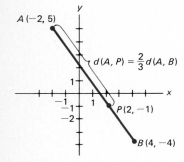

has coordinates given by

$$x = (1 - t)x_1 + tx_2$$
$$y = (1 - t)y_1 + ty_2$$

Note that $t = \frac{1}{2}$ gives the midpoint formula. Use the formula to find the point $P(x, y)$ two-thirds of the way from $A(-2, 5)$ to $B(4, -4)$.

Solution.   Substitution of $t = \frac{2}{3}$ gives

$$x = \frac{1}{3}(-2) + \frac{2}{3}(4) = \frac{6}{3} = 2$$

$$y = \frac{1}{3}(5) + \frac{2}{3}(-4) = -\frac{3}{3} = -1$$

The required point is $(2, -1)$.

Use the point-of-division formula to find $P(x, y)$ for the given value of t and given points A and B. Plot the points A, P, and B.

34. $A(5, -8)$, $B(11, 4)$, $t = \frac{1}{3}$

35. $A(5, -8)$, $B(11, 4)$, $t = \frac{2}{3}$

36. $A(5, -8)$, $B(11, 4)$, $t = \frac{5}{6}$

37. $A(4, 9)$, $B(104, 209)$, $t = \frac{13}{100}$

Miscellaneous Problems

38. Let A and B be points on the number line with coordinates -3 and 5, respectively.
    (a) Find the coordinate of the midpoint of AB.
    (b) Find the coordinate of C if B is the midpoint of the segment AC.

39. Let A and B be the points on the number line with coordinates -4 and 6, respectively. Find the coordinates of the three points which divide AB into 4 equal parts.

40. Find the lengths of the sides of the triangle with vertices $A(2, -4)$, $B(3, 2)$, and $C(-5, 1)$.

41. Given the points $A(2, 3)$ and $B(-4, 7)$.
    (a) Find the coordinates of the midpoint of the segment AB.
    (b) Find the coordinates of C if B is the midpoint of the segment AC.

42. Find x such that the distance between the points $(3, 0)$ and $(x, 3)$ is 5.

43. Find the area of the triangle with vertices $(0, 0)$, $(4, 0)$, and $(2, 6)$.

44. Consider the triangle with vertices $A(0, 0)$, $B(4, 0)$, and $C(0, 5)$.
    (a) Find the area of this triangle.
    (b) Find the length of BC.
    (c) Find the length of the altitude from A to BC.

45. Given $A(-4, 3)$ and $B(21, 38)$, find the coordinates of the four points that divide AB into five equal parts. (*Hint:* For the point closest to A, use $t = \frac{1}{5}$.)

46. Consider the figure in the margin.
    (a) Find the coordinates of the midpoints L, M, and N of the sides of the triangle OAB.
    (b) The segment OM is called the median from vertex O. Find the coordinates of the point on OM that is $\frac{2}{3}$ of the way from O to M.
    (c) Answer the question in part (b) for the medians from A to L and from B to N.
    (d) What fact about the medians of a triangle is illustrated by your results in parts (b) and (c)?

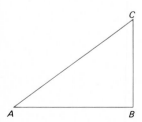

c 47. Cities at A, B, and C are vertices of a right triangle, as shown in the margin. Also, AB and BC are roads of lengths 214 and 179 miles, respectively. An airline flies route AC, which is not a road. It costs \$3.71 per mile to ship a certain product by truck and \$4.82 per mile by air. Calculate the costs for both methods of transportation from A to C in order to see which is cheaper.

C 48. City B is 10 miles downstream from city A and on the opposite side of a river $\frac{1}{2}$ mile wide. Tom O'Shanter will run from city A along the river for 6 miles, then swim diagonally to city B. If he runs at 8 miles per hour and swims at 3 miles per hour, how long will it take him to get from city A to city B?

Algebra   Geometry

"As long as algebra and geometry traveled separate paths, their advance was slow and their applications limited. But when the two sciences joined company, they drew from each other fresh vitality and thenceforward marched on at a rapid pace toward perfection."

Joseph-Louis Lagrange

4-2
Algebra and Geometry United

The Greeks were preeminent geometers but poor algebraists. Though they were able to solve a host of geometry problems, their limited algebraic skills kept others beyond their grasp. By 1600, geometry was a mature and eligible bachelor. Algebra was a young woman only recently come of age. Fermat and Descartes were the matchmakers; they brought the two together. The resulting union is called **analytic geometry**, or coordinate geometry.

THE GRAPH OF AN EQUATION

An equation is an algebraic object. By means of a coordinate system, it can be transformed into a curve, a geometric object. Here is how it is done.

Consider the equation $y = x^2 - 3$. Its set of solutions is the set of ordered pairs (x, y) that satisfy the equation. These ordered pairs are the coordinates of points in the plane. The set of all such points is called the **graph** of the equation. The graph of an equation in two variables x and y will usually be a curve.

To obtain this graph, we follow a definite procedure.

1. Obtain the coordinates of a few points.
2. Plot those points in the plane.
3. Connect the points with a smooth curve in the order of increasing x values.

The best way to do step 1 is to make a **table of values**. Assign values to one of the variables, say x, determine the corresponding values of the other, and then list the pairs of values in tabular form. The whole three-step procedure is illustrated below for the previously mentioned equation, $y = x^2 - 3$.

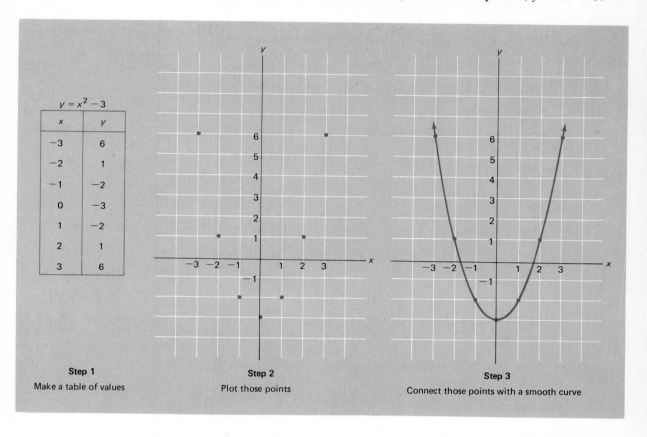

$y = x^2 - 3$	
x	y
-3	6
-2	1
-1	-2
0	-3
1	-2
2	1
3	6

Step 1
Make a table of values

Step 2
Plot those points

Step 3
Connect those points with a smooth curve

Of course, you need to use common sense and even a little faith. When you connect the points you have plotted with a smooth curve, you are assuming that the curve behaves nicely between consecutive points; that is faith. This is why you should plot enough points so the outline of the curve seems very clear; the more points you plot, the less faith you will need. Also you should recognize that you can seldom display the whole curve. In our example, the curve has infinitely long arms opening wider and wider. But our graph does show the essential features. That is what we always aim to do—show enough of the graph so the essential features are visible.

SYMMETRY OF A GRAPH

The graph of $y = x^2 - 3$, drawn above and again in the margin, has a nice property of symmetry. If the coordinate plane were folded along the y-axis the two branches would coincide. For example, $(3, 6)$ would coincide with

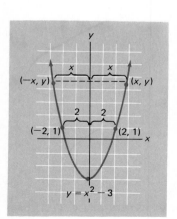

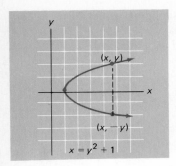

$x = y^2 + 1$

$(-3, 6)$; $(2, 1)$ would coincide with $(-2, 1)$; and, more generally, (x, y) would coincide with $(-x, y)$. Algebraically, this corresponds to the fact that we may replace x by $-x$ in the equation $y = x^2 - 3$ without changing it. More precisely, $y = x^2 - 3$ and $y = (-x)^2 - 3$ are equivalent equations.

Whenever an equation is unchanged by replacing (x, y) with $(-x, y)$, the graph of the equation is **symmetric with respect to the y-axis.** Likewise, if the equation is unchanged when (x, y) is replaced by $(x, -y)$, its graph is **symmetric with respect to the x-axis.** The equation $x = 1 + y^2$ is of the latter type; its graph is shown in the margin.

A third type of symmetry is **symmetry with respect to the origin.** It occurs whenever replacing (x, y) by $(-x, -y)$ produces no change in the equation. The equation $y = x^3$ is a good example as $-y = (-x)^3$ is equivalent to $y = x^3$. The graph is shown below. Note that the dotted line segment from $(-x, -y)$ to (x, y) is bisected by the origin.

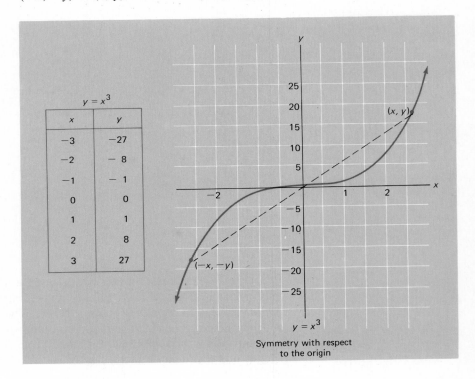

Symmetry with respect
to the origin

In graphing $y = x^3$, we used a smaller scale on the y-axis than on the x-axis. This made it possible to show a larger portion of the graph. We suggest that before putting scales on the two axes, you should examine your table of values. Choose scales so that all of your points can be plotted and still keep your graph of reasonable size.

Graphing an equation is an extremely important operation. It gives us a picture to look at. Most of us can absorb qualitative information from a picture much more easily than from symbols. But if we want precise quantitative information, then symbols are better; they are easier to manipulate. That is why we must be able to reverse the process just described, which is our next topic.

THE EQUATION OF A GRAPH

A graph is a geometric object, a picture. How can we turn it into an algebraic object, an equation? Sometimes it is easy, but not always. As an example, consider a circle of radius 3. It consists of all points 3 units from a fixed point, called the center. The picture in the margin shows this circle with its center at the origin of a coordinate system.

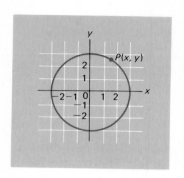

Take *any* point P on the circle and label its coordinates (x, y). It must satisfy the equation

$$d(P, O) = 3$$

From the distance formula of the previous section, we have

$$\sqrt{(x - 0)^2 + (y - 0)^2} = 3$$

or equivalently (after squaring both sides)

$$x^2 + y^2 = 9$$

This is the equation we sought.

We could move to other types of curves, attempting to find their equations. In fact, we will do exactly that in later sections. Right now, we shall consider more general circles, that is, circles with arbitrary radii and arbitrary centers.

THE STANDARD EQUATION OF A CIRCLE

Consider a circle of radius r with center at (a, b). To find its equation, take an arbitrary point on the circle with coordinates (x, y). According to the distance formula, it must satisfy the equation

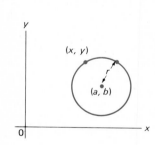

$$\sqrt{(x - a)^2 + (y - b)^2} = r$$

or, equivalently,

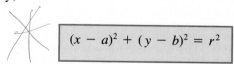

$$\boxed{(x - a)^2 + (y - b)^2 = r^2}$$

Every circle has an equation of this form; we call it the **standard equation of a circle.**

As an example, let us find the equation of a circle of radius 6 centered at $(2, -1)$. We use the boxed equation with $a = 2$, $b = -1$, and $r = 6$. This gives

$$(x - 2)^2 + (y + 1)^2 = 36$$

Let us go the other way. What circle has the equation

$$(x + 3)^2 + (y - 4)^2 = 49$$

The answer is the circle with center at $(-3, 4)$ and radius 7.

Consider this last equation again. Notice that it could be written as

$$x^2 + 6x + 9 + y^2 - 8y + 16 = 49$$

or equivalently as

$$x^2 + y^2 + 6x - 8y = 24$$

A natural question to ask is whether every equation of the form

$$x^2 + y^2 + Dx + Ey = F$$

is the equation of a circle. Take

$$x^2 + y^2 - 6y + 16x = 8$$

as an example. Recalling a skill we learned in Section 3-4 (completing the square), we may rewrite this as

$$(x^2 + 16x + \quad) + (y^2 - 6y + \quad) = 8$$

or

$$(x^2 + 16x + 64) + (y^2 - 6y + 9) = 8 + 64 + 9$$

or

$$(x + 8)^2 + (y - 3)^2 = 81$$

We recognize this to be the equation of a circle with center at $(-8, 3)$ and radius 9. In fact, we will always have a circle unless the number on the right side in the last step is negative or zero.

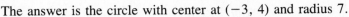

$(x - 4)^2 + (y + 2)^2 = 11$
Center : $(-4, 2)$
Radius : 11

$(x - 4)^2 + (y + 2)^2 = 11$
Center : $(4, -2)$
Radius : $\sqrt{11}$

Problem Set 4-2

Graph each of the following equations, showing enough of the graph to bring out its essential features. Note any of the three kinds of symmetry discussed in the text.

1. $y = 3x - 2$ 2. $y = 2x + 1$ 3. $y = -x^2 + 4$
4. $y = -x^2 - 2x$ 5. $y = x^2 - 4x$ 6. $y = x^3 + 2$
7. $y = -x^3$ 8. $y = \dfrac{12}{x^2 + 4}$ 9. $y = \dfrac{4}{x^2 + 1}$
10. $y = x^3 + x$

EXAMPLE (More graphing)   Graph the equation $x = y^2 - 2y + 4$.

Solution.   Assign values to y and calculate the corresponding values of x.
Note that this graph is symmetric with respect to the line $y = 1$.

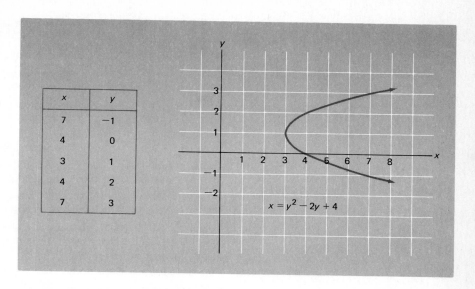

x	y
7	−1
4	0
3	1
4	2
7	3

$x = y^2 - 2y + 4$

Graph each of the following equations.

11. $x = 2y - 1$

12. $x = -3y + 1$

13. $x = -2y^2$

14. $x = 2y - y^2$

15. $x = y^3$

16. $x = 8 - y^3$

Write the equation of the circle with the given center and radius.

17. Center $(0, 0)$, radius 6.

18. Center $(2, 3)$, radius 3.

19. Center $(4, 1)$, radius 5.

20. Center $(2, -1)$, radius $\sqrt{7}$.

21. Center $(-2, 1)$, radius $\sqrt{3}$.

22. Center $(\pi, \frac{3}{4})$, radius $\frac{1}{2}$.

Graph the following equations.

23. $(x - 2)^2 + y^2 = 16$

24. $(x - 2)^2 + (y + 2)^2 = 25$

25. $(x + 1)^2 + (y - 3)^2 = 64$

26. $(x + 4)^2 + (y + 6)^2 = \frac{49}{4}$

Find the center and radius of each of the following circles. (Hint: Complete the squares.)

27. $x^2 + y^2 + 2x - 10y + 25 = 0$

28. $x^2 + y^2 - 6y = 16$

29. $x^2 + y^2 - 12x + 35 = 0$

30. $x^2 + y^2 - 10x + 10y = 0$

31. $4x^2 + 4y^2 + 4x - 12y + 1 = 0$

32. $3x^2 + 3y^2 - 2x + 4y = \frac{20}{3}$

Miscellaneous Problems

Sketch the graphs of the equations in Problems 33–37.

33. $y = 12/x$

34. $y = 3 + 4/x^2$

35. $y = 2(x - 1)^2$

36. $x = -y^2 + 8$

37. $x = 4y - y^2$

38. Which of the graphs in 33–37 are symmetric with respect to the y-axis? The x-axis? The origin?

39. Write the equation of the circle with center at $(0, 0)$ and passing through $(4, 5)$.

40. Write the equation of the circle with center at $(2, -3)$ and passing through $(5, 2)$.

41. Write the equation of the circle with AB as diameter where A and B have coordinates $(-3, 7)$ and $(3, -7)$.

42. Write the equation of the circle with diameter AB where A and B have coordinates $(3, -1)$ and $(-5, 7)$.

43. Find the equation of the circle with center on the line $y = x$ and tangent to the y-axis at $(0, 6)$.

44. Find the equation of the circle pictured in the margin.

45. The points $(2, 3)$, $(6, 3)$, $(6, -1)$, and $(2, -1)$ are the corners of a square. Find the equation of the inscribed circle of the square (the circle which is tangent to all four sides).

46. Find the equation of the circumscribed circle of the square in Problem 45. (This circle passes through all four corner points.)

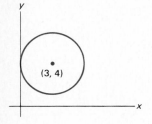

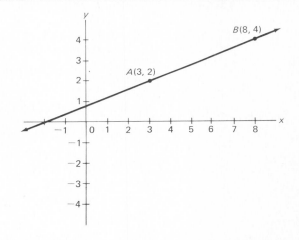

A point is that which has no part.

A line is breadthless length.

A straight line is a line which lies evenly with the points of itself.

Euclid
300 B.C.

4-3
The Straight Line

Euclid's definition of a straight line is not very helpful, but neither are most of the alternatives we have heard. Fortunately, we all know what we mean by a straight line even if we cannot seem to describe it in terms of more primitive ideas. There is one thing on which we must agree: Given two points (for example, A and B above), there is one and only one straight line that passes through them. And contrary to Euclid, let us agree that the word *line* shall always mean straight line.

A line is a geometric object. When it is placed in a coordinate system, it ought to have an equation just as a circle does. How do we find the equation of a line? To answer this question we will need the notion of slope.

THE SLOPE OF A LINE

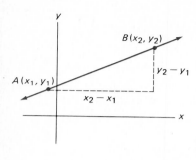

Consider the line in our opening diagram. From point A to point B, there is a **rise** (vertical change) of 2 units and a **run** (horizontal change) of 5 units. We say that the line has a slope of $\frac{2}{5}$. In general, for a line through $A(x_1, y_1)$ and $B(x_2, y_2)$, where $x_1 \neq x_2$, we define the **slope** m of that line by

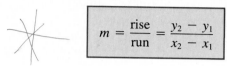

$$m = \frac{\text{rise}}{\text{run}} = \frac{y_2 - y_1}{x_2 - x_1}$$

You should immediately raise a question. A line has many points. Does the value we get for the slope depend on which pair of points we use for A and B? The similar triangles in the marginal diagram show us that

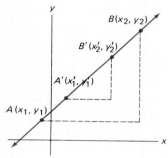

$$\frac{y_2' - y_1'}{x_2' - x_1'} = \frac{y_2 - y_1}{x_2 - x_1}$$

Thus, points A' and B' would do just as well as A and B. It does not even matter whether A is to the left or right of B since

$$\frac{y_1 - y_2}{x_1 - x_2} = \frac{y_2 - y_1}{x_2 - x_1}$$

All that matters is that we subtract the coordinates in the same order in numerator and denominator.

The slope m is a measure of the steepness of a line, as the diagram below illustrates. Notice that a horizontal line has zero slope and a line that rises to the right has positive slope. The larger this positive slope is, the more steeply the line rises. A line that falls to the right has negative slope. The concept of slope for a vertical line makes no sense since it would involve division by zero. Therefore the notion of slope for a vertical line is left undefined.

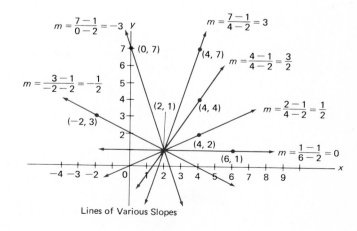

Lines of Various Slopes

THE POINT-SLOPE FORM

Consider again the line of our opening diagram; it is reproduced in the margin. We know:

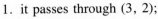

1. it passes through (3, 2);
2. it has slope $\frac{2}{5}$.

Take any other point on that line, such as one with coordinates (x, y). If we use this point together with (3, 2) to measure slope, we must get $\frac{2}{5}$; that is,

$$\frac{y - 2}{x - 3} = \frac{2}{5}$$

or, after multiplying by $x - 3$,

$$y - 2 = \frac{2}{5}(x - 3)$$

Notice that this last equation is satisfied by all points on the line, even by (3, 2). Moreover, no points not on the line can satisfy this equation.

What we have just done in an example can be done in general. The line passing through the (fixed) point (x_1, y_1) with slope m has equation

$$\boxed{y - y_1 = m(x - x_1)}$$

We call it the **point-slope** form of the equation of a line.

Consider once more the line of our example. That line passes through (8, 4) as well as (3, 2). If we use (8, 4) as (x_1, y_1), we get the equation

$$y - 4 = \frac{2}{5}(x - 8)$$

which looks quite different from

$$y - 2 = \frac{2}{5}(x - 3)$$

However, both can be simplified to $5y - 2x = 4$; they are equivalent.

THE SLOPE-INTERCEPT FORM

The equation of a line can be expressed in various forms. Suppose we are given the slope m for a line and the y-intercept b (that is, the line intersects the y-axis at $(0, b)$). Choosing $(0, b)$ as (x_1, y_1) and applying the point-slope form, we get

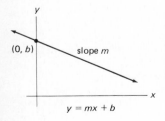

$$y - b = m(x - 0)$$

which we can rewrite as

$$y = mx + b$$

The latter is called the **slope-intercept** form.

Why get excited about that, you ask? Because any time we see an equation written this way, we recognize it as the equation of a line and can immediately read its slope and y-intercept. For example, consider the equation

$$3x - 2y + 4 = 0$$

If we solve for y, we get

$$y = \frac{3}{2}x + 2$$

It is the equation of a line with slope $\frac{3}{2}$ and y-intercept 2.

EQUATION OF A VERTICAL LINE

Vertical lines do not fit within the discussion above; they do not have slopes. But they do have equations, very simple ones. The line in the marginal picture has equation $x = \frac{5}{2}$, since every point on the line satisfies this equation. The equation of any vertical line can be put in the form

$$x = k$$

where k is a constant. It should be noted that the equation of a horizontal line can be written in the form $y = k$.

THE FORM $Ax + By + C = 0$

It would be nice to have a form that covered all lines including vertical lines. Consider for example,

(i) $\qquad y - 2 = -4(x + 2)$

(ii) $\qquad y = 5x - 3$

(iii) $\qquad x = 5$

These can be rewritten (by taking everything to the left side) as follows:

(i) $\qquad 4x + y + 6 = 0$

(ii) $\qquad -5x + y + 3 = 0$

(iii) $\qquad x + 0y - 5 = 0$

All are of the form

$$Ax + By + C = 0$$

which we call the **general linear equation.** It takes only a moment's thought to see that the equation of any line can be put in this form. Conversely, the graph of $Ax + By + C = 0$ is always a line (if A and B are not both zero (see Problem 64)).

PARALLEL LINES

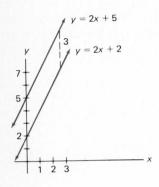

If two lines have the same slope, they are parallel. Thus $y = 2x + 2$ and $y = 2x + 5$ represent parallel lines; both have a slope of 2. The second line is 3 units above the first for every value of x.

Similarly, the lines with equations $-2x + 3y + 12 = 0$ and $4x - 6y = 5$ are parallel. To see this, solve these equations for y (that is, find the slope-intercept form); you get $y = \frac{2}{3}x - 4$ and $y = \frac{2}{3}x - \frac{5}{6}$, respectively. Both have slope $\frac{2}{3}$; they are parallel.

We may summarize by stating that *two nonvertical lines are parallel if and only if they have the same slope.*

PERPENDICULAR LINES

Is there a simple slope condition which characterizes perpendicular lines? Yes; *two nonvertical lines are perpendicular if and only if their slopes are negative reciprocals of each other.* We are not going to prove this, but an example will help explain why it is true. The slopes of the lines $y = \frac{3}{4}x$ and $y = -\frac{4}{3}x$ are negative reciprocals of each other. Both lines pass through the origin. The points $(4, 3)$ and $(3, -4)$ are on the first and second lines, respectively. The two right triangles shown in the diagram in the margin are congruent with $\angle\alpha = \angle\delta$ and $\angle\beta = \angle\gamma$. But

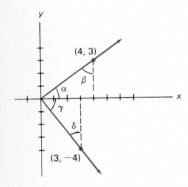

$$\angle\alpha + \angle\beta = 90°$$

and therefore

$$\angle\alpha + \angle\gamma = 90°$$

That says the two lines are perpendicular to each other.

The lines $2x - 3y = 5$ and $3x + 2y = -4$ are also perpendicular, since—after solving them for y—we see that the first has slope $\frac{2}{3}$ and the second has slope $-\frac{3}{2}$.

Problem Set 4-3

Find the slope of the line containing the given two points.

1. $(2, 3)$ and $(4, 8)$
2. $(4, 1)$ and $(8, 2)$
3. $(-4, 2)$ and $(3, 0)$
4. $(2, -4)$ and $(0, -6)$
5. $(3, 0)$ and $(0, 5)$
6. $(-6, 0)$ and $(0, 6)$
7. $(-1.732, 5.014)$ and $(4.315, 6.175)$
8. $(\pi, \sqrt{3})$ and $(1.642, \sqrt{2})$

Find an equation for each of the following lines. Then write your answer in the form $Ax + By + C = 0$.

9. Through (2, 3) with slope 4.
10. Through (4, 2) with slope 3.
11. Through (3, −4) with slope −2.
12. Through (−5, 2) with slope −1.
13. With y-intercept 4 and slope −2.
14. With y-intercept −3 and slope 1.
15. With y-intercept 5 and slope 0.
16. With y-intercept 1 and slope −1.
17. Through (2, 3) and (4, 8).
18. Through (4, 1) and (8, 2).
19. Through (3, 0) and (0, 5).
20. Through (−6, 0) and (0, 6).
[c] 21. Through $(\sqrt{3}, \sqrt{7})$ and $(\sqrt{2}, \pi)$.
[c] 22. Through $(\pi, \sqrt{3})$ and $(\pi + 1, 2\sqrt{3})$.
23. Through (2, −3) and (2, 5).
24. Through (−5, 0) and (−5, 4).

In Problems 25–32, find the slope and y-intercept of each line.

25. $y = 3x + 5$
26. $y = 6x + 2$
27. $3y = 2x - 4$
28. $2y = 5x + 2$
29. $2x + 3y = 6$
30. $4x + 5y = -20$
31. $y + 2 = -4(x - 1)$
32. $y - 3 = 5(x + 2)$

33. Write the equation of the line through (3, −3):
    (a) parallel to the line $y = 2x + 5$;
    (b) perpendicular to the line $y = 2x + 5$;
    (c) parallel to the line $2x + 3y = 6$;
    (d) perpendicular to the line $2x + 3y = 6$;
    (e) parallel to the line through (−1, 2) and (3, −1);
    (f) parallel to the line $x = 8$;
    (g) perpendicular to the line $x = 8$.

34. Find the value of k for which the line $4x + ky = 5$:
    (a) passes through the point (2, 1);
    (b) is parallel to the y-axis;
    (c) is parallel to the line $6x - 9y = 10$;
    (d) has equal x- and y-intercepts;
    (e) is perpendicular to the line $y - 2 = 2(x + 1)$.

35. Write the equation of the line through (0, −4) that is perpendicular to the line $y + 2 = -\frac{1}{2}(x - 1)$.

36. Find the value of k such that the line $kx - 3y = 10$:
    (a) is parallel to the line $y = 2x + 4$;
    (b) is perpendicular to the line $y = 2x + 4$;
    (c) is perpendicular to the line $2x + 3y = 6$.

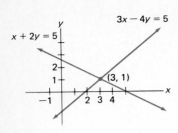

EXAMPLE A (Intersection of two lines)   Find the coordinates of the point of intersection of the lines $3x - 4y = 5$ and $x + 2y = 5$.

Solution.   We simply solve the two equations simultaneously (see Section 3-3). Multiply the second equation by -3 and then add the two equations.

$$
\begin{aligned}
3x - 4y &= 5 \\
-3x - 6y &= -15 \\
\hline
-10y &= -10 \\
y &= 1 \\
x &= 3
\end{aligned}
$$

Find the coordinates of the point of intersection in each problem below. Then write the equation of the line through that point perpendicular to the line given first.

37. $2x + 3y = 4$
       $-3x + y = 5$

38. $4x - 5y = 8$
       $2x + y = -10$

39. $3x - 4y = 5$
      $2x + 3y = 9$

40. $5x - 2y = 5$
      $2x + 3y = 6$

EXAMPLE B (Distance from a point to a line)   It can be shown that the distance d from the point (x_1, y_1) to the line $Ax + By + C = 0$ is

$$d = \frac{|Ax_1 + By_1 + C|}{\sqrt{A^2 + B^2}}$$

Find the distance from $(1, 2)$ to $3x - 4y = 5$.

Solution.   First write the equation as $3x - 4y - 5 = 0$. The formula gives

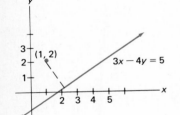

$$d = \frac{|3 \cdot 1 - 4 \cdot 2 - 5|}{\sqrt{3^2 + (-4)^2}} = \frac{|-10|}{5} = 2$$

In each case, find the distance from the given point to the given line.

41. $(-3, 2)$, $3x + 4y = 6$

42. $(4, -1)$, $2x - 2y + 4 = 0$

43. $(-2, -1)$, $5y = 12x + 1$

44. $(3, -1)$, $y = 2x - 5$

Find the (perpendicular) distance between the given parallel lines. (Hint: First find a point on one of the lines.)

45. $3x + 4y = 6$, $3x + 4y = 12$

46. $5x + 12y = 2$, $5x + 12y = 7$

Miscellaneous Problems

47. Find the slope of the line containing the points $(2, 1)$ and $(-3, 5)$.

48. Find the equation of the line through the point $(4, 1)$ that:
    (a) passes through the origin;
    (b) is parallel to the line $y = 3x - 4$;
    (c) is perpendicular to the line $x + 2y = 6$;

(d) is parallel to the y-axis;

(e) has y-intercept 2.

49. Find the equation of the line that passes through the intersection of the lines $2x - 3y = 2$ and $5x + 3y = 26$ and is parallel to the line $x + y = 4$.

50. Consider the line segment AB where A and B have coordinates $(2, -1)$ and $(6, 5)$. Find the equation of the line which bisects AB at right angles.

51. Which pairs of lines below are parallel, which are perpendicular, and which are neither?

(a) $y = 4x - 7$; $y + 3 = 4(x + 1)$

(b) $x + 2y = 4$; $x - 2y = -4$

(c) $3x + 2y = 4$; $2x - 3y = 0$

(d) $x = 4$; $x = -2$

(e) $2x + 5y = 0$; $5x = 2y$

52. Find the distance between the parallel lines $3x - 4y = 12$ and $3x - 4y = -13$.

53. The Celsius and Fahrenheit temperature scales are related by the equation $F = \frac{9}{5}C + 32$.

(a) If C increases by 1, how much does F increase?

(b) The graph of $F = \frac{9}{5}C + 32$ is a line. Assuming the F-axis to be vertical, what is the slope of the line? What is its F-intercept?

(c) Solve the system of equations

$$F = \frac{9}{5}C + 32$$

$$F = C$$

Interpret your answer.

54. Suppose that you borrow \$20,000 today at 8 percent simple interest per year. Find a formula for the amount A you will owe after t years (see Problem 64 in Section 3-1).

55. Graph the equation you obtained in Problem 54 using a vertical A-axis. What is the slope of the graph? What is its A-intercept?

56. A heavy piece of road equipment costs \$120,000 and each year it depreciates at 8 percent of its original value. Find a formula for the value V of the item after t years.

57. The graph of the answer to Problem 56 is a straight line. What is its slope, assuming the t-axis to be horizontal? Interpret the slope.

58. Past experience indicates that egg production in Matlin County is growing linearly. In 1960, it was 700,000 cases and in 1970, it was 820,000 cases. Write a formula for the number N of cases produced n years after 1960 and use it to predict egg production in the year 2000.

59. A piece of equipment purchased today for \$80,000 will depreciate linearly to a scrap value of \$2000 after 20 years. Write a formula for its value V after n years.

60. Suppose that the profit P that a company realizes in selling x items of a certain commodity is given by $P = 450x - 2000$.

(a) Interpret the value of P when $x = 0$.

(b) Find the slope of the graph of the above equation. This slope is called the **marginal profit.** What is its economic interpretation?

61. The cost C of producing x items of a certain commodity is given by $C = .75x + 200$. The slope of its graph is called the **marginal cost.** Find it and give an economic interpretation.

62. Does $(3, 9)$ lie above or below the line $y = 3x - 1$?

63. Show that that the equation of the line with x-intercept a and y-intercept b is

$$\frac{x}{a} + \frac{y}{b} = 1$$

64. Show that the graph of $Ax + By + C = 0$ is always a line (provided A and B are not both 0). (*Hint:* Consider two cases: (i) $B = 0$ and (ii) $B \neq 0$.)

65. Find the equation of the line through $(2, 3)$ which has equal x- and y-intercepts.

66. Show that for each value of k, the equation

$$2x - y + 4 + k(x + 3y - 6) = 0$$

represents a line through the intersection of the two lines $2x - y + 4 = 0$ and $x + 3y - 6 = 0$. (*Hint:* It is not necessary to find the point of intersection, (x_0, y_0).)

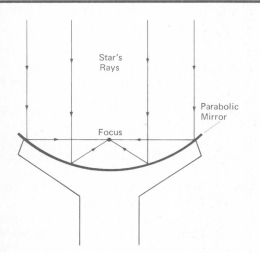

Star's Rays

Parabolic Mirror

Focus

Reflecting Telescopes

A reflecting telescope, such as the one at Mount Palomar in California, makes use of a parabolic mirror. The rays of light from a star (parallel lines) are focused by the mirror at a single point. This is a characteristic property of the mathematical curve that we call a parabola.

4.4
The Parabola

We have seen that the graph $y = ax + b$ is always a line. Now we want to study the graph of $y = ax^2 + bx + c\,(a \neq 0)$. As we shall discover, this graph is always a smooth, cup-shaped curve something like the cross section of the mirror shown in the opening diagram. We call it a **parabola.**

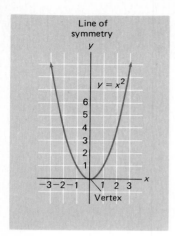

Line of symmetry

$y = x^2$

Vertex

SOME SIMPLE CASES

The simplest case of all is $y = x^2$. The graph of this equation is shown in the margin. Two important features should be noted.

1. The curve is symmetric about the y-axis. This follows from the fact that the equation $y = x^2$ is not changed if we replace (x, y) by $(-x, y)$.
2. The curve reaches its lowest point at $(0, 0)$, the point where the curve intersects the line of symmetry. We call this point the **vertex** of the parabola.

Next we consider how the graph of $y = x^2$ is modified as we look successively at $y = ax^2$, $y = x^2 + k$, $y = (x - h)^2$, and $y = a(x - h)^2 + k$.

THE GRAPH OF $y = ax^2$

In the diagrams below, we show the graphs of $y = x^2$, $y = 3x^2$, $y = -2x^2$, and $y = \frac{1}{2}x^2$. They suggest the following general facts. The graph of $y = ax^2$, $a \neq 0$, is a parabola with vertex at the origin, opening upward if $a > 0$ and downward if $a < 0$. Increasing $|a|$ makes the graph narrower.

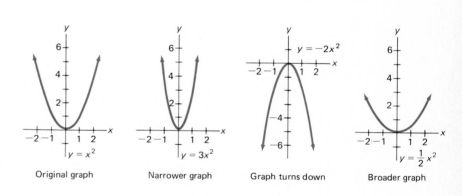

Original graph   Narrower graph   Graph turns down   Broader graph

THE GRAPHS OF $y = x^2 + k$ AND $y = (x - h)^2$

The graphs of $y = x^2 + 4$, $y = x^2 - 6$, $y = (x - 2)^2$, and $y = (x + 1)^2$ can all be obtained by shifting (translating) the graph of $y = x^2$, while maintaining its shape. They are shown at the top of the next page. After studying these graphs carefully, you will understand the general situation we now describe.

The graph of $y = x^2 + k$ is obtained by shifting the graph of $y = x^2$ vertically $|k|$ units, upward if $k > 0$ and downward if $k < 0$. The vertex is at $(0, k)$.

The graph of $y = (x - h)^2$ is obtained by a horizontal shift of $|h|$ units, to the right if $h > 0$ and to the left if $h < 0$. The vertex is at $(h, 0)$.

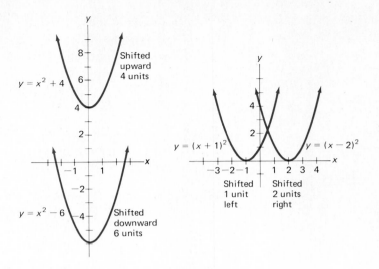

THE GRAPH OF $y = a(x - h)^2 + k$ OR $y - k = a(x - h)^2$

We can get the graph of $y = 2(x - 3)^2 - 4$ by shifting the graph of $y = 2x^2$ three units to the right and four units down. This puts the vertex at $(3, -4)$. More generally, the graph of $y = a(x - h)^2 + k$ is the graph of $y = ax^2$ shifted horizontally $|h|$ units and vertically $|k|$ units, so that the vertex is at (h, k). The graphs below illustrate these facts.

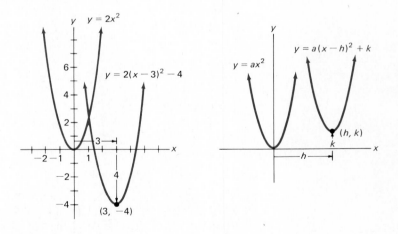

What about the graph of $y = -2(x - 4)^2 + 6$? It has the same shape as the graph of $y = 2x^2$, but turns down with its vertex at $(4, 6)$.

For the circle $x^2 + y^2 = r^2$, recall that replacing x by $x - h$ and y by $y - k$ to obtain $(x - h)^2 + (y - k)^2 = r^2$ had the effect of shifting the center from $(0, 0)$ to (h, k). A similar thing happens to the parabola $y = ax^2$. Replacing x by $x - h$ and y by $y - k$ changes $y = ax^2$ to $y - k = a(x - h)^2$

(which you will note is equivalent to $y = a(x - h)^2 + k$) and correspondingly shifts the vertex from $(0,0)$ to (h, k).

THE GRAPH OF $y = ax^2 + bx + c$

The most general equation considered so far is $y = a(x - h)^2 + k$. If we expand $a(x - h)^2$ and collect terms on the right side, the equation takes the form $y = ax^2 + bx + c$. Conversely, $y = ax^2 + bx + c \, (a \neq 0)$ always represents a parabola with a vertical line of symmetry, as we shall now show. We use the method of completing the square to rewrite $y = ax^2 + bx + c$ as follows.

$$y = a\left(x^2 + \frac{b}{a}x + \quad \right) + c$$

$$= a\left[x^2 + \frac{b}{a}x + \left(\frac{b}{2a}\right)^2\right] + c - a\left(\frac{b}{2a}\right)^2$$

$$= a\left(x + \frac{b}{2a}\right)^2 + \left(c - \frac{b^2}{4a}\right)$$

This is the equation of a parabola with vertex at $(-b/2a, \, c - b^2/4a)$ and line of symmetry $x = -b/2a$.

What we have just said leads to an important conclusion. The graph of the equation $y = ax^2 + bx + c \, (a \neq 0)$ is always a parabola with vertical axis of symmetry. Moreover, we have the following facts.

1. The x-coordinate of the vertex is $-b/2a$; the y-coordinate is easily found by substitution in the equation.
2. The parabola turns upward if $a > 0$ and downward if $a < 0$. It is a fat or thin parabola, according as $|a|$ is small or large.

As an example, consider $y = 2x^2 + 8x + 3$. Its graph is a parabola with vertex at $x = -b/2a = -8/4 = -2$. The y-coordinate of the vertex, obtained by substituting $x = -2$ in the equation, is $y = -5$. The parabola turns upward and it is rather thin. A sketch is shown in the margin.

APPLICATIONS

Our opening panel hinted at an important application of the parabola related to its optical properties. When light rays, parallel to the axis of a parabolic mirror, hit the mirror, they are reflected to a single point called the *focus*. Conversely, if a light source is placed at the focus of a parabolic mirror, the reflected rays of light are parallel to the axis, a principle used in flashlights.

The parabola is used also in the design of suspension bridges and load-bearing arches. If equal weights are placed along a line, equally spaced, and suspended from a thin flexible cable, the cable will assume a shape that closely approximates a parabola.

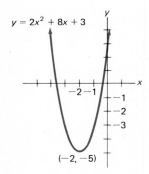

$y = 2x^2 + 8x + 3$

$(-2, -5)$

Cross section of a parabolic mirror with light source at focus

We turn to a very different kind of application. From physics, we learn that the path of a projectile is a parabola. It is known, for example, that a projectile fired at an angle of 45° from the horizontal with an initial speed of $320\sqrt{2}$ feet per second follows a curve with equation

$$y = -\frac{1}{6400}x^2 + x$$

where the coordinate axes are placed as shown in the diagram in the margin. Taking this for granted, we may ask two questions

1. What is the maximum height attained by the projectile?
2. What is the range (horizontal distance traveled) of the projectile?

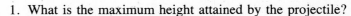

To find the maximum height is simply to find the y-coordinate of the vertex. First we find the x-coordinate.

$$x = \frac{-b}{2a} = -\frac{1}{-2/6400} = 3200$$

When we substitute this value in the equation, we get

$$y = -\frac{1}{6400}(3200)^2 + 3200 = -1600 + 3200 = 1600$$

The greatest height is thus 1600 feet.

The range of the projectile is the x-coordinate of the point where it lands. By symmetry, this is simply twice the x-coordinate of the vertex; that is,

$$\text{range} = 2(3200) = 6400 \text{ feet}$$

This value could also be obtained by solving the quadratic equation

$$-\frac{1}{6400}x^2 + x = 0$$

since the x-coordinate of the landing point is the value of x when $y = 0$.

Problem Set 4-4

The equations in Problems 1–10 represent parabolas. Sketch the graph of each parabola, indicating the coordinates of the vertex.

1. $y = 3x^2$
2. $y = -2x^2$
3. $y = x^2 + 5$
4. $y = 2x^2 - 4$
5. $y = (x - 4)^2$
6. $y = -(x + 3)^2$
7. $y = 2(x - 1)^2 + 5$
8. $y = 3(x + 2)^2 - 4$
9. $y = -4(x - 2)^2 + 1$
10. $y = \frac{1}{2}(x + 3)^2 + 3$

Write each of the following in the form $y = ax^2 + bx + c$.

11. $y = 2(x - 1)^2 + 7$

12. $y = -3(x + 2)^2 + 5$

13. $-2y + 5 = (x - 5)^2$

14. $3y + 6 = (x + 3)^2$

Sketch the graph of each equation. Begin by plotting the vertex and at least one point on each side of the vertex. Recall that the x-coordinate of the vertex for $y = ax^2 + bx + c$ is $x = -b/2a$.

15. $y = x^2 + 2x$

16. $y = 3x^2 - 6x$

17. $y = -2x^2 + 8x + 1$

18. $y = -3x^2 + 6x + 4$

EXAMPLE A (Horizontal parabolas) Sketch the parabolas (a) $x = 2y^2$; (b) $x = 2(y - 3)^2 - 4$.

Solution. Note that the roles of x and y are interchanged when compared to earlier examples. The line of symmetry will therefore be horizontal. The vertex in (a) is $(0, 0)$; in (b), it is $(-4, 3)$. The second graph can be obtained by shifting the first one 4 units to the left and 3 units upward.

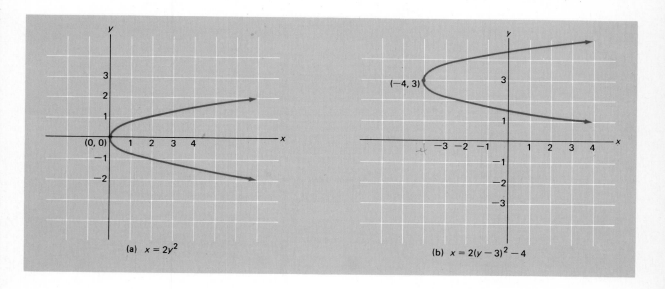

(a) $x = 2y^2$

(b) $x = 2(y - 3)^2 - 4$

In Problems 19–24, sketch the graph of the equation and indicate the coordinates of the vertex.

19. $x = -2y^2$

20. $x = -2y^2 + 8$

21. $x = -2(y + 2)^2 + 8$

22. $x = 3(y - 1)^2 + 6$

23. $x = y^2 + 4y + 2$. (*Note:* The y-coordinate of the vertex is at $-b/2a$.)

24. $x = 4y^2 - 8y + 10$

EXAMPLE B (Intersection of a line and a parabola)   Find the points of intersection of the line $y = -2x + 2$ and the parabola $y = 2x^2 - 4x - 2$.

Solution.   We must solve the two equations simultaneously. This is easy to do by equating the two expressions for y and then solving the resulting equation for x.

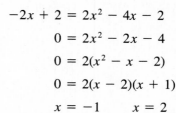

$$-2x + 2 = 2x^2 - 4x - 2$$
$$0 = 2x^2 - 2x - 4$$
$$0 = 2(x^2 - x - 2)$$
$$0 = 2(x - 2)(x + 1)$$
$$x = -1 \qquad x = 2$$

By substitution, we find the corresponding values of y to be 4 and -2; the intersection points are therefore $(-1, 4)$ and $(2, -2)$.

Find the points of intersection for the given line and parabola.

25. $y = -x + 1$
    $y = x^2 + 2x + 1$
26. $y = -x + 4$
    $y = -x^2 + 2x + 4$
27. $y = -2x + 1$
    $y = -x^2 - x + 3$
28. $y = -3x + 15$
    $y = 3x^2 - 3x + 12$
ⓒ 29. $y = 1.5x + 3.2$
    $y = x^2 - 2.9x$
ⓒ 30. $y = 2.1x - 6.4$
    $y = -1.2x^2 + 4.3$

EXAMPLE C (Using side conditions)   Find the equation of the vertical parabola with vertex $(2, -3)$ and passing through the point $(9, -10)$.

Solution.   The equation must have the form

$$y + 3 = a(x - 2)^2$$

To find a, we substitute $x = 9$ and $y = -10$.

$$-7 = a(7)^2$$

Thus $a = -\frac{1}{7}$, and the required equation is $y + 3 = -\frac{1}{7}(x - 2)^2$.

In Problems 31–34, find the equation of the vertical parabola that satisfies the given conditions.

31. Vertex $(0, 0)$, passing through $(-6, 3)$.
32. Vertex $(0, 0)$, passing through $(3, -6)$.
33. Vertex $(-2, 0)$, passing through $(6, -8)$.
34. Vertex $(3, -1)$, passing through $(-2, 5)$.
35. Find the equation of the parabola passing through $(1, 1)$ and $(2, 7)$ and having the y-axis as the line of symmetry. (*Hint:* The equation has the form $y = ax^2 + c$.)

36. Find the equation of the parabola passing through $(1, 2)$ and $(-2, -7)$ and having the y-axis as the line of symmetry.

37. Sketch the following parabolas on the same coordinate system.
    (a) $y = 2x^2 - 8x + 4$ (b) $y = 2x^2 - 8x + 8$
    (c) $y = 2x^2 - 8x + 11$

38. Sketch the following parabolas on the same coordinate system.
    (a) $y = -3x^2 + 6x + 9$ (b) $y = -3x^2 + 6x - 3$
    (c) $y = -3x^2 + 6x - 9$

39. Calculate the value of the discriminant $b^2 - 4ac$ for each of the parabolas in Problem 37. Make a conjecture about the number of x-intercepts of the parabola $y = ax^2 + bx + c$ if $b^2 - 4ac > 0$, if $b^2 - 4ac = 0$, and if $b^2 - 4ac < 0$.

40. For what values of k does the parabola $y = x^2 - 6x + k$ have two x-intercepts?

41. Find the value of a for which the parabola $y = ax^2$ passes through the point $(2, 5)$.

42. Find the points of intersection of the parabolas $y = 4 - x^2$ and $y = x^2 - 14$.

43. Find the points where the line $y = 2x - 4$ intersects the parabola $y = x^2 - 12$.

44. Use algebra to show that the parabola $y = x^2 - 4x + 6$ has just one point in common with the line $y = 2x - 3$. Then sketch both equations on the same coordinate system.

45. Use algebra to show that the parabola $y = -x^2 + 2x + 4$ and the line $y = -2x + 9$ have no point in common. Then sketch both equations on the same coordinate system.

46. For any $a \neq 0$, the equation $y = a(x - 2)(x - 8)$ represents a parabola.
    (a) Find its x-intercepts.
    (b) Use part (a) to determine the x-coordinate of the vertex.
    (c) Find the value of a if the parabola passes through $(10, 40)$.

47. If the shape of the hanging rope shown in the margin is part of a parabola, find the distance PQ.

48. Starting at $(0, 0)$, a projectile travels along the path $y = -\frac{1}{256}x^2 + \sqrt{3}x$. Sketch the path, find the maximum height, and find the range.

49. Find the equation of the vertical parabola passing through $(-1, 2)$, $(1, -1)$, and $(2, 1)$.

50. A retailer has learned from experience that if she charges x dollars apiece for a toy truck, she can sell $300 - 100x$ of them. The trucks cost her \$2 each. Write a formula for her total profit P in terms of x. Then determine what she should charge to maximize her profit.

51. A company that makes fancy golf carts will have overhead of \$12,000 per year and direct costs (labor and materials) of \$80 per cart. It sells all its carts to a certain retailer and has a standard price of \$120 each. The company will

15 ft   15 ft
10 ft
5 ft   5 ft
Q
P
20 ft

give a discount of 1 percent for 100 carts, 2 percent for 200 carts, and in general $x/100$ percent for x carts. Its maximum possible production is 2500 carts.

(a) Write a formula for C, the cost of producing x carts.

(b) Show that its total receipts R in dollars is $R = 120x - .012x^2$

c (c) What is the smallest number of carts it can produce and still break even?

c (d) What number of carts will produce a maximum profit and what is this profit?

52. Starting at $(0, 0)$, a ball travels along the path $y = ax^2 + bx$. Find a and b if the ball reaches a height of 75 at $x = 50$ and if it reaches its greatest height at $x = 100$.

53. Let $P(x, y)$ move so that its distance from the point $F(0, 3)$ is always equal to its perpendicular distance from the line $y = -3$. Derive and simplify the equation of the path. It should turn out to be a parabola with vertex at the origin.

54. Suppose in Problem 53 that the fixed point is $F(0, p)$ and that the given line is $y = -p$. Now let $P(x, y)$ move so that its distance from F is always equal to its perpendicular distance from the given line. Show that the equation of the path is $x^2 = 4py$.

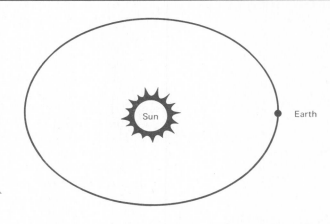

Planetary Orbits

"Kepler discovered by the analysis of astronomical observations, and Newton proved mathematically on the basis of the inverse square law of gravitational attraction, that the planets move in ellipses. The geometry of ancient Greece thus became the cornerstone of modern astronomy."

J. L. Synge

4-5
Ellipses and Hyperbolas (Optional)

Roughly speaking, an ellipse is a flattened circle. In the case of the earth's orbit about the sun, there is very little flattening (less than the opening panel indicates). But all ellipses, be they nearly circular or very flat, have important properties in common. As with parabolas, we shall begin by discussing equations of ellipses.

EQUATIONS OF ELLIPSES

Consider the equation

$$\frac{x^2}{25} + \frac{y^2}{16} = 1$$

Because x and y can be replaced by $-x$ and $-y$, respectively, without changing the equation, the graph is symmetric with respect to both axes and the origin. To find the x-intercepts, we let $y = 0$ and solve for x.

$$\frac{x^2}{25} = 1$$
$$x^2 = 25$$
$$x = \pm 5$$

Thus the graph intersects the x-axis at $(\pm 5, 0)$. By a similar procedure (letting $x = 0$), we find that the graph intersects the y-axis at $(0, \pm 4)$. Plotting these points and a few others leads to the graph below.

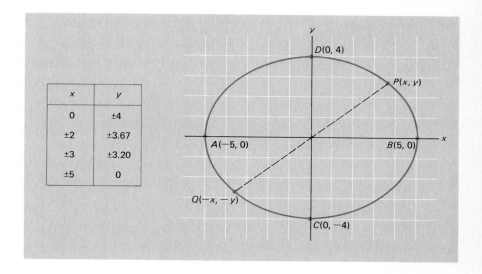

x	y
0	±4
±2	±3.67
±3	±3.20
±5	0

This curve is an example of an ellipse. The dotted line segment PQ with endpoints on the ellipse and passing through the origin is called a **diameter**. The longest diameter, AB, is the **major diameter** (sometimes called the major axis) and the shortest one, CD, is the **minor diameter**. The origin, which is a bisector of every diameter, is appropriately called the **center** of the ellipse. The endpoints of the major diameter are called the **vertices** of the ellipse.

More generally, if a and b are any positive numbers, the equation

$$\frac{x^2}{a^2} + \frac{y^2}{b^2} = 1$$

represents an ellipse with center at the origin and intersecting the x- and y-axes at $(\pm a, 0)$ and $(0, \pm b)$, respectively. If $a > b$, it is called a *horizontal* ellipse (because the major diameter is horizontal); if $a = b$, it is a *circle* of radius a; and if $a < b$, it is called a *vertical* ellipse.

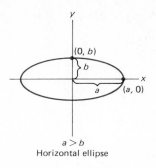 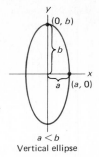 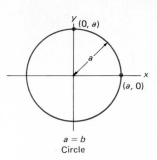

| $a > b$ | $a < b$ | $a = b$ |
| Horizontal ellipse | Vertical ellipse | Circle |

The graph of $\dfrac{x^2}{a^2} + \dfrac{y^2}{b^2} = 1$

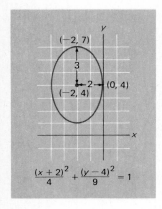

$$\frac{(x + 2)^2}{4} + \frac{(y - 4)^2}{9} = 1$$

TRANSLATING THE ELLIPSE

If we move the ellipse $x^2/a^2 + y^2/b^2 = 1$ (without turning it) so that its center is at (h, k) rather than the origin, its equation takes the form

$$\frac{(x - h)^2}{a^2} + \frac{(y - k)^2}{b^2} = 1$$

This is called the **standard form** for the equation of an ellipse. Again, the relative sizes of a and b determine whether it is a horizontal ellipse, a vertical ellipse, or a circle. For example, the graph of

$$\frac{(x + 2)^2}{4} + \frac{(y - 4)^2}{9} = 1$$

is a vertical ellipse centered at $(-2, 4)$ with major diameter of length $2 \cdot 3 = 6$ and minor diameter of length $2 \cdot 2 = 4$ (see the diagram in the margin).

EQUATIONS OF HYPERBOLAS

What a difference a change in sign can make! The graphs of

$$\frac{x^2}{25} + \frac{y^2}{16} = 1$$

and

$$\frac{x^2}{25} - \frac{y^2}{16} = 1$$

are as different as night and day. The first is an ellipse; the second is a hyperbola. Let us see what we can find out about this hyperbola.

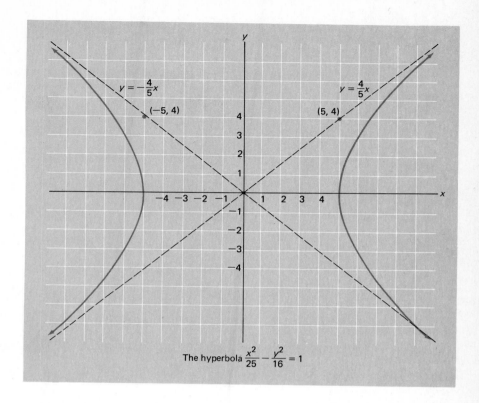

The hyperbola $\dfrac{x^2}{25} - \dfrac{y^2}{16} = 1$

First note that it has x-intercepts ± 5, but no y-intercepts (setting $x = 0$ in the equation yields $y^2 = -16$). Since x can be replaced by $-x$ and y can be replaced by $-y$ without changing the equation, the graph is symmetric with respect to both axes and the origin. This makes it appropriate to call the origin the **center** of the hyperbola. If we solve for y in terms of x, we get

$$y = \pm\tfrac{4}{5}\sqrt{x^2 - 25}$$

This implies first that we must have $|x| \geq 5$; so the graph has no points between $x = -5$ and $x = 5$. Second, since for large $|x|$, $\tfrac{4}{5}\sqrt{x^2 - 25}$ behaves very much like $\tfrac{4}{5}x$, the hyperbola must draw closer and closer to the lines $y = \pm\tfrac{4}{5}x$. These lines are called **asymptotes** of the graph.

When we put all of this information together and plot a few points, we are led to the graph above. You can see from it why $(-5, 0)$ and $(5, 0)$ are called **vertices** of the hyperbola.

The example analyzed above suggests the general situation. If a and b are positive numbers, the graphs of

$$\frac{x^2}{a^2} - \frac{y^2}{b^2} = 1 \quad \text{or} \quad \frac{y^2}{b^2} - \frac{x^2}{a^2} = 1$$

are hyperbolas. In the first case, the hyperbola is said to be *horizontal*, since the vertices are at $(\pm a, 0)$, on a horizontal line. In the second case, the hyperbola is *vertical* with vertices $(0, \pm b)$. In both cases, the asymptotes are the lines $y = \pm(b/a)x$. Note in the diagrams below how the numbers a and b determine a rectangle with the asymptotes as diagonals.

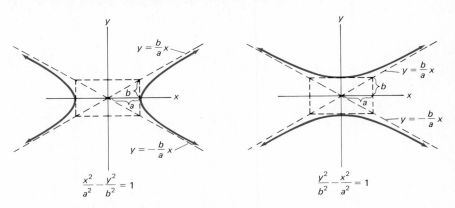

$$\frac{x^2}{a^2} - \frac{y^2}{b^2} = 1 \qquad\qquad \frac{y^2}{b^2} - \frac{x^2}{a^2} = 1$$

As an example of a vertical hyperbola, consider

$$\frac{y^2}{16} - \frac{x^2}{25} = 1$$

The vertices are at $(0, \pm 4)$ and the asymptotes have equations $y = \pm\frac{4}{5}x$.

TRANSLATING THE HYPERBOLA

If we translate the general hyperbolas discussed above so that their centers are at (h, k) rather than the origin, their equations take the form

HORIZONTAL CASE	VERTICAL CASE
$\dfrac{(x-h)^2}{a^2} - \dfrac{(y-k)^2}{b^2} = 1$	$\dfrac{(y-k)^2}{b^2} - \dfrac{(x-h)^2}{a^2} = 1$

These are called the **standard forms** for the equation of a hyperbola.

As a final example, consider

$$\frac{(y-1)^2}{9} - \frac{(x+2)^2}{4} = 1$$

This is the equation of a vertical hyperbola with center at $(-2, 1)$ and vertices 3 units above and below the center—that is, $(-2, 4)$ and $(-2, -2)$. The graph, including the asymptotes, is shown in the margin.

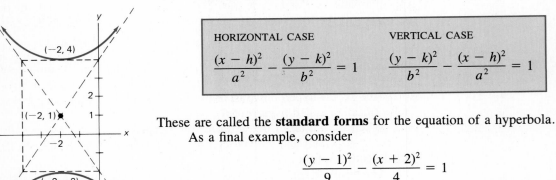

THE CONIC SECTIONS

The Greeks called the four curves—circles, parabolas, ellipses, and hyperbolas—*conic sections*. If you take a cone with two nappes and pass planes through it at various angles, the curve of intersection in each instance is one of these four curves (assuming that the intersecting plane does not pass through the apex of the cone). The diagrams below illustrate this important fact.

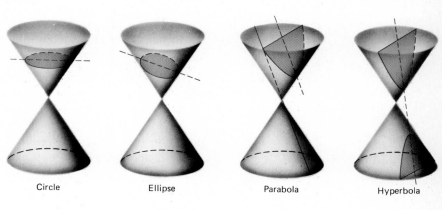

Circle   Ellipse   Parabola   Hyperbola

Problem Set 4-5

Each of the following equations represents an ellipse. Find its center and the endpoints of the major and minor diameters and graph the ellipse.

1. $\dfrac{x^2}{25} + \dfrac{y^2}{9} = 1$

2. $\dfrac{x^2}{36} + \dfrac{y^2}{16} = 1$

3. $\dfrac{x^2}{9} + \dfrac{y^2}{25} = 1$

4. $\dfrac{x^2}{4} + \dfrac{y^2}{25} = 1$

5. $\dfrac{(x - 2)^2}{25} + \dfrac{(y + 1)^2}{9} = 1$

6. $\dfrac{(x - 3)^2}{36} + \dfrac{(y - 4)^2}{25} = 1$

7. $\dfrac{(x + 3)^2}{9} + \dfrac{y^2}{16} = 1$

8. $\dfrac{x^2}{4} + \dfrac{(y - 4)^2}{49} = 1$

In Problems 9–14, find the center of the ellipse with AB and CD as major and minor diameters, respectively. Then write the equation of the ellipse.

9. $A(6, 0), B(-6, 0), C(0, 3), D(0, -3)$

10. $A(5, 0), B(-5, 0), C(0, 1), D(0, -1)$

11. $A(0, 6), B(0, -6), C(4, 0), D(-4, 0)$

12. $A(0, 3), B(0, -3), C(2, 0), D(-2, 0)$

13. $A(-4, 3), B(8, 3), C(2, 1), D(2, 5)$
    (*Hint:* Use the midpoint formula to find the center.)

14. $A(3, -4), B(11, -4), C(7, -7), D(7, -1)$

Each of the following equations represents a hyperbola. Find its center and its vertices. Then graph the equation.

15. $\dfrac{x^2}{16} - \dfrac{y^2}{9} = 1$

16. $\dfrac{x^2}{36} - \dfrac{y^2}{16} = 1$

17. $\dfrac{y^2}{9} - \dfrac{x^2}{16} = 1$

18. $\dfrac{y^2}{16} - \dfrac{x^2}{36} = 1$

19. $\dfrac{(x-3)^2}{9} - \dfrac{(y+2)^2}{16} = 1$

20. $\dfrac{(x+1)^2}{16} - \dfrac{(y-4)^2}{36} = 1$

21. $\dfrac{(y+3)^2}{4} - \dfrac{x^2}{25} = 1$

22. $\dfrac{y^2}{49} - \dfrac{(x-2)^2}{9} = 1$

EXAMPLE A (Hyperbolas with side conditions)   Find the equation of the hyperbola satisfying the given conditions.

(a) Its vertices are $(4, 0)$ and $(-4, 0)$ and one of its asymptotes has slope $\frac{3}{2}$.

(b) The center is at $(2, 1)$, one vertex is at $(2, -4)$, and the equation of one asymptote is $5x - 7y = 3$.

Solution.

(a) This is a horizontal hyperbola with center at $(0, 0)$. Since $\frac{3}{2} = b/a = b/4$, we get $b = 6$. The required equation is

$$\frac{x^2}{16} - \frac{y^2}{36} = 1$$

(b) Since the vertex $(2, -4)$ is 5 units directly below the center, the hyperbola is vertical and $b = 5$. Solving $5x - 7y = 3$ for y in terms of x gives $y = \frac{5}{7}x - \frac{3}{7}$, so the given asymptote has slope $\frac{5}{7}$. Thus $\frac{5}{7} = b/a = 5/a$, so $a = 7$. The required equation is

$$\frac{(y-1)^2}{25} - \frac{(x-2)^2}{49} = 1$$

In Problems 23–28, find the equation of the hyperbola satisfying the given conditions.

23. The vertices are $(4, 0)$ and $(-4, 0)$ and one asymptote has slope $\frac{5}{4}$.

24. The vertices are $(7, 0)$ and $(-7, 0)$ and one asymptote has slope $\frac{2}{7}$.

25. The center is $(6, 3)$, one vertex is $(8, 3)$, and one asymptote has slope 1.

26. The center is $(3, -3)$, one vertex is $(-2, -3)$, and one asymptote has slope $\frac{6}{5}$.

27. The vertices are $(4, 3)$ and $(4, 15)$ and the equation of one asymptote is $3x - 2y + 6 = 0$.

28. The vertices are $(5, 0)$ and $(5, 14)$ and the equation of one asymptote is $7x - 5y = 0$.

EXAMPLE B (Changing to standard form)   Change each of the following equations to standard form. Decide whether the corresponding curve is an ellipse or a hyperbola and whether it is horizontal or vertical. Find the center and the vertices.

(a) $x^2 + 4y^2 - 8x + 16y = -28$

(b) $4x^2 - 9y^2 + 24x + 36y + 36 = 0$

Solution.

(a) We use the familiar process of completing the square.

$$(x^2 - 8x + \quad) + 4(y^2 + 4y + \quad) = -28$$
$$(x^2 - 8x + 16) + 4(y^2 + 4y + 4) = -28 + 16 + 16$$
$$(x - 4)^2 + 4(y + 2)^2 = 4$$
$$\frac{(x - 4)^2}{4} + \frac{(y + 2)^2}{1} = 1$$

We recognize this to be a horizontal ellipse (the larger denominator is in the x-term). Its center is at $(4, -2)$ and its vertices are at $(2, -2)$ and $(6, -2)$.

(b) Again we complete the squares.

$$4(x^2 + 6x + \quad) - 9(y^2 - 4y + \quad) = -36$$
$$4(x^2 + 6x + 9) - 9(y^2 - 4y + 4) = -36 + 36 - 36$$
$$4(x + 3)^2 - 9(y - 2)^2 = -36$$
$$\frac{(y - 2)^2}{4} - \frac{(x + 3)^2}{9} = 1$$

This is a vertical hyperbola (the y-term is positive). The center is at $(-3, 2)$ and the vertices are at $(-3, 0)$ and $(-3, 4)$, 2 units below and above the center.

For each of the following, change the equation to standard form and decide whether the corresponding curve is an ellipse or a hyperbola and whether it is horizontal or vertical. Find the center and the vertices.

29. $9x^2 + 16y^2 + 36x - 96y = -36$

30. $x^2 + 4y^2 - 2x + 32y = -61$

31. $4x^2 - 9y^2 - 16x - 18y - 29 = 0$

32. $9x^2 - y^2 + 90x + 8y + 200 = 0$

33. $25x^2 + y^2 - 4y = 96$

34. $25x^2 + 9y^2 - 100x - 125 = 0$

35. $4x^2 - y^2 - 32x - 4y + 69 = 0$

36. $x^2 - y^2 + 8x + 4y + 13 = 0$

Miscellaneous Problems

In Problems 37–44, decide whether the graph of the given equation is a circle, an ellipse, or a hyperbola. Then sketch the graph.

37. $\dfrac{x^2}{64} + \dfrac{y^2}{16} = 1$

38. $\dfrac{x^2}{16} + \dfrac{y^2}{16} = 1$

39. $\dfrac{x^2}{64} - \dfrac{y^2}{16} = 1$ 　　　　　40. $\dfrac{x^2}{4} - \dfrac{y^2}{9} = 1$

41. $\dfrac{(x-2)^2}{25} + \dfrac{(y-1)^2}{4} = 1$ 　　42. $\dfrac{(x-2)^2}{4} - \dfrac{(y+2)^2}{9} = 1$

43. $x^2 - 2y^2 - 6x + 8y = 1$ 　　　44. $x^2 + 2y^2 - 6x + 8y = 1$

45. Find the equation of the ellipse whose vertices are $(-4, 2)$ and $(10, 2)$ and whose minor diameter has length 10.

46. Find the equation of the ellipse that passes through the point $(3, 2)$ and has its vertices at $(\pm 5, 0)$.

47. Find the equation of the hyperbola having vertices $(\pm 6, 0)$ and one of its asymptotes passing through $(3, 2)$.

48. Find the equation of the hyperbola with vertices at $(0, \pm 2)$ and passing through the point $(2, 4)$.

ⓒ 49. Find the y-coordinates of the points for which $x = \pm 2.5$ on the ellipse

$$\frac{x^2}{24} + \frac{y^2}{19} = 1$$

ⓒ 50. Find the y-coordinates of the points for which $x = \pm 5$ on the hyperbola

$$\frac{x^2}{17} - \frac{y^2}{11} = 1$$

ⓒ 51. The area of the ellipse $x^2/a^2 + y^2/b^2 = 1$ is πab. Find the areas (correct to two decimal places) of the following ellipses.

(a) $\dfrac{x^2}{7} + \dfrac{y^2}{11} = 1$ 　　(b) $\dfrac{x^2}{111} + y^2 = 1$

ⓒ 52. Find the radius (correct to two decimal places) of the circle which has the same area as the given ellipse.

(a) $\dfrac{x^2}{256} + \dfrac{y^2}{89} = 1$ 　　(b) $\dfrac{x^2}{50} + \dfrac{y^2}{19} = 1$

53. Let $F_1(-3, 0)$ and $F_2(3, 0)$ be two given points and let $P(x, y)$ move so that

$$d(P, F_1) + d(P, F_2) = 10$$

Derive and simplify the equation of the path. It should turn out to be an ellipse.

54. Let $F_1(-5, 0)$ and $F_2(5, 0)$ be two given points and let $P(x, y)$ move so that

$$\left| d(P, F_1) - d(P, F_2) \right| = 8$$

Derive and simplify the equation of the path. It should turn out to be a hyperbola.

55. Suppose in Problem 53 that the points F_1 and F_2 have coordinates $(-c, 0)$ and $(c, 0)$. Now let $P(x, y)$ move so that

$$d(P, F_1) + d(P, F_2) = 2a$$

where $a > c$. Show that the equation of the path is $x^2/a^2 + y^2/b^2 = 1$, where $b^2 = a^2 - c^2$. The given distance condition is often used as the definition of the ellipse.

56. Suppose in Problem 54 that the points F_1 and F_2 have coordinates $(-c, 0)$ and $(c, 0)$. Now let $P(x, y)$ move so that

$$|d(P, F_1) - d(P, F_2)| = 2a$$

where $a < c$. Show that the equation of the path is $x^2/a^2 - y^2/b^2 = 1$, where $b^2 = c^2 - a^2$. The given distance condition is often used as the definition of the hyperbola.

CHAPTER SUMMARY

Like a city planner, we introduce in the plane two main streets, one vertical (the **y-axis**) and the other horizontal (the **x-axis**). Relative to these axes, we can specify any point by giving its address (x, y). The numbers x and y, called **Cartesian coordinates**, measure the directed distances from the vertical and horizontal axes, respectively. And given two points A and B with addresses (x_1, y_1) and (x_2, y_2), we may calculate the distance between them from the **distance formula:**

$$d(A, B) = \sqrt{(x_2 - x_1)^2 + (y_2 - y_1)^2}$$

In **analytic geometry**, we use the notion of coordinates to combine algebra and geometry. Thus we may graph the equation $y = x^2$ (algebra), thereby turning it into a curve (geometry). Conversely, we may take a circle with radius 6 and center $(-1, 4)$ and give it the equation

$$(x + 1)^2 + (y - 4)^2 = 36$$

The simplest of all curves is a **line**. If a line passes through (x_1, y_1) and (x_2, y_2) with $x_1 \neq x_2$, then its **slope** m is given by

$$m = \frac{\text{rise}}{\text{run}} = \frac{y_2 - y_1}{x_2 - x_1}$$

There are two important forms for the equation of a nonvertical line.

point-slope form:	$y - y_1 = m(x - x_1)$
slope-intercept form:	$y = mx + b$

Vertical lines do not have slope; their equations take the form $x = k$. All lines (vertical and nonvertical) can be written in the form of the **general linear equation.**

$$Ax + By + C = 0$$

The distance d from the point (x_1, y_1) to the line $Ax + By + C = 0$ is

$$d = \frac{|Ax_1 + By_1 + C|}{\sqrt{A^2 + B^2}}$$

Nonvertical lines are parallel if their slopes are equal, and they are perpendicular if their slopes are negative reciprocals.

Somewhat more complicated curves are the *conic sections:* **parabolas, circles, ellipses,** and **hyperbolas.** Here are typical equations for them.

parabola: $\qquad y - k = a(x - h)^2$

circle: $\qquad (x - h)^2 + (y - k)^2 = r^2$

ellipse: $\qquad \dfrac{(x - h)^2}{a^2} + \dfrac{(y - k)^2}{b^2} = 1$

hyperbola: $\qquad \dfrac{(x - h)^2}{a^2} - \dfrac{(y - k)^2}{b^2} = 1$

CHAPTER REVIEW PROBLEM SET

1. Name the graph of each of the following.

(a) $y = -4x + 3$

(b) $x^2 + y^2 = 25$

(c) $2x - 3y = 0$

(d) $y = x^2$

(e) $(y - 2)^2 = -3(x + 1)$

(f) $(x - 1)^2 + (y + 2)^2 = 9$

(g) $\dfrac{x^2}{9} + \dfrac{y^2}{49} = 1$

(h) $\dfrac{x^2}{9} - \dfrac{y^2}{49} = 1$

(i) $x^2 - 4y = 16$

(j) $x^2 + 4y^2 = 16$

(k) $x^2 - 4y^2 = 16$

(l) $x^2 - 4y^2 = -16$

2. Sketch the graph of each of the following equations. For some, you will need to make a table of values.

(a) $y = -3x + 4$

(b) $x^2 + y^2 = 16$

(c) $y = x^2 + \dfrac{1}{x}$

(d) $y = x^3 - 4x$

(e) $y = \sqrt{x} + 2$

(f) $\dfrac{x^2}{25} + \dfrac{y^2}{9} = 1$

3. Consider the triangle determined by the points $A(-2, 3)$, $B(3, 5)$ and $C(1, 9)$.

(a) Sketch the triangle.

(b) Find the lengths of the three sides.

(c) Find the slopes of the lines which contain the three sides.

(d) Write an equation for each of these lines.

(e) Find the midpoints of AB, BC, and CA..

(f) Find the equation of the line through A parallel to BC.

(g) Find the equation of the line through A perpendicular to BC.

(h) Find the length of the altitude from A to side BC.

(i) Calculate the area of the triangle.

(j) Show that the line segment joining the midpoints of AB and AC is parallel to BC and is one-half the length of BC.

4. Write the equation of the line satisfying the given conditions in the form $Ax + By + C = 0$.

(a) It is vertical and passes through $(4, -1)$.

(b) It is horizontal and passes through $(4, -1)$.

(c) It is parallel to the line $3x + 2y = 4$ and passes through $(4, -1)$.

(d) It passes through $(2, -5)$ and has x-intercept 3.

(e) It is perpendicular to the line $2y = -3x$ and passes through $(-2, 3)$.

(f) It is parallel to the lines $y = 4x + 5$ and $y = 4x - 7$ and is midway between them.

(g) It passes through $(4, 6)$ and has equal x- and y-intercepts.

(h) It is tangent to the circle $x^2 + y^2 = 169$ at the point $(12, 5)$.

5. Find the vertex of each of the following parabolas and decide which way it opens (up, down, to the right, or to the left).

(a) $x^2 = -6y$

(b) $y^2 = 16x$

(c) $y - 1 = \frac{1}{10}(x + 2)^2$

(d) $x - \frac{1}{4} = -\frac{1}{3}(y + \frac{1}{2})^2$

(e) $y = x^2 - 4x$

(f) $2y^2 + 12y = x - 1$

6. Write the equation of the parabola satisfying the following conditions.

(a) Vertex at $(0, 0)$, y-axis as line of symmetry, and passing through $(10, -5)$.

(b) Vertex at $(3, 0)$, axis of symmetry $y = 0$, and passing through $(1, 12)$.

(c) Vertex at $(-3, 5)$, horizontal line of symmetry, and passing through $(-1, 6)$.

7. Find the points of intersection of the parabola $y = x^2 + x$ and the line $6x - y - 4 = 0$.

8. By completing the squares, identify each of the following conic sections and give pertinent information about it.

(a) $x^2 + 2x + y^2 - 4y = 2$

(b) $2x^2 + y^2 + 4x - 4y = 14$

(c) $2x^2 - y^2 + 4x - 4y = 14$

(d) $2x^2 + 4x - y = 14$

9. Find the equation of the ellipse satisfying the given conditions.

(a) Vertices $(0, \pm 5)$; passing through $(2, 5\sqrt{5}/3)$.

(b) Vertices $(-2, 0)$ and $(6, 0)$; length of minor diameter 4.

10. Find the equation of the hyperbola with vertices $(-2, 5)$ and $(-2, 11)$ and with one of its asymptotes passing through $(0, 14)$.

5

FUNCTIONS
AND THEIR GRAPHS

Mathematicians do not deal in objects, but in relations between objects; thus, they are free to replace some objects by others so long as the relations remain unchanged. Content to them is irrelevant: they are interested in form only.

Henri Poincaré

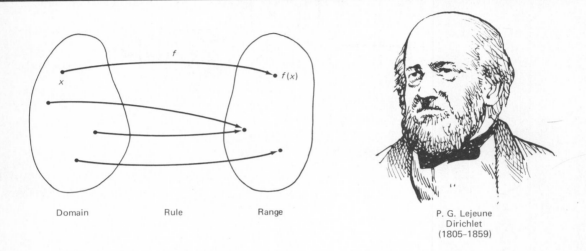

Domain   Rule   Range

P. G. Lejeune
Dirichlet
(1805–1859)

5-1
Functions

One of the most important ideas in mathematics is that of a function. For a long time, mathematicians and scientists wanted a precise way to describe the relationships that may exist between two variables. It is somewhat surprising that it took so long for the idea to crystallize into a clear, unambiguous concept. The French mathematician P. G. Lejeune Dirichlet (1805–1859) is credited with the modern definition of function.

DEFINITION

*A **function** is a rule which assigns to each element in one set (called the **domain** of the function) exactly one value from another set. The set of all assigned values is called the **range** of the function.*

Three examples will help clarify this idea. When a grocer puts a price tag on each of the watermelons for sale, a function is determined. Its domain is the set of watermelons, its range is the set of prices, and the rule is the procedure the grocer uses in assigning prices (perhaps a specified amount per pound.)

$1.98

$2.21

$2.30

Watermelon Price Function

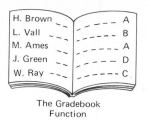

The Gradebook
Function

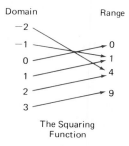

The Squaring
Function

When a professor assigns a grade to each student in a class, he or she is determining a function. The domain is the set of students and the range is the set of grades, but who can say what the rule is? It varies from professor to professor; some may even prefer to keep it a secret.

A much more typical function from our point of view is the *squaring* function displayed in the margin. It takes a number from the domain $\{-2, -1, 0, 1, 2, 3\}$ and squares it, producing a number in the range $\{0, 1, 4, 9\}$. This function is typical for two reasons: Both the domain and range are sets of numbers and the rule can be specified by giving an algebraic formula. Most functions in this book will be of this type.

FUNCTIONAL NOTATION

Long ago, mathematicians introduced a special notation for functions. A single letter like f (or g or h) is used to name the function. Then $f(x)$, read f *of* x or f *at* x, denotes the value that f assigns to x. Thus, if f names the squaring function,

$$f(-2) = 4 \qquad f(2) = 4 \qquad f(-1) = 1$$

and, in general,

$$f(x) = x^2$$

We call this last result the *formula* for the function f. It tells us in a concise algebraic way what f does to any number. Notice that the given formula and

$$f(y) = y^2 \qquad f(z) = z^2$$

all say the same thing; the letter used for the domain variable is a matter of no significance, though it does happen that we shall usually use x. Many functions do not have simple formulas (see Problems 17 and 18), but in this book, most of them do.

For a further example, consider the function that cubes a number and then subtracts 1 from the result. If we name this function g, then

$$g(2) = 2^3 - 1 = 7$$
$$g(-1) = (-1)^3 - 1 = -2$$
$$g(.5) = (.5)^3 - 1 = -.875$$
$$g(\pi) = \pi^3 - 1 \approx 30$$

and, in general,

$$g(x) = x^3 - 1$$

Few students would have trouble using this formula when x is replaced by a specific number. However, it is important to be able to use it when x is replaced by anything whatever, even an algebraic expression. Be sure you understand the following calculations.

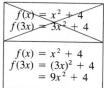

$$f(x) = x^2 + 4$$
$$f(3x) = 3x^2 + 4$$

$$f(x) = x^2 + 4$$
$$f(3x) = (3x)^2 + 4$$
$$\qquad = 9x^2 + 4$$

$$g(a) = a^3 - 1$$

$$g(y^2) = (y^2)^3 - 1 = y^6 - 1$$

$$g\left(\frac{1}{z}\right) = \left(\frac{1}{z}\right)^3 - 1 = \frac{1}{z^3} - 1$$

$$g(2 + h) = (2 + h)^3 - 1 = 8 + 12h + 6h^2 + h^3 - 1$$

$$= h^3 + 6h^2 + 12h + 7$$

$$g(x + h) = (x + h)^3 - 1 = x^3 + 3x^2h + 3xh^2 + h^3 - 1$$

DOMAIN AND RANGE

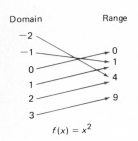

Domain   Range

-2

-1 0

0 1

1 4

2 9

3

$f(x) = x^2$

The rule of correspondence is the heart of a function, but a function is not completely determined until its domain is given. Recall that the **domain** is the set of elements to which the function assigns values. In the case of the squaring function f (reproduced in the margin), we gave the domain as the set $\{-2, -1, 0, 1, 2, 3\}$. We could just as well have specified the domain as the set of all real numbers; the formula $f(x) = x^2$ would still make perfectly good sense. In fact, there is a common agreement that if no domain is specified, it is understood to be the largest set of real numbers for which the rule for the function makes sense and gives real number values. We call it the **natural domain** of the function. Thus if no domain is specified for the function with formula $g(x) = x^3 - 1$, it is assumed to be the set of all real numbers. Similarly, for

$$h(x) = \frac{1}{x - 1}$$

we would take the natural domain to consist of all real numbers except 1. Here, the number 1 is excluded to avoid division by zero.

Once the domain is understood and the rule of correspondence is given, the **range** of the function is determined. It is the set of values of the function. Here are several examples.

RULE	DOMAIN	RANGE
$F(x) = 4x$	All reals	All reals
$G(x) = \sqrt{x - 3}$	$\{x : x \geq 3\}$	Nonnegative reals
$H(x) = \dfrac{1}{(x - 2)^2}$	$\{x : x \neq 2\}$	Positive reals

INDEPENDENT AND DEPENDENT VARIABLES

Scientists like to use the language of variables in talking about functions. Let us illustrate by referring to an object falling under the influence of gravity near the earth's surface. If the object is very dense (so air resistance can be neglected), it will fall according to the formula

$$d = 16t^2$$

where d represents the distance in feet the object falls during the first t seconds. Since time moves along quite independently of everything else, a physicist would call t the **independent variable.** The distance d depends on t; it is called the **dependent variable.** And d is said to be a function of t.

VARIATION

Sometimes we describe the relationship of a dependent variable to one or more independent variables using the language of variation. To say that y **varies directly** as x means that $y = kx$ for some fixed number k. To say that y **varies inversely** as x means that $y = k/x$ for some fixed number k. Finally, to say that z **varies jointly** as x and y means that $z = kxy$ for a fixed number k.

In variation problems, we are often not only given the form of the relationship, but also some set of corresponding values of the variables involved. Then we can evaluate the constant k and obtain an explicit formula for the dependent variable. Here is an illustration.

It is known that y varies directly as x and that $y = 10$ when $x = -2$. Find an explicit formula for y.

We substitute the given values in the equation

$$y = kx$$

and get $10 = k(-2)$. Thus $k = -5$ and we can write the explicit formula

$$y = -5x$$

Problem Set 5-1

1. If $f(x) = x^2 - 4$, evaluate each expression.
   (a) $f(-2)$ (b) $f(0)$ (c) $f(\frac{1}{2})$ (d) $f(.1)$
   (e) $f(\sqrt{2})$ (f) $f(a)$ (g) $f(1/x)$ (h) $f(x + 1)$

2. If $f(x) = (x - 4)^2$, evaluate each expression in Problem 1.

3. If $f(x) = 1/(x - 4)$, evaluate each expression.
   (a) $f(8)$ (b) $f(2)$ (c) $f(\frac{9}{2})$ (d) $f(\frac{31}{8})$
   (e) $f(4)$ (f) $f(4.01)$ (g) $f(1/x)$ (h) $f(x^2)$
   (i) $f(2 + h)$ (j) $f(2 - h)$

4. If $f(x) = x^2$ and $g(x) = 2/x$, evaluate each expression.
   (a) $f(-7)$ (b) $g(-4)$ (c) $f(\frac{1}{4})$ (d) $1/f(4)$
   (e) $g(\frac{1}{4})$ (f) $1/g(4)$ (g) $g(0)$ (h) $g(1)f(1)$
   (i) $f(g(1))$ (j) $f(1)/g(1)$

EXAMPLE A (Finding natural domains) Find the natural domain of
(a) $f(x) = 4x/[(x + 2)(x - 3)]$; (b) $g(x) = \sqrt{x^2 - 4}$.

Solution. We recall that the natural domain is the largest set of real numbers for which the formula makes sense and gives real number values. Thus in part (a), the domain consists of all real numbers except -2 and 3; in part (b) we must have $x^2 \geq 4$, which is equivalent to $|x| \geq 2$. Notice that if $|x| < 2$, we would be taking the square root of a negative number, so the result would not be a real number.

In Problems 5–16, find the natural domain of the given function.

5. $f(x) = x^2 - 4$

6. $f(x) = (x - 4)^2$

7. $g(x) = \dfrac{1}{x^2 - 4}$

8. $g(x) = \dfrac{1}{9 - x^2}$

9. $h(x) = \dfrac{2}{x^2 - x - 6}$

10. $h(x) = \dfrac{1}{2x^2 + 3x - 2}$

11. $F(x) = \dfrac{1}{x^2 + 4}$

12. $F(x) = \dfrac{1}{9 + x^2}$

13. $G(x) = \sqrt{x - 2}$

14. $G(x) = \sqrt{x + 2}$

15. $H(x) = \dfrac{1}{5 - \sqrt{x}}$

16. $H(x) = \dfrac{1}{\sqrt{x + 1} - 2}$

17. Not all functions arising in mathematics have rules given by simple algebraic formulas. Let $f(n)$ be the nth digit in the decimal expansion of

$$\pi = 3.14159265358979323846 \ldots$$

Thus $f(1) = 3$ and $f(3) = 4$. Find (a) $f(6)$; (b) $f(9)$; (c) $f(16)$. What is the natural domain for this function?

18. Let g be the function which assigns to each positive integer the number of factors in its prime factorization. Thus

$$g(2) = 1$$
$$g(4) = g(2 \cdot 2) = 2$$
$$g(36) = g(2 \cdot 2 \cdot 3 \cdot 3) = 4$$

Find (a) $g(24)$; (b) $g(37)$; (c) $g(64)$; (d) $g(162)$. Can you find a formula for this function?

EXAMPLE B (Functions generated on a calculator) Show how the function $f(x) = 2\sqrt{x} + 5$ can be generated on a calculator. Then calculate $f(\pi)$.

Solution. To avoid confusion with the times sign, we use # rather than x for an arbitrary number. Then on a typical algebraic logic calculator,

$$f(\#) = 2 \;\boxed{\times}\; \# \;\boxed{\sqrt{x}}\; \boxed{+}\; 5 \;\boxed{=}$$

In particular,

$$f(\pi) = 2 \;\boxed{\times}\; \pi \;\boxed{\sqrt{x}}\; \boxed{+}\; 5 \;\boxed{=}$$

which yields the value 8.544907.

$\boxed{\text{c}}$ *In Problems 19–26 on page 177, write the sequence of keys that will generate the given function, using # for an arbitrary number. Then use the sequence to calculate f(2.9).*

19. $f(x) = (x + 2)^2$

20. $f(x) = \sqrt{x + 3}$

21. $f(x) = 3(x + 2)^2 - 4$

22. $f(x) = 4\sqrt{x + 3} - 11$

23. $f(x) = \left(3x + \dfrac{2}{\sqrt{x}}\right)^3$

24. $f(x) = \left(\dfrac{x}{3} + \dfrac{3}{x}\right)^5$

25. $f(x) = \dfrac{\sqrt{x^5 - 4}}{2 + 1/x}$

26. $f(x) = \dfrac{(\sqrt{x} - 1)^3}{x^2 + 4}$

$\boxed{\text{c}}$ *In Problems 27–32, write the algebraic formula for the function that is generated by the given sequence of calculator keys.*

27. $\#\ \boxed{+}\ 5\ \boxed{\sqrt{x}}$

28. $\boxed{(}\ \#\ \boxed{+}\ 5\ \boxed{)}\ \boxed{\sqrt{x}}$

29. $2\ \boxed{\times}\ \#\ \boxed{x^2}\ \boxed{+}\ 3\ \boxed{\times}\ \#$

30. $3\ \boxed{+/-}\ \boxed{\times}\ \#\ \boxed{x^2}\ \boxed{-}\ 7$

31. $\boxed{(}\ 3\ \boxed{\times}\ \boxed{(}\ \#\ \boxed{-}\ 2\ \boxed{)}\ \boxed{x^2}\ \boxed{+}\ 9\ \boxed{)}\ \boxed{\sqrt{x}}$

32. $\boxed{(}\ \#\ \boxed{1/x}\ \boxed{\times}\ 2\ \boxed{+}\ 4\ \boxed{)}\ \boxed{\sqrt{x}}$

EXAMPLE C (Functions of two variables)   The formula $f(x, y) = x^2 + 2y^2$ determines a function of two variables. It assigns a number to each ordered pair of numbers (x, y). For example.

$$f(2, -3) = 2^2 + 2(-3)^2 = 22$$

$$f(3, 8) = 3^2 + 2(8)^2 = 137$$

Let $g(x, y) = 3xy - 5x$ and $G(x, y) = (5x + 3y)/(2x - y)$. Find each of the following.

33. $g(2, 5)$ 34. $g(-1, 3)$ 35. $g(5, 2)$ 36. $g(3, -1)$

37. $G(1, 1)$ 38. $G(3, 3)$ 39. $G(\tfrac{1}{2}, 1)$ 40. $G(5, 0)$

41. $g(2x, 3y)$ 42. $G(2x, 4y)$ 43. $g(x, 1/x)$ 44. $G(x - y, y)$

EXAMPLE D (More on variation)   Suppose that w varies jointly as x and the square root of y and that $w = 14$ when $x = 2$ and $y = 4$. Find an explicit formula for w.

Solution.   We translate the first statement into mathematical symbols as

$$w = kx\sqrt{y}$$

To evaluate k, we substitute the given values for w, x, and y.

$$14 = k \cdot 2\sqrt{4} = 4k$$

or

$$k = \frac{14}{4} = \frac{7}{2}$$

Thus the explicit formula for w is

$$w = \frac{7}{2}x\sqrt{y}$$

In Problems 45–50, find an explicit formula for the dependent variable.

45. y varies directly as x, and $y = 12$ when $x = 3$.

46. y varies directly as x^2, and $y = 4$ when $x = 0.1$.

47. y varies inversely as x (that is, $y = k/x$), and $y = 5$ when $x = \frac{1}{5}$.

48. V varies jointly as r^2 and h, and $V = 75$ when $r = 5$ and $h = 9$.

49. I varies directly as s and inversely as d^2, and $I = 9$ when $s = 4$ and $d = 12$.

50. W varies directly as x and inversely as the square root of yz, and $W = 5$ when $x = 7.5$, $y = 2$, and $z = 18$.

51. The maximum range of a projectile varies as the square of the initial velocity. If the range is 16,000 feet when the initial velocity is 600 feet per second,
    (a) write an explicit formula for R in terms of v, where R is the range in feet and v is the initial velocity in feet per second;
    (b) use this formula to find the range when the initial velocity is 800 feet per second.

52. Suppose that the amount of gasoline used by a car varies jointly as the distance traveled and the square root of the average speed. If a car used 8 gallons on a 100-mile trip going at an average speed of 64 miles per hour, how many gallons would that car use on a 160-mile trip at an average speed of 25 miles per hour?

Miscellaneous Problems

53. If $f(x) = -3x^2 + (12/x)$, evaluate each expression.
    (a) $f(2)$ (b) $f(-1)$ (c) $f(\frac{3}{2})$
    (d) $f(1/x)$ (e) $f(\sqrt{3})$ (f) $f(a + b)$

54. If $g(x) = [5/(x - 2)^2] + 4x$, evaluate each expression.
    (a) $g(0)$ (b) $g(3)$ (c) $g(1)$
    (d) $g(\frac{3}{2})$ (e) $g(2.1)$ (f) $g(2 + h)$

55. If $f(x, y) = (2x^2 + xy)/(x - y)$, evaluate each expression.
    (a) $f(2, 1)$ (b) $f(3, 0)$ (c) $f(2\sqrt{2}, \sqrt{2})$

In Problems 56–59, find the natural domain of the function.

56. $g(x) = \dfrac{x - 2}{(x^2 + 1)(x - 4)}$

57. $g(x) = \dfrac{3x + 8}{(2 - x)(x^2 + 7)}$

58. $h(x) = \dfrac{2x}{2 - \sqrt{x}}$

59. $h(x) = \dfrac{3x^2 + 2}{1 - \sqrt{2x}}$

ⓒ 60. Write a sequence of calculator keys for calculating $f(\#)$ if
$$f(x) = \frac{(-5 + \sqrt{3x^2 + 1}\,)^3}{(x + 1)^2}$$

ⓒ 61. Write an algebraic formula for the function that is calculated by using the following sequence of keys.

$$\boxed{(}\ 4\ \boxed{+/-}\ \boxed{+}\ \#\ \boxed{x^2}\ \boxed{1/x}\ \boxed{)}\ \boxed{y^x}\ 3$$

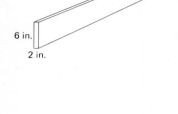

10 ft.

6 in.

2 in.

In Problems 62 and 63, find an explicit formula for y and evaluate y when x = 4.

62. $y = kx^3$, and $y = 4$ when $x = 0.1$.

63. y varies inversely as x^2, and $y = 48$ when $x = \frac{1}{4}$.

64. The safe load of a horizontal beam supported at both ends varies directly as its breadth and the square of its depth and inversely as its length. If a 2 inch by 6 inch white pine joist 10 feet long holds 1000 pounds when placed edgewise, what is its safe load when placed flatwise?

65. What is the safe load of a beam of the same material as in Problem 64 if it is 20 feet long, 4 inches wide, 1 foot thick, and placed edgewise?

☐c 66. The density $D(T)$ of dry air at a pressure of 76 centimeters of mercury and at a temperature of T degrees Celsius is given by

$$D(T) = \frac{.001293}{1 + (.00367)T}$$

in kilograms per liter.
(a) What is the density at 13.6 degrees Celsius?
(b) Does the density increase or decrease as the temperature increases?

67. A water tank has the shape of a right circular cone with vertex downward. The radius of the top is 20 feet and the height is 40 feet. Express the volume $V(d)$ of the water as a function of its depth.

68. A plant has the capacity to produce from 1 to 100 refrigerators per day. The daily overhead for the plant is $2200 and the direct cost (labor and material) of producing one refrigerator is $151. Write a formula for $T(x)$, the total cost of producing x refrigerators per day, and also the unit cost $U(x)$. What are the domains for these functions?

69. It cost the ABC company $800 + 20\sqrt{x}$ dollars to make x toy ovens. Each oven sells for $6. Express the total profit $T(x)$ and average profit $A(x)$ as functions of x. What are the domains for these functions?

$$y = f(x) = \frac{2}{3}x + 2$$

$$y = f(x) = x^2 + 2x - 3$$

$$y = f(x) = x^3 - 3x + 4$$

$$y = f(x) = x^4 - x^3 - 8x^2 + 16x + 20$$

Geometry, however, supplies sustenance and meaning to bare formulas . . . . One can still believe Plato's statement that "geometry draws the soul toward truth."

Morris Kline

5-2
Graphs of Functions

We have said that functions are usually specified by giving formulas. Formulas are fine for manipulation and almost essential for exact quantitative information, but to grasp the overall qualitative features of a function, we need a picture. The best picture of a function is its graph. And the **graph of a function** f is simply the graph of the equation $y = f(x)$. We learned how to graph equations in the previous chapter.

POLYNOMIAL FUNCTIONS

We look first at polynomial functions, that is, functions of the form

$$f(x) = a_n x^n + a_{n-1} x^{n-1} + \cdots + a_1 x + a_0$$

Four typical graphs are shown above. We know from the last chapter that the graph of $f(x) = ax + b$ is always a straight line and that, if $a \neq 0$, the graph of $f(x) = ax^2 + bx + c$ is necessarily a parabola.

The graphs of higher degree polynomial functions are harder to describe, but after we have studied two examples, we can offer some general guidelines. Consider first the cubic function

$$f(x) = x^3 - 3x + 4$$

With the help of a table of values, we sketch its graph (top of page 181).

Notice that for large positive values of x, the values of y are large and positive; similarly, for large negative values of x, y is large and negative. This is due to the dominance of the leading term x^3 for large $|x|$. This dominance is responsible for a drooping left arm and a right arm held high on the graph.

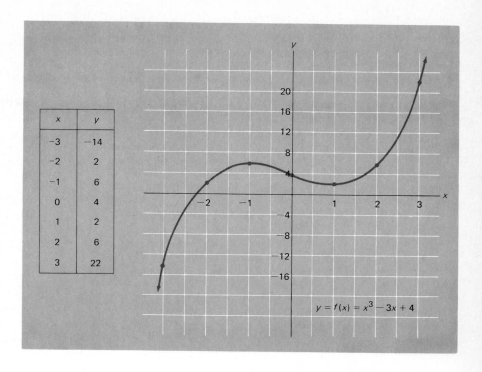

x	y
-3	-14
-2	2
-1	6
0	4
1	2
2	6
3	22

$$y = f(x) = x^3 - 3x + 4$$

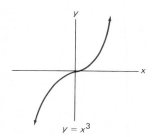

$y = x^3$

Notice also that the graph has one hill and one valley. This is typical of the graph of a cubic function, though it is possible for it to have no hills or valleys. The graph of $y = x^3$ illustrates this latter behavior.

Next consider a typical fourth degree polynomial function.

$$f(x) = -x^4 + 4x^3 + 2x^2 - 12x - 3$$

A table of values and the graph are shown at the top of page 182.

The leading term $-x^4$, which is negative for all values of x, determines that the graph has two drooping arms. Note that there are two hills and one valley.

In general, we can make the following statements about the graph of

$$f(x) = a_n x^n + a_{n-1} x^{n-1} + \cdots + a_1 x + a_0, \qquad a_n \neq 0$$

1. If n is even and $a_n < 0$, the graph will have two drooping arms; if n is even and $a_n > 0$, it will have both arms raised. This is due to the dominance of $a_n x^n$ for large values of $|x|$.

2. If n is odd, one arm droops and the other points upward. Again, this is dictated by the dominance of $a_n x^n$.

3. The combined number of hills and valleys cannot exceed $n - 1$, although it can be less.

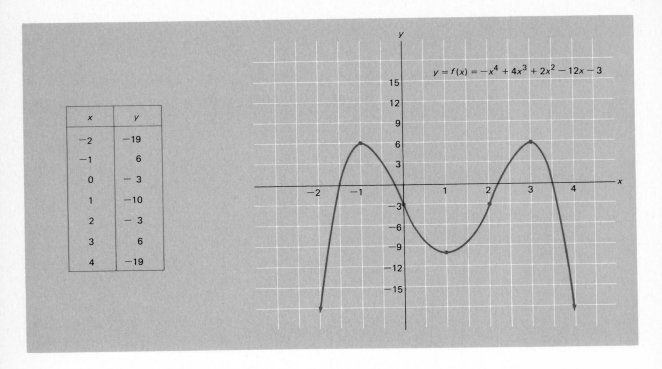

x	y
-2	-19
-1	6
0	-3
1	-10
2	-3
3	6
4	-19

$y = f(x) = -x^4 + 4x^3 + 2x^2 - 12x - 3$

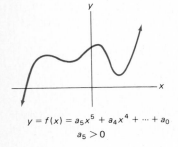

$y = f(x) = a_5 x^5 + a_4 x^4 + \cdots + a_0$

$a_5 > 0$

Based on these facts, we expect the graph of a fifth degree polynomial function with positive leading coefficient to look something like the graph in the margin.

FACTORED POLYNOMIAL FUNCTIONS

The task of graphing can be simplified considerably if our polynomial is factored. The real solutions of $f(x) = 0$ correspond to the x-intercepts of the graph of $y = f(x)$—that is, to the x-coordinates of the points where the graph intersects the x-axis. If the polynomial is factored, these intercepts are easy to find. Consider as an example.

$$y = f(x) = x(x + 3)(x - 1)$$

The solutions of $f(x) = 0$ are 0, -3, and 1; these are the x-intercepts of the graph. Clearly, $f(x)$ cannot change signs between adjacent x-intercepts since only at these points can any of the linear factors change sign. The signs of $f(x)$ on the four intervals determined by $x = -3$, $x = 0$, and $x = 1$ are shown below (to check this, try substituting an x-value from each of these intervals, as in the split-point method of Section 3-5).

With this information and a few plotted points, we can easily sketch the graph.

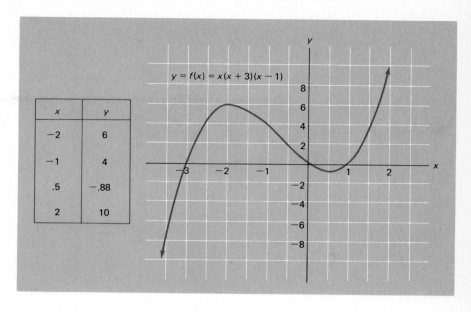

x	y
−2	6
−1	4
.5	−.88
2	10

FUNCTIONS WITH MULTI-PART RULES

Sometimes a function has polynomial components even though it is not a polynomial function. Especially notable is the absolute value function $f(x) = |x|$, which has the two-part rule

$$f(x) = \begin{cases} -x & \text{if } x < 0 \\ x & \text{if } x \geq 0 \end{cases}$$

For $x < 0$, the graph of this function coincides with the line $y = -x$; for $x \geq 0$, it coincides with the line $y = x$. Note the sharp corner at the origin.

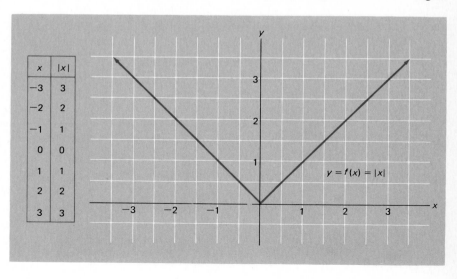

x	|x|
−3	3
−2	2
−1	1
0	0
1	1
2	2
3	3

Here is a more complicated example.

$$g(x) = \begin{cases} x + 2 & \text{if } x < 0 \\ x^2 & \text{if } 0 \le x \le 2 \\ 4 & \text{if } x > 2 \end{cases}$$

Though this way of describing a function may seem strange, it is not at all unusual in more advanced courses. The graph of g consists of three pieces.

1. A part of the line $y = x + 2$.
2. A part of the parabola $y = x^2$.
3. A part of the horizontal line $y = 4$.

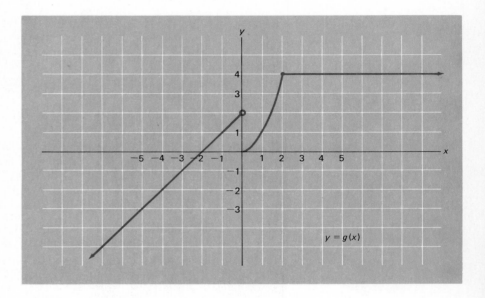

Note the use of the open circle at $(0, 2)$ to indicate that this point is not part of the graph.

Problem Set 5-2

Graph each of the following polynomial functions. The first two are called constant functions.

1. $f(x) = 5$
2. $f(x) = -4$
3. $f(x) = -3x + 5$
4. $f(x) = 4x - 3$
5. $f(x) = x^2 - 5x + 4$
6. $f(x) = x^2 + 2x - 3$
7. $f(x) = x^3 - 9x$
8. $f(x) = x^3 - 16x$
9. $f(x) = 2.12x^3 - 4.13x + 2$
10. $f(x) = -1.2x^3 + 2.3x^2 - 1.4x$

Graph each of the following functions.

11. $f(x) = 2|x|$

12. $f(x) = |x| - 2$

13. $f(x) = |x - 2|$

14. $f(x) = |x| + 2$

15. $f(x) = \begin{cases} x & \text{if } x < 0 \\ 2 & \text{if } x \geq 0 \end{cases}$

16. $f(x) = \begin{cases} -1 & \text{if } x \leq 0 \\ 2x & \text{if } x > 0 \end{cases}$

17. $f(x) = \begin{cases} -5 & \text{if } x \leq -3 \\ 4 - x^2 & \text{if } -3 < x \leq 3 \\ -5 & \text{if } x > 3 \end{cases}$

18. $f(x) = \begin{cases} 9 & \text{if } x < 0 \\ 9 - x^2 & \text{if } 0 \leq x \leq 3 \\ x^2 - 9 & \text{if } x > 3 \end{cases}$

EXAMPLE A (Symmetry properties) A function f is called an **even function** if $f(-x) = f(x)$ for all x in its domain. The graph of an even function is symmetric with respect to the y-axis. A function g is called an **odd function** if $g(-x) = -g(x)$ for all x in its domain; its graph is symmetric with respect to the origin. Graph the following two functions, observing their symmetries.

(a) $f(x) = x^4 + x^2 - 3$ (b) $g(x) = x^3 + 2x$

Solution. Notice that

$$f(-x) = (-x)^4 + (-x)^2 - 3 = x^4 + x^2 - 3 = f(x)$$

$$g(-x) = (-x)^3 + 2(-x) = -x^3 - 2x = -g(x)$$

Thus f is even and g is odd. Their graphs are sketched below.

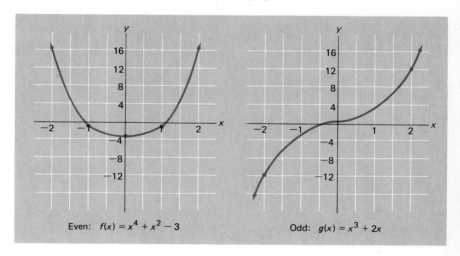

Even: $f(x) = x^4 + x^2 - 3$ Odd: $g(x) = x^3 + 2x$

Note that a polynomial function involving only even powers of x is even, while one involving only odd powers of x is odd.

Determine which of the following are even functions, which are odd functions, and which are neither. Then sketch the graphs of those that are even or odd, making use of the symmetry properties.

19. $f(x) = 2x^2 - 5$
20. $f(x) = -3x^2 + 2$
21. $f(x) = x^2 - x + 1$
22. $f(x) = -2x^3$
23. $f(x) = 4x^3 - x$
24. $f(x) = x^3 + x^2$
25. $f(x) = 2x^4 - 5x^2$
25. $f(x) = 3x^4 + x^2$

EXAMPLE B (More on factored polynomials)   Graph

$$f(x) = (x - 1)^2(x - 3)(x + 2)$$

Solution.   The x-intercepts are at 1, 3, and -2. The new feature is that $x - 1$ occurs as a square. The factor $(x - 1)^2$ never changes sign, so the graph does not cross the x-axis at $x = 1$; it merely touches the axis there. Note the entries in the table of values corresponding to $x = 0.9$ and $x = 1.1$.

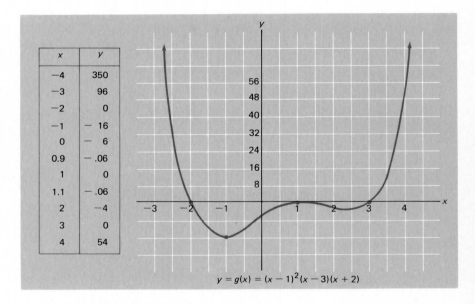

x	y
−4	350
−3	96
−2	0
−1	− 16
0	− 6
0.9	− .06
1	0
1.1	− .06
2	−4
3	0
4	54

$y = g(x) = (x - 1)^2(x - 3)(x + 2)$

Sketch the graph of each of the following.

27. $f(x) = (x + 1)(x - 1)(x - 3)$
28. $f(x) = x(x - 2)(x - 4)$
29. $f(x) = x^2(x - 4)$
30. $f(x) = x(x + 2)^2$
31. $f(x) = (x + 2)^2(x - 2)^2$
32. $f(x) = x(x - 1)^3$

Miscellaneous Problems

Graph each of the functions in Problems 33–42.

33. $f(x) = -3$
34. $f(x) = 2x - 3$

35. $f(x) = x^2 - 1$

36. $f(x) = x^3 - 3x$

37. $f(x) = (x + 2)(x - 1)^2(x - 3)$

38. $f(x) = x(x - 2)^3$

39. $f(x) = \begin{cases} 4 - x^2 & \text{if } -2 \leq x < 2 \\ x - 1 & \text{if } 2 \leq x \leq 4 \end{cases}$

40. $f(x) = \begin{cases} |x| & \text{if } -2 \leq x \leq 1 \\ x^2 & \text{if } 1 < x \leq 2 \\ 4 & \text{if } 2 < x \leq 4 \end{cases}$

C 41. $f(x) = x^4 + x^2 + 2,\ -2 \leq x \leq 2$

C 42. $f(x) = x^5 + x^3 + x,\ -1 \leq x \leq 1$

43. The graphs of the three functions $f(x) = x^2$, $g(x) = x^4$, and $h(x) = x^6$ all pass through the points $(-1, 1)$, $(0, 0)$ and $(1, 1)$. Draw careful sketches of these three functions using the same axes. Be sure to show clearly how they differ for $-1 < x < 1$.

C 44. Sketch the graph of $f(x) = x^{50}$ for $-1 \leq x \leq 1$. Be sure to calculate $f(.5)$ and $f(.9)$. What simple figure does the graph resemble?

C 45. Notice that $f(x) = 3x^4 + 2x^3 - 3x^2 + x + 1$ can be written as

$$f(x) = [((3x + 2)x - 3)x + 1]x + 1$$

It is now easy to calculate $f(2), f(1.3), f(4.2)$ on a calculator. Do so.

C 46. Use the trick described in Problem 45 to evaluate $f(3), f(4.3)$, and $f(-1.6)$ for

$$f(x) = 4x^5 - 3x^4 + 2x^3 - x^2 + 7x - 3$$

47. The function $f(x) = [x]$ is called the *greatest integer function*. It assigns to each real number x the largest integer which is less than or equal to x. For example. $[\frac{5}{2}] = 2$, $[13] = 13$, and $[-14.25] = -15$. Graph this function on the interval $-2 \leq x \leq 6$.

48. Graph each of the following functions on the interval $-2 \leq x \leq 8$.
    (a) $f(x) = 2[x]$ (b) $g(x) = 2 + [x]$
    (c) $h(x) = [x - 2]$ (d) $k(x) = x - [x]$

49. A machine purchased for $8000 is expected to have a scrap value of $1500 after 12 years. If the machine is depreciated linearly, what will its value $V(x)$ be x years after purchase? Graph this function.

50. It costs the XYZ company $1000 + 10\sqrt{x}$ dollars to make x toy dolls, which sell for $8 each. Express the total profit $T(x)$ as a function of x and graph $T(x)$.

51. Suppose that the cost of shipping a package is 15 cents for anything less than one ounce and 10 cents for each additional ounce or fraction thereof. Write a formula for the cost $C(x)$ of shipping a package weighing x ounces (using [ ]) and graph it.

52. An open box is made from a piece of 12 inch by 18 inch cardboard by cutting a square of side x inches from each corner and turning up the sides. Express the volume $V(x)$ in terms of x and graph the resulting function. What is the appropriate domain for this function?

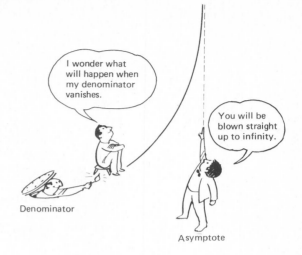

Rational Functions

The graph of the rational function

$$f(x) = \frac{p(x)}{q(x)}$$

exhibits spectacular behavior whenever the denominator $q(x)$ nears 0. It must either blow up to plus infinity or down to minus infinity.

5-3
Graphing Rational Functions

If $f(x)$ is given by

$$f(x) = \frac{p(x)}{q(x)}$$

where $p(x)$ and $q(x)$ are polynomials, then f is called a **rational function.** For simplicity, we shall assume that $f(x)$ is in reduced form, that is, that $p(x)$ and $q(x)$ have no common nontrivial factors. Typical examples of rational functions are

$$f(x) = \frac{x + 1}{x^2 - x + 6} = \frac{x + 1}{(x - 3)(x + 2)}$$

$$g(x) = \frac{(x + 2)(x - 5)}{(x + 3)^3}$$

Graphing a rational function can be tricky, primarily because of the denominator $q(x)$. Whenever it is zero, something dramatic is sure to happen to the graph. That is the point of our opening cartoon.

THE GRAPHS OF $1/x$ AND $1/x^2$
Let us consider two simple cases.

$$f(x) = \frac{1}{x} \qquad g(x) = \frac{1}{x^2}$$

Notice that f is an odd function ($f(-x) = -f(x)$), while g is even ($g(-x) = g(x)$). These facts imply that the graph of f is symmetric with respect to the origin, and that the graph of g is symmetric with respect to the y-axis. Thus we need to use only positive values of x to calculate y-values. Each calculation yields two points on the graph. Observe particularly the behavior of each graph near $x = 0$.

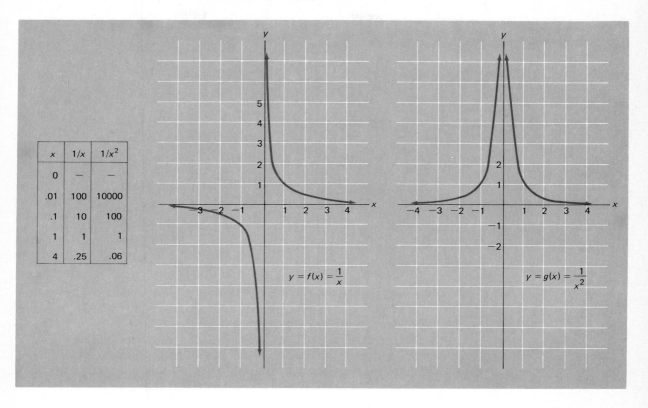

x	$1/x$	$1/x^2$
0	—	—
.01	100	10000
.1	10	100
1	1	1
4	.25	.06

$$y = f(x) = \frac{1}{x}$$

$$y = g(x) = \frac{1}{x^2}$$

In both cases, the x- and y-axis play special roles; we call them *asymptotes* for the graphs. If, as a point moves away from the origin along a curve, the distance between it and a line becomes closer and closer to zero, then that line is called an **asymptote** for the curve. Clearly the line $x = 0$ is a vertical asymptote for both of our curves and the line $y = 0$ is a horizontal asymptote for both of them.

THE GRAPHS OF $1/(x - 2)$ AND $1/(x - 2)^2$

If we replace x by $x - 2$ in our two functions, we get two new functions.

$$h(x) = \frac{1}{x - 2} \qquad k(x) = \frac{1}{(x - 2)^2}$$

Their graphs are just like those of f and g except that they are moved two units to the right, as you can see at the top of the next page.

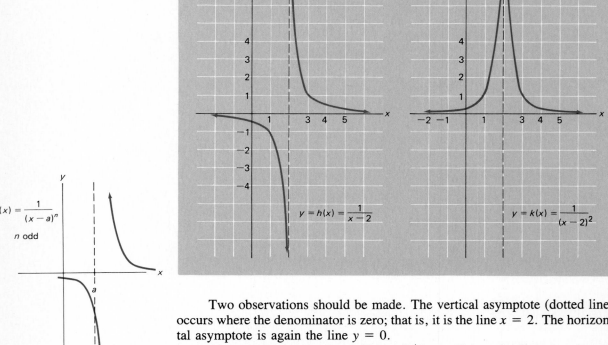

$$f(x) = \frac{1}{(x-a)^n}$$

n odd

$$f(x) = \frac{1}{(x-a)^n}$$

n even

$$y = h(x) = \frac{1}{x-2}$$

$$y = k(x) = \frac{1}{(x-2)^2}$$

Two observations should be made. The vertical asymptote (dotted line) occurs where the denominator is zero; that is, it is the line $x = 2$. The horizontal asymptote is again the line $y = 0$.

In general, the graph of $f(x) = 1/(x - a)^n$ has the line $y = 0$ as a horizontal asymptote and the line $x = a$ as a vertical asymptote. The behavior of the graph near $x = a$ for n even and n odd is illustrated in the margin.

MORE COMPLICATED EXAMPLES

Consider next the rational function determined by

$$y = f(x) = \frac{x}{x^2 + x - 6} = \frac{x}{(x - 2)(x + 3)}$$

We expect its graph to have vertical asymptotes at $x = 2$ and $x = -3$. Again, the line $y = 0$ will be a horizontal asymptote since, as $|x|$ gets large, the term x^2 in the denominator will dominate, so that y will behave much like x/x^2 or $1/x$ and will thus approach zero. The graph crosses the x-axis where the numerator is zero, namely, at $x = 0$. Finally, with the help of a table of values, we sketch the graph, shown at the top of page 191.

Lastly we consider

$$y = f(x) = \frac{2x^2 + 2x}{x^2 - 4x + 4} = \frac{2x(x + 1)}{(x - 2)^2}$$

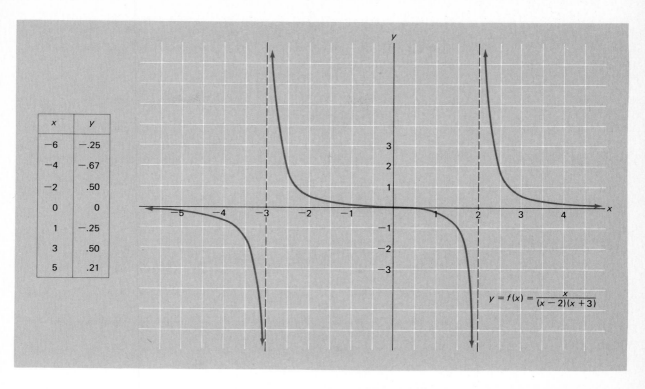

x	y
−6	−.25
−4	−.67
−2	.50
0	0
1	−.25
3	.50
5	.21

$$y = f(x) = \frac{x}{(x-2)(x+3)}$$

The graph will have one vertical asymptote, at $x = 2$. To check on a horizontal asymptote, we note that for large $|x|$, the numerator behaves like $2x^2$ and the denominator behaves like x^2. It follows that $y = 2$ is a horizontal asymptote. The graph crosses the x-axis where the numerator $2x(x+1)$ is zero, namely, at $x = 0$ and $x = -1$. A good approximation to the graph is shown below.

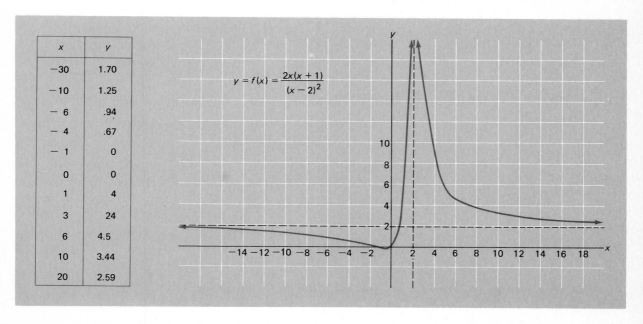

x	y
−30	1.70
−10	1.25
−6	.94
−4	.67
−1	0
0	0
1	4
3	24
6	4.5
10	3.44
20	2.59

$$y = f(x) = \frac{2x(x+1)}{(x-2)^2}$$

A GENERAL PROCEDURE

Here is an outline of the procedure for graphing a rational function which is in reduced form.

1. Check for symmetry with respect to the y-axis and the origin.
2. Factor the numerator and denominator.
3. Determine the vertical asymptotes (if any) by checking where the denominator is zero. Draw a dotted line for each asymptote. Be sure to examine the behavior of the graph near a vertical asymptote.
4. Determine the horizontal asymptote (if any) by examining the quotient of the leading terms for large $|x|$. Indicate any horizontal asymptote with a dotted line.
5. Determine the x-intercepts (if any). These occur where the numerator is zero.
6. Make a small table of values and plot corresponding points.
7. Sketch the graph.

Problem Set 5-3

Sketch the graph of each of the following functions.

1. $f(x) = \dfrac{2}{x + 2}$ 2. $f(x) = \dfrac{-1}{x + 2}$

3. $f(x) = \dfrac{2}{(x + 2)^2}$ 4. $f(x) = \dfrac{1}{(x - 3)^2}$

5. $f(x) = \dfrac{2x}{x + 2}$ 6. $f(x) = \dfrac{x + 2}{x - 3}$

7. $f(x) = \dfrac{1}{(x + 2)(x - 1)}$ 8. $f(x) = \dfrac{3}{x^2 - 9}$

9. $f(x) = \dfrac{x + 1}{(x + 2)(x - 1)}$ 10. $f(x) = \dfrac{3x}{x^2 - 9}$

11. $f(x) = \dfrac{2x^2}{(x + 2)(x - 1)}$ 12. $f(x) = \dfrac{x^2 - 4}{x^2 - 9}$

EXAMPLE A (No vertical asymptotes) Sketch the graph of

$$f(x) = \frac{x^2 - 4}{x^2 + 1} = \frac{(x - 2)(x + 2)}{x^2 + 1}$$

Solution. Note that f is an even function, so the graph will be symmetric with respect to the y-axis. The denominator is not zero for any real x, so there are no vertical asymptotes. The line $y = 1$ is a horizontal asymptote, since for large $|x|$, $f(x)$ behaves like x^2/x^2. The x-intercepts are $x = 2$ and $x = -2$. The graph is shown at the top of page 193.

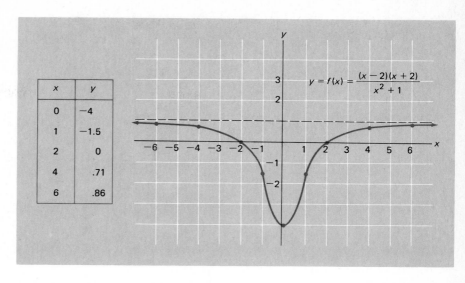

x	y
0	-4
1	-1.5
2	0
4	.71
6	.86

$y = f(x) = \dfrac{(x-2)(x+2)}{x^2+1}$

Sketch the graph of each of the following.

13. $f(x) = \dfrac{1}{x^2+2}$

14. $f(x) = \dfrac{x^2-2}{x^2+2}$

15. $f(x) = \dfrac{x}{x^2+2}$

16. $f(x) = \dfrac{x^3}{x^2+2}$

EXAMPLE B (Rational functions that are not in reduced form)   Sketch the graph of

$$f(x) = \frac{x^2+x-6}{x-2}$$

Solution.   Notice that

$$f(x) = \frac{(x+3)(x-2)}{x-2}$$

You have the right to expect that we will cancel the factor $x - 2$ from numerator and denominator and graph

$$g(x) = x + 3$$

But note that 2 is in the domain of g but not in the domain of f. Thus f and g and their graphs are exactly alike except at one point, namely, at $x = 2$. Both graphs are shown at the top of page 194. You will notice the hole in the graph of $y = f(x)$ at $x = 2$. This technical distinction is occasionally important.

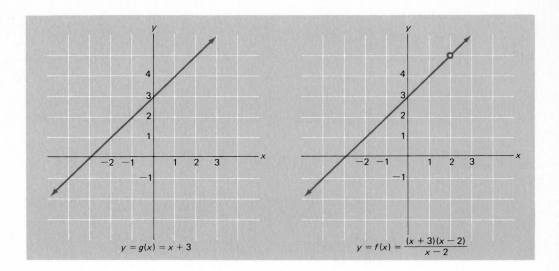

$y = g(x) = x + 3$

$y = f(x) = \dfrac{(x + 3)(x - 2)}{x - 2}$

Sketch the graph of each of the following rational functions, which, you will note, are not in reduced form.

17. $f(x) = \dfrac{(x + 2)(x - 4)}{x + 2}$

18. $f(x) = \dfrac{x^2 - 4}{x - 2}$

19. $f(x) = \dfrac{x^3 - x^2 - 12x}{x + 3}$

20. $f(x) = \dfrac{x^3 - 4x}{x^2 - 2x}$

Miscellaneous Problems

Sketch the graphs of the rational functions in Problems 21–26.

21. $f(x) = \dfrac{x}{x + 5}$

22. $f(x) = \dfrac{x - 2}{x + 3}$

23. $f(x) = \dfrac{x^2 - 9}{x^2 - x - 2}$

24. $f(x) = \dfrac{x - 2}{(x + 3)^2}$

25. $f(x) = \dfrac{x^2 - 9}{x^2 - x - 6}$

26. $f(x) = \dfrac{x - 2}{x^2 + 3}$

27. Sketch the graph of $f(x) = x^n/(x^2 + 1)$ for $n = 1$, 2, and 3. Which graph does not have a horizontal asymptote? Which has the x-axis as horizontal asymptote?

28. Let $f(x) = x^4/(x^n + 2x + 3)$.
    (a) Show (by reasoning) that the graph of f has horizontal asymptote $y = 1$ if $n = 4$.
    (b) Show that the graph does not have a horizontal asymptote if $n < 4$.
    (c) Does the graph have a horizontal asymptote if $n > 4$ and, if so, what is it?

29. A manufacturer of gizmos has overhead of $20,000 per year and direct costs (labor and material) of $50 per gizmo. Write an expression for $U(x)$, the average cost per unit, if the company makes x gizmos this year. Graph the resulting function.

30. A cylindrical tin can is to contain 10π cubic inches. Write a formula for $S(r)$, the total surface area, in terms of the radius r. Graph the resulting function.

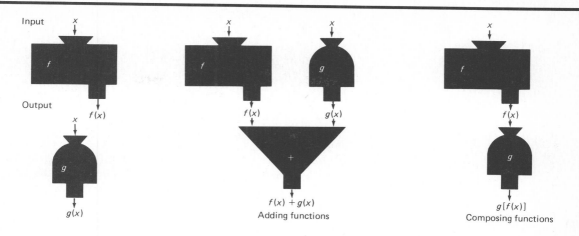

Input

Output

x $f(x)$

x

$g(x)$

x x

$f(x)$ $g(x)$

$+$

$f(x) + g(x)$

Adding functions

x

$f(x)$

$g[f(x)]$

Composing functions

There is still another way to visualize a function. Think of the function named f as a machine. It accepts a number x as input, operates on it, and then presents the number $f(x)$ as output. Machines can be hooked together to make more complicated machines; similarly, functions can be combined to produce more complicated functions. That is the subject of this section.

SUMS, DIFFERENCES, PRODUCTS, AND QUOTIENTS

The simplest way to make new functions from old ones is to use the four arithmetic operations on them. Suppose, for example, that the functions f and g have the formulas

$$f(x) = \frac{x - 3}{2} \qquad g(x) = \sqrt{x}$$

We can make a new function $f + g$ by having it assign to x the value $(x - 3)/2 + \sqrt{x}$; that is,

$$(f + g)(x) = f(x) + g(x) = \frac{x - 3}{2} + \sqrt{x}$$

Of course, we must be a little careful about domains. Clearly, x must be a number on which both f and g can operate. In other words, the domain of $f + g$ is the intersection (common part) of the domains of f and g.

The functions $f - g, f \cdot g$, and f/g are defined in a completely analogous way. Assuming that f and g have their respective natural domains—namely, all reals and the nonnegative reals, respectively—we have the following.

FORMULA   *DOMAIN*

$$(f + g)(x) = f(x) + g(x) = \frac{x - 3}{2} + \sqrt{x} \qquad x \geq 0$$

$$(f - g)(x) = f(x) - g(x) = \frac{x - 3}{2} - \sqrt{x} \qquad x \geq 0$$

$$(f \cdot g)(x) = f(x) \cdot g(x) = \frac{x - 3}{2} \sqrt{x} \qquad x \geq 0$$

$$(f/g)(x) = f(x)/g(x) = \frac{x - 3}{2\sqrt{x}} \qquad x > 0$$

To graph the function $f + g$, it is often best to graph f and g separately in the same coordinate plane and then add the y-coordinates together along vertical lines. We illustrate this method (called **addition of ordinates**) below.

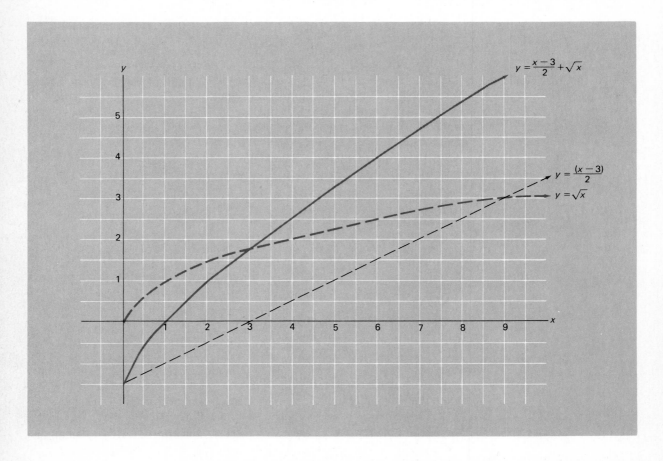

The graph of $f - g$ can be handled similarly. Simply graph f and g in the same coordinate plane and subtract ordinates. We can even graph $f \cdot g$ and f/g in the same manner, but that is harder.

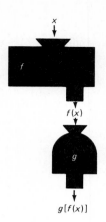

COMPOSITION OF FUNCTIONS

To compose functions is to string them together in tandem. Part of our opening display (reproduced in the margin) shows how this is done. If f operates on x to produce $f(x)$ and then g operates on $f(x)$ to produce $g(f(x))$, we say that we have composed g and f. The resulting function, called the **composite of g with f,** is denoted by $g \circ f$. Thus

$$(g \circ f)(x) = g(f(x))$$

Recall our earlier examples, $f(x) = (x - 3)/2$ and $g(x) = \sqrt{x}$. We may compose them in two ways.

$$(g \circ f)(x) = g(f(x)) = g\left(\frac{x-3}{2}\right) = \sqrt{\frac{x-3}{2}}$$

$$(f \circ g)(x) = f(g(x)) = f(\sqrt{x}) = \frac{\sqrt{x}-3}{2}$$

We note one thing right away: Composition of functions is not commutative; $g \circ f$ and $f \circ g$ are not the same. We must also be careful in describing the domain of a composite function. The domain of $g \circ f$ is that part of the domain of f for which g can acccept $f(x)$ as input. In our example, the domain of $g \circ f$ is $x \geq 3$, not all x or $x \geq 0$ as we might have thought at first glance. The diagram below offers another view of these matters. The shaded portion of the domain of f is not in the domain of $g \circ f$; for x in this portion, $f(x)$ is outside the domain of g.

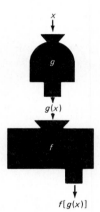

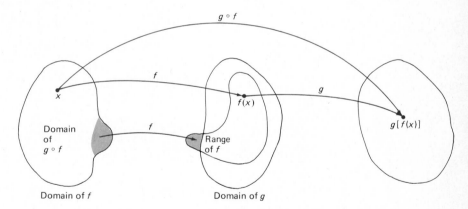

In calculus, we shall often need to take a given function and decompose it, that is, break it into composite pieces. Usually, this can be done in several ways. Take $p(x) = \sqrt{x^2 + 3}$ for example. We may think of it as

$$p(x) = g(f(x)) \quad \text{where} \quad g(x) = \sqrt{x} \quad \text{and} \quad f(x) = x^2 + 3$$

or as

$$p(x) = g(f(x)) \quad \text{where} \quad g(x) = \sqrt{x + 3} \quad \text{and} \quad f(x) = x^2$$

TRANSLATIONS

Observing how a function is built up from simpler ones can be a big aid in graphing. We may ask this question: How are the graphs of

$$y = f(x) \qquad y = f(x - 3) \qquad y = f(x) + 2 \qquad y = f(x - 3) + 2$$

related to each other? Consider $f(x) = |x|$ as an example. The corresponding four graphs are displayed below.

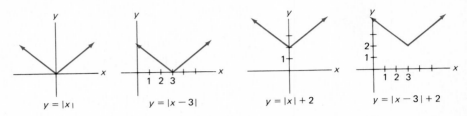

What happened with $f(x) = |x|$ is typical. Notice that all four graphs have the same shape; the last three are just translations of the first. Replacing x by $x - 3$ translates the graph 3 units to the right; adding 2 translates it upward by 2 units.

Here is another illustration of these principles for the function $f(x) = x^3 + x^2$.

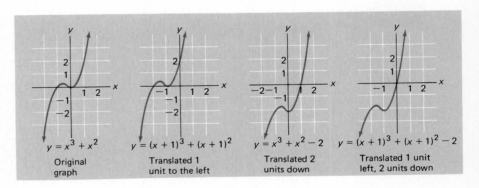

Exactly the same principles apply in the general situation, here illustrated for h and k positive.

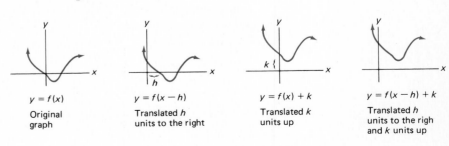

If $h < 0$, the translation is to the left; if $k < 0$, the translation is downward.

Problem Set 5-4

1. Let $f(x) = x^2 - 2x + 2$ and $g(x) = 2/x$. Calculate each of the following.
   - (a) $(f + g)(2)$
   - (b) $(f + g)(0)$
   - (c) $(f - g)(1)$
   - (d) $(f \cdot g)(-1)$
   - (e) $(f/g)(2)$
   - (f) $(g/f)(2)$
   - (g) $(f \circ g)(-1)$
   - (h) $(g \circ f)(-1)$
   - (i) $(g \circ g)(3)$

2. Let $f(x) = 3x + 5$ and $g(x) = |x - 2|$. Perform the calculations in Problem 1 for these functions.

In each of the following, write the formulas for $(f + g)(x)$, $(f - g)(x)$, $(f \cdot g)(x)$, and $(f/g)(x)$ and give the domains of these four functions.

3. $f(x) = x^2$, $g(x) = x - 2$

4. $f(x) = x^3 - 1$, $g(x) = x + 3$

5. $f(x) = x^2$, $g(x) = \sqrt{x}$

6. $f(x) = 2x^2 + 5$, $g(x) = \dfrac{1}{x}$

7. $f(x) = \dfrac{1}{x - 2}$, $g(x) = \dfrac{x}{x - 3}$

8. $f(x) = \dfrac{1}{x^2}$, $g(x) = \dfrac{1}{5 - x}$

For each of the following, write the formulas for $(g \circ f)(x)$ and $(f \circ g)(x)$ and give the domains of these composite functions.

9. $f(x) = x^2$, $g(x) = x - 2$

10. $f(x) = x^3 - 1$, $g(x) = x + 3$

11. $f(x) = \dfrac{1}{x}$, $g(x) = x + 3$

12. $f(x) = 2x^2 + 5$, $g(x) = \dfrac{1}{x}$

13. $f(x) = \sqrt{x - 2}$, $g(x) = x^2 - 2$

14. $f(x) = \sqrt{2x}$, $g(x) = x^2 + 1$

15. $f(x) = 2x - 3$, $g(x) = \frac{1}{2}(x + 3)$

16. $f(x) = x^3 + 1$, $g(x) = \sqrt[3]{x - 1}$

EXAMPLE (Decomposing functions) In each of the following, H can be thought of as a composite function $g \circ f$. Write formulas for $f(x)$ and $g(x)$.

(a) $H(x) = (2 + 3x)^2$

(b) $H(x) = \dfrac{1}{(x^2 + 4)^3}$

Solution.

(a) Think of how you might calculate $H(x)$. You would first calculate $2 + 3x$ and then square the result. That suggests

$$f(x) = 2 + 3x \qquad g(x) = x^2$$

(b) Here there are two obvious ways to proceed. One way would be to let

$$f(x) = x^2 + 4 \qquad g(x) = \dfrac{1}{x^3}$$

Another selection, which is just as good, is

$$f(x) = \dfrac{1}{x^2 + 4} \qquad g(x) = x^3$$

We could actually think of H as the composite of four functions. Let

$$f(x) = x^2 \qquad g(x) = x + 4 \qquad h(x) = x^3 \qquad j(x) = \frac{1}{x}$$

Then

$$H = j \circ h \circ g \circ f$$

You should check this result.

In each of the following, write formulas for $g(x)$ and $f(x)$ so that $H = g \circ f$. The answer is not unique.

17. $H(x) = (x + 4)^3$ 18. $H(x) = (2x + 1)^3$

19. $H(x) = \sqrt{x + 2}$ 20. $H(x) = \sqrt[3]{2x + 1}$

21. $H(x) = \dfrac{1}{(2x + 5)^3}$ 22. $H(x) = \dfrac{6}{(x + 4)^3}$

23. $H(x) = |x^3 - 4|$ 24. $H(x) = |4 - x - x^2|$

Use the method of addition or subtraction of ordinates to graph each of the following. That is, graph $y = f(x)$ and $y = g(x)$ in the same coordinate plane and then obtain the graph of $f + g$ or $f - g$ by adding or subtracting ordinates.

25. $f + g$ where $f(x) = x^2$ and $g(x) = x - 2$
26. $f + g$ where $f(x) = |x|$ and $g(x) = x$
27. $f - g$ where $f(x) = 1/x$ and $g(x) = x$
28. $f - g$ where $f(x) = x^3$ and $g(x) = -x + 1$

In each of the following, graph the function f carefully and then use translations to sketch the graphs of the functions g, h, and j.

29. $f(x) = x^2$, $g(x) = (x - 2)^2$, $h(x) = x^2 - 4$, and $j(x) = (x - 2)^2 + 1$
30. $f(x) = x^3$, $g(x) = (x + 2)^3$, $h(x) = x^3 + 4$, and $j(x) = (x + 2)^3 - 2$
31. $f(x) = \sqrt{x}$, $g(x) = \sqrt{x - 3}$, $h(x) = \sqrt{x} + 2$, and $j(x) = \sqrt{x - 3} - 2$
32. $f(x) = \dfrac{1}{x}$, $g(x) = \dfrac{1}{x - 4}$, $h(x) = \dfrac{1}{x} + 3$, and $j(x) = \dfrac{1}{x - 4} - 5$

Miscellaneous Problems

33. Let $f(x) = 2x + 3$ and $g(x) = x^3$. Write formulas for each of the following.
   (a) $(f + g)(x)$ (b) $(g - f)(x)$ (c) $(f \cdot g)(x)$ (d) $(f/g)(x)$
   (e) $(f \circ g)(x)$ (f) $(g \circ f)(x)$ (g) $(f \circ f)(x)$ (h) $(g \circ g \circ g)(x)$
34. If $f(x) = 1/(x - 1)$ and $g(x) = \sqrt{x + 1}$, write formulas for $(f \circ g)(x)$ and $(g \circ f)(x)$ and give the domains of these composite functions.
35. If $f(x) = x^2 - 4$, $g(x) = |x|$, and $h(x) = 1/x$, write a formula for $(h \circ g \circ f)(x)$ and indicate its domain.
36. In general, how many different functions can be obtained by composing three different functions f, g, and h in different orders?
37. In calculus, the *difference quotient*

$$\frac{f(x + h) - f(x)}{h}$$

arises repeatedly. Calculate this expression and simplify it for each of the following.

(a) $f(x) = x^2$ (b) $f(x) = 2x + 3$
(c) $f(x) = 1/x$ (d) $f(x) = 2/(x - 2)$

38. Calculate $[g(x - h) - g(x)]/h$ for each of the following. Simplify your answer.

(a) $g(x) = 4x - 9$ (b) $g(x) = x^2 + 2x$
(c) $g(x) = x + 1/x$ (d) $g(x) = x^3$

$\boxed{c}$ 39. Let $f(x) = (1 + \sqrt{x})^3/(3x^2 + 1)$. Calculate each of the following.
   (a) $f(3.1)$ (b) $f(.03)$

$\boxed{c}$ 40. Let $g(x) = (3 + 1/x)^2\sqrt{x^3 + 1}$. Calculate each of the following.
   (a) $g(4.2)$ (b) $g(-.91)$

41. The relationship between the price (in cents) for a certain product and the demand (in thousands of units) appears to satisfy

$$P = \sqrt{29 - 3D + D^2}$$

On the other hand, the demand has risen over the past t years according to $D = 2 + \sqrt{t}$.

(a) Express P as a function of t. $\boxed{c}$ (b) Evaluate P when $t = 15$.

42. After being in business x years, a certain tractor manufacturer is making $100 + x + 2x^2$ units each year. The sales price (in dollars) per unit has risen according to the formula $P = 500 + 60x$. Assuming all this is true, write a formula for the manufacturer's yearly revenue $R(x)$ after x years.

A one-to-one function   has an inverse.

5-5
Inverse Functions

Some processes are reversible; most are not. If I take off my shoes, I may put them back on again. The second operation undoes the first one and brings things back to the original state. But if I throw my shoes in the fire, I will have a hard time undoing the damage I have done.

A function f operates on a number x to produce a number $y = f(x)$. It may be that we can find a function g that will operate on y and give back x. For example, if

$$y = f(x) = 2x + 1$$

then

$$g(x) = \frac{1}{2}(x - 1)$$

is such a function, since

$$g(y) = g(f(x)) = \frac{1}{2}(2x + 1 - 1) = x$$

When we can find such a function g, we call it the *inverse* of f. Not all functions have inverses. Whether they do or not has to do with a concept called one-to-oneness.

ONE-TO-ONE FUNCTIONS

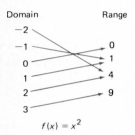

$f(x) = x^2$

In the margin, we have reproduced an example we studied earlier, the squaring function with domain $\{-2, -1, 0, 1, 2, 3\}$. It is a perfectly fine function, but it does have one troublesome feature. It may assign the same value to two different x's. In particular, $f(-2) = 4$ and $f(2) = 4$. Such a function cannot possibly have an inverse g. For what would g do with 4? It would not know whether to give back -2 or 2 as the value.

In contrast, consider $f(x) = 2x + 1$, also pictured in the margin. Notice that this function never assigns the same value to two different values of x. Therefore there is an unambiguous way of undoing it.

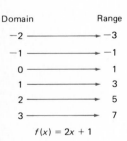

$f(x) = 2x + 1$

We say that a function f is **one-to-one** if $x_1 \neq x_2$ implies $f(x_1) \neq f(x_2)$, that is, if different values for x always result in different values for $f(x)$. Some functions are one-to-one; some are not. It would be nice to have a graphical criterion for deciding.

Consider the functions $f(x) = x^2$ and $f(x) = 2x + 1$ again, but now let the domains be the set of all real numbers. Their graphs appear below. In the

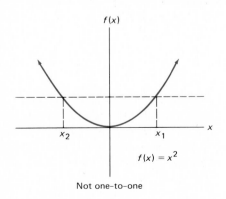

Not one-to-one

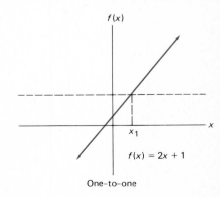

One-to-one

first case, certain horizontal lines (those which are above the x-axis) meet the graph in two points; in the second case, every horizontal line meets the graph in exactly one point. Notice on the first graph that $f(x_1) = f(x_2)$ even though $x_1 \neq x_2$. On the second graph, this cannot happen. Thus we have the important fact that *if every horizontal line meets the graph of a function f in at most one point, then f is one-to-one*.

INVERSE FUNCTIONS

Now we are ready to give a formal definition of the main idea of this section.

DEFINITION

Let f be a one-to-one function with domain X and range Y. Then the function g with domain Y and range X which satisfies

$$g(f(x)) = x$$

*for all x in X is called the **inverse of f**.*

We make several important observations. First, the boxed formula simply says that g undoes what f did. Second, if g undoes f, then f will undo g, that is,

$$f(g(y)) = y$$

for all y in Y. Third, the function g is usually denoted by the symbol f^{-1}. You are cautioned to remember that f^{-1} does *not* mean $1/f$, as you have the right to expect. Mathematicians decided long ago that f^{-1} should stand for the inverse function (the undoing function). Thus

$$(f^{-1} \circ f)(x) = x \quad \text{and} \quad (f \circ f^{-1})(y) = y$$

For example, if $f(x) = 4x$, then $f^{-1}(y) = \frac{1}{4}y$ since

$$(f^{-1} \circ f)(x) = f^{-1}(f(x)) = f^{-1}(4x) = \tfrac{1}{4}(4x) = x$$

and

$$(f \circ f^{-1})(y) = f(f^{-1}(y)) = f(\tfrac{1}{4}y) = 4(\tfrac{1}{4}y) = y$$

The boxed results are illustrated in the diagrams below.

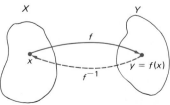

$$f^{-1}[f(x)] = x$$

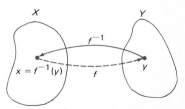

$$f[f^{-1}(y)] = y$$

FINDING A FORMULA FOR f^{-1}

If f adds 2, then f^{-1} ought to subtract 2. To say it in symbols, if $f(x) = x + 2$, then we might expect $f^{-1}(y) = y - 2$. And we are right, for

$$f^{-1}[f(x)] = f^{-1}(x + 2) = x + 2 - 2 = x$$

If f divides by 3 and then subtracts 4, we expect f^{-1} to add 4 and multiply by 3. Symbolically, if $f(x) = x/3 - 4$, then we expect $f^{-1}(y) = 3(y + 4)$. Again we are right, for

$$f^{-1}[f(x)] = f^{-1}\left(\frac{x}{3} - 4\right) = 3\left(\frac{x}{3} - 4 + 4\right) = x$$

Note that you must undo things in the reverse order in which you did them (that is, we divided by 3 and then subtracted 4, so to undo this, we first add 4 and then multiply by 3).

When we get to more complicated functions, it is not always easy to find the formula for the inverse function. Here is an important way to look at it.

$$x = f^{-1}(y) \quad \text{if and only if} \quad y = f(x)$$

That means that we can get the formula for f^{-1} by solving the equation $y = f(x)$ for x. Here is an example. Let $y = f(x) = 3/(x - 2)$. Follow the steps below.

$$y = \frac{3}{x - 2}$$

$$(x - 2)y = 3$$

$$xy - 2y = 3$$

$$xy = 3 + 2y$$

$$x = \frac{3 + 2y}{y}$$

Thus

$$f^{-1}(y) = \frac{3 + 2y}{y}$$

In the formula for f^{-1} just derived, there is no need to use y as the variable. We might use u or t or even x. The formulas

$$f^{-1}(u) = \frac{3 + 2u}{u}$$

$$f^{-1}(t) = \frac{3 + 2t}{t}$$

$$f^{-1}(x) = \frac{3 + 2x}{x}$$

all say the same thing in the sense that they give the same rule. It is conventional to give formulas for functions using x as the variable, and so we would write $f^{-1}(x) = (3 + 2x)/x$ as our answer. Let us summarize. To find the formula for $f^{-1}(x)$, use the following steps.

THREE-STEP PROCEDURE FOR FINDING $f^{-1}(x)$

1. Solve $y = f(x)$ for x in terms of y.
2. Use $f^{-1}(y)$ to name the resulting expression in y.
3. Replace y by x to get the formula for $f^{-1}(x)$.

THE GRAPHS OF f AND f^{-1}

Since $y = f(x)$ and $x = f^{-1}(y)$ are equivalent, the graphs of these two equations are the same. Suppose we want to compare the graphs of $y = f(x)$ and $y = f^{-1}(x)$ (where, you will note, we have used x as the domain variable in both cases). To get $y = f^{-1}(x)$ from $x = f^{-1}(y)$, we interchange the roles of x and y. Graphically, this corresponds to folding (reflecting) the graph across the 45° line—that is, across the line $y = x$. This is the same as saying that if the point (a, b) is on one graph, then (b, a) is on the other.

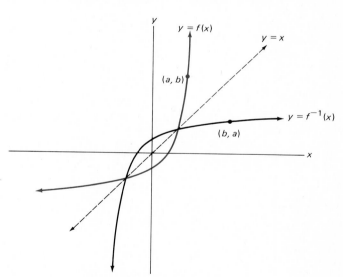

Reflecting a graph across the line $y = x$

Here is a simple example. Let $f(x) = x^3$; then $f^{-1}(x) = \sqrt[3]{x}$, the cube root of x. The graphs of $y = x^3$ and $y = \sqrt[3]{x}$ are shown at the top of page 206, first separately and then on the same coordinate plane. Note that $f(2) = 8$ and $f^{-1}(8) = 2$.

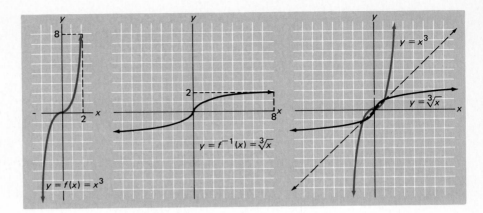

Problem Set 5-5

1. Examine the graphs below.
   (a) Which of these are the graphs of functions with x as domain variable?
   (b) Which of these functions are one-to-one?
   (c) Which of them have inverses?

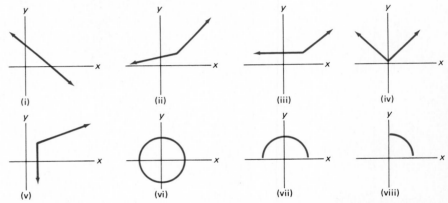

(i)   (ii)   (iii)   (iv)

(v)   (vi)   (vii)   (viii)

2. Let each of the following functions have their natural domains. Which of them are one-to-one? (*Hint:* Consider their graphs.)

   (a) $f(x) = x^4$ (b) $f(x) = x^3$ (c) $f(x) = \dfrac{1}{x}$

   (d) $f(x) = \dfrac{1}{x^2}$ (e) $f(x) = x^2 + 2x + 3$ (f) $f(x) = |x|$

   (g) $f(x) = \sqrt{x}$ (h) $f(x) = -3x + 2$

3. Let $f(x) = 3x - 2$. To find $f^{-1}(2)$, note that $f^{-1}(2) = a$ if $f(a) = 2$, that is, if $3a - 2 = 2$; so $f^{-1}(2) = a = \frac{4}{3}$. Find each of the following.
   (a) $f^{-1}(1)$ (b) $f^{-1}(-3)$ (c) $f^{-1}(14)$

4. Let $g(x) = 1/(x - 1)$. Find each of the following.
   (a) $g^{-1}(1)$ (b) $g^{-1}(-1)$ (c) $g^{-1}(14)$

EXAMPLE A (Finding $f^{-1}(x)$) Use the three-step procedure to find $f^{-1}(x)$ if $f(x) = 2x^3 - 1$. Check your result by calculating $f(f^{-1}(x))$.

Solution.

    Step 1   We solve $y = 2x^3 - 1$ for x in terms of y.

$$2x^3 = y + 1$$

$$x^3 = \frac{y + 1}{2}$$

$$x = \sqrt[3]{\frac{y + 1}{2}}$$

    Step 2   Call the result $f^{-1}(y)$.

$$f^{-1}(y) = \sqrt[3]{\frac{y + 1}{2}}$$

    Step 3   Replace y by x.

$$f^{-1}(x) = \sqrt[3]{\frac{x + 1}{2}}$$

Check: $f(f^{-1}(x)) = 2\left(\sqrt[3]{\frac{x + 1}{2}}\right)^3 - 1 = 2\left(\frac{x + 1}{2}\right) - 1 = x$

Each of the functions in Problems 5–14 has an inverse (using its natural domain). Find the formula for $f^{-1}(x)$. Then check your result by calculating $f(f^{-1}(x))$.

  5. $f(x) = 5x$ 6. $f(x) = -4x$ 7. $f(x) = 2x - 7$

  8. $f(x) = -3x + 2$ 9. $f(x) = \sqrt{x} + 2$ 10. $f(x) = 2\sqrt{x} - 6$

  11. $f(x) = \dfrac{x}{x - 3}$ 12. $f(x) = \dfrac{x - 3}{x}$ 13. $f(x) = (x - 2)^3 + 2$

  14. $f(x) = \frac{1}{3}x^5 - 2$

  15. In the same coordinate plane, sketch the graphs of $y = f(x)$ and $y = f^{-1}(x)$ for $f(x) = \sqrt{x} + 2$ (see Problem 9).

  16. In the same coordinate plane, sketch the graphs of $y = f(x)$ and $y = f^{-1}(x)$ for $f(x) = x/(x - 3)$ (see Problem 11).

  17. Sketch the graph of $y = f^{-1}(x)$ if the graph of $y = f(x)$ is as shown below.

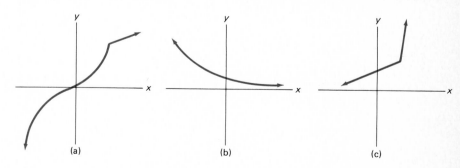

(a)   (b)   (c)

18. Show that $f(x) = 2x/(x - 1)$ and $g(x) = x/(x - 2)$ are inverses of each other by calculating $f(g(x))$ and $g(f(x))$.

19. Show that $f(x) = 3x/(x + 2)$ and $g(x) = 2x/(3 - x)$ are inverses of each other.

20. Sketch the graph of $f(x) = x^3 + 1$ and note that f is one-to-one. Find a formula for $f^{-1}(x)$.

EXAMPLE B (Restricting the domain)   The function $f(x) = x^2$ does not have an inverse if we use its natural domain (all real numbers). However, if we restrict its domain to $x \geq 0$ so that we are considering only its right branch (see graphs in margin), then it has an inverse, $f^{-1}(x) = \sqrt{x}$. Use the same idea to show that $g(x) = x^2 - 2x - 1$ has an inverse when its domain is appropriately restricted. Find $g^{-1}(x)$.

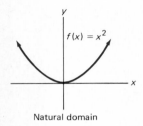

$f(x) = x^2$

Natural domain

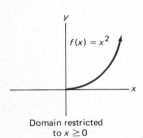

$f(x) = x^2$

Domain restricted to $x \geq 0$

Solution.   The graph of $g(x)$ is shown in the margin; it is a parabola with vertex at $x = 1$. Accordingly, we can restrict the domain to $x \geq 1$. To find the formula for $g^{-1}(x)$, we first solve $y = x^2 - 2x - 1$ for x using an old trick, completing the square.

$$y + 1 = x^2 - 2x$$
$$y + 1 + 1 = x^2 - 2x + 1$$
$$y + 2 = (x - 1)^2$$
$$\pm\sqrt{y + 2} = x - 1$$
$$1 \pm \sqrt{y + 2} = x$$

Notice that there are two expressions for x; they correspond to the two halves of the parabola. We chose to make $x \geq 1$, so $x = 1 + \sqrt{y + 2}$ is the correct expression for $g^{-1}(y)$. Thus

$$g^{-1}(x) = 1 + \sqrt{x + 2}$$

If we had chosen to make $x \leq 1$, the correct answer would have been $g^{-1}(x) = 1 - \sqrt{x + 2}$.

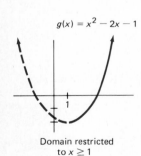

$g(x) = x^2 - 2x - 1$

Domain restricted to $x \geq 1$

In each of the following, restrict the domain so that f has an inverse. Describe the restricted domain and find a formula for $f^{-1}(x)$. Note: Different restrictions of the domain are possible.

21. $f(x) = (x - 1)^2$

22. $f(x) = (x + 3)^2$

23. $f(x) = (x + 1)^2 - 4$

24. $f(x) = (x - 2)^2 + 3$

25. $f(x) = x^2 + 6x + 7$

26. $f(x) = x^2 - 4x + 9$

27. $f(x) = |x + 2|$

28. $f(x) = 2|x - 3|$

29. $f(x) = \dfrac{(x - 1)^2}{1 + 2x - x^2}$

30. $f(x) = \dfrac{-1}{x^2 + 4x + 3}$

31. Sketch the graph of $f(x) = 1/(x - 1)$. Is f one-to-one? Calculate each of the following.

    (a) $f(3)$ (b) $f^{-1}(\frac{1}{2})$ (c) $f(0)$

    (d) $f^{-1}(-1)$ (e) $f^{-1}(3)$ (f) $f^{-1}(-2)$

32. If $f(x) = x/(x - 2)$, find the formula for $f^{-1}(x)$.

33. Find the formula for $f^{-1}(x)$ if $f(x) = 1/(x - 1)$ and sketch the graph of $y = f^{-1}(x)$. Compare this graph with the graph of $y = f(x)$ that you sketched in Problem 31.

34. The functions $f(x) = 3/(x - 2)$ and $f^{-1}(x) = (3 + 2x)/x$ were shown to be inverses in the text. Sketch the graphs of $y = f(x)$ and $y = f^{-1}(x)$ using the same coordinate axes.

35. Sketch the graph of $f(x) = x^2 - 2x - 3$ and observe that is is not one-to-one. Restrict its domain so it is and then find a formula for $f^{-1}(x)$.

36. Show that $f(x) = x/(x - 1)$ is its own inverse by showing that $f(f(x)) = x$. Can you give other examples of functions with this property? What must be true about their graphs?

In Problems 37–38, find formulas for $f^{-1}(x)$, $g^{-1}(x)$, $(g \circ f)(x)$, $(g \circ f)^{-1}(x)$, and $(f^{-1} \circ g^{-1})(x)$.

37. $f(x) = 2x + 5$ and $g(x) = 1/x$

38. $f(x) = x - 4$ and $g(x) = x^3 + 1$

39. From your results in Problems 37–38, what do you conjecture about the relationship between $(g \circ f)^{-1}$ and $f^{-1} \circ g^{-1}$?

CHAPTER SUMMARY

A **function** f is a rule which assigns to each element x in one set (called the **domain**) a value $f(x)$ from another set. The set of all these values is called the **range** of the function. Numerical functions are usually specified by formulas (for example, $g(x) = (x^2 + 1)/(x + 1)$). The **natural domain** for such a function is the largest set of real numbers for which the formula makes sense and gives real values (thus, the natural domain for g consists of all real numbers except $x = -1$). Related to the notion of function is that of **variation.**

The **graph** of a function is simply the graph of the equation $y = f(x)$. Of special interest are the graphs of **polynomial functions** and **rational functions.** In graphing them, we should show the hills, the valleys, the x-**intercepts,** and, in the case of rational functions, the vertical and horizontal **asymptotes.**

Functions can be combined in many ways. Of these, composition is perhaps the most significant. The **composite** of f with g is defined by $(f \circ g)(x) = f(g(x))$.

Some functions are **one-to-one;** some are not. Those that are one-to-one have undoing functions called inverses. The **inverse** of f, denoted by f^{-1}, satisfies $f^{-1}(f(x)) = x$. Finding a formula for $f^{-1}(x)$ can be tricky; therefore, we described a definite procedure for doing it.

CHAPTER REVIEW PROBLEM SET

1. Let $f(x) = x^2 - 1$ and $g(x) = 2/x$. Calculate if possible.
   (a) $f(4)$ (b) $g(\frac{1}{2})$ (c) $g(0)$
   (d) $f(1)/g(3)$ (e) $f(g(4))$ (f) $g(f(4))$

2. Find the natural domain of f if $f(x) = \sqrt{x} + 1/(x - 1)$.

3. If y varies directly as the cube of x and $y = 1$ when $x = 2$, find an explicit formula for y in terms of x.

4. If z varies directly as x and inversely as the square of y, and if $z = 1$ when x and y are both 3, find z when $x = 16$ and $y = 2$.

5. Graph each f.
   (a) $f(x) = (x - 2)^2$ (b) $f(x) = x^3 + 2x$

   (c) $f(x) = \dfrac{1}{x^2 - x - 2}$ (d) $f(x) = \begin{cases} 0 & \text{if } x \le 0 \\ x^2 & \text{if } 0 < x < 1 \\ 1 & \text{if } x \ge 1 \end{cases}$

6. Suppose that g is an even function satisfying $g(x) = \sqrt{x}$ for $x \ge 0$. Sketch its graph on $-4 \le x \le 4$.

7. Sketch the graph of $h(x) = x^2 + 1/x$ by first graphing $y = x^2$ and $y = 1/x$ and then adding ordinates.

8. If $f(x) = x^3 + 1$ and $g(x) = x + 2$, give formulas for each of the following.
   (a) $f(g(x))$ (b) $g(f(x))$ (c) $f(f(x))$
   (d) $g(g(x))$ (e) $f^{-1}(x)$ (f) $g^{-1}(x)$
   (g) $g(x + h)$ (h) $[g(x + h) - g(x)]/h$ (i) $f(3x)$

9. How does the graph of $y = f(x - 2) + 3$ relate to the graph of $y = f(x)$?

10. Which of the following functions are even? Odd? One-to-one?
    (a) $f(x) = 1/(x^2 - 1)$ (b) $f(x) = 1/x$
    (c) $f(x) = |x|$ (d) $f(x) = 3x + 4$

11. Let $f(x) = x/(x - 2)$. Find a formula for $f^{-1}(x)$. Graph $y = f(x)$ and $y = f^{-1}(x)$ using the same coordinate axes.

12. How could we restrict the domain of $f(x) = (x + 2)^2$ so that f has an inverse?

6

EXPONENTIAL AND LOGARITHMIC FUNCTIONS

The method of logarithms, by reducing to a few days the labor of many months, doubles as it were, the life of the astronomer, besides freeing him from the errors and disgust inseparable from long calculation.

P. S. Laplace

Radicals on Top of Radicals

The discovery of the quadratic formula led to a long search for a corresponding formula for the cubic equation. Success had to wait until the sixteenth century, when the Italian school of mathematicians at Bologna found formulas for both the cubic and quartic equations. A typical result is Cardano's solution to $x^3 + ax = b$ which takes the form

$$x = \sqrt[3]{\sqrt{\frac{a^3}{27} + \frac{b^2}{4}} + \frac{b}{2}} - \sqrt[3]{\sqrt{\frac{a^3}{27} + \frac{b^2}{4}} - \frac{b}{2}}$$

Hieronimo Cardano
1501–1576

6-1
Radicals

Historically, interest in radicals has been associated with the desire to solve equations. Even the general cubic equation leads to very complicated radical expressions. Today, powerful iterative methods make results like Cardano's solution historical curiosities. Yet the need for radicals continues; it is important that we know something about them.

Raising a number to the 3rd power (or cubing it) is a process which can be undone. The inverse process—taking the 3rd root—is denoted by $\sqrt[3]{\ }$. We call $\sqrt[3]{a}$ a *radical* and read it "the cube root of a." Thus $\sqrt[3]{8} = 2$ and $\sqrt[3]{-125} = -5$ since $2^3 = 8$ and $(-5)^3 = -125$.

Our first goal is to give meaning to the symbol $\sqrt[n]{a}$ when n is any positive integer. Naturally, we require that $\sqrt[n]{a}$ be a number which yields a when raised to the nth power; that is

$$(\sqrt[n]{a})^n = a$$

When n is odd, that is all we need to say, since for any real number a, there is exactly one real number whose nth power is a.

When n is even, we face two serious problems, problems that are already apparent when $n = 2$. We have already discussed square roots, using the symbol $\sqrt{}$ rather than $\sqrt[2]{}$ (see Section 3-4). Recall that if $a < 0$, then $\sqrt{a}$ is not a real number (for example, $\sqrt{-4} = 2i$). Even if $a > 0$, we are in trouble since there are always two real numbers with squares equal to a. For example, both -3 and 3 have squares equal to 9. We agree that in this ambiguous case, $\sqrt{a}$ shall always denote the positive square root of a. Thus $\sqrt{9}$ is equal to 3, not -3.

We make a similar agreement about $\sqrt[n]{a}$ for n an even number greater than 2. First, we shall avoid the case $a < 0$. Second, when $a \geq 0$, $\sqrt[n]{a}$ will always denote the nonnegative number whose nth power is a. Thus $\sqrt[4]{81} = 3$, $\sqrt[4]{16} = 2$, and $\sqrt[4]{0} = 0$; however, the symbol $\sqrt[4]{-16}$ will be assigned no meaning in this book (even roots of negative numbers are discussed in more advanced books). Let us summarize.

If n is odd, $\sqrt[n]{a}$ is the unique real number satisfying $(\sqrt[n]{a})^n = a$.
If n is even and $a \geq 0$, $\sqrt[n]{a}$ is the unique nonnegative real number satisfying $(\sqrt[n]{a})^n = a$.

The symbol $\sqrt[n]{a}$, as we have defined it, is called the **principal nth root of a;** for brevity, we often drop the adjective *principal*.

RULES FOR RADICALS

Radicals, like exponents, obey certain rules. The most important ones are listed below, where it is assumed that all radicals name real numbers.

RULES FOR RADICALS

1. $(\sqrt[n]{a})^n = a$
2. $\sqrt[n]{a^n} = a \qquad (a \geq 0)$
3. $\sqrt[n]{ab} = \sqrt[n]{a}\,\sqrt[n]{b}$
4. $\sqrt[n]{\dfrac{a}{b}} = \dfrac{\sqrt[n]{a}}{\sqrt[n]{b}}$

Rule 2 holds also for $a < 0$ if n is odd; for example, $\sqrt[5]{(-2)^5} = -2$.

These rules can all be proved, but we believe that the following illustrations will be more helpful to you than proofs.

$$(\sqrt[4]{7})^4 = 7$$

$$\sqrt[14]{3^{14}} = 3$$

$$\sqrt{2} \cdot \sqrt{18} = \sqrt{36} = 6$$

$$\frac{\sqrt[3]{750}}{\sqrt[3]{6}} = \sqrt[3]{\frac{750}{6}} = \sqrt[3]{125} = 5$$

SIMPLIFYING RADICALS

One use of the four rules given above is to simplify radicals. Here are two examples.

$$\sqrt[3]{54x^4y^6} \qquad \sqrt[4]{x^8 + x^4y^4}$$

We assume that x and y represent positive numbers.

In the first example, we start by factoring out the largest possible third power.

$$\sqrt[3]{54x^4y^6} = \sqrt[3]{(27x^3y^6)(2x)}$$
$$= \sqrt[3]{(3xy^2)^3(2x)}$$
$$= \sqrt[3]{(3xy^2)^3} \sqrt[3]{2x} \qquad \text{(Rule 3)}$$
$$= 3xy^2 \sqrt[3]{2x} \qquad \text{(Rule 2)}$$

In the second example, it is tempting to write $\sqrt[4]{x^8 + x^4y^4} = x^2 + xy$, thereby pretending that $\sqrt[4]{a^4 + b^4} = a + b$. This is wrong, because $(a + b)^4 \neq a^4 + b^4$. Here is what we can do.

$$\sqrt[4]{x^8 + x^4y^4} = \sqrt[4]{x^4(x^4 + y^4)}$$
$$= \sqrt[4]{x^4} \sqrt[4]{x^4 + y^4} \qquad \text{(Rule 3)}$$
$$= x\sqrt[4]{x^4 + y^4} \qquad \text{(Rule 2)}$$

We were able to take x^4 out of the radical because it is a 4th power and a factor of $x^8 + x^4y^4$.

RATIONALIZING DENOMINATORS

For some purposes (including hand calculations), fractions with radicals in their denominators are considered to be needlessly complicated. Fortunately, we can usually rewrite a fraction so that its denominator is free of radicals. The process we go through is called **rationalizing the denominator.** Here are two examples.

$$\frac{1}{\sqrt[5]{x}} \qquad \frac{x}{\sqrt{x} + \sqrt{y}}$$

In the first case, we multiply numerator and denominator by $\sqrt[5]{x^4}$, which gives the 5th root of a 5th power in the denominator.

$$\frac{1}{\sqrt[5]{x}} = \frac{1 \cdot \sqrt[5]{x^4}}{\sqrt[5]{x} \cdot \sqrt[5]{x^4}} = \frac{\sqrt[5]{x^4}}{\sqrt[5]{x \cdot x^4}} = \frac{\sqrt[5]{x^4}}{\sqrt[5]{x^5}} = \frac{\sqrt[5]{x^4}}{x}$$

In the second case, we make use of the identity $(a + b)(a - b) = a^2 - b^2$. If we multiply numerator and denominator of the fraction by $\sqrt{x} - \sqrt{y}$, the radicals in the denominator disappear

$$\frac{x}{\sqrt{x} + \sqrt{y}} = \frac{x(\sqrt{x} - \sqrt{y})}{(\sqrt{x} + \sqrt{y})(\sqrt{x} - \sqrt{y})} = \frac{x\sqrt{x} - x\sqrt{y}}{x - y}$$

We should point out that this manipulation is valid provided $x \neq y$.

Problem Set 6-1

Simplify the following radical expressions. This will involve removing perfect powers from radicals and rationalizing denominators. Assume that all letters represent positive numbers.

$$\sqrt{a^4 + a^4b^2} = a^2 + a^2b$$

$$\sqrt{a^4 + a^4b^2} = \sqrt{a^4(1 + b^2)}$$
$$= a^2\sqrt{1 + b^2}$$

1. $\sqrt{9}$

2. $\sqrt[3]{-8}$

3. $\sqrt[5]{32}$

4. $\sqrt[4]{16}$

5. $(\sqrt[3]{7})^3$

6. $(\sqrt{\pi})^2$

7. $\sqrt[3]{(\frac{3}{2})^3}$

8. $\sqrt[5]{(-2/7)^5}$

9. $(\sqrt{5})^4$

10. $(\sqrt[3]{5})^6$

11. $\sqrt{3}\sqrt{27}$

12. $\sqrt{2}\sqrt{32}$

13. $\sqrt[3]{16}/\sqrt[3]{2}$

14. $\sqrt[4]{48}/\sqrt[4]{3}$

15. $\sqrt[3]{10^{-6}}$

16. $\sqrt[4]{10^8}$

17. $1/\sqrt{2}$

18. $1/\sqrt{3}$

19. $\sqrt{10}/\sqrt{2}$

20. $\sqrt{6}/\sqrt{3}$

21. $\sqrt[3]{54x^4y^5}$

22. $\sqrt[3]{-16x^3y^8}$

23. $\sqrt[4]{(x + 2)^4y^7}$

24. $\sqrt[4]{x^5(y - 1)^8}$

25. $\sqrt{x^2 + x^2y^2}$

26. $\sqrt{25 + 50y^4}$

27. $\sqrt[3]{x^6 - 9x^3y}$

28. $\sqrt[4]{16x^{12} + 64x^8}$

29. $\sqrt[3]{x^4y^{-6}z^6}$

30. $\sqrt[4]{32x^{-4}y^9}$

31. $\dfrac{2}{\sqrt{x} + 3}$

32. $\dfrac{4}{\sqrt{x} - 2}$

33. $\dfrac{2}{\sqrt{x + 3}}$

34. $\dfrac{4}{\sqrt{x - 2}}$

35. $\dfrac{1}{\sqrt[4]{8x^3}}$

36. $\dfrac{1}{\sqrt[3]{5x^2y^4}}$

37. $\sqrt[3]{2x^{-2}y^4}\sqrt[3]{4xy^{-1}}$

38. $\sqrt[4]{125x^5y^3}\sqrt[4]{5x^{-9}y^5}$

39. $\sqrt{50} - 2\sqrt{18} + \sqrt{8}$

40. $\sqrt[3]{24} + \sqrt[3]{375}$

EXAMPLE A (Equations involving radicals)   Solve the following equations.

(a) $\sqrt[3]{x - 2} = 3$ (b) $x = \sqrt{2 - x}$

Solution.

(a) Raise both sides to the 3rd power and solve for x.

$$(\sqrt[3]{x - 2})^3 = 3^3$$

$$x - 2 = 27$$

$$x = 29$$

(b) Square both sides and solve for x.

$$x^2 = 2 - x$$

$$x^2 + x - 2 = 0$$

$$(x - 1)(x + 2) = 0$$

$$x = 1 \qquad x = -2$$

Let us check our answers in part (b) by substituting them in the original equation. When we substitute these numbers for x in $x = \sqrt{2 - x}$, we find that 1 works but -2 does not.

$$1 = \sqrt{2 - 1} \qquad -2 \neq \sqrt{2 - (-2)}$$

In squaring both sides of $x = \sqrt{2 - x}$, we introduced an extraneous solution. That happened because $a = b$ and $a^2 = b^2$ are not equivalent statements. Whenever you square both sides of an equation (or raise both sides of an equation to any even power), be sure to check your answers.

Solve each of the following equations.

41. $\sqrt{x - 1} = 5$ 42. $\sqrt{x + 2} = 3$ 43. $\sqrt[3]{2x - 1} = 2$

44. $\sqrt[3]{1 - 5x} = 6$ 45. $\sqrt{\dfrac{x}{x + 2}} = 4$ 46. $\sqrt[3]{\dfrac{x - 2}{x + 1}} = -2$

47. $\sqrt{x^2 + 4} = x + 2$ 48. $\sqrt{x^2 + 9} = x - 3$

49. $\sqrt{2x + 1} = x - 1$ 50. $\sqrt{x} = 12 - x$

EXAMPLE B (Combining fractions involving radicals) Sums and differences of fractions involving radicals occur often in calculus. It is usually desirable to combine these fractions. Do so in

(a) $\dfrac{1}{\sqrt[3]{x + h}} - \dfrac{1}{\sqrt[3]{x}}$;   (b) $\dfrac{x}{\sqrt{x^2 + 4}} - \dfrac{\sqrt{x^2 + 4}}{x}$.

Solution.

(a) $\dfrac{1}{\sqrt[3]{x + h}} - \dfrac{1}{\sqrt[3]{x}} = \dfrac{\sqrt[3]{x}}{\sqrt[3]{x}\sqrt[3]{x + h}} - \dfrac{\sqrt[3]{x + h}}{\sqrt[3]{x}\sqrt[3]{x + h}} = \dfrac{\sqrt[3]{x} - \sqrt[3]{x + h}}{\sqrt[3]{x}\sqrt[3]{x + h}}$

(b) $\dfrac{x}{\sqrt{x^2 + 4}} - \dfrac{\sqrt{x^2 + 4}}{x} = \dfrac{x^2}{x\sqrt{x^2 + 4}} - \dfrac{\sqrt{x^2 + 4}\sqrt{x^2 + 4}}{x\sqrt{x^2 + 4}}$

$\qquad\qquad\qquad\qquad = \dfrac{x^2 - (x^2 + 4)}{x\sqrt{x^2 + 4}} = \dfrac{-4}{x\sqrt{x^2 + 4}}$

Combine the fractions in each of the following. Do not bother to rationalize denominators.

51. $\dfrac{2}{\sqrt{x + h}} - \dfrac{2}{\sqrt{x}}$ 52. $\dfrac{\sqrt{x}}{\sqrt{x + 2}} - \dfrac{1}{\sqrt{x}}$

53. $\dfrac{1}{\sqrt{x + 6}} + \sqrt{x + 6}$ 54. $\dfrac{\sqrt{x + 1}}{\sqrt{x + 3}} - \dfrac{\sqrt{x + 3}}{x + 1}$

55. $\dfrac{\sqrt[3]{(x + 2)^2}}{2} - \dfrac{1}{\sqrt[3]{x + 2}}$ 56. $\dfrac{\sqrt{x + 7}}{\sqrt{x - 2}} - \dfrac{\sqrt{x - 2}}{x + 7}$

57. $\dfrac{1}{\sqrt{x^2 + 9}} - \dfrac{\sqrt{x^2 + 9}}{x^2}$ 58. $\dfrac{x}{\sqrt{x^2 + 3}} + \dfrac{\sqrt{x^2 + 3}}{x}$

Miscellaneous Problems

Simplify the following. Assume that all letters represent positive numbers.

59. $(\sqrt[3]{2})^3$ 60. $\sqrt[3]{b^3}$ 61. $\sqrt[4]{16y^8}$

62. $\sqrt{\dfrac{9a^2}{b^4}}$ 63. $\sqrt[7]{\left(\dfrac{-3x}{y}\right)^7}$ 64. $\sqrt{27}\sqrt{3}$

65. $\sqrt{54b^3c^4}$ 　　　　66. $\sqrt[3]{250x^5y^9}$ 　　　　67. $\sqrt[4]{x^4 + x^8y^4}$

68. $\dfrac{2}{\sqrt{a} - 1}$ 　　　　69. $\dfrac{2}{\sqrt{a - 1}}$ 　　　　70. $\dfrac{1}{\sqrt{7bc^3}}$

71. $\sqrt{a}\left(\sqrt{a} + \dfrac{1}{a\sqrt{a}}\right)$ 　　72. $\sqrt{12} + \sqrt{48} - \sqrt{27}$

Solve the equations in Problems 73–78.

73. $\sqrt{3x - 2} = 4$ 　　　　74. $\sqrt{2 - 3x} = \sqrt{5}$ 　　　　75. $\sqrt[3]{1 - 5x} = -4$

76. $\sqrt{x + 2} = \sqrt{2x - 3}$ 　　　77. $\sqrt{4x + 1} = x + 1$ 　　　78. $2\sqrt{x} = 3x - 1$

C 79. Taking a root is the inverse of raising to a power. Thus, on some calculators, you may take roots by using the $\boxed{\text{INV}}$ and $\boxed{y^x}$ keys. For example, to find $\sqrt[4]{31}$, press 31 $\boxed{\text{INV}}$ $\boxed{y^x}$ 4 $\boxed{=}$. You should get 2.3596111. Use your calculator to find each of the following.

(a) $\sqrt[5]{87}$ 　　(b) $\sqrt[3]{213}$ 　　(c) $\sqrt[50]{100}$ 　　(d) $\sqrt[8]{390,625}$

C 80. Calculate.

(a) $\dfrac{4}{\sqrt[6]{11}}$ 　　(b) $\sqrt[5]{12} - \sqrt[3]{2}$ 　　(c) $\dfrac{\sqrt[4]{29} + \sqrt[3]{6}}{\sqrt[3]{14}}$

(d) $\sqrt[8]{.012}(\sqrt{130} + 4)^2$

81. Recall that for any real x (positive or negative) $\sqrt{x^2} = |x|$. Use this to simplify each expression.

(a) $\sqrt{x^4 + 4x^2}$ 　　(b) $\sqrt{x^{-2}y^5}$

82. Show that, if x is any real number and n is even, $\sqrt[n]{x^n} = |x|$.

83. We know that $f(x) = x^5$ and $g(x) = \sqrt[5]{x}$ are inverse functions. What does this mean about their graphs? Draw both graphs in the same coordinate plane.

84. Draw the graphs of $f(x) = x^4$ and $g(x) = \sqrt[4]{x}$ for $x \geq 0$ in the same coordinate plane.

85. Use algebra to solve $\sqrt{x} \leq \sqrt[3]{x}$. (*Hint:* To what power can you raise both sides to clear the equation of radicals?)

86. Show that there are only two integral values of x between 1 and 1000 for which $\sqrt{x}$ and $\sqrt[3]{x}$ are both integers.

C 87. Here is a well-known iterative method for finding the square root of a positive number A to any desired degree of accuracy. Use the formulas

$$x_n = \frac{x_{n-1} + A/x_{n-1}}{2} \qquad n = 2, 3, 4, \ldots$$

where x_1 is a first approximation to $\sqrt{A}$. As an example, let $A = 37$ and take $x_1 = 6$. Then

$$x_2 = \frac{x_1 + 37/x_1}{2} = \frac{6 + 37/6}{2} \approx 6.08$$

$$x_3 = \frac{x_2 + 37/x_2}{2} = \frac{6.08 + 37/6.08}{2} \approx 6.083$$

and so on. Find the second and third approximations, x_2 and x_3, of each of the following square roots using the given first approximation x_1.

(a) $\sqrt{15}$, $x_1 = 4$ (b) $\sqrt{40}$, $x_1 = 6$ (c) $\sqrt{40}$, $x_1 = 7$

88. Show that $\sqrt[3]{2}$ is a solution of the equation

$$2x^9 - 4x^6 + 5x^3 - 10 = 0$$

89. Show that

$$\frac{1/\sqrt{x + h} - 1/\sqrt{x}}{h} = \frac{-1}{\sqrt{x + h}\sqrt{x}(\sqrt{x} + \sqrt{x + h})}$$

90. Show that

$$\frac{\sqrt{x + h} - \sqrt{x}}{h} = \frac{1}{\sqrt{x + h} + \sqrt{x}}$$

Some Sense—Some Nonsense

One of the authors once asked a student to write the definition of $2^{4.6}$ on the blackboard. After thinking deeply for a minute, he wrote:

$$2^2 = 2 \cdot 2$$
$$2^3 = 2 \cdot 2 \cdot 2$$
$$2^4 = 2 \cdot 2 \cdot 2 \cdot 2$$
$$2^{4.6} = 2 \cdot 2 \cdot 2 \cdot 2 \cdot \zeta$$

6-2
Exponents and Exponential Functions

After you have criticized the student mentioned above, ask yourself how you would define $2^{4.6}$. Of course, integral powers of 2 make perfectly good sense, although 2^{-3} and 2^0 became meaningful only after we had *defined* a^{-n} to be $1/a^n$ and a^0 to be 1 (see Section 2-1). Those were good definitions because they were consistent with the familiar rules of exponents. Now we ask what meaning we can give to powers like $2^{1/2}$, $2^{4.6}$, and even 2^π so that these familiar rules still hold.

RATIONAL EXPONENTS

We assume throughout this section that $a > 0$. If n is any positive integer, we want

$$(a^{1/n})^n = a^{(1/n)\cdot n} = a^1 = a$$

But we know that $(\sqrt[n]{a})^n = a$. Thus we define

$$a^{1/n} = \sqrt[n]{a}$$

Rules for Exponents

1. $a^m a^n = a^{m+n}$

2. $\dfrac{a^m}{a^n} = a^{m-n}$

3. $(a^m)^n = a^{mn}$

For example, $2^{1/2} = \sqrt{2}$, $27^{1/3} = \sqrt[3]{27} = 3$, and $(16)^{1/4} = \sqrt[4]{16} = 2$.

Next, if m and n are positive integers, we want

$$(a^{1/n})^m = a^{m/n} \quad \text{and} \quad (a^m)^{1/n} = a^{m/n}$$

This forces us to define

$$a^{m/n} = (\sqrt[n]{a})^m = \sqrt[n]{a^m}$$

Accordingly,

$$2^{3/2} = (\sqrt{2})^3 = \sqrt{2}\sqrt{2}\sqrt{2} = 2\sqrt{2}$$

and

$$27^{2/3} = (\sqrt[3]{27})^2 = 3^2 = 9$$

Lastly, we define

$$a^{-m/n} = \frac{1}{a^{m/n}}$$

so that

$$2^{-1/2} = \frac{1}{2^{1/2}} = \frac{1}{\sqrt{2}}$$

and

$$4^{-3/2} = \frac{1}{4^{3/2}} = \frac{1}{(\sqrt{4})^3} = \frac{1}{8}$$

We have just succeeded in defining a^x for all rational numbers x (recall that a rational number is a ratio of two integers). What is more important is that we have done it in such a way that the rules of exponents still hold. Incidentally, we can now answer the question in our opening display.

$$2^{4.6} = 2^4 2^{.6} = 2^4 2^{6/10} = 16(\sqrt[10]{2})^6$$

For simplicity, we have assumed that a is positive in our discussion of $a^{m/n}$. But we should point out that the definition of $a^{m/n}$ given above is also appropriate for the case in which a is negative and n is odd. For example,

$$(-27)^{2/3} = (\sqrt[3]{-27})^2 = (-3)^2 = 9$$

REAL EXPONENTS

Irrational powers such as 2^π and $3^{\sqrt{2}}$ are intrinsically more difficult to define than are rational powers. Rather than attempt a technical definition, we ask you to consider what 2^π might mean. The decimal expansion of π is 3.14159 . . . . Thus we could look at the sequence of rational powers

$$2^3,\ 2^{3.1},\ 2^{3.14},\ 2^{3.141},\ 2^{3.1415},\ 2^{3.14159},\ \ldots$$

As you should suspect, when the exponents get closer and closer to π, the corresponding powers of 2 get closer and closer to a definite number. We shall call the number 2^π.

The process of starting with integral exponents and then extending to rational exponents and finally to real exponents can be clarified by means of three graphs. Note the table of values in the margin.

x	2^x
-3	$\frac{1}{8}$
-2	$\frac{1}{4}$
-1	$\frac{1}{2}$
0	1
$\frac{1}{2}$	$\sqrt{2} \approx 1.4$
1	2
$\frac{3}{2}$	$2\sqrt{2} \approx 2.8$
2	4
3	8

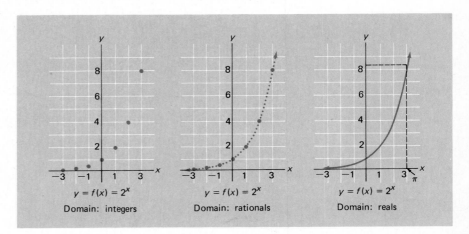

$y = f(x) = 2^x$
Domain: integers

$y = f(x) = 2^x$
Domain: rationals

$y = f(x) = 2^x$
Domain: reals

The first graph suggests a curve rising from left to right. The second graph makes the suggestion stronger. The third graph leaves nothing to the imagination; it is a continuous curve and it shows 2^x for all values of x, rational and irrational. As x increases in the positive direction, the values of 2^x increase without bound; in the negative direction, the values of 2^x approach 0. Notice that 2^π is a little less than 9; its value correct to seven decimal places is

$$2^\pi = 8.8249778$$

See if your calculator gives this value.

EXPONENTIAL FUNCTIONS

The function $f(x) = 2^x$, graphed above, is one example of an exponential function. But what has been done with 2 can be done with any positive real number a. In general, the formula

$$f(x) = a^x$$

determines a function called an **exponential function with base a.** Its domain is the set of all real numbers and its range is the set of positive numbers.

Let us see what effect the size of a has on the graph of $f(x) = a^x$. We choose $a = 3$ and $a = \frac{1}{3}$, showing both graphs below.

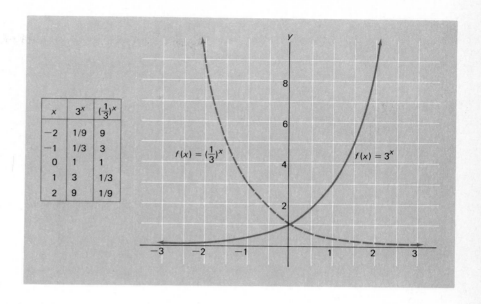

The graph of $f(x) = 3^x$ looks much like the graph of $f(x) = 2^x$, although it rises more rapidly. The graph of $f(x) = 10^x$ would be even steeper. All three functions are *increasing functions,* meaning that the values of $f(x)$ increase as x increases; more formally, $x_2 > x_1$ implies $f(x_2) > f(x_1)$. The function $f(x) = (\frac{1}{3})^x$, on the other hand, is a *decreasing function.* In fact, you can get the graph of $f(x) = (\frac{1}{3})^x$ by reflecting the graph of $f(x) = 3^x$ about the y-axis. This is because $(\frac{1}{3})^x = 3^{-x}$.

We can summarize what is suggested by our discussion as follows.

If $a > 1$, $f(x) = a^x$ is an increasing function.
If $0 < a < 1$, $f(x) = a^x$ is a decreasing function.

In both of these cases, the graph of f has the x-axis as an asymptote. The case $a = 1$ is not very interesting since it yields the constant function $f(x) = 1$.

PROPERTIES OF EXPONENTIAL FUNCTIONS

It is easy to describe the main properties of exponential functions, since they obey the rules we learned in Section 2-1. Perhaps it is worth repeating them, since we do want to emphasize that they now hold for all *real* exponents x and y (at least for the case where a and b are both positive).

$$1.\ a^x a^y = a^{x+y}$$

$$2.\ \frac{a^x}{a^y} = a^{x-y}$$

$$3.\ (a^x)^y = a^{xy}$$

$$4.\ (ab)^x = a^x b^x$$

$$5.\ \left(\frac{a}{b}\right)^x = \frac{a^x}{b^x}$$

Here are a number of examples that are worth studying.

$$3^{1/2} 3^{3/4} = 3^{1/2+3/4} = 3^{5/4}$$

$$\frac{\pi^4}{\pi^{5/2}} = \pi^{4-5/2} = \pi^{3/2}$$

$$(2^{\sqrt{3}})^4 = 2^{4\sqrt{3}}$$

$$(5^{\sqrt{2}} 5^{1-\sqrt{2}})^{-3} = (5^1)^{-3} = 5^{-3}$$

Problem Set 6-2

Write each of the following as a power of 7.

1. $\sqrt[3]{7}$ 2. $\sqrt[5]{7}$ 3. $\sqrt[3]{7^2}$ 4. $\sqrt[5]{7^3}$ 5. $\dfrac{1}{\sqrt[3]{7}}$

6. $\dfrac{1}{\sqrt[5]{7}}$ 7. $\dfrac{1}{\sqrt[3]{7^2}}$ 8. $\dfrac{1}{\sqrt[3]{7^3}}$ 9. $7\sqrt[3]{7}$ 10. $7\sqrt[5]{7}$

Rewrite each of the following using exponents instead of radicals. For example, $\sqrt[5]{x^3} = x^{3/5}$.

11. $\sqrt[3]{x^2}$ 12. $\sqrt[4]{x^3}$ 13. $x^2\sqrt{x}$ 14. $x\sqrt[3]{x}$
15. $\sqrt{(x+y)^3}$ 16. $\sqrt[3]{(x+y)^2}$ 17. $\sqrt{x^2+y^2}$ 18. $\sqrt[3]{x^3+8}$

Rewrite each of the following using radicals instead of fractional exponents. For example, $(xy^2)^{3/7} = \sqrt[7]{x^3 y^6}$

19. $4^{2/3}$ 20. $10^{3/4}$ 21. $8^{-3/2}$
22. $12^{-5/6}$ 23. $(x^4 + y^4)^{1/4}$ 24. $(x^2 + xy)^{1/2}$
25. $(x^2 y^3)^{2/5}$ 26. $(3ab^2)^{2/3}$ 27. $(x^{1/2} + y^{1/2})^{1/2}$
28. $(x^{1/3} + y^{2/3})^{1/3}$

Simplify each of the following. Give your answer without any exponents.

29. $25^{1/2}$ 30. $27^{1/3}$
31. $8^{2/3}$ 32. $16^{3/2}$

33. $9^{-3/2}$

34. $64^{-2/3}$

35. $(-.008)^{2/3}$

36. $(-.027)^{5/3}$

37. $(.0025)^{3/2}$

38. $(1.44)^{3/2}$

39. $5^{2/3}5^{-5/3}$

40. $4^{3/4}4^{-1/4}$

41. $16^{7/6}16^{-5/6}16^{-4/3}$

42. $9^2 9^{2/3} 9^{-7/6}$

43. $(8^2)^{-2/3}$

44. $(4^{-3})^{3/2}$

EXAMPLE A (Simplifying expressions involving exponents) Simplify and write the answer without negative exponents.

(a) $\dfrac{x^{1/3}(8x)^{-2/3}}{x^{-3/4}}$

(b) $\left(\dfrac{2x^{-1/2}}{y}\right)^4 \left(\dfrac{x}{y}\right)^{-1}(3x^{10/3})$

Solution.

(a) $\dfrac{x^{1/3}(8x)^{-2/3}}{x^{-3/4}} = x^{1/3}8^{-2/3}x^{-2/3}x^{3/4} = \dfrac{x^{1/3-2/3+3/4}}{8^{2/3}} = \dfrac{x^{5/12}}{4}$

(b) $\left(\dfrac{2x^{-1/2}}{y}\right)^4 \left(\dfrac{x}{y}\right)^{-1}(3x^{10/3}) = \left(\dfrac{16x^{-2}}{y^4}\right)\left(\dfrac{y}{x}\right)(3x^{10/3})$

$$= \dfrac{48x^{-2-1+10/3}}{y^{4-1}} = \dfrac{48x^{1/3}}{y^3}$$

Simplify, writing your answer without negative exponents.

45. $(3a^{1/2})(-2a^{3/2})$

46. $(2x^{3/4})(5x^{-3/4})$

47. $(2^{1/2}x^{-2/3})^6$

48. $(\sqrt{3}x^{-1/4}y^{3/4})^4$

49. $(xy^{-2/3})^3(x^{1/2}y)^2$

50. $(a^2b^{-1/4})^2(a^{-1/3}b^{1/2})^3$

51. $\dfrac{(2x^{-1}y^{2/3})^2}{x^2y^{-2/3}}$

52. $\left(\dfrac{a^{1/2}b^{1/3}}{c^{5/6}}\right)^{12}$

53. $\left(\dfrac{x^{-2}y^{3/4}}{x^{1/2}}\right)^{12}$

54. $\dfrac{x^{1/3}y^{-3/4}}{x^{-2/3}y^{1/2}}$

55. $y^{2/3}(2y^{4/3} - y^{-5/3})$

56. $x^{-3/4}\left(-x^{7/4} + \dfrac{2}{\sqrt[4]{x}}\right)$

57. $(x^{1/2} + y^{1/2})^2$

58. $(a^{3/2} + \pi)^2$

EXAMPLE B (Combining fractions) Perform the following addition.

$$\dfrac{(x + 1)^{2/3}}{x} + \dfrac{1}{(x + 1)^{1/3}}$$

Solution.

$$\dfrac{(x + 1)^{2/3}}{x} + \dfrac{1}{(x + 1)^{1/3}} = \dfrac{(x + 1)^{2/3}(x + 1)^{1/3}}{x(x + 1)^{1/3}} + \dfrac{x}{x(x + 1)^{1/3}}$$

$$= \dfrac{x + 1 + x}{x(x + 1)^{1/3}} = \dfrac{2x + 1}{x(x + 1)^{1/3}}$$

Combine the fractions in each of the following.

59. $\dfrac{(x + 2)^{4/5}}{3} + \dfrac{2x}{(x + 2)^{1/5}}$

60. $\dfrac{(x - 3)^{1/3}}{4} - \dfrac{1}{(x - 3)^{2/3}}$

61. $(x^2 + 1)^{1/3} - \dfrac{2x^2}{(x^2 + 1)^{2/3}}$

62. $(x^2 + 2)^{1/4} + \dfrac{x^2}{(x^2 + 2)^{3/4}}$

EXAMPLE C (Mixing radicals of different orders) Express $\sqrt{2}\,\sqrt[3]{5}$ using just one radical.

Solution. Square roots and cube roots mix about as well as oil and water, but exponents can serve as a blender. They allow us to write both $\sqrt{2}$ and $\sqrt[3]{5}$ as sixth roots.

$$\sqrt{2}\,\sqrt[3]{5} = 2^{1/2} \cdot 5^{1/3}$$
$$= 2^{3/6} \cdot 5^{2/6}$$
$$= (2^3 \cdot 5^2)^{1/6}$$
$$= \sqrt[6]{200}$$

Express each of the following in terms of at most one radical in simplest form.

63. $\sqrt{2}\,\sqrt[3]{2}$ 64. $\sqrt[3]{2}\,\sqrt[4]{2}$ 65. $\sqrt[4]{2}\,\sqrt[6]{x}$

66. $\sqrt[3]{5}\,\sqrt{x}$ 67. $\sqrt[3]{x}\sqrt{x}$ 68. $\sqrt{x}\sqrt[3]{x}$

© *Use a calculator to find an approximate value of each of the following.*

69. $2^{1.34}$ 70. $2^{-.79}$ 71. $\pi^{1.34}$ 72. π^{π}

73. $(1.46)^{\sqrt{2}}$ 74. $\pi^{\sqrt{2}}$ 75. $(.9)^{50.2}$ 76. $(1.01)^{50.2}$

Sketch the graph of each of the following functions.

77. $f(x) = 4^x$ 78. $f(x) = 4^{-x}$ 79. $f(x) = (\tfrac{2}{3})^x$

80. $f(x) = (\tfrac{2}{3})^{-x}$ 81. $f(x) = \pi^x$ 82. $f(x) = (\sqrt{2})^x$

Miscellaneous Problems

Rewrite each of the following using exponents instead of radicals.

83. $\sqrt[3]{a^2}$ 84. $\sqrt[6]{x^3}$ 85. $\sqrt[5]{(a + 2b)^3}$ 86. $\sqrt{x^2 + 9}$

Rewrite using radicals instead of exponents.

87. $6^{3/4}$ 88. $12^{-2/3}$ 89. $(4 + x^{1/2})^{-1/2}$ 90. $(16a^2 b^3)^{3/4}$

Simplify.

91. $8^{1/3}$ 92. $8^{2/3}$

93. $8^{-2/3}$ 94. $(.04)^{-1/2}$

95. $(.125)^{4/3}$

96. $7^{2/3}7^{1/2}7^{-1/6}$

97. $5^{10}[(25)^3]^{-3/2}$

98. $(y^2 z^{-4})^{1/2}$

99. $x^{1/4}(x^{-5/4} + x^{3/4})$

100. $(x^{3/2} + x^{-3/2})^2$

Sketch the graphs of the functions in 101–104.

101. $f(x) = 5^x$

102. $f(x) = (\frac{1}{5})^x$

[c] 103. $f(x) = (.9)^x$

[c] 104. $f(x) = (1.1)^x$

105. Rank the numbers $5^{5/3}$, $5^{7/4}$, and $5^{\sqrt{3}}$ from least to greatest.

106. How would you define $2^{\sqrt{2}}$?

107. Solve $x^{2/3} - 3x^{1/3} + 2 = 0$.

A Packing Problem

World population is growing at about 2 percent per year. If this continues indefinitely, how long will it be until we are all packed together like sardines? In answering, assume that "sardine packing" for humans is one person per square foot of land area.

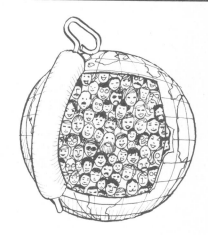

6-3
Exponential Growth and Decay

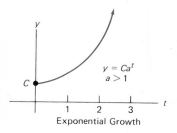

$y = Ca^t$
$a > 1$

Exponential Growth

The phrase *exponential growth* is used repeatedly by professors, politicians, and pessimists. Population, energy use, mining of ores, pollution, and the number of books about these things are all said to be growing exponentially. Most people probably do not know what exponential growth means, except that they have heard it guarantees alarming consequences. For students of this book, it is easy to explain its meaning. For y to grow exponentially with time t means that it satisfies the relationship

$$y = Ca^t$$

for constants C and a, with $C > 0$ and $a > 1$. Why should so many ingredients of modern society behave this way? The basic cause is population growth.

POPULATION GROWTH

Simple organisms reproduce by cell division. If, for example, there is one cell today, that cell may split so that there are two cells tomorrow. Then each of

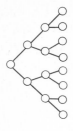

those cells may divide giving four cells the following day. As this process continues, the numbers of cells on successive days form the sequence

$$1, 2, 4, 8, 16, 32, \ldots$$

If we start with 100 cells and let $f(t)$ denote the number present t days from now, we have the results indicated in the table below.

t	0	1	2	3	4	5
$f(t)$	100	200	400	800	1600	3200

It seems that

$$f(t) = (100)2^t$$

A perceptive reader will ask if this formula is really valid. Does it give the right answer when $t = 5.7$? Is not population growth a discrete process, occurring in unit amounts at distinct times, rather than a continuous process as the formula implies? The answer is that the exponential growth model provides a very good approximation to the growth of simple organisms, provided the initial population is large.

The mechanism of reproduction is different (and more interesting) for people, but the pattern of population growth is similar. World population is presently growing at about 2 percent per year. In 1975, there were about 4 billion people. Accordingly, the population in 1976 in billions was $4 + 4(.02) = 4(1.02)$, in 1977 it was $4(1.02)^2$, in 1978 it was $4(1.02)^3$, and so on. If this trend continues, there will be $4(1.02)^{20}$ billion people in the world in 1995, that is, 20 years after 1975. It appears that world population obeys the formula

$$p(t) = 4(1.02)^t$$

where $p(t)$ represents the number of people (in billions) t years after 1975.

In general, if $A(t)$ is the amount at time t of a quantity growing exponentially at the rate of r percent per year, then

$$A(t) = A(0)\left(1 + \frac{r}{100}\right)^t$$

Here $A(0)$ is the initial amount—that is, the amount present at $t = 0$.

DOUBLING TIMES

One way to get a feeling for the spectacular nature of exponential growth is via the concept of **doubling time;** this is the length of time required for an exponentially growing quantity to double in size. It is easy to show that if a quantity doubles in an initial time interval of length T, it will double in size in

t	$(1.02)^t$
5	1.104
10	1.219
15	1.346
20	1.486
25	1.641
30	1.811
35	2.000
40	2.208
45	2.438
50	2.692
55	2.972
60	3.281
65	3.623
70	4.000
75	4.416
80	4.875
85	5.383
90	5.943

any time interval of length T. Consider the world population problem as an example. By the table in the margin, $(1.02)^{35} \approx 2$, so world population doubles in 35 years. Since it is 4 billion in 1975, it should be 8 billion in 2010, 16 billion in 2045, and so on. This alarming information is displayed on the graph below.

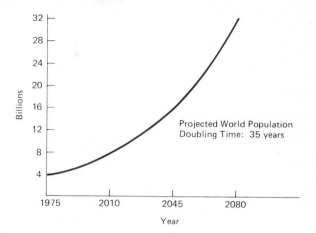

Now we can answer the question about sardine packing in our opening display. There are slightly more than 1,000,000 billion square feet of land area on the surface of the earth. Sardine packing for humans is about 1 square foot per person. Thus we are asking when $4(1.02)^t$ billion will equal 1,000,000 billion. This leads to the equation

$$(1.02)^t = 250,000$$

To solve this exponential equation, we use the following approximations

$$(1.02)^{35} \approx 2 \qquad 250,000 \approx 2^{18}$$

Our equation can then be rewritten as

$$[(1.02)^{35}]^{t/35} = 2^{18}$$

or

$$2^{t/35} = 2^{18}$$

We conclude that

$$\frac{t}{35} = 18$$

$$t = (18)(35) = 630$$

Thus, after about 630 years, we will be packed together like sardines. If it is any comfort, war, famine, or birth control will change population growth patterns before then.

COMPOUND INTEREST

One of the best practical illustrations of exponential growth is money earning compound interest. Suppose that Amy puts $1000 in a bank today at 8 percent interest compounded annually. Then at the end of one year the bank adds the interest of (.08)(1000) = $80 to her $1000, giving her a total of $1080. But note that 1080 = 1000(1.08). During the second year, $1080 draws interest. At the end of that year, the bank adds (.08)(1080) to the account, bringing the total to

$$1080 + (.08)(1080) = (1080)(1.08)$$
$$= 1000(1.08)(1.08)$$
$$= 1000(1.08)^2$$

Continuing in this way, we see that Amy's account will have grown to $1000(1.08)^3$ by the end of 3 years, $1000(1.08)^4$ by the end of 4 years, and so on. By the end of 15 years, it will have grown to

$$1000(1.08)^{15} \approx 1000(3.172169)$$
$$= \$3172.17$$

To calculate $(1.08)^{15}$, we used the table at the end of the problem set. We could have used a calculator.

How long would it take for Amy's money to double—that is, when will

$$1000(1.08)^t = 2000$$

This will occur when $(1.08)^t = 2$. According to the table just mentioned, this happens at $t \approx 9$, or in about 9 years.

EXPONENTIAL DECAY

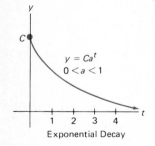

$y = Ca^t$
$0 < a < 1$

1 2 3 4

Exponential Decay

Fortunately, not all things grow; some decline or decay. In fact, some things—notably the radioactive elements—decay exponentially. This means that the amount y present at time t satisfies

$$y = Ca^t$$

for some constants C and a with $C > 0$ and $0 < a < 1$.

Here an important idea is that of **half-life,** the time required for half of a substance to disappear. For example, radium decays with a half-life of 1620 years. Thus if 1000 grams of radium are present now, 1620 years from now 500 grams will be present, 2(1620) = 3240 years from now only 250 grams will be present, and so on.

The precise nature of radioactive decay is used to date old objects. If an object contains radium and lead (the product to which radium decays) in the ratio 1 to 3, then it is believed that an original amount of pure radium has decayed to $\frac{1}{4}$ its original size. The object must be two half-lives, or 3240 years, old. Two important assumptions have been made: (1) decay of radium is exactly exponential over long periods of time; and (2) no lead was originally present. Recent research raises some question about the correctness of such assumptions.

Problem Set 6-3

1. In each of the following, indicate whether y grows exponentially or decays exponentially with t.
   (a) $y = 128\left(\dfrac{1}{2}\right)^t$ (b) $y = 5\left(\dfrac{5}{3}\right)^t$
   (c) $y = 4(10)^9(1.03)^t$ (d) $y = 1000(.99)^t$

2. Find the values of y corresponding to $t = 0$, $t = 1$, and $t = 2$ for each case in Problem 1.

3. Use the table at the end of the problem set or a calculator with a $\boxed{y^x}$ key to find each value.
   (a) $(1.08)^{20}$ (b) $(1.12)^{25}$ (c) $1000(1.04)^{40}$ (d) $2000(1.02)^{80}$

4. Evaluate
   (a) $(1.01)^{100}$ (b) $(1.02)^{40}$ (c) $100(1.12)^{50}$ (d) $500(1.04)^{30}$

5. Silver City's present population of 1000 is expected to grow exponentially over the next 10 years at 4 percent per year. How many people will it have at the end of that time? (*Hint:* $A(t) = A(0)[1 + (r/100)]^t$.)

6. The value of houses in Longview is said to be growing exponentially at 12 percent per year. What will a house valued at $100,000 today be worth after 8 years?

7. Under the assumptions concerning world population used in this section, what will be the approximate number of people on earth in each year?
   (a) 1990 (that is, 15 years after 1975)
   (b) 2000
   (c) 2065

8. A certain radioactive substance has a half-life of 40 minutes. What fraction of an initial amount of this substance will remain after 1 hour, 20 minutes (that is, after 2 half-lives)? After 2 hours, 40 minutes?

EXAMPLE A (Compound interest)   Roger put $1000 in a money market fund at 15 percent interest compounded annually. How much was it worth after 4 years?

Solution.   We will have to use a calculator since $(1.15)^n$ is not in our table. The answer is

$$1000(1.15)^4 = \$1749.01$$

9. If you put $100 in the bank for 8 years, how much will it be worth at the end of that time at
   (a) 8 percent compounded annually;
   (b) 12 percent compounded annually?

10. If you invest $500 in the bank today, how much will it be worth after 25 years at
    (a) 8 percent compounded annually;

(b) 4 percent compounded annually;

(c) 12 percent compounded annually?

11. If you put $3500 in the bank today, how much will it be worth after 40 years at

    (a) 8 percent compounded annually;

    (b) 12 percent compounded annually?

12. Approximately how long will it take for money to accumulate to twice its value if

    (a) it is invested at 8 percent compounded annually;

    (b) it is invested at 12 percent compounded annually?

13. Suppose that you invest p dollars at r percent compounded annually. Write an expression for the amount accumulated after n years.

EXAMPLE B (More on compound interest)   If $1000 is invested at 8 percent compounded quarterly, find the accumulated amount after 15 years.

Solution.   Interest calculated at 2 percent ($\frac{1}{4}$ of 8 percent) is converted to principal every 3 months. By the end of the first 3-month period, the account has grown to $1000(1.02) = \$1020$; by the end of the second 3-month period, it has grown to $1000(1.02)^2$; and so on. The accumulated amount after 15 years, or 60 conversion periods, is

$$1000(1.02)^{60} \approx 1000(3.28103)$$

$$= \$3281.03$$

Suppose more generally that P dollars is invested at a rate r (written as a decimal), which is compounded m times per year. Then the accumulated amount A after t years is given by

$$A = P\left(1 + \frac{r}{m}\right)^{tm}$$

In our example

$$A = 1000\left(1 + \frac{.08}{4}\right)^{15 \cdot 4} = 1000(1.02)^{60}$$

Find the accumulated amount for the indicated initial principal, compound interest rate, and total time period.

14. $2000; 8 percent compounded annually; 15 years

15. $5000; 8 percent compounded semiannually; 5 years

16. $5000; 12 percent compounded monthly; 5 years

$\boxed{c}$ 17. $3000; 9 percent compounded annually; 10 years

$\boxed{c}$ 18. $3000; 9 percent compounded semiannually; 10 years

$\boxed{c}$ 19. $3000; 9 percent compounded quarterly; 10 years

$\boxed{c}$ 20. $3000; 9 percent compounded monthly; 10 years

$\boxed{c}$ 21. $1000; 8 percent compounded monthly; 10 years

$1000; 8 percent compounded daily; 10 years   (*Hint:* Assume there are 365 days in a year, so that the interest rate per day is .08/365.)

23. If $y = 1600(\frac{3}{4})^t$, find the values of y corresponding to $t = 0$, $t = 1$, $t = 2$, and $t = 3$.

24. If $f(t) = 2500(\frac{6}{5})^t$, find $f(0)$, $f(1)$, $f(2)$, $f(3)$, and $f(4)$.

25. If $(1.034)^T = 2$, find the value of $3000(1.034)^{3T}$. (*Hint:* $a^{3T} = (a^T)^3$.)

26. If $(.63)^H = \frac{1}{2}$, find the value of $640(.63)^{4H}$.

27. If $4000 is invested today, how much will it be worth after 10 years at 8 percent interest if interest is
    (a) compounded annually;
    (b) compounded quarterly;
    ☐ (c) compounded monthly;
    ☐ (d) compounded daily?
    (*Hint:* In part (d), the interest rate per day is .08/365.)

28. If $2500 is invested today, how much will it be worth after 5 years at 12 percent interest if interest is
    (a) compounded annually;
    (b) compounded monthly;
    ☐ (c) compounded daily?

☐ 29. Find the accumulated amount for the indicated initial principal, compound interest rate, and total time period.
    (a) $2500; 11.25 percent compounded quarterly; 4 years and 6 months.
    (b) $4175; 10.76 percent compounded monthly; 11 years and 9 months.

☐ 30. Suppose that the population of a certain city obeys the formula

$$p(t) = 4600(1.016)^t$$

where $p(t)$ is the population t years after 1980.
    (a) What is the population in 2020? In 2080?
    (b) What is the doubling time for this population? (*Hint:* Use the trial-and-error method, beginning with $t = 50$.)

31. The number of bacteria in a certain culture triples every hour. Suppose that the count at 12:00 noon is 162,000. What was the count at 11:00 A.M.? at 10:00 A.M.? at 8:00 A.M.?

32. Assuming that the half-life of a certain radioactive substance is 1690 years, what fraction of an initial amount of the substance will remain after 3380 years? After 5070 years?

33. Assume that the number of bacteria in a certain culture t hours from now is given by $Q(t) = 5000(2^t)$. Find the number of bacteria
    (a) at the present time;   (b) after 30 minutes;
    (c) after 3 hours;   (d) after 90 minutes.

34. Assume in Problem 32 that the quantity of the substance in milligrams after t years is given by $q(t) = 30(\frac{1}{2})^{kt}$.
    (a) Find the value of k. (Use the half-life.)
    ☐ (b) How much of the substance will remain after 2385 years?

35. Carbon 14 has a half-life of 5720 years. If initially there were 10 grams, how much will there be after 17,160 years?

36. About how long does it take a population to double if it is growing at
    (a) 1 percent per year:
    (b) 4 percent per year;
    [c] (c) 1.5 percent per year;
    [c] (d) .5 percent per year?

[c] 37. If $f(t) = 1000(.983)^t$ is the formula for an exponential decay, find
    (a) $f(10)$;   (b) $f(40)$;   (c) the approximate half-life.

[c] 38. If $1 is invested at 10 percent compounded annually, in how many years will it accumulate to $1,000,000,000? (*Hint:* Use the trial-and-error method, beginning with $t = 1000$.)

[c] 39. The amount A (in dollars) of the average life insurance coverage of families in the United States satisfies (approximately)

$$A = 7,040(1.074)^t$$

where t is the number of years after 1955. Assume that is valid for the future and find the average coverage of a family in 1995.

TABLE 2   Compound Interest Table

n	$(1.01)^n$	$(1.02)^n$	$(1.04)^n$	$(1.08)^n$	$(1.12)^n$
1	1.01000000	1.02000000	1.04000000	1.08000000	1.12000000
2	1.02010000	1.04040000	1.08160000	1.16640000	1.25440000
3	1.03030100	1.06120800	1.12486400	1.25971200	1.40492800
4	1.04060401	1.08243216	1.16985856	1.36048896	1.57351936
5	1.05101005	1.10408080	1.21665290	1.46932808	1.76234168
6	1.06152015	1.12616242	1.26531902	1.58687432	1.97382269
7	1.07213535	1.14868567	1.31593178	1.71382427	2.21068141
8	1.08285671	1.17165938	1.36856905	1.85093021	2.47596318
9	1.09368527	1.19509257	1.42331181	1.99900463	2.77307876
10	1.10462213	1.21899442	1.48024428	2.15892500	3.10584821
11	1.11566835	1.24337431	1.53945406	2.33163900	3.47854999
12	1.12682503	1.26824179	1.60103222	2.51817012	3.89597599
15	1.16096896	1.34586834	1.80094351	3.17216911	5.47356576
20	1.22019004	1.48594740	2.19112314	4.66095714	9.64629309
25	1.28243200	1.64060599	2.66583633	6.84847520	17.00006441
30	1.34784892	1.81136158	3.24339751	10.06265689	29.95992212
35	1.41660276	1.99988955	3.94608899	14.78534429	52.79961958
40	1.48886373	2.20803966	4.80102063	21.72452150	93.05097044
45	1.56481075	2.43785421	5.84117568	31.92044939	163.98760387
50	1.64463182	2.69158803	7.10668335	46.90161251	289.00218983
55	1.72852457	2.97173067	8.64636692	68.91385611	509.32060567
60	1.81669670	3.28103079	10.51962741	101.25706367	897.59693349
65	1.90936649	3.62252311	12.79873522	148.77984662	1581.87249060
70	2.00676337	3.99955822	15.57161835	218.60640590	2787.79982770
75	2.10912847	4.41583546	18.94525466	321.20452996	4913.05584077
80	2.21671522	4.87543916	23.04979907	471.95483426	8658.48310008
85	2.32978997	5.38287878	28.04360494	693.45648897	15259.20568055
90	2.44863267	5.94313313	34.11933334	1018.91508928	26891.93422336
95	2.57353755	6.56169920	41.51138594	1497.12054855	47392.77662369
100	2.70481383	7.24464612	50.50494818	2199.76125634	83522.26572652

40. A manufacturer of radial tires has found that the percentage of tires P still usable after being driven m miles is given by

$$P = 100(2.71)^{-.000025m}$$

(a) Graph this equation.

(b) What percentage of the tires are still usable at 80,000 miles?

41. One method of depreciation allowed by IRS is the double-declining-balance method. If this is done over N years, the original value C is depreciated each year by $100(2/N)$ percent of its value at the beginning of that year. Thus, the value after n years is

$$V = C\left(1 - \frac{2}{N}\right)^n$$

(a) If a piece of equipment costing $8000 is depreciated by this method over 10 years, what is its value after 5 years?

(b) Answer the same question if the equipment in part (a) is depreciated to zero linearly over 10 years.

(c) Does the value ever become zero by the double-declining-balance method?

An active participant in the political and religious battles of his day, the Scot John Napier amused himself by studying mathematics and science. In 1614 he published a book containing the idea that made him famous. He gave it the name *logarithm*.

John Napier
(1550–1617)

6-4
Logarithms and Logarithmic Functions

Napier's approach to logarithms is out of style, but the goal he had in mind is still worth considering. He hoped to replace multiplications by additions. He thought additions were easier to do, and he was right.

Consider the exponential function $f(x) = 2^x$ and recall that

$$2^x \cdot 2^y = 2^{x+y}$$

On the left, we have a multiplication and on the right, an addition. If we are to fulfill Napier's objective, we want logarithms to behave like exponents. That suggests a definition. The logarithms of N to the base 2 is the exponent to which 2 must be raised to yield N. That is,

$$\log_2 N = x \quad \text{if and only if} \quad 2^x = N$$

Thus

$$\log_2 4 = 2 \quad \text{since} \quad 2^2 = 4$$
$$\log_2 8 = 3 \quad \text{since} \quad 2^3 = 8$$
$$\log_2 \sqrt{2} = \tfrac{1}{2} \quad \text{since} \quad 2^{1/2} = \sqrt{2}$$

and, in general,

$$\log_2(2^x) = x \quad \text{since} \quad 2^x = 2^x$$

Has Napier's goal been achieved? Does the logarithm turn a product into a sum? Yes, for note that

$$\log_2(2^x \cdot 2^y) = \log_2(2^{x+y}) \qquad \text{(property of exponents)}$$
$$= x + y \qquad \text{(definition of } \log_2)$$
$$= \log_2(2^x) + \log_2(2^y)$$

Thus

$$\log_2(2^x \cdot 2^y) = \log_2(2^x) + \log_2(2^y)$$

which has the form

$$\log_2(M \cdot N) = \log_2 M + \log_2 N$$

THE GENERAL DEFINITION

What has been done for 2 can be done for any base $a > 1$. The **logarithm of N to the base a** is the exponent x to which a must be raised to yield N. Thus

$$\log_a N = x \quad \text{if and only if} \quad a^x = N$$

Now we can calculate many kinds of logarithms.

$$\log_4 16 = 2 \quad \text{since} \quad 4^2 = 16$$
$$\log_{10} 1000 = 3 \quad \text{since} \quad 10^3 = 1000$$
$$\log_{10}(.001) = -3 \quad \text{since} \quad 10^{-3} = \frac{1}{1000} = .001$$

What is $\log_{10} 7$? We are not ready to answer that yet, except to say it is a number x satisfying $10^x = 7$ (see Section 6-6).

We point out that negative numbers and zero do not have logarithms. Suppose -4 and 0 did have logarithms, that is, suppose

$$\log_a(-4) = m \quad \text{and} \quad \log_a 0 = n$$

Then

$$a^m = -4 \quad \text{and} \quad a^n = 0$$

But that is impossible; we learned earlier that a^x is always positive.

PROPERTIES OF LOGARITHMS

There are three main properties of logarithms.

> **PROPERTIES OF LOGARITHMS**
> 1. $\log_a(M \cdot N) = \log_a M + \log_a N$
> 2. $\log_a(M/N) = \log_a M - \log_a N$
> 3. $\log_a(M^p) = p \log_a M$

To establish Property 1, let

$$x = \log_a M \quad \text{and} \quad y = \log_a N$$

Then, by definition,

$$M = a^x \quad \text{and} \quad N = a^y$$

so that

$$M \cdot N = a^x \cdot a^y = a^{x+y}$$

Thus $x + y$ is the exponent to which a must be raised to yield $M \cdot N$, that is,

$$\log_a(M \cdot N) = x + y = \log_a M + \log_a N$$

Properties 2 and 3 are demonstrated in a similar fashion.

THE LOGARITHMIC FUNCTION

The function determined by

$$g(x) = \log_a x$$

is called the **logarithmic function with base a.** We can get a feeling for the behavior of this function by drawing its graph for $a = 2$ (top of next page).

Several properties of $y = \log_2 x$ are apparent from this graph. The domain consists of all positive real numbers. If $0 < x < 1$, $\log_2 x$ is negative; if $x > 1$, $\log_2 x$ is positive. The y-axis is a vertical asymptote of the graph since

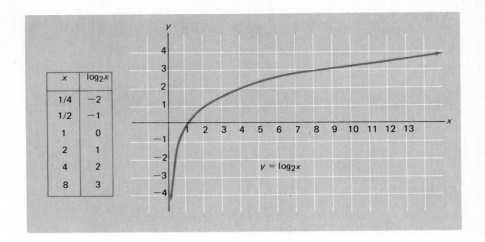

very small positive x values yield large negative y values. Although $\log_2 x$ continues to increase as x increases, even this small part of the complete graph indicates how slowly it grows for large x. In fact, by the time x reaches 1,000,000, $\log_2 x$ is still loafing along at about 20. In this sense, it behaves in a manner opposite to the exponential function 2^x, which grows more and more rapidly as x increases. There is a good reason for this opposite behavior; the two functions are inverses of each other.

INVERSE FUNCTIONS

We begin by emphasizing two facts that you must not forget.

$$a^{\log_a x} = x$$
$$\log_a(a^x) = x$$

For example, $2^{\log_2 7} = 7$ and $\log_2(2^{-19}) = -19$. Both of these facts are direct consequences of the definition of logarithms; the second is also a special case of Property 3, stated earlier. What these facts tell us is that the logarithmic and exponential functions undo each other.

Let us put it in the language of Section 5-5. If $f(x) = a^x$ and $g(x) = \log_a x$, then

$$f(g(x)) = f(\log_a x) = a^{\log_a x} = x$$

and

$$g(f(x)) = g(a^x) = \log_a(a^x) = x$$

Thus g is really f^{-1}. This fact also tells us something about the graphs of g and f: They are simply reflections of each other about the line $y = x$.

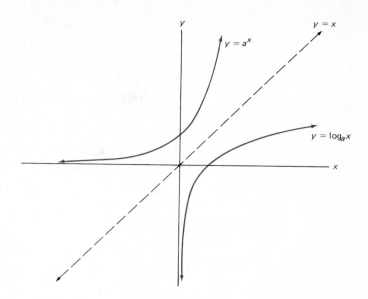

Note finally that $f(x) = a^x$ has the set of all real numbers as its domain and the positive real numbers as its range. Thus its inverse $f^{-1}(x) = \log_a x$ has domain consisting of the positive real numbers and range consisting of all real numbers. We emphasize again a fact that is important to remember. *Negative numbers and zero do not have logarithms.*

Problem Set 6-4

Write each of the following in logarithmic form. For example, $3^4 = 81$ can be written as $\log_3 81 = 4$.

1. $4^3 = 64$ 2. $7^3 = 343$ 3. $27^{1/3} = 3$

4. $16^{1/4} = 2$ 5. $4^0 = 1$ 6. $81^{-1/2} = \frac{1}{9}$

7. $125^{-2/3} = \frac{1}{25}$ 8. $2^{9/2} = 16\sqrt{2}$ 9. $10^{\sqrt{3}} = a$

10. $5^{\sqrt{2}} = b$ 11. $10^a = \sqrt{3}$ 12. $b^x = y$

Write each of the following in exponential form. For example, $\log_5 125 = 3$ can be written as $5^3 = 125$.

13. $\log_5 625 = 4$ 14. $\log_6 216 = 3$ 15. $\log_4 8 = \frac{3}{2}$

16. $\log_{27} 9 = \frac{2}{3}$ 17. $\log_{10}(.01) = -2$ 18. $\log_3(\frac{1}{27}) = -3$

19. $\log_c c = 1$ 20. $\log_b N = x$ 21. $\log_c Q = y$

Determine the value of each of the following logarithms.

22. $\log_4 16$ 23. $\log_5 25$ 24. $\log_7 \frac{1}{7}$

25. $\log_3 \frac{1}{3}$ 26. $\log_4 2$ 27. $\log_{27} 3$

28. $\log_{10}(10^{-6})$ 29. $\log_{10}(.0001)$ 30. $\log_8 1$

31. $\log_3 1$ 32. $\log_{100} 1000$ 33. $\log_8 16$

Find the value of c in each of the following.

34. $\log_c 25 = 2$ 35. $\log_c 8 = 3$ 36. $\log_4 c = -\frac{1}{2}$

37. $\log_9 c = -\frac{3}{2}$ 38. $\log_2(2^{5.6}) = c$ 39. $\log_3(3^{-2.9}) = c$

40. $8^{\log_8 11} = c$ 41. $5^{2 \log_5 7} = c$ 42. $3^{4 \log_3 2} = c$

Given $\log_{10} 2 = .301$ and $\log_{10} 3 = .477$, calculate each of the following without the use of tables. For example, in Problem 43, $\log_{10} 6 = \log_{10} 2 \cdot 3 = \log_{10} 2 + \log_{10} 3$, and in Problem 50, $\log_{10} 54 = \log_{10} 2 \cdot 3^3 = \log_{10} 2 + \log_{10} 3^3 = \log_{10} 2 + 3 \log_{10} 3$.

43. $\log_{10} 6$ 44. $\log_{10} \frac{3}{2}$ 45. $\log_{10} 16$

46. $\log_{10} 27$ 47. $\log_{10} \frac{1}{4}$ 48. $\log_{10} \frac{1}{27}$

49. $\log_{10} 24$ 50. $\log_{10} 54$ 51. $\log_{10} \frac{8}{9}$

52. $\log_{10} \frac{3}{8}$ 53. $\log_{10} 5$ 54. $\log_{10} \sqrt[3]{3}$

© *Your scientific calculator has a $\log_{10}$ key (which may be abbreviated log). Use it to find each of the following.*

55. $\log_{10} 34$ 56. $\log_{10} 1417$ 57. $\log_{10}(.0123)$

58. $\log_{10}(.3215)$ 59. $\log_{10} 9723$ 60. $\log_{10}(\frac{21}{312})$

EXAMPLE A (Combining logarithms)   Write the following expression as a single logarithm.

$$2 \log_{10} x + 3 \log_{10}(x + 2) - \log_{10}(x^2 + 5)$$

Solution.   We use the properties of logarithms to rewrite this as

$$\log_{10} x^2 + \log_{10}(x + 2)^3 - \log_{10}(x^2 + 5) \qquad \text{(Property 3)}$$

$$= \log_{10} x^2(x + 2)^3 - \log_{10}(x^2 + 5) \qquad \text{(Property 1)}$$

$$= \log_{10}\left[\frac{x^2(x + 2)^3}{x^2 + 5}\right] \qquad \text{(Property 2)}$$

Write each of the following as a single logarithm.

61. $3 \log_{10}(x + 1) + \log_{10}(4x + 7)$

62. $\log_{10}(x^2 + 1) + 5 \log_{10} x$

63. $3 \log_2(x + 2) + \log_2 8x - 2 \log_2(x + 8)$

64. $2 \log_5 x - 3 \log_5(2x + 1) + \log_5(x - 4)$

65. $\frac{1}{2} \log_6 x + \frac{1}{3} \log_6(x^3 + 3)$

66. $-\frac{2}{3} \log_3 x + \frac{5}{2} \log_3(2x^2 + 3)$

EXAMPLE B (Solving logarithmic equations)   Solve the equation

$$\log_2 x + \log_2(x + 2) = 3$$

Solution. First we note that we must have $x > 0$ so that both logarithms exist. Next we rewrite the equation using the first property of logarithms and then the definition of a logarithm.

$$\log_2 x(x + 2) = 3$$

$$x(x + 2) = 2^3$$

$$x^2 + 2x - 8 = 0$$

$$(x + 4)(x - 2) = 0$$

We reject $x = -4$ (because $-4 < 0$) and keep $x = 2$. To make sure that 2 is a solution, we substitute 2 for x in the original equation.

$$\log_2 2 + \log_2(2 + 2) \stackrel{?}{=} 3$$

$$1 + 2 = 3$$

Solve each of the following equations.

67. $\log_7(x + 2) = 2$
68. $\log_5(3x + 2) = 1$
69. $\log_2(x + 3) = -2$
70. $\log_4(\frac{1}{64}x + 1) = -3$
71. $\log_2 x - \log_2(x - 2) = 3$
72. $\log_3 x - \log_3(2x + 3) = -2$
73. $\log_2(x - 4) + \log_2(x - 3) = 1$
74. $\log_{10} x + \log_{10}(x - 3) = 1$

EXAMPLE C (Change of base) In Problems 89 and 90, you will be asked to establish the **change-of-base formula**

$$\log_b x = \frac{\log_a x}{\log_a b}$$

Use this formula and the $\boxed{\log_{10}}$ key on a calculator to find $\log_2 13$.

Solution.

$$\log_2 13 = \frac{\log_{10} 13}{\log_{10} 2} \approx \frac{1.1139433}{.30103} \approx 3.7004$$

Ⓒ *Use the method of Example C to find the following.*

75. $\log_2 128$
76. $\log_3 128$
77. $\log_3 82$
78. $\log_5 110$
79. $\log_6 39$
80. $\log_2(.26)$

Miscellaneous Problems

81. Write in logarithmic form.
    (a) $16^{-3/4} = \frac{1}{8}$ (b) $y = 10^{3.24}$ (c) $b = (1.4)^a$
82. Write in exponential form.
    (a) $\log_3 \frac{1}{9} = -2$ (b) $\log_8 4 = \frac{2}{3}$
    (c) $\log_8 c = b$ (d) $\log_b N = x$
83. Find the value of x in each of the following.
    (a) $x = \log_6 36$ (b) $x = \log_4 2$

(c) $\log_{25} x = \frac{3}{2}$ (d) $\log_4 x = \frac{5}{2}$

(e) $\log_x 10\sqrt{10} = \frac{3}{2}$ (f) $\log_x \frac{1}{8} = -\frac{3}{2}$

84. Write each of the following as a single logarithm.
    (a) $3 \log_2 5 - 2 \log_2 7$
    (b) $\frac{1}{2} \log_5 64 + \frac{1}{3} \log_5 27 - \log_5(x^2 + 4)$
    (c) $\frac{2}{3} \log_{10}(x + 5) + 4 \log_{10} x - 2 \log_{10}(x - 3)$

85. What is the domain of the following function?

$$f(x) = \frac{2}{3} \log_{10}(x + 5) + 4 \log_{10} x - 2 \log_{10}(x - 3)$$

86. Solve each of the following equations for x.
    (a) $\log_5(2x - 1) = 2$
    (b) $\log_4\left(\dfrac{x - 2}{2x + 3}\right) = 0$
    (c) $\log_4(x - 2) - \log_4(2x + 3) = 0$
    (d) $\log_{10} x + \log_{10}(x - 15) = 2$

87. Derive the second property of logarithms, that is, show that $\log_a (M/N) = \log_a M - \log_a N$. (*Hint:* Use the method given in the text to derive Property 1.)

88. Show that $\log_a (M^p) = p \log_a M$.

89. Show that $\log_2 x = \log_{10} x \cdot \log_2 10$, where $x > 0$. (*Hint:* Let $\log_{10} x = c$. Then $x = 10^c$. Next take $\log_2$ of both sides.)

90. Use the technique outlined in Problem 89 to show that

$$\log_a x = \log_b x \cdot \log_a b$$

This is equivalent to the change-of-base formula of Example C.

91. Show that $\log_a b = 1/\log_b a$, where $a, b > 0$. (*Hint:* Use the change-of-base formula and the fact that $1 = \log_a a$.)

92. Graph the equations $y = 2^x$ and $y = \log_2 x$ on the same coordinate plane.

93. Graph the equations $y = 3^x$ and $y = \log_3 x$ on the same coordinate plane.

94. Give the value(s) of x for which
    (a) $2^x = 3^x$; (b) $2^x > 3^x$; (c) $2^x < 3^x$.

95. Give the value(s) of x for which
    (a) $\log_2 x = \log_3 x$; (b) $\log_2 x > \log_3 x$; (c) $\log_2 x < \log_3 x$.

96. Give the solution set for each of the following inequalities.
    (a) $\log_2 x < 0$ (b) $\log_2 x > 3$
    (c) $\log_{10} x < -1$ (d) $\log_{10} x \geq 2$
    (e) $2 < \log_3 x < 3$ (f) $0 \leq \log_{10} x \leq 2$
    (g) $-3 < \log_2 x < -2$ (h) $-2 \leq \log_{10} x \leq -1$

97. Sketch the graph of each of the following functions using the same coordinate system.
    (a) $f(x) = \log_2 x$ (b) $g(x) = \log_2(x + 1)$ (c) $h(x) = 3 + \log_2 x$

The collected works of this brilliant Swiss mathematician will fill 74 volumes when completed. No other person has written so profusely on mathematical topics. Remarkably, 400 of his research papers were written after he was totally blind. One of his contributions was the introduction of the number $e = 2.71828 \ldots$ as the base for natural logarithms.

Leonhard Euler
1707-1783

6-5
Applications of Logarithms

Napier invented logarithms to simplify arithmetic calculations. Computers and calculators have reduced that application to minor significance, though we shall discuss such a use of logarithms later in this chapter. Here we have in mind deeper applications such as solving exponential equations, defining power functions, and modeling physical phenomena.

In order to make any progress, we shall need an easy way to calculate logarithms. Fortunately, this has been done for us as tables of logarithms to several bases are available. For our purposes in this section, one base is as good as another. Base 10 would be an appropriate choice, but we would rather defer discussion of logarithms to base 10 (common logarithms) until Section 6-6. We have chosen rather to introduce you to the number e (after Euler), which is used as a base of logarithms in all advanced mathematics courses. You will see the importance of logarithms to this base (**natural logarithms**) when you study calculus. An approximate value of e is

$$e \approx 2.71828$$

and, like π, e is an irrational number. Page 243 shows a table of values of natural logarithms, which we shall denote by ln instead of $\log_e$.

Since ln denotes a genuine logarithm function, we have as in Section 6-4

$$\ln N = x \quad \text{if and only if} \quad e^x = N$$

and consequently

$$\ln e^x = x \quad \text{and} \quad e^{\ln N} = N$$

e and interest

Suppose one dollar is invested at 100 percent interest compounded n times per year. Then it will be worth $(1 + \frac{1}{n})^n$ dollars at the end of the year. As n increases without bound (that is, as compounding occurs more and more often)

$$(1 + \frac{1}{n})^n \to e$$

and we have what is called continuous compounding. Under this plan, $1 grows to $2.72 in one year.

Moreover the three properties of logarithms hold.

1. $\ln(MN) = \ln M + \ln N$
2. $\ln(M/N) = \ln M - \ln N$
3. $\ln(N^p) = p \ln N$

SOLVING EXPONENTIAL EQUATIONS

Consider first the simple equation

$$5^x = 1.7$$

We call it an *exponential equation* because the unknown is in the exponent. To solve it, we take natural logarithms of both sides.

$$5^x = 1.7$$

$$\ln(5^x) = \ln 1.7$$

$$x \ln 5 = \ln 1.7 \qquad \text{(Property 3)}$$

$$x = \frac{\ln 1.7}{\ln 5}$$

$$x \approx \frac{.531}{1.609} \approx .330 \qquad \text{(table of ln values)}$$

We point out that the last step can also be done on a scientific calculator, which has a key for calculating natural logarithms.

Here is a more complicated example.

$$5^{2x-1} = 7^{x+2}$$

Begin by taking natural logarithms of both sides and then solve for x.

$$\ln(5^{2x-1}) = \ln(7^{x+2})$$

$$(2x - 1) \ln 5 = (x + 2) \ln 7 \qquad \text{(Property 3)}$$

$$2x \ln 5 - \ln 5 = x \ln 7 + 2 \ln 7$$

$$2x \ln 5 - x \ln 7 = \ln 5 + 2 \ln 7$$

$$x(2 \ln 5 - \ln 7) = \ln 5 + 2 \ln 7$$

$$x = \frac{\ln 5 + 2 \ln 7}{2 \ln 5 - \ln 7}$$

$$\approx \frac{1.609 + 2(1.946)}{2(1.609) - 1.946} \qquad \begin{array}{l}\text{(table of ln values}\\ \text{or a calculator)}\end{array}$$

$$\approx 4.325$$

You get a more accurate answer, 4.321, if you use a calculator all the way.

TABLE 3   Table of Natural Logarithms

x	$\ln x$	x	$\ln x$	x	$\ln x$
		4.0	1.386	8.0	2.079
0.1	−2.303	4.1	1.411	8.1	2.092
0.2	−1.609	4.2	1.435	8.2	2.104
0.3	−1.204	4.3	1.459	8.3	2.116
0.4	−0.916	4.4	1.482	8.4	2.128
0.5	−0.693	4.5	1.504	8.5	2.140
0.6	−0.511	4.6	1.526	8.6	2.152
0.7	−0.357	4.7	1.548	8.7	2.163
0.8	−0.223	4.8	1.569	8.8	2.175
0.9	−0.105	4.9	1.589	8.9	2.186
1.0	0.000	5.0	1.609	9.0	2.197
1.1	0.095	5.1	1.629	9.1	2.208
1.2	0.182	5.2	1.649	9.2	2.219
1.3	0.262	5.3	1.668	9.3	2.230
1.4	0.336	5.4	1.686	9.4	2.241
1.5	0.405	5.5	1.705	9.5	2.251
1.6	0.470	5.6	1.723	9.6	2.262
1.7	0.531	5.7	1.740	9.7	2.272
1.8	0.588	5.8	1.758	9.8	2.282
1.9	0.642	5.9	1.775	9.9	2.293
2.0	0.693	6.0	1.792	10	2.303
2.1	0.742	6.1	1.808	20	2.996
2.2	0.788	6.2	1.825	30	3.401
2.3	0.833	6.3	1.841	40	3.689
2.4	0.875	6.4	1.856	50	3.912
2.5	0.916	6.5	1.872	60	4.094
2.6	0.956	6.6	1.887	70	4.248
2.7	0.993	6.7	1.902	80	4.382
2.8	1.030	6.8	1.917	90	4.500
2.9	1.065	6.9	1.932	100	4.605
3.0	1.099	7.0	1.946		
3.1	1.131	7.1	1.960		
3.2	1.163	7.2	1.974	e	1.000
3.3	1.194	7.3	1.988		
3.4	1.224	7.4	2.001	π	1.145
3.5	1.253	7.5	2.015		
3.6	1.281	7.6	2.028		
3.7	1.308	7.7	2.041		
3.8	1.335	7.8	2.054		
3.9	1.361	7.9	2.067		

To find the natural logarithm of a number N which is either smaller than 0.1 or larger than 10, write N in scientific notation, that is, write $N = c \times 10^k$. Then $\ln N = \ln c + k \ln 10 = \ln c + k(2.303)$. A more complete table of natural logarithms appears as Table A of the Appendix.

THE GRAPHS OF $\ln x$ AND e^x

We have already pointed out that $\ln e^x = x$ and $e^{\ln x} = x$. Thus $f(x) = \ln x$ and $g(x) = e^x$ are inverse functions, which means that their graphs are reflections of each other across the line $y = x$. They are shown on the following page.

x	$\ln x$
.1	-2.3
.5	$-.7$
1	0
2	.7
3	1.1
4	1.4
5	1.6

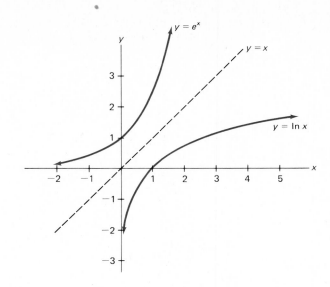

EXPONENTIAL FUNCTIONS VERSUS POWER FUNCTIONS

Look closely at the formulas below.

$$f(x) = 2^x \qquad f(x) = x^2$$

They are very different, yet easily confused. The first is an exponential function, while the second is called a power function. Both grow rapidly for large x, but the exponential function ultimately gets far ahead (see the graphs in the margin).

The situation described above is a special instance of two very general classes of functions.

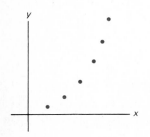

EXPONENTIAL FUNCTIONS

$$f(x) = ba^x$$

POWER FUNCTIONS

$$f(x) = bx^a$$

While these functions are very different in character, they are related, as you will be asked to show in Problem 78.

CURVE FITTING

A recurring theme in science is to fit a mathematical curve to a set of experimental data. Suppose that a scientist, studying the relationship between two variables x and y, obtained the data plotted in the margin. In searching for curves to fit these data, the scientist naturally thought of exponential curves and power curves. How did he or she decide if either was appropriate? The scientist took logarithms. Let us see why.

MODEL 1	MODEL 2

$$y = ba^x$$

$$\ln y = \ln b + x \ln a$$

$$Y = B + Ax$$

$$y = bx^a$$

$$\ln y = \ln b + a \ln x$$

$$Y = B + aX$$

Here the scientist made the substitutions $Y = \ln y$, $B = \ln b$, $A = \ln a$, and $X = \ln x$.

In both cases, the final result is a linear equation. But note the difference. In the first case, $\ln y$ is a linear function of x, while in the second case, $\ln y$ is a linear function of $\ln x$. These considerations suggest the following procedures. Make two additional plots of the data. In the first, plot $\ln y$ against x, and in the second, plot $\ln y$ against $\ln x$. If the first plotting gives data nearly along a straight line, Model 1 is appropriate; if the second does, then Model 2 is appropriate. If neither plot approximates a straight line, our scientist should look for a different and perhaps more complicated model.

We have used natural logarithms in the discussion above; we could also have used common logarithms (logarithms to the base 10). In the latter case, special kinds of graph paper are available to simplify the curve fitting process. On semilog paper, the vertical axis has a logarithmic scale; on log-log paper, both axes have logarithmic scales. The xy-data can be plotted *directly* on this paper. If semilog paper gives an (approximately) straight line, Model 1 is indicated; if log-log paper does so, then Model 2 is appropriate. You will have ample opportunity to use these special kinds of graph paper in your science courses.

LOGARITHMS AND PHYSIOLOGY

The human body appears to have a built-in logarithmic calculator. What do we mean by this statement?

In 1834, the German physiologist E. Weber noticed an interesting fact. Two heavy objects must differ in weight by considerably more than two light objects if a person is to perceive a difference between them. Other scientists noted the same phenomenon when human subjects tried to differentiate loudness of sounds, pitches of musical tones, brightness of light, and so on. Experiments suggested that people react to stimuli on a logarithmic scale, a result formulated as the Weber-Fechner law.

$$S = C \ln\left(\frac{R}{r}\right)$$

Here R is the actual intensity of the stimulus, r is the threshold value (smallest value at which the stimulus is observed), C is a constant depending on the type

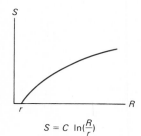

$S = C \ln(\frac{R}{r})$

of stimulus, and S is the perceived intensity of the stimulus. Note that a change in R is not as perceptible for large R as for small R because as R increases, the graph of the logarithmic functions gets steadily flatter.

Problem Set 6-5

In Problems 1–8, find the value of each natural logarithm.

1. $\ln e$ 2. $\ln(e^2)$ 3. $\ln 1$ 4. $\ln\left(\dfrac{1}{e}\right)$

5. $\ln\sqrt{e}$ 6. $\ln(e^{1.1})$ 7. $\ln\left(\dfrac{1}{e^3}\right)$ 8. $\ln(e^n)$

For Problems 9–14, find each value. Assume that $\ln a = 2.5$ and $\ln b = -.4$.

9. $\ln(ae)$ 10. $\ln\left(\dfrac{e}{b}\right)$ 11. $\ln\sqrt{b}$

12. $\ln(a^2 b^{10})$ 13. $\ln\left(\dfrac{1}{a^3}\right)$ 14. $\ln(a^{4/5})$

Use the table of natural logarithms to calculate the values in Problems 15–20.

15. $\ln 120 = \ln 60 + \ln 2$ 16. $\ln 150$ 17. $\ln 690$
18. $\ln 84$ 19. $\ln \frac{6}{5}$ 20. $\ln 20{,}000$

$$\ln \frac{4}{3} = \frac{\ln 4}{\ln 3}$$
$$= \frac{1.386}{1.099} = 1.261$$

$$\ln \frac{4}{3} = \ln 4 - \ln 3$$
$$= 1.386 - 1.099 = .287$$

In Problems 21–26, use the natural logarithm table to find N.

21. $\ln N = 2.208$ 22. $\ln N = 1.808$ 23. $\ln N = -.105$
24. $\ln N = -.916$ 25. $\ln N = 4.500$ 26. $\ln N = 9.000$

© *Use your calculator ($\boxed{\ln}$ or $\boxed{\log_e}$ key) to find each of the following.*

27. $\ln 4.31$ 28. $\ln 517$ 29. $\ln(.127)$ 30. $\ln(.00424)$

31. $\ln\left(\dfrac{6.71}{42.3}\right)$ 32. $\ln\sqrt{457}$ 33. $\dfrac{\ln 6.71}{\ln 42.3}$ 34. $\sqrt{\ln 457}$

35. $\ln(51.4)^3$ 36. $\ln(31.2 + 43.1)$ 37. $(\ln 51.4)^3$ 38. $3 \ln 51.4$

© *Use your calculator to find N in each of the following. (Hint: $\ln N = 5.1$ if and only if $N = e^{5.1}$. On some calculators, there is an $\boxed{e^x}$ key; on others, you use the two keys $\boxed{\text{inv}}$ $\boxed{\ln}$.)*

39. $\ln N = 2.12$ 40. $\ln N = 5.63$ 41. $\ln N = -.125$
42. $\ln N = .00257$ 43. $\ln\sqrt{N} = 3.41$ 44. $\ln N^3 = .415$

EXAMPLE A (Exponential equations)   Solve $4^{3x-2} = 15$ for x.

Solution.   Take natural logarithms of both sides and then solve for x. Complete the solution using the ln table or a calculator.

$$\ln 4^{3x-2} = \ln 15$$

$$(3x - 2)\ln 4 = \ln 15$$

$$3x \ln 4 - 2 \ln 4 = \ln 15$$

$$3x \ln 4 = 2 \ln 4 + \ln 15$$

$$x = \frac{2 \ln 4 + \ln 15}{3 \ln 4}$$

$$x \approx 1.32$$

[c] *Solve for x using the method above.*

45. $3^x = 20$ 46. $5^x = 40$ 47. $2^{x-1} = .3$

48. $4^x = 3^{2x-1}$ 49. $(1.4)^{x+2} = 19.6$ 50. $5^x = \frac{1}{2}(4x)$

EXAMPLE B (Doubling time)   How long will it take money to double in value if it is invested at 9.5 percent interest compounded annually?

Solution.   For convenience, consider investing \$1. From Section 3-3, we know that this dollar will grow to $(1.095)^t$ dollars after t years. Thus we must solve the exponential equation $(1.095)^t = 2$. This we do by the method of Example A, using a calculator. The result is

$$t = \frac{\ln 2}{\ln 1.095} \approx 7.64$$

[c] 51. How long would it take money to double at 12 percent compounded annually?

[c] 52. How long would it take money to double at 12 percent compounded monthly? (*Hint:* After t months, \$1 is worth $(1.01)^t$ dollars.)

[c] 53. How long would it take money to double at 15 percent compounded quarterly?

[c] 54. A certain substance decays according to the formula $y = 100e^{-.135t}$, where t is in years. Find its half-life.

[c] 55. By finding the natural logarithm of the numbers in each pair, determine which is larger.
(a) $10^5, 5^{10}$ (b) $10^9, 9^{10}$
(c) $10^{20}, 20^{10}$ (d) $10^{1000}, 1000^{10}$

56. What do your answers in Problem 55 confirm about the growth of 10^x and x^{10} for large x?

57. On the same coordinate plane, graph $y = 3^x$ and $y = x^3$ for $0 \le x \le 4$.

58. By means of a change of variable(s) (as explained in the text), transform each equation below to a linear equation. Find the slope and Y-intercept of the resulting line.
(a) $y = 3e^{2x}$ (b) $y = 2x^3$ (c) $xy = 12$
(d) $y = x^e$ (e) $y = 5(3^x)$ (f) $y = ex^{1.1}$

t	N
0	100
1	700
2	5000
3	40,000

EXAMPLE C (Curve fitting)   The table in the margin shows the number N of bacteria in a certain culture found after t hours. Which is a better description of these data,

$$N = ba^t \quad \text{or} \quad N = bt^a$$

Find a and b.

Solution. Following the discussion of curve fitting in the text, we begin by plotting ln N against t. If the resulting points lie along a line, we choose $N = ba^t$ as the appropriate model. If not, we will plot ln N against ln t to check on the second model.

t	ln N
0	4.6
1	6.6
2	8.5
3	10.6

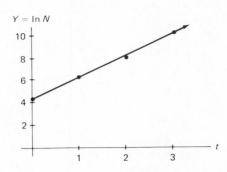

Since the fit to a line is quite good, we accept $N = ba^t$ as our model. To find a and b, we write $N = ba^t$ in the form

$$\ln N = \ln b + t \ln a \quad \text{or} \quad Y = \ln b + (\ln a)t$$

Examination of the line shows that it has a Y-intercept of 4.6 and a slope of about 2; so for its equation we write $Y = 4.6 + 2t$. Comparing this with $Y = \ln b + (\ln a)t$ gives

$$\ln b = 4.6 \qquad \ln a = 2$$

Finally, we use the natural logarithm table or a calculator to find

$$b \approx 100 \qquad a \approx 7.4$$

Thus the original data are described reasonably well by the equation

$$N = 100(7.4)^t$$

For the data sets below, decide whether $y = ba^x$ or $y = bx^a$ is the better model. Then determine a and b.

59.

x	1	2	3	4
y	96	145	216	325

61.

x	1	2	3	5
y	12	190	975	7490

60.

x	0	1	2	4
y	243	162	108	48

62.

x	1	4	9
y	16	128	432

Miscellaneous Problems

In Problems 63–70, find the value of each expression using the natural logarithm table where necessary.

63. $\ln(e^{3.5})$

64. $\ln(e^{-1.2})$

65. $3 \ln(e^{2/3})$

66. $\ln(1/\sqrt{e})$

67. $\ln(10e)$

68. $\ln 4300$

69. $\ln(80^2/\sqrt{6.3})$ 70. $\ln(4.1 \times 10^4)$

[c] 71. Solve for x.
   (a) $10^{2x+5} = 800$ (b) $e^{2x} = 8^{x-1}$

72. By means of a change of variables, transform each of the following equations to a linear equation. Find the slope and y-intercept of the resulting line.
   (a) $xy^2 = 40$ (b) $y = 9(2^{2x})$

73. The Weber-Fechner law discussed in the text says that $S = C \ln(R/r)$.
   (a) What is S when $R = r$?
   (b) What is S when $R = r^3$?
   (c) What is S when $R = 2r$?

74. If the perceived intensity S of the stimulus in the Weber-Fechner law has doubled, what has happened to R/r?

75. A certain substance decays according to the formula

$$y = 100e^{-3t}$$

where y is the amount present (in grams) after t years. Find its half-life.

[c] 76. Suppose that the number of bacteria in a certain culture t hours from now will be $200(1.5)^t$.
   (a) What will the bacteria count be 2.3 hours from now?
   (b) When will the count reach 10,000?

[c] 77. A radioactive substance decays so that the amount present is A grams after t years where

$$A = 50e^{-.0135t}$$

When will there be only 10 grams left?

78. We know that $N = e^{\ln N}$ for any $N > 0$. By letting x^a take the place of N, show that

$$x^a = e^{a \ln x}, \qquad x > 0$$

This equation is sometimes used as the definition of x^a, especially when a is irrational.

79. Along the lines of Problem 78, how would you define x^x?

80. How would you define $(x^2 + 5)^{2x}$?

81. How would you define $(x - 5)^{1/x}$? How must x be restricted for your definition to make sense?

[c] 82. In calculus it is shown that

$$e^x \approx 1 + x + \frac{x^2}{2} + \frac{x^3}{6} + \frac{x^4}{24} + \frac{x^5}{120}$$

Use this to approximate e and $e^{-.5}$.

[c] 83. If $f(x) = [\ln(x^2 + \sqrt{x} + 1/x)]^3$, calculate
   (a) $f(2.1)$;   (b) $f(.12)$.

[c] 84. Calculate.
   (a) $250\left(1 + \dfrac{.14}{12}\right)^{120}$

   (b) $250\left(1 + \dfrac{.14}{365}\right)^{3650}$

   (c) $250e^{1.4}$

ⓒ 85. It is shown in calculus that

$$\left(1 + \frac{r}{n}\right)^n \longrightarrow e^r$$

as n gets larger and larger. If P dollars is invested at the rate r compounded n times per year, it will grow to

$$P\left(1 + \frac{r}{n}\right)^{nt}$$

dollars at the end of t years; if interest is compounded continuously (see the box **e and interest** on page 241), P dollars will grow to Pe^{rt} dollars after t years. Use these facts to calculate the value of $100 after 10 years if interest is 12 percent (that is, $r = .12$) and is compounded
(a) monthly;   (b) daily;   (c) continuously.

86. Interpret your answers to Problem 84 in light of Problem 85.

Henry Briggs 1561–1631

When 10 is used as the base for logarithms, some very nice things happen as the display at the right shows. Realizing this, John Napier and his friend Henry Briggs decided to produce a table of these logarithms. The job was accomplished by Briggs after Napier's death and published in 1624 in the famous book *Arithmetica Logarithmica*.

$\log .0853 = .9309 - 2$

$\log .853 \ \ = .9309 - 1$

$\log 8.53 \ \ = .9309$

$\log 85.3 \ \ = .9309 + 1$

$\log 853 \ \ = .9309 + 2$

$\log 8530 = .9309 + 3$

6-6 Common Logarithms (Optional)

In Section 6-4, we defined the logarithm of a positive number N to the base a as follows.

$$\log_a N = x \quad \text{if and only if} \quad a^x = N$$

Then in Section 6-5, we introduced base e, calling the result natural logarithms. These are the logarithms that are most important in advanced branches of mathematics such as calculus.

In this section, we shall consider logarithms to the base 10, often called **common logarithms,** or Briggsian logarithms. They have been studied by high school and college students for centuries as an aid to computation, a subject we take up in the next section. We shall write $\log N$ instead of $\log_{10} N$ (just as we write $\ln N$ instead of $\log_e N$). Note that

$$\log N = x \quad \text{if and only if} \quad 10^x = N$$

In other words, the common logarithm of any power of 10 is simply its exponent. Thus

$$\log 100 = \log 10^2 = 2$$

$$\log 1 = \log 10^0 = 0$$

$$\log .001 = \log 10^{-3} = -3$$

But how do we find common logarithms of numbers that are not integral powers of 10, such as 8.53 or 14,600? That is the next topic.

FINDING COMMON LOGARITHMS

We know that $\log 1 = 0$ and $\log 10 = 1$. If $1 < N < 10$, we correctly expect $\log N$ to be between 0 and 1. Table B (in the appendix) gives us four-place approximations of the common logarithms of all three-digit numbers between 1 and 10. For example,

$$\log 8.53 = .9309$$

We find this value by locating 8.5 in the left column and then moving right to the entry with 3 as heading. Similarly,

$$\log 1.08 = .0334$$

and

$$\log 9.69 = .9863$$

You should check these values.

You may have noticed in the opening box that log 8.53, log 8530, and log .0853 all have the same positive fractional part, .9309, called the **mantissa** of the logarithm. They differ only in the integer part, called the **characteristic** of the logarithm. To see why this is so, recall that

$$\log(M \cdot N) = \log M + \log N$$

Thus

$$\log 8530 = \log(8.53 \times 10^3) = \log 8.53 + \log 10^3$$

$$= .9309 + 3$$

$$\log .0853 = \log(8.53 \times 10^{-2}) = \log 8.53 + \log 10^{-2}$$

$$= .9309 - 2$$

Clearly the mantissa .9309 is determined by the sequence of digits 8, 5, 3, while the characteristic is determined by the position of the decimal point. Let us say that the decimal point is in **standard position** when it occurs immediately after the first nonzero digit. The characteristic for the logarithm of a number with the decimal point in standard position is 0. The characteristic is k if the decimal point is k places to the right of standard position; it is $-k$ if it is k places to the left of standard position.

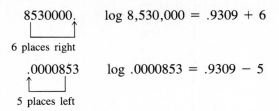

$$\log 8,530,000 = .9309 + 6$$

6 places right

$$\log .0000853 = .9309 - 5$$

5 places left

Here is another way of describing the mantissa and characteristic. If $c \times 10^n$ is the scientific notation for N, then $\log c$ is the mantissa and n is the characteristic of $\log N$.

FINDING ANTILOGARITHMS

If we are to make significant use of common logarithms, we must know how to find a number when its logarithm is given. This process is called finding the inverse logarithm, or the **antilogarithm.** The process is simple: Use the mantissa to find the sequence of digits and then let the characteristic tell you where to put the decimal point.

Suppose, for example, that you are given

$$\log N = .4031 - 4$$

Locate .4031 in the body of Table B. You will find it across from 2.5 and below 3. Thus the number N must have 2, 5, 3 as its sequence of digits. Since the characteristic is -4, put the decimal point 4 places to the left of standard position. The result is

$$N = .000253$$

As a second example, let us find antilog 5.9547. The mantissa .9547 gives us the digits 9, 0, 1. Since the characteristic is 5,

$$\text{antilog } 5.9547 = 901,000$$

LINEAR INTERPOLATION

Suppose that for some function f we know $f(a)$ and $f(b)$, but we want $f(c)$, where c is between a and b (see the diagrams at the top of the next page). As a reasonable approximation, we may pretend that the graph of f is a straight line between a and b. Then

$$f(c) \approx f(a) + d$$

where, by similarity of triangles,

$$\frac{d}{f(b) - f(a)} = \frac{c - a}{b - a}$$

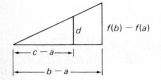

That is,

$$d = \frac{f(b) - f(a)}{b - a} (c - a)$$

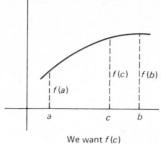

We want $f(c)$

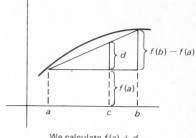

We calculate $f(a) + d$

The process just described is called **linear interpolation.** The process of linear interpolation for the logarithm function is explained in Examples A and B of the problem set, as well as at the beginning of the logarithm tables at the back of the book.

Problem Set 6-6

Find the common logarithm of each of the following numbers.

1. 10,000
2. 1,000,000
3. .01
4. .0001
5. $10^4 10^{3/2}$
6. $10^3 10^{4/3}$
7. $(10^3)^{-5}$
8. $(10^5)^{-1/10}$

Find N in each case.

9. $\log N = 4$
10. $\log N = 6$
11. $\log N = -2$
12. $\log N = -5$
13. $\log N = \frac{3}{2}$
14. $\log N = \frac{1}{3}$
15. $\log N = -\frac{3}{4}$
16. $\log N = -\frac{1}{6}$

Use Table B to find the following logarithms.

17. log 4.32
18. log 3.09
19. log 158
20. log 47.3
21. log .0329
22. log .0715
23. log 563,000
24. log 420,000
25. $\log(9.23 \times 10^8)$
26. $\log(2.83 \times 10^{-11})$

Find N in each case.

27. $\log N = 1.5159$
28. $\log N = 3.9015$
29. $\log N = .0043 - 2$
30. $\log N = .8627 - 4$
31. $\log N = 8.5999$
32. $\log N = 4.7427$

Find the antilogarithm of each number.

33. 2.2201
34. 3.8639
35. $.9232 - 1$
36. $.8500 - 5$

EXAMPLE A (Linear interpolation in finding logarithms)   Find log 34.67.

Solution.   Our table gives the logarithms of 34.6 and 34.7, so we use linear interpolation to get an intermediate value. Here is how we arrange our work.

$$.10\left[.07\left[\begin{array}{l}\log 34.60 = 1.5391 \\ \log 34.67 = \quad ? \\ \log 34.70 = 1.5403\end{array}\right]d\right].0012$$

$$\frac{d}{.0012} = \frac{.07}{.10} = \frac{7}{10}$$

$$d = \frac{7}{10}(.0012) \approx .0008$$

$$\log 34.67 \approx \log 34.60 + d \approx 1.5391 + .0008 = 1.5399$$

Use linear interpolation in Table B to find each value.

37. log 5.237   38. log 9.826   39. log 7234

40. log 68.04   41. log .001234   42. log .09876

EXAMPLE B (Interpolation in finding antilogarithms)   Find antilog 2.5285.

Solution.   We find .5285 sandwiched between .5276 and .5289 in the body of Table B.

$$.0013\left[.0009\left[\begin{array}{l}\text{antilog } 2.5276 = 337.0 \\ \text{antilog } 2.5285 = \quad ? \\ \text{antilog } 2.5289 = 338.0\end{array}\right]d\right]1.0$$

$$\frac{d}{1.0} = \frac{.0009}{.0013} = \frac{9}{13}$$

$$d = \frac{9}{13}(1.0) \approx .7$$

$$\text{antilog } 2.5285 \approx 337.0 + .7 = 337.7$$

Find the antilogarithm of each of the following using linear interpolation.

43. 0.8497   44. 0.8516   45. 3.9130

46. 1.9849   47. .6004 − 2   48. .4946 − 4

Miscellaneous Problems

Find the common logarithm of each of the following.

49. $10^{3/2}10^{-1/4}$ 50. $(.0001)^{1/3}$ 51. $\sqrt{10}\sqrt[3]{10}$ 52. $10^{\log .001}$

Find N in each of the following.

53. $\log N = 0$ 54. $\log N = -2$ 55. $\log N = \frac{2}{3}$ 56. $\log N = 10$

Use Table B to find the following logarithms.

57. log 9.83   58. log 46.3   59. log .00219

60. log 492.7   61. log 9.623   62. log .04705

Use Table B to find N in Problems 63–66, using interpolation to find the fourth digit.

63. $\log N = 2.9325$
64. $\log N = .2695 - 3$
65. $\log N = 0.7300$
66. $\log N = .8619 - 1$
67. Find N if $\log N = -2.4473$ (*Hint:* First write -2.4473 as an integer plus a positive decimal fraction: $-2.4473 = (-2.4473 + 3) - 3 = .5527 - 3$.)
68. Find N.
    (a) $\log N = -1.0074$ (b) $\log N = -4.0729$
69. If $b = 100a$ and $\log a = .75$, find $\log b$.
70. If $b = .001a$ and $\log a = 5.5$, find $\log b$.
71. Find antilog(log .78).
72. Find log(antilog(.4275 − 2)).
73. If $.000001 < N < .00001$, what is the characteristic of $\log N$?
74. Find $\log\left(\dfrac{471 + 328}{471 \times 328}\right)$.

The Laws of Logs

If $M > 0$ and $N > 0$, then

1. $\log M \cdot N = \log M + \log N$

2. $\log \dfrac{M}{N} = \log M - \log N$

3. $\log M^n = n \log M$

"The miraculous powers of modern calculation are due to three inventions: the Arabic Notation, Decimal Fractions, and Logarithms."

F. Cajori, 1897

"Electronic calculators make calculations with logarithms as obsolete as whale oil lamps."

Anonymous Reviewer, 1982

6-7
Calculations with Logarithms (Optional)

For 300 years, scientists depended on logarithms to reduce the drudgery associated with long computations. The invention of electronic computers and calculators has diminished the importance of this long-established technique. Still, we think that any student of algebra should know how products, quotients, powers, and roots can be calculated by means of common logarithms.

About all you need are the three laws stated above and Appendix Table B. A little common sense and the ability to organize your work will help.

PRODUCTS

Suppose you want to calculate $(.00872)(95,300)$. Call this number x. Then by Law 1 and Table B,

$$\log x = \log .00872 + \log 95{,}300$$
$$= (.9405 - 3) + 4.9791$$
$$= 5.9196 - 3$$
$$= 2.9196$$

Now use Table B backwards to find that antilog $.9196 = 8.31$, so $x = $ antilog $2.9196 = 831$.

Here is a good way to organize your work in a compact systematic way.

$$x = (.00872)(95{,}300)$$

$$
\begin{array}{rl}
\log .00872 = & .9405 - 3 \\
(+)\ \underline{\log 95300 = } & \underline{4.9791} \\
\log x = & 5.9196 - 3 \\
\end{array}
$$

$$x = 831$$

QUOTIENTS

Suppose we want to calculate $x = .4362/91.84$. Then by Law 2,

$$\log x = \log .4362 - \log 91.84$$
$$= (.6397 - 1) - 1.9630$$
$$= .6397 - 2.9630$$
$$= -2.3233$$

What we have done is correct; however, it is poor strategy. The result we found for log x is not in characteristic-mantissa form and therefore is not usable. Remember that the mantissa must be positive. We can bring this about by adding and subtracting 3.

$$-2.3233 = (-2.3233 + 3) - 3 = .6767 - 3$$

Actually it is better to anticipate the need for doing this and arrange the work as follows.

$$x = \frac{.4362}{91.84}$$

$$
\begin{array}{rll}
\log .4362 = & .6397 - 1 = & 2.6397 - 3 \\
(-)\ \underline{\log 91.48 = } & & \underline{1.9630} \\
\log x = & & .6767 - 3 \\
\end{array}
$$

$$x = .00475$$

POWERS OR ROOTS

Here the main tool is Law 3. We illustrate with two examples.

$$x = (31.4)^{11}$$

$$\log x = 11 \log 31.4$$

$$\log 31.4 = 1.4969$$

$$11 \log 31.4 = 16.4659$$

$$\log x = 16.4659$$

$$x = 29{,}230{,}000{,}000{,}000{,}000$$

$$= 2.923 \times 10^{16}$$

$$x = \sqrt[4]{.427} = (.427)^{1/4}$$

$$\log x = \frac{1}{4} \log .427$$

$$\log .427 = .6304 - 1$$

$$\frac{1}{4} \log .427 = \frac{1}{4}(.6304 - 1) = \frac{1}{4}(3.6304 - 4) = .9076 - 1$$

$$\log x = .9076 - 1$$

$$x = .8084$$

Notice in the second example that we wrote $3.6304 - 4$ in place of $.6304 - 1$, so that multiplication by $\frac{1}{4}$ gave the logarithm in characteristic-mantissa form.

Problem Set 6-7

Use logarithms and Table B without interpolation to find approximate values for each of the following.

1. $(46.3)(2.76)$
2. $(378)(9.63)$
3. $\dfrac{46.3}{483}$

4. $\dfrac{437}{92300}$
5. $\dfrac{.00912}{.439}$
6. $\dfrac{.0429}{15.7}$

7. $(37.2)^5$
8. $(113)^3$
9. $\sqrt[3]{42.9}$

10. $\sqrt[4]{312}$
11. $\sqrt[5]{.918}$
12. $\sqrt[3]{.0307}$

13. $(14.9)^{2/3}$
14. $(98.6)^{3/4}$

Use logarithms and Table B with interpolation to approximate each of the following.

15. $(31.96)(149)$
16. $(6236)(.00108)$
17. $\dfrac{43.98}{7.16}$

18. $\dfrac{115}{4.623}$
19. $(.1234)^6$
20. $(92.83)^3$

EXAMPLE A (More complicated calculations)   Use logarithms, without interpolation, to calculate

$$\frac{(31.4)^3(.982)}{(.0463)(824)}$$

Solution.   Let N denote the entire numerator, D the entire denominator, and x the fraction N/D. Then

$$\log x = \log N - \log D$$

where

$$\log N = 3 \log 31.4 + \log .982$$

$$\log D = \log .0463 + \log 824$$

Here is a good way to organize the work.

$$
\begin{array}{rl}
\log 31.4 = & 1.4969 \\
3 \log 31.4 = & 4.4907 \\
(+) \quad \log .982 = & .9921 - 1 \\
\hline
\log N = & 5.4828 - 1 \\
= & 4.4828
\end{array}
\qquad
\begin{array}{rl}
\log .0463 = & .6656 - 2 \\
(+) \quad \log 824 = & 2.9159 \\
\hline
\log D = & 3.5815 - 2 \\
= & 1.5815
\end{array}
$$

$$
\begin{array}{rl}
\log N = & 4.4828 \\
(-) \; \log D = & 1.5815 \\
\hline
\log x = & 2.9013 \\
x = & 797
\end{array}
$$

Carry out the following calculations using logarithms without interpolation.

21. $\dfrac{(.56)^2(619)}{21.8}$

22. $\dfrac{.413}{(4.9)^2(.724)}$

23. $\dfrac{(14.3)\sqrt{92.3}}{\sqrt[3]{432}}$

24. $\dfrac{(91)(41.3)^{2/3}}{42.6}$

EXAMPLE B (Solving exponential equations)   Solve the equation

$$2^{2x-1} = 13$$

Solution.   We begin by taking logarithms of both sides and then solving for x.

$$(2x - 1)\log 2 = \log 13$$

$$2x - 1 = \frac{\log 13}{\log 2} = \frac{1.1139}{.3010}$$

$$\log(2x - 1) = \log 1.1139 - \log .3010$$

$$= (1.0469 - 1) - (.4786 - 1)$$

$$= .5683$$

$$2x - 1 = \text{antilog } .5683 = 3.70$$

$$2x = 4.70$$

$$x = 2.35$$

Notice that we did the division of (log 13)/(log 2) by means of logarithms. We could have done it by long division (or on a calculator) if we preferred.

Use logarithms to solve the following exponential equations. You need not interpolate.

25. $3^x = 300$ 26. $5^x = 14$ 27. $10^{2-3x} = 6240$

28. $10^{5x-1} = .00425$ 29. $2^{3x} = 3^{x+2}$ 30. $2^{x^2} = 3^x$

Miscellaneous Problems

Use common logarithms and Table B to approximate the following.

31. $(86.4)(.000139)$ 32. $(14.8)^3$ 33. $\dfrac{38.9}{6150}$

34. $\sqrt[3]{.0427}$ 35. $\dfrac{(42.9)^2(.983)}{\sqrt{323}}$ 36. $\dfrac{10^{6.42}}{8^{7.2}}$

Use common logarithms to solve the following equations. You need not interpolate.

37. $4^{2x} = 150$ 38. $(.975)^x = .5$

In Problems 39–42, solve for x. (Hint: First combine the logarithms.)

39. $\log(x + 2) - \log x = 1$ 40. $\log(2x + 1) - \log(x + 3) = 0$

41. $\log \dfrac{x + 3}{x} + \log 2x^2 = \log 8$.

42. $\log(x + 3) + \log(x - 1) = \log 4x$.

43. Suppose that the amount Q of a radioactive substance (in grams) remaining after t years is

$$Q = (42)2^{-.017t}$$

    (a) Use logarithms to calculate the amount remaining after 20 years.
    (b) After how many years will the amount remaining be only .42 grams?

44. Suppose that the number of bacteria in a certain culture t hours from now is $(800)3^t$.
    (a) What will the bacteria count be 3.12 hours from now?
    (b) When will the bacteria count reach 100,000?

45. Assume the 1977 population of the earth was 4.19 billion and that the growth rate is 2 percent per year. Then the population t years after 1977 will be $4.19(10^9)(1.02)^t$.
    (a) What will the population of the earth be in the year 2000?
    (b) When will the population reach 8.63 billion?

46. Answer the two questions in Problem 45 for a growth rate of 3 percent and then for a growth rate of 1 percent.

47. The volume of a sphere is given by $V = \frac{4}{3}\pi r^3$, where r is the radius of the sphere. Find the radius of a sphere whose volume is 42,900 cubic feet.

CHAPTER SUMMARY

The symbol $\sqrt[n]{a}$, the principal nth root of a, denotes one of the numbers whose nth power is a. For odd n, that is all that needs to be said. For n even and $a > 0$, we specify that $\sqrt[n]{a}$ signifies the positive nth root. Thus $\sqrt[3]{-8} = -2$ and $\sqrt{16} = \sqrt[2]{16} = 4$. (It is wrong to write $\sqrt{16} = -4$.) These symbols are also called **radicals.** These radicals obey four carefully prescribed rules (page 213). These rules allow us to simplify complicated radical expressions, in particular, to **rationalize denominators.**

The key to understanding **rational exponents** is the definition $a^{1/n} = \sqrt[n]{a}$, which implies $a^{m/n} = (\sqrt[n]{a})^m$. Thus $16^{5/4} = (\sqrt[4]{16})^5 = 2^5 = 32$. The meaning of **real exponents** is determined by considering rational approximations. For example, 2^π is the number that the sequence $2^3, 3^{3.1}, 2^{3.14}, \ldots$ approaches. The function $f(x) = a^x$ (and more generally $f(x) = b \cdot a^x$) is called an **exponential function.**

A variable y is **growing exponentially** or **decaying exponentially** according as $a > 1$ or $0 < a < 1$ in the equation $y = b \cdot a^x$. Typical of the former are biological populations; of the latter, radioactive elements. Corresponding key ideas are **doubling times** and **half-lives.**

Logarithms are exponents. In fact, $\log_a N = x$ means $a^x = N$, that is, $a^{\log_a N} = N$. The functions $f(x) = \log_a x$ and $g(x) = a^x$ are **inverses** of each other. Logarithms have three primary properties (page 235). **Natural logarithms** correspond to the choice of base $a = e = 2.71828 \ldots$ and play a fundamental role in advanced courses. **Common logarithms** correspond to base 10 and have historically been used to simplify arithmetic calculations.

CHAPTER REVIEW PROBLEM SET

1. Simplify, rationalizing all denominators. Assume letters represent positive numbers.
   (a) $\sqrt[3]{\dfrac{-8y^6}{z^{14}}}$ (b) $\sqrt[4]{32x^5y^8}$ (c) $\sqrt{4\sqrt[3]{5}}$
   (d) $\dfrac{2}{\sqrt{x} - \sqrt{y}}$ (e) $\sqrt{50 + 25x^2}$ (f) $\sqrt{32} + \sqrt{8}$

2. Solve the equations.
   (a) $\sqrt{x - 3} = 3$ (b) $\sqrt{x} = 6 - x$

3. Simplify, writing your answer in exponential form with all positive exponents.
   (a) $(25a^2)^{3/2}$ (b) $(a^{-1/2}aa^{-3/4})^2$ (c) $\dfrac{1}{5\sqrt[4]{5^3}}$
   (d) $\dfrac{(3x^{-2}y^{3/4})^2}{3x^2y^{-2/3}}$ (e) $(x^{1/2} - y^{1/2})^2$ (f) $\sqrt[3]{4}\sqrt[4]{4}$

4. Sketch the graph of $y = (\tfrac{3}{2})^x$ and use your graph to estimate the value of $(\tfrac{3}{2})^\pi$.

5. A certain radioactive substance has a half-life of 3 days. What fraction of an initial amount will be left after 243 days?

6. A population grows so that its doubling time is 40 years. If this population is 1 million today, what will it be after 160 years?

7. If $100 is put in the bank at 8 percent interest compounded quarterly, what will it be worth at the end of 10 years?

8. Find x in each of the following.
   (a) $\log_4 64 = x$ (b) $\log_2 x = -3$
   (c) $\log_x 49 = 2$ (d) $\log_4 x = 0$
   (e) $\log_9 27 = x$ (f) $\log_{10} x + \log_{10}(x - 3) = 1$
   (g) $a^{\log_a 10} = x$ (h) $x = \log_a(a^{1.14})$

9. Write as a single logarithm.

$$2 \log_4(3x + 1) - \frac{1}{2} \log_4 x + \log_4(x - 1)$$

10. Evaluate.
    (a) $\log_{10} \sqrt{1000}$ (b) $\log_{27} 81$
    (c) $\ln \sqrt{e}$ (d) $\ln(1/2^4)$

11. Use the table of natural logarithms to determine each of the following.
    (a) $\ln(\frac{7}{4})^3$ (b) N if $\ln N = 2.230$
    (c) $\ln[(3.4)(9.9)]$ (d) N if $\ln N = -0.105$

12. By taking ln of both sides, solve $2^{x+1} = 7$.

13. A certain substance decays according to the formula $y = y_0 e^{-.05t}$, where t is measured in years. Find its half-life.

14. A substance initially weighing 100 grams decays exponentially according to the formula $y = 100e^{-kt}$. If 30 grams are left after 10 days, determine k.

15. Sketch the graphs of $y = \log_3 x$ and $y = 3^x$ using the same coordinate axes.

16. Use common logarithms to calculate

$$\frac{(13.2)^4 \sqrt{15.2}}{29.6}$$

7

THEORY OF POLYNOMIAL EQUATIONS

He who loves practice without theory is like the sailor who boards ship without a rudder and compass and never knows where he may cast.

Leonardo da Vinci

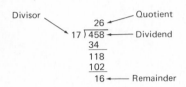

Division of Integers

Divisor ⟍ ⟋ Quotient
26
17)‾458‾ ⟵ Dividend
34
118
102
16 ⟵ Remainder

$$\frac{458}{17} = 26 + \frac{16}{17}$$

Division of Polynomials

Divisor ⟍ ⟋ Quotient
3x − 1 ⟵
$2x^2 - x + 3$)‾$6x^3 - 5x^2 + x - 4$‾ ⟵ Dividend
$6x^3 - 3x^2 + 9x$

$- 2x^2 - 8x - 4$
$- 2x^2 + x - 3$
$- 9x - 1$ ⟵ Remainder

$$\frac{6x^3 - 5x^2 + x - 4}{2x^2 - x + 3} = 3x - 1 + \frac{-9x - 1}{2x^2 - x + 3}$$

7-1
Division
of Polynomials

In Section 2-4, we learned that polynomials in x can always be added, sub-tracted, and multiplied; the result in every case is a polynomial. For example,

$$(3x^2 + x - 2) + (3x - 2) = 3x^2 + 4x - 4$$

$$(3x^2 + x - 2) - (3x - 2) = 3x^2 - 2x$$

$$(3x^2 + x - 2) \cdot (3x - 2) = 9x^3 - 3x^2 - 8x + 4$$

Now we are going to study division of polynomials. Occasionally the division is exact.

$$(3x^2 + x - 2) \div (3x - 2) = \frac{(3x - 2)(x + 1)}{3x - 2} = x + 1$$

We say $3x - 2$ is an exact divisor, or factor, of $3x^2 + x - 2$. More often than not, the division is inexact and there is a nonzero remainder.

THE DIVISION ALGORITHM

The opening display shows that the process of division for polynomials is much the same as for integers. Both processes (we call them *algorithms*) involve subtraction. The first one shows how many times 17 can be subtracted from 458 before obtaining a remainder less than 17. The answer is 26 times, with a remainder of 16. The second algorithm shows how many times $2x^2 - x + 3$ can be subtracted from $6x^3 - 5x^2 + x - 4$ before obtaining a remainder of lower degree than $2x^2 - x + 3$. The answer is $3x - 1$ times, with a remainder of $-9x - 1$. Thus in these two examples we have

$$458 - 26(17) = 16$$

$$(6x^3 - 5x^2 + x - 4) - (3x - 1)(2x^2 - x + 3) = -9x - 1$$

Of course, we can also write these equalities as

$$458 = (17)(26) + 16$$

$$6x^3 - 5x^2 + x - 4 = (2x^2 - x + 3)(3x - 1) + (-9x - 1)$$

Notice that both of them can be summarized in the words

$$\text{dividend} = (\text{divisor}) \cdot (\text{quotient}) + \text{remainder}$$

This statement is very important and is worth stating again for polynomials in very precise language.

If $P(x)$ and $D(x)$ are any two nonconstant polynomials, then there are unique polynomials $Q(x)$ and $R(x)$ such that

$$\boxed{P(x) = D(x)Q(x) + R(x)}$$

where $R(x)$ is either zero or it is of lower degree than $D(x)$.

Here you should think of $P(x)$ as the dividend, $D(x)$ as the divisor, $Q(x)$ as the quotient, and $R(x)$ as the remainder.

The algorithm we use to find $Q(x)$ and $R(x)$ was illustrated in the opening display. Here is another illustration, this time for polynomials with some nonreal coefficients. Notice that we always arrange the divisor and dividend in descending powers of x before we start the division process.

$$
\begin{array}{r}
x - i \\
3x + i \overline{\smash{\big)}\, 3x^2 - 2ix + 7} \\
\underline{3x^2 + ix} \\
-3ix + 7 \\
\underline{-3ix + 1} \\
6
\end{array}
\qquad
\begin{array}{l}
Q(x) = x - i \\
R(x) = 6
\end{array}
$$

SYNTHETIC DIVISION

It is often necessary to divide by polynomials of the form $x - c$. For such a division, there is a shortcut called *synthetic division*. We illustrate how it works for

$$(2x^3 - x^2 + 5) \div (x - 2)$$

Certainly the result depends on the coefficients; the powers of x serve mainly to determine the placement of the coefficients. Below, we show the division in its usual form and then in a skeletal form with the x's omitted. Note that we leave a blank space for the missing first degree term in the long division but indicate it with a 0 in the skeletal form.

LONG DIVISION

$$
\begin{array}{r}
2x^2 + 3x + 6 \\
x - 2 \overline{\smash)\ 2x^3 - x^2 \qquad + 5} \\
\underline{2x^3 - 4x^2} \\
3x^2 \\
\underline{3x^2 - 6x} \\
6x + 5 \\
\underline{6x - 12} \\
17
\end{array}
$$

FIRST CONDENSATION

$$
\begin{array}{r}
②\quad ③\quad ⑥ \\
① - 2 \overline{\smash)\ 2 \quad -1 \quad 0 \quad\quad 5} \\
2 \quad -4 \\
3 \quad\quad ⓪ \\
③ - 6 \\
6 \quad\quad ⑤ \\
⑥ - 12 \\
17
\end{array}
$$

We can condense things still more by discarding all of the circled digits. The coefficients of the quotient, 2, 3, and 6, the remainder, 17, and the numbers from which they were calculated remain. All of the important numbers appear in the diagram below, on the left. On the right, we show the final modification. There we have changed the divisor from -2 to 2 to allow us to do addition rather than subtraction at each stage.

SECOND CONDENSATION

$$
\begin{array}{r}
-2 \,\underline{\big|\ 2 \quad -1 \quad\ 0 \quad\ 5} \\
-4 \quad -6 \quad -12 \\
\hline
2 \quad\ 3 \quad\ 6 \quad\ 17
\end{array}
$$

SYNTHETIC DIVISION

$$
\begin{array}{r}
2 \,\underline{\big|\ 2 \quad -1 \quad\ 0 \quad\ 5} \\
4 \quad\ 6 \quad\ 12 \\
\hline
2 \quad\ 3 \quad\ 6 \quad\ 17
\end{array}
$$

The process shown in the final format is called **synthetic division.** We can describe it by a series of steps.

1. To divide by $x - 2$, use 2 as the synthetic divisor.
2. Write down the coefficients of the dividend. Be sure to write zeros for missing powers.
3. Bring down the first coefficient.

4. Follow the arrows, first multiplying by 2 (the divisor), then adding, multiplying the sum by 2, adding, and so on.
5. The last number in the third row is the remainder and the others are the coefficients of the quotient.

Here is another example. We use synthetic division to divide $3x^3 - x^2 + 15x - 5$ by $x - \frac{1}{3}$.

$$\frac{1}{3} \begin{array}{|rrrr} 3 & -1 & 15 & -5 \\ & 1 & 0 & 5 \\ \hline 3 & 0 & 15 & 0 \end{array}$$

Since the remainder is 0, the division is exact. We conclude that

$$3x^3 - x^2 + 15x - 5 = (x - \tfrac{1}{3})(3x^2 + 15)$$

PROPER AND IMPROPER RATIONAL EXPRESSIONS

A rational expression (ratio of two polynomials) is said to be **proper** if the degree of its numerator is smaller than that of its denominator. Thus

$$\frac{x + 1}{x^2 - 3x + 2}$$

is a proper rational expression, but

$$\frac{2x^3}{x^2 - 3}$$

is improper. The division law, $P(x) = D(x)Q(x) + R(x)$, can be written as

$$\frac{P(x)}{D(x)} = Q(x) + \frac{R(x)}{D(x)}$$

It implies that any improper rational expression can be rewritten as the sum of a polynomial and a proper rational expression. For example,

$$\frac{2x^3}{x^2 - 3} = 2x + \frac{6x}{x^2 - 3}$$

a result we obtained by dividing $x^2 - 3$ into $2x^3$.

Problem Set 7-1

In Problems 1–8, find the quotient and the remainder if the first polynomial is divided by the second.

1. $x^3 - x^2 + x + 3$; $x^2 - 2x + 3$
2. $x^3 - 2x^2 - 7x - 4$; $x^2 + 2x + 1$
3. $6x^3 + 7x^2 - 18x + 15$; $2x^2 + 3x - 5$

4. $10x^3 + 13x^2 + 5x + 12; 5x^2 - x + 4$
5. $4x^4 - x^2 - 6x - 9; 2x^2 - x - 3$
6. $25x^4 - 20x^3 + 4x^2 - 4; 5x^2 - 2x + 2$
7. $2x^5 - 2x^4 + 9x^3 - 12x^2 + 4x - 16; 2x^3 - 2x^2 + x - 4$
8. $3x^5 - x^4 - 8x^3 - x^2 - 3x + 12; 3x^3 - x^2 + x - 4$

In Problems 9–14, write each rational expression as the sum of a polynomial and a proper rational expression.

9. $\dfrac{x^3 + 2x^2 + 5}{x^2}$

10. $\dfrac{2x^3 - 4x^2 - 3}{x^2 + 1}$

11. $\dfrac{x^3 - 4x + 5}{x^2 + x - 2}$

12. $\dfrac{2x^3 + x - 8}{x^2 - x + 4}$

13. $\dfrac{2x^2 - 4x + 5}{x^2 + 1}$

14. $\dfrac{5x^3 - 6x + 11}{x^3 - x}$

In Problems 15–20, use synthetic division to find the quotient and remainder if the first polynomial is divided by the second.

15. $2x^3 - x^2 + x - 4; x - 1$

16. $3x^3 + 2x^2 - 4x + 5; x - 2$

17. $3x^3 + 5x^2 + 2x - 10; x - 1$

18. $2x^3 - 5x^2 + 4x - 4; x - 2$

19. $x^4 - 2x^2 - 1; x - 3$

20. $x^4 + 3x^2 - 340; x - 4$

EXAMPLE A (Synthetic division by $x + c$)   Find the quotient and remainder when $2x^3 - 2x^2 + 5$ is divided by $x + 3$.

Solution.   Since $x + 3 = x - (-3)$, we may use synthetic division with -3 as the synthetic divisor.

$$
\begin{array}{r|rrrr}
-3 & 2 & -2 & 0 & 5 \\
   &   & -6 & 24 & -72 \\
\hline
   & 2 & -8 & 24 & -67
\end{array}
$$

The quotient is $2x^2 - 8x + 24$ and the remainder is -67.

Find the quotient and the remainder when the first polynomial is divided by the second.

21. $x^3 + 2x^2 - 3x + 2; x + 1$
22. $x^3 - x^2 + 11x - 1; x + 1$
23. $x^4 + 2x^3 + 4x^2 + 7x + 4; x + 1$
24. $2x^4 + x^3 + x^2 + 10x - 8; x + 2$

EXAMPLE B (Division by $x - (a + bi)$)   Find the quotient and remainder when $x^3 - 2x^2 - 6ix + 18$ is divided by $x - 2 - 3i$.

Solution.   Synthetic division works just fine even when some or all of the coefficients are nonreal. Since $x - 2 - 3i = x - (2 + 3i)$, we use $2 + 3i$ as the synthetic divisor.

$$\begin{array}{r|rrrr} 2 + 3i & 1 & -2 & -6i & 18 \\ & & 2 + 3i & -9 + 6i & -18 - 27i \\ \hline & 1 & 3i & -9 & -27i \end{array}$$

Quotient: $x^2 + 3ix - 9$; remainder: $-27i$.

Use synthetic division to find the quotient and remainder when the first polynomial is divided by the second.

☐ 25. $x^3 - 2x^2 + 5x + 30$; $x - 2 + 3i$
☐ 26. $2x^3 - 11x^2 + 44x + 35$; $x - 3 - 4i$
☐ 27. $x^4 - 17$; $x - 2i$
☐ 28. $x^4 + 18x^2 + 90$; $x - 3i$

Miscellaneous Problems

29. Find the quotient and remainder when $x^4 + 6x^3 - 2x^2 + 4x - 15$ is divided by $x^2 - 2x + 3$.
30. Express $(3x^2 + 4x)/(x - 2)$ as the sum of a polynomial and a proper rational expression.

In Problems 31–34, find the quotient and remainder when the first polynomial is divided by the second.

31. $2x^3 + 5x - 2$; $x + 2$
32. $x^4 - 4x^3 + 27$; $x - 3$
33. $x^5 - 32$; $x - 2$
☐ 34. $x^3 + (2i - 2)x^2 - (9 + 10i)x + 20 + 12i$; $x - 3$

In Problems 35–38, use division to show that the second polynomial is a factor of the first. You may use synthetic division where it applies.

35. $x^5 + x^4 - 16x - 16$; $x - 2$
☐ 36. $x^5 + x^4 - 16x - 16$; $x - 2i$
☐ 37. $x^5 + x^4 - 16x - 16$; $x + 2i$
38. $x^5 - 32$; $x^4 + 2x^3 + 4x^2 + 8x + 16$

In Problems 39–42, determine by inspection the remainder when the first polynomial is divided by the second.

39. $(x - 2)^3 - 3(x - 2)^2 + 5(x - 2) + 11$; $x - 2$
    (*Hint:* The first polynomial can be written as $(x - 2)[(x - 2)^2 - 3(x - 2) + 5] + 11$.)
40. $2(x + 3)^2 + 118(x + 3) - 14$; $x + 3$
41. $(x - 4)^6 + 14(x - 4)^3 + 25$; $(x - 4)^3$
42. $(x + b)^3 - 10(x + b)^2 + 4(x + b) - 11$; $x + b$

Carl F. Gauss (1777–1855)
"The Prince of Mathematicians"

Young Scholar: And does $x^4 + 99x^3 + 21$ have a zero?
Carl Gauss: Yes.
Young Scholar: How about $\pi x^{67} - \sqrt{3}\, x^{19} + 4i$?
Carl Gauss: It does.
Young Scholar: How can you be sure?
Carl Gauss: When I was young, 22 I think, I proved that every non-constant polynomial, no matter how complicated, has at least one zero. Many people call it the *Fundamental Theorem of Algebra.*

7-2
Factorization Theory for Polynomials

Our young scholar could have asked a harder question. How do you find the zeros of a polynomial? Even the eminent Gauss would have had trouble with that question. You see, it is one thing to know a polynomial has zeros; it is quite another thing to find them.

Even though it is a difficult task and one at which we will have only limited success, our goal for this and the next two sections is to develop methods for finding zeros of polynomials. Remember that a polynomial is an expression of the form

$$P(x) = a_n x^n + a_{n-1} x^{n-1} + \cdots + a_1 x + a_0$$

Unless otherwise specified, the coefficients (the a_i's) are allowed to be *complex* numbers. And by a **zero** of $P(x)$, we mean any complex number c (real or nonreal) such that $P(c) = 0$. The number c is also called a **solution,** or a **root,** of the equation $P(x) = 0$. Note the use of words: Polynomials have zeros, but polynomial equations have solutions.

THE REMAINDER AND FACTOR THEOREMS

Recall the division law from Section 7-1, which had as its conclusion

$$P(x) = D(x)Q(x) + R(x)$$

If $D(x)$ has the form $x - c$, this becomes

$$P(x) = (x - c)Q(x) + R$$

where R, which is of lower degree than $x - c$, must be a constant. This last equation is an identity; it is true for all values of x, including $x = c$. Thus

$$P(c) = (c - c)Q(c) + R = 0 + R$$

We have just proved an important result.

REMAINDER THEOREM

If a polynomial $P(x)$ is divided by $x - c$, then the constant remainder R is given by $R = P(c)$.

Here is a nice example. Suppose we want to know the remainder R when $P(x) = x^{1000} + x^{22} - 15$ is divided by $x - 1$. We could, of course, divide it out—but what a waste of energy, especially since we know the remainder theorem. From it we learn that

$$R = P(1) = 1^{1000} + 1^{22} - 15 = -13$$

Much more important than the mere calculation of remainders is a consequence called the *factor theorem*. Since $R = P(c)$, as we have just seen, we may rewrite the division law as

$$P(x) = (x - c)Q(x) + P(c)$$

It is plain to see that $P(c) = 0$ if and only if the division of $P(x)$ by $x - c$ is exact; that is, if and only if $x - c$ is a factor of $P(x)$.

FACTOR THEOREM

A polynomial $P(x)$ has c as a zero if and only if it has $x - c$ as a factor.

Sometimes it is easy to spot one zero of a polynomial. If so, the factor theorem may help us find the other zeros. Consider the polynomial

$$P(x) = 3x^3 - 8x^2 + 3x + 2$$

Notice that

$$P(1) = 3 - 8 + 3 + 2 = 0$$

so 1 is a zero. By the factor theorem, $x - 1$ is a factor of $P(x)$. We can use synthetic division to find the other factor.

$$
\begin{array}{r|rrrr}
1 & 3 & -8 & 3 & 2 \\
  &   & 3 & -5 & -2 \\
\hline
  & 3 & -5 & -2 & 0
\end{array}
$$

The remainder is 0 as we expected, and

$$P(x) = (x - 1)(3x^2 - 5x - 2)$$

Using the quadratic formula, we find the zeros of $3x^2 - 5x - 2$ to be $(5 \pm \sqrt{49})/6$, which simplify to 2 and $-\frac{1}{3}$. Thus $P(x)$ has 1, 2, and $-\frac{1}{3}$ as its three zeros.

COMPLETE FACTORIZATION OF POLYNOMIALS

In the example above, we did not really need the quadratic formula. If we had been clever, we would have factored $3x^2 - 5x - 2$.

$$3x^2 - 5x - 2 = (3x + 1)(x - 2)$$
$$= 3(x + \tfrac{1}{3})(x - 2)$$

Thus $P(x)$, our original polynomial, may be written as

$$P(x) = 3(x - 1)(x + \tfrac{1}{3})(x - 2)$$

from which all three of the zeros are immediately evident.

But now we make another key observation. Notice that $P(x)$ can be factored as a product of its leading coefficient and three factors of the form $(x - c)$, where the c's are the zeros of $P(x)$. This holds true in general.

COMPLETE FACTORIZATION THEOREM

If

$$P(x) = a_n x^n + a_{n-1} x^{n-1} + \cdots + a_1 x + a_0$$

is an nth degree polynomial with $n > 0$, then there are n numbers c_1, c_2, . . . , c_n, not necessarily distinct, such that

$$P(x) = a_n(x - c_1)(x - c_2) \cdots (x - c_n)$$

The c's are the zeros of $P(x)$; they may or may not be real numbers.

To prove the complete factorization theorem, we must go back to Carl Gauss and our opening display. In his doctoral dissertation in 1799, Gauss gave a proof of the following important theorem, a proof that unfortunately is beyond the scope of this book.

FUNDAMENTAL THEOREM OF ALGEBRA

Every nonconstant polynomial has at least one zero.

Now let $P(x)$ be any polynomial of degree $n > 0$. By the fundamental theorem, it has a zero, which we may call c_1. By the factor theorem, $x - c_1$ is a factor of $P(x)$; that is,

$$P(x) = (x - c_1)P_1(x)$$

where $P_1(x)$ is a polynomial of degree $n - 1$ and with the same leading coefficient as $P(x)$—namely, a_n.

If $n - 1 > 0$, we may repeat the argument on $P_1(x)$. It has a zero c_2 and hence a factor $x - c_2$; that is,

$$P_1(x) = (x - c_2)P_2(x)$$

where $P_2(x)$ has degree $n - 2$. For our original polynomial $P(x)$, we may now write

$$P(x) = (x - c_1)(x - c_2)P_2(x)$$

Continuing in the pattern now established, we eventually get

$$P(x) = (x - c_1)(x - c_2) \cdots (x - c_n)P_n$$

where P_n has degree zero; that is, P_n is a constant. In fact, $P_n = a_n$, since the leading coefficient stayed the same at each step. This establishes the complete factorization theorem.

ABOUT THE NUMBER OF ZEROS

Each of the numbers c_i in

$$P(x) = a_n(x - c_1)(x - c_2) \cdots (x - c_n)$$

is a zero of $P(x)$. Are there any other zeros? No, for if d is any number different from each of the c_i's, then

$$P(d) = a_n(d - c_1)(d - c_2) \cdots (d - c_n) \neq 0$$

All of this tempts us to say that a polynomial of degree n has exactly n zeros. But hold on! The numbers $c_1, c_2, \ldots, c_n$ need not all be different. For example, the sixth degree polynomial

$$P(x) = 4(x - 2)^3(x + 1)(x - 4)^2$$

has only three distinct zeros, 2, -1, and 4. We have to settle for the following statement.

An nth degree polynomial has at most n distinct zeros.

There is a way in which we can say that there are exactly n zeros. Call c a **zero of multiplicity k** of $P(x)$ if $x - c$ appears k times in its complete factorization. For example, in

$$P(x) = 4(x - 2)^3(x + 1)(x - 4)^2$$

the zeros 2, -1, and 4 have multiplicities 3, 1, and 2, respectively. A zero of multiplicity 1 is called a **simple zero.** Notice in our example that the multiplicities add to 6, the degree of the polynomial. In general, we may say that an nth degree polynomial has exactly n zeros if we count a zero of multiplicity k as k zeros.

Problem Set 7-2

Use the remainder theorem to find P(c). Check your answer by substituting c for x.

1. $P(x) = 2x^3 - 5x^2 + 3x - 4; c = 2$
2. $P(x) = x^3 + 4x^2 - 11x - 5; c = 3$
3. $P(x) = 8x^4 - 3x^2 - 2; c = \frac{1}{2}$
4. $P(x) = 2x^4 + \frac{3}{4}x + \frac{3}{2}; c = -\frac{1}{2}$

Find the remainder if the first polynomial is divided by the second. Do it without actually dividing.

5. $x^{10} - 15x + 8$; $x - 1$

6. $2x^{20} + 5$; $x + 1$

7. $64x^6 + 13$; $x + \frac{1}{2}$

8. $81x^3 + 9x^2 - 2$; $x - \frac{1}{3}$

Find all of the zeros of the given polynomial and give their multiplicities.

9. $(x - 1)(x + 2)(x - 3)$

10. $(x + 2)(x + 5)(x - 7)$

11. $(2x - 1)(x - 2)^2 x^3$

12. $(3x + 1)(x + 1)^3 x^2$

⊡ 13. $3(x - 1 - 2i)(x + \frac{2}{3})$

14. $5(x - 2 + \sqrt{5})(x - \frac{4}{5})$

In Problems 15–18, show that $x - c$ is a factor of $P(x)$.

15. $P(x) = 2x^3 - 7x^2 + 9x - 4$; $c = 1$

16. $P(x) = 3x^3 + 4x^2 - 6x - 1$; $c = 1$

17. $P(x) = x^3 - 7x^2 + 16x - 12$; $c = 3$

18. $P(x) = x^3 - 8x^2 + 13x + 10$; $c = 5$

19. In Problem 15, you know that $P(x)$ has 1 as a zero. Find the other zeros.

20. Find all of the zeros of $P(x)$ in Problem 16. Then factor $P(x)$ completely.

21. Find all of the zeros of $P(x)$ in Problem 17.

22. Find all of the zeros of $P(x)$ in Problem 18.

In Problems 23–26, factor the given polynomial into linear factors. You should be able to do it by inspection.

23. $x^2 - 5x + 6$

24. $2x - 14x + 24$

25. $x^4 - 5x^2 + 4$

26. $x^4 - 13x^2 + 36$

In Problems 27–30, factor $P(x)$ into linear factors given that c is a zero of $P(x)$.

27. $P(x) = x^3 - 3x^2 - 28x + 60$; $c = 2$

28. $P(x) = x^3 - 2x^2 - 29x - 42$; $c = -2$

29. $P(x) = x^3 + 3x^2 - 10x - 12$; $c = -1$

30. $P(x) = x^3 + 11x^2 - 5x - 55$; $c = -11$

EXAMPLE A (Finding a polynomial from its zeros)
(a) Find a cubic polynomial having simple zeros 3, $2i$, and $-2i$.
(b) Find a polynomial $P(x)$ with integral coefficients and having $\frac{1}{2}$ and $-\frac{2}{3}$ as simple zeros and 1 as a zero of multiplicity 2.

Solution.
(a) Let us call the required polynomial $P(x)$. Then

$$P(x) = a(x - 3)(x - 2i)(x + 2i)$$

where a can be any nonzero number. Choosing $a = 1$ and multiplying, we have

$$P(x) = (x - 3)(x^2 + 4) = x^3 - 3x^2 + 4x - 12$$

(b) $P(x) = a(x - \frac{1}{2})(x + \frac{2}{3})(x - 1)^2$

We choose $a = 6$ to eliminate fractions.

$$P(x) = 6(x - \tfrac{1}{2})(x + \tfrac{2}{3})(x - 1)^2$$
$$= 2(x - \tfrac{1}{2})3(x + \tfrac{2}{3})(x - 1)^2$$
$$= (2x - 1)(3x + 2)(x^2 - 2x + 1)$$
$$= 6x^4 - 11x^3 + 2x^2 + 5x - 2$$

In Problems 31–38, find a polynomial $P(x)$ with integral coefficients having the given zeros. Assume each zero to be simple (multiplicity 1) unless otherwise indicated.

31. $2, 1,$ and -4 32. $3, -2,$ and 5

33. $\frac{1}{2}, -\frac{5}{6}$ 34. $\frac{3}{7}, \frac{3}{4}$

35. $2, \sqrt{5}, -\sqrt{5}$ 36. $-3, \sqrt{7}, -\sqrt{7}$

37. $\frac{1}{2}$ (multiplicity 2), -2 (multiplicity 3)   38. $0, -2, \frac{3}{4}$ (multiplicity 3)

In Problems 39–42, find a polynomial $P(x)$ having only the indicated simple zeros.

⊡ 39. $2, -2, i, -i$ ⊡ 40. $2i, -2i, 3i, -3i$

⊡ 41. $2, -5, 2 + 3i, 2 - 3i$ ⊡ 42. $-3, 2, 1 - 4i, 1 + 4i$

EXAMPLE B (More on zeros of polynomials)   Show that 2 is a zero of multiplicity 3 of the polynomial

$$P(x) = 2x^5 - 17x^4 + 51x^3 - 58x^2 + 4x + 24$$

and find the remaining zeros.

Solution.   We must show that $x - 2$ appears as a factor 3 times in the factored form of $P(x)$. Synthetic division can be used successively, as shown below.

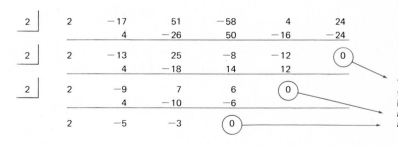

CAUTION

$$(x - 3i)(x + 3i) = x^2 - 9$$

$$(x - 3i)(x + 3i) = x^2 - 9i^2$$
$$= x^2 + 9$$

The final quotient $2x^2 - 5x - 3$ factors as $(2x + 1)(x - 3)$. Therefore the remaining two zeros are $-\frac{1}{2}$ and 3. The factored form of $P(x)$ is

$$2(x - 2)^3(x + \tfrac{1}{2})(x - 3)$$

43. Show that 1 is a zero of multiplicity 3 of the polynomial $x^5 + 2x^4 - 6x^3 - 4x^2 + 13x - 6$, find the remaining zeros, and factor completely.

44. Show that the polynomial $x^5 - 11x^4 + 46x^3 - 90x^2 + 81x - 27$ has 3 as a zero of multiplicity 3, find the remaining zeros, and factor completely.

45. Show that the polynomial

$$x^6 - 8x^5 + 7x^4 + 32x^3 + 31x^2 + 40x + 25$$

has -1 and 5 as zeros of multiplicity 2 and find the remaining zeros.

46. Show that the polynomial

$$x^6 + 3x^5 - 9x^4 - 50x^3 - 84x^2 - 72x - 32$$

has 4 as a simple zero and -2 as a zero of multiplicity 3. Find the remaining zeros.

EXAMPLE C (Factoring a polynomial with a given nonreal zero) Show that $1 + 2i$ is a zero of $P(x) = x^3 - (1 + 2i)x^2 - 4x + 4 + 8i$. Then factor $P(x)$ into linear factors.

Solution. We start by using synthetic division.

$$
\begin{array}{r|rrrr}
1 + 2i & 1 & -1 - 2i & -4 & 4 + 8i \\
       &   & 1 + 2i & 0 & -4 - 8i \\
\hline
       & 1 & 0   & -4 & 0
\end{array}
$$

Therefore $1 + 2i$ is a zero and $x - 1 - 2i$ is a factor of $P(x)$, and

$$P(x) = (x - 1 - 2i)(x^2 - 4)$$
$$= (x - 1 - 2i)(x + 2)(x - 2)$$

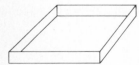

In each of the following, factor P(x) into linear factors given that c is a zero of P(x).

47. $P(x) = x^3 - 2ix^2 - 9x + 18i; \; c = 2i$

48. $P(x) = x^3 + 3ix^2 - 4x - 12i; \; c = -3i$

49. $P(x) = x^3 + (1 - i)x^2 - (1 + 2i)x - 1 - i; \; c = 1 + i$

50. $P(x) = x^3 - 3ix^2 + (3i - 3)x - 2 + 6i; \; c = -1 + 3i$

Miscellaneous Problems

51. Without actually dividing, find the remainder when $x^5 - 10$ is divided by $x - 2$.

In Problems 52-54, find all the zeros of the given polynomials and give their multiplicities.

52. $(x^2 - 4)^3$ 53. $(x^2 - 3x + 2)^2$ 54. $(x^2 + 2x + 4)(x + 2)^2$

55. Find a polynomial $P(x)$ with integral coefficients that has the given zeros.
    (a) $\frac{3}{4}, 2, -\frac{2}{3}$ (b) $3, -2$ (multiplicity 2)

56. Find the polynomial $P(x) = ax^2 + bx + c$ which has 1 and -2 as zeros and which satisfies $P(3) = 40$. (*Hint:* First find a from the factored form of $P(x)$.)

57. Sketch the graph of $y = x^3 + 4x^2 - 2x$. Now find the three values of x for which $y = 8$. (*Hint:* One of these is an integer.)

58. Refer to the graph of Problem 57 and note that there is only one real x (namely, $x = 2$) for which $y = 20$. What does this tell you about the other two solutions of $x^3 + 4x^2 - 2x = 20$? Find the other two solutions.

59. A tray is to be constructed from a piece of sheet metal 16 inches square by cutting equal squares from the corners and then folding up the flaps (see the diagrams in the margin). How large must these squares be if the volume is to be 300 cubic inches? (*Hint:* There are two possible answers, one of which is an integer.)

60. Take a number, add 2, multiply the result by the original number, and then add 9. Now multiply this result by the original number, subtract 2, again multiply by the original number, and finally subtract 10. What must your original number be if the final outcome is 0?

61. Find $P(x) = a_4x^4 + a_3x^3 + a_2x^2 + a_1x + a_0$ if $P(x)$ has $\frac{1}{2}$ as a zero of multiplicity 4 and $P(0) = 1$.

62. Use the factor theorem to show that
    (a) $x^n - a^n$ has $x - a$ as a factor for any positive integer n, where a is any complex number;
    (b) $x^n + a^n$ has $x + a$ as a factor if n is odd;
    (c) $x^n + a^n$ does not have $x + a$ as a factor if n is even.

63. Show that the polynomial

$$P(x) = x^{40} + 10x^{22} + 4x^8 + 16x^2 + 15$$

has no real zeros.

64. Find the zeros of $(x^2 + 1)^4$ and their multiplicities.

65. Show that if

$$P(x) = a_nx^n + a_{n-1}x^{n-1} + \cdots + a_1x + a_0$$

has $n + 1$ distinct zeros, then all of the coefficients have to be zero.

66. Let

$$P_1(x) = x^n + a_{n-1}x^{n-1} + a_{n-2}x^{n-2} + \cdots + a_1x + a_0$$

and

$$P_2(x) = x^n + b_{n-1}x^{n-1} + b_{n-2}x^{n-2} + \cdots + b_1x + b_0$$

Show that if $P_1(x) = P_2(x)$ for n distinct values of x, then the two polynomials are identical; that is, $a_i = b_i$ for each i. (*Hint:* Let $P(x) = P_1(x) - P_2(x)$ and apply Problem 65.)

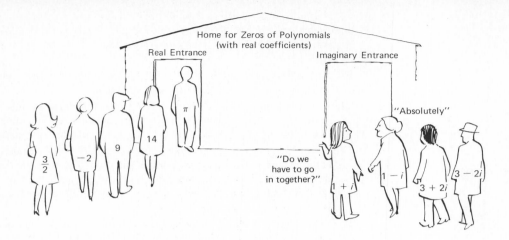

7-3
Polynomial Equations with Real Coefficients

Finding the zeros of a polynomial $P(x)$ is the same as finding the solutions of the equation $P(x) = 0$. We have solved polynomial equations before, especially in Chapter 3. In particular, recall that the two solutions of a quadratic equation with real coefficients are either both real or both nonreal. For example, the equation $x^2 - 7x + 12 = 0$ has the real solutions 3 and 4. On the other hand, $x^2 - 8x + 17 = 0$ has two nonreal solutions $4 + i$ and $4 - i$.

In this section, we shall see that if a polynomial equation has real coefficients, then its nonreal solutions (if any) must occur in pairs—conjugate pairs. In the opening display, $1 + i$ and $1 - i$ must enter together or not at all; so must $3 + 2i$ and $3 - 2i$.

PROPERTIES OF CONJUGATES

We indicate the conjugate of a complex number by putting a bar over it. If $u = a + bi$, then $\bar{u} = a - bi$. For example,

$$\overline{1 + i} = 1 - i$$
$$\overline{3 - 2i} = 3 + 2i$$
$$\overline{4} = 4$$
$$\overline{2i} = -2i$$

The operation of taking the conjugate behaves nicely in both addition and multiplication. There are two pertinent properties, stated first in words.

1. The conjugate of a sum is the sum of the conjugates.
2. The conjugate of a product is the product of the conjugates.

In symbols these properties become

1. $\overline{u_1 + u_2 + \cdots + u_n} = \overline{u}_1 + \overline{u}_2 + \overline{u}_3 + \cdots + \overline{u}_n$
2. $\overline{u_1 u_2 u_3 \cdots u_n} = \overline{u}_1 \cdot \overline{u}_2 \cdot \overline{u}_3 \cdots \overline{u}_n$

A third property follows from the second property if we set all u_i's equal to u.

3. $\overline{u^n} = (\overline{u})^n$

Rather than prove these properties, we shall illustrate them.

1. $\overline{(2 + 3i) + (1 - 4i)} = \overline{2 + 3i} + \overline{1 - 4i}$
2. $\overline{(2 + 3i)(1 - 4i)} = \overline{(2 + 3i)}\,\overline{(1 - 4i)}$
3. $\overline{(2 + 3i)^3} = (\overline{2 + 3i})^3$

Let us check that the second statement is correct. We will do it by computing both sides independently.

$$\overline{(2 + 3i)(1 - 4i)} = \overline{2 + 12 + (-8 + 3)i} = \overline{14 - 5i} = 14 + 5i$$

$$\overline{(2 + 3i)}\,\overline{(1 - 4i)} = (2 - 3i)(1 + 4i) = 2 + 12 + (8 - 3)i = 14 + 5i$$

We can use the three properties of conjugates to demonstrate some very important results. For example, we can show that if u is a solution of the equation

$$ax^3 + bx^2 + cx + d = 0$$

where a, b, c, and d are real, then $\overline{u}$ is also a solution. We need to show that

$$a\overline{u}^3 + b\overline{u}^2 + c\overline{u} + d = 0$$

Since u is a solution of the equation, we know that

$$au^3 + bu^2 + cu + d = 0$$

Now take the conjugate of both sides, using the three properties of conjugates and the fact that a real number is its own conjugate.

$$\overline{au^3 + bu^2 + cu + d} = \overline{0}$$

$$\overline{au^3} + \overline{bu^2} + \overline{cu} + \overline{d} = 0$$

$$\overline{a}\,\overline{u^3} + \overline{b}\,\overline{u^2} + \overline{c}\,\overline{u} + \overline{d} = 0$$

$$a\overline{u}^3 + b\overline{u}^2 + c\overline{u} + d = 0$$

NONREAL SOLUTIONS OCCUR IN PAIRS

We are ready to state the main theorem of this section.

CONJUGATE PAIR THEOREM

Let

$$a_n x^n + a_{n-1} x^{n-1} + \cdots + a_1 x + a_0 = 0$$

be a polynomial equation with real coefficients. If u is a solution, its conjugate $\bar{u}$ is also a solution.

We feel confident that you will be willing to accept the truth of this theorem without further argument. The formal proof would mimic the proof given above for the cubic equation.

As an illustration of one use of this theorem, suppose that we know that $3 + 4i$ is a solution of the equation

$$x^3 - 8x^2 + 37x - 50 = 0$$

Then we know that $3 - 4i$ is also a solution. We can easily find the third solution, which incidentally must be real. (Why?) Here is how we do it.

$$
\begin{array}{r|rrrr}
3 + 4i & 1 & -8 & 37 & -50 \\
& & 3 + 4i & -31 - 8i & 50 \\
\hline
3 - 4i & 1 & -5 + 4i & 6 - 8i & 0 \\
& & 3 - 4i & -6 + 8i & \\
\hline
& 1 & -2 & 0 &
\end{array}
$$

From this it follows that

$$x^3 - 8x^2 + 37x - 50 = [x - (3 + 4i)][x - (3 - 4i)][x - 2]$$

and that 2 is the third solution of our equation.

Notice something special about the product of the first two factors displayed above.

$$[x - (3 + 4i)][x - (3 - 4i)]$$

$$= x^2 - (3 + 4i)x - (3 - 4i)x + (3 + 4i)(3 - 4i)$$

$$= x^2 - 6x + 25$$

The product is a quadratic polynomial with *real* coefficients. This is not an accident. If u is any complex number, then

$$(x - u)(x - \bar{u}) = x^2 - (u + \bar{u})x + u\bar{u}$$

is a real quadratic polynomial, since both $u + \bar{u}$ and $u\bar{u}$ are real (see Problem 42). Thus the conjugate pair theorem, when combined with the complete factorization theorem of Section 7-2, has the following consequence.

REAL FACTORS THEOREM

Any polynomial with real coefficients can be factored into a product of linear and quadratic polynomials having real coefficients, where the quadratic polynomials have no real zeros.

RATIONAL SOLUTIONS *skip*

How does one get started on solving an equation of high degree? So far, all we can suggest is to guess. If you are lucky and find a solution, you can use synthetic division to reduce the degree of the equation to be solved. Eventually you may get it down to a quadratic equation, for which we have the quadratic formula.

Guessing would not be so bad if there were not so many possibilities to consider. Is there an intelligent way to guess? There is, but unfortunately it works only if the coefficients are integers, and then it only helps us find rational solutions.

Consider

$$3x^3 + 13x^2 - x - 6 = 0$$

which, as you will note, has integral coefficients. Suppose it has a rational solution c/d which is in reduced form (that is, c and d are integers without common divisors greater than 1 and $d > 0$). Then

$$3 \cdot \frac{c^3}{d^3} + 13 \cdot \frac{c^2}{d^2} - \frac{c}{d} - 6 = 0$$

or, after multiplying by d^3,

$$3c^3 + 13c^2d - cd^2 - 6d^3 = 0$$

We can rewrite this as

$$c(3c^2 + 13cd - d^2) = 6d^3$$

and also as

$$d(13c^2 - cd - 6d^2) = -3c^3$$

The first of these tells us that c divides $6d^3$ and the second that d divides $-3c^3$. But c and d have no common divisors. Therefore c must divide 6 and d must divide 3.

The only possibilities for c are ± 1, ± 2, ± 3, and ± 6; for d, the only possibilities are 1 and 3. Thus the possible rational solutions must come from the list below.

$$\frac{c}{d}: \quad \pm 1, \ \pm 2, \ \pm 3, \ \pm 6, \ \pm \tfrac{1}{3}, \ \pm \tfrac{2}{3}$$

Upon checking all 12 numbers (which takes time, but a bit less time than checking *all* numbers would take!) we find that only $\tfrac{2}{3}$ works.

$$\begin{array}{r|rrrr}
\tfrac{2}{3} & 3 & 13 & -1 & -6 \\
& & 2 & 10 & 6 \\
\hline
& 3 & 15 & 9 & 0
\end{array}$$

We could prove the following theorem by using similar reasoning.

RATIONAL SOLUTION THEOREM (RATIONAL ROOT THEOREM)

Let

$$a_n x^n + a_{n-1} x^{n-1} + \cdots + a_1 x + a_0 = 0$$

have integral coefficients. If c/d is a rational solution in reduced form, then c divides a_0 and d divides a_n.

Problem Set 7-3

In Problems 1–10, write the conjugate of the number.

1. $2 + 3i$
2. $3 - 5i$
3. $4i$
4. $-6i$
5. $4 + \sqrt{6}$
6. $3 - \sqrt{5}$
7. $(2 - 3i)^8$
8. $(3 + 4i)^{12}$
9. $2(1 + 2i)^3 - 3(1 + 2i)^2 + 5$
10. $4(6 - i)^4 + 11(6 - i) - 23$
11. If $P(x)$ is a cubic polynomial with real coefficients and has -3 and $5 - i$ as zeros, what other zero does it have?
12. If $P(x)$ is a cubic polynomial with real coefficients and has 0 and $\sqrt{2} + 3i$ as zeros, what other zero does it have?
13. Suppose that $P(x)$ has real coefficients and is of the fourth degree. If it has $3 - 2i$ and $5 + 4i$ as two of its zeros, what other zeros does it have?
14. If $P(x)$ is a fourth degree polynomial with real coefficients and has $5 + 6i$ as a zero of multiplicity 2, what are its other zeros?

EXAMPLE A (Solving an equation given some solutions)   Given that -1 and $1 + 2i$ are solutions of the equation

$$2x^4 - 5x^3 + 9x^2 + x - 15 = 0$$

find the other solutions.

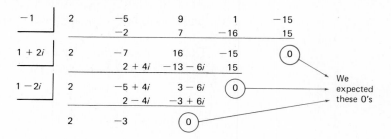

Last quotient: $2x - 3$

The 4th solution: $\dfrac{3}{2}$

Solution. Since the coefficients are real, $1 - 2i$ is a solution. The fourth solution is found by progressively using synthetic division. This division is shown at the bottom of the previous page.

In Problems 15–18, one or more solutions of the specified equations are given. Find the other solutions.

☐ 15. $2x^3 - x^2 + 2x - 1 = 0$; i
☐ 16. $x^3 - 3x^2 + 4x - 12 = 0$; $2i$
☐ 17. $x^4 + x^3 + 6x^2 + 26x + 20 = 0$; $1 + 3i$
☐ 18. $x^6 + 2x^5 + 4x^4 + 4x^3 - 4x^2 - 16x - 16 = 0$; $2i$, $-1 + i$

EXAMPLE B (Obtaining a polynomial given some of its zeros) Find a cubic polynomial with real coefficients that has 3 and $2 + 3i$ as zeros. Make the leading coefficient 1.

Solution. The third zero has to be $2 - 3i$. In factored form, our polynomial is

$$(x - 3)(x - 2 - 3i)(x - 2 + 3i)$$

We multiply this out in stages.

$$(x - 3)[(x - 2)^2 + 9]$$
$$(x - 3)(x^2 - 4x + 13)$$
$$x^3 - 7x^2 + 25x - 39$$

Find a polynomial with real coefficients that has the indicated degree and the given zero(s). Make the leading coefficient 1.

☐ 19. Degree 2; zero: $2 + 5i$.
☐ 20. Degree 2; zero: $\sqrt{6}i$.
☐ 21. Degree 3; zeros: -3, $2i$.
☐ 22. Degree 3; zeros: 5, $-3i$.
☐ 23. Degree 5; zeros: 2, $3i$ (multiplicity 2).
☐ 24. Degree 5; zeros: 1, $1 - i$ (multiplicity 2).

EXAMPLE C (Finding rational solutions) Find the rational solutions of the equation

$$3x^4 + 2x^3 + 2x^2 + 2x - 1 = 0$$

Then find the remaining solutions.

Solution. The only way that c/d (in reduced form) can be a solution is for c to be 1 or -1 and for d to be 1 or 3. This means that the possibilities for c/d are ± 1 and $\pm \frac{1}{3}$. Synthetic division shows that -1 and $\frac{1}{3}$ work.

$$\begin{array}{r|rrrrr}
-1 & 3 & 2 & 2 & 2 & -1 \\
  &   & -3 & 1 & -3 & 1 \\
\hline
\tfrac{1}{3} & 3 & -1 & 3 & -1 & 0 \\
  &   & 1 & 0 & 1 & \\
\hline
  & 3 & 0 & 3 & 0 &
\end{array}$$

Setting the final quotient, $3x^2 + 3$, equal to zero and solving, we get

$$3x^2 = -3$$
$$x^2 = -1$$
$$x = \pm i$$

The complete solution set is $\{-1, \tfrac{1}{3}, i, -i\}$.

In Problems 25–30, find the rational solutions of each equation. If possible, find the other solutions.

25. $x^3 - 3x^2 - x + 3 = 0$
26. $x^3 + 3x^2 - 4x - 12 = 0$
27. $2x^3 + 3x^2 - 4x + 1 = 0$
28. $5x^3 - x^2 + 5x - 1 = 0$
29. $\tfrac{1}{3}x^3 - \tfrac{1}{2}x^2 - \tfrac{1}{6}x + \tfrac{1}{6} = 0$ (*Hint:* Clear the equation of fractions.)
30. $\tfrac{2}{3}x^3 - \tfrac{1}{2}x^2 + \tfrac{2}{3}x - \tfrac{1}{2} = 0$

Miscellaneous Problems

31. The equation $2x^3 - 11x^2 + 20x + 13 = 0$ has $3 - 2i$ as a solution. Find the other solutions.
32. Find the fourth degree polynomial with real coefficients and leading coefficient 1 that has 2, -4, and $1 + 2i$ as three of its zeros.
33. Find all the solutions of

$$x^4 - 3x^3 - 20x^2 - 24x - 8 = 0$$

Start by looking for rational solutions.
34. Find all the solutions of

$$2x^4 - x^3 + x^2 - x - 1 = 0$$

35. Answer *true* or *false*.
    (a) Every nth degree polynomial has n distinct zeros.
    (b) Every quadratic polynomial with real coefficients has either 2 distinct real zeros or 2 distinct nonreal zeros.
    (c) If $2 + 3i$ is a solution of a polynomial equation, so is $2 - 3i$.
    (d) It is possible for a sixth degree polynomial to have only 3 distinct zeros.
    (e) There is exactly one fourth degree polynomial which has 3, -2, and $2 \pm 5i$ as zeros.
36. Show that a polynomial equation with real coefficients and of odd degree has at least one real solution.

37. Show by means of an example that the statement in Problem 36 is not necessarily true if the coefficients of the equation are complex.

38. A cubic equation with real coefficients has either 3 real solutions (not necessarily distinct), or 1 real solution and a pair of nonreal solutions. State all of the possibilities for:
    (a) A fourth degree equation with real coefficients;
    (b) A fifth degree equation with real coefficients.

39. Suppose that a polynomial has integral coefficients and a leading coefficient of 1. Show that if this equation has a rational solution, it must be an integer.

40. Show that the following equation has no rational solution.

$$x^4 + x^3 - 4x^2 - 5x - 5 = 0$$

41. Show that the following equation has 1 as a solution of multiplicity 4. Find the remaining solution.

$$x^5 + 6x^4 - 34x^3 + 56x^2 - 39x + 10 = 0$$

42. (a) Let $u = -4 + 3i$. Calculate $u + \bar{u}$ and $u\bar{u}$.
    (b) Show that if u is any complex number, then $u + \bar{u}$ and $u\bar{u}$ are real numbers. (*Hint:* Let $u = a + bi$.)

43. Write $x^4 + 3x^3 + 3x^2 - 3x - 4$ as a product of linear and quadratic factors with real coefficients as guaranteed by the real factors theorem. (*Hint:* This polynomial has two integer zeros.)

44. Find all the zeros of

$$x^6 - 9x^5 + 38x^4 - 106x^3 + 181x^2 - 205x + 100$$

given that $1 + 2i$ is a zero of multiplicity 2. Then write this polynomial as a product of real linear and quadratic factors.

45. (a) Solve the equation $x^4 + 1 = 0$. (*Hint:* First show that $(\sqrt{2} + \sqrt{2}i)/2$ is a solution; then find remaining solutions.)
    (b) Find the rational solution(s) of $x^5 - 2x^4 + x - 2 = 0$. Then find the remaining solutions.

46. Show that the sum and product of the zeros of the polynomial

$$a_n x^n + a_{n-1} x^{n-1} + \cdots + a_1 x + a_0$$

are given by $-a_{n-1}/a_n$ and $(-1)^n a_0/a_n$, respectively. (*Hint:* Multiply the factors in the complete factorization theorem.)

47. Verify the truth of the result in Problem 46 for the polynomial

$$4x^4 + 3x^2 - 1 = (4x^2 - 1)(x^2 + 1)$$

Isaac Newton
1642–1727

7-4
The Method of Successive Approximations

ship

Even with the theory developed so far, we are often unable to get started on solving an equation of high degree. Imagine being given a fifth degree equation whose true solutions are $\sqrt{2} + 5i$, $\sqrt{2} - 5i$, $\sqrt[3]{19}$, 1.597, and 3π. How would you ever find them? Nothing you have learned until now would be of much help.

There is a general method of solving problems known to all resourceful people. We call it "muddling through" or "trial and error." Given a cup of tea, we add sugar a bit at a time until it tastes just right. Given a stopper too large for a hole, we whittle it down until it fits. We change the solution a step at a time, continually improving the accuracy until we are satisfied. Mathematicians call it a **method of successive approximations.**

We explore two such methods in this section. The first is a graphical method; the second is Newton's algebraic method. Both are designed to find the real solutions of polynomial equations with real coefficients. Both require many computations. We suggest that you keep your pocket calculator handy.

METHOD OF SUCCESSIVE ENLARGEMENTS

Consider the equation

$$x^3 - 3x - 5 = 0$$

We begin by graphing

$$y = x^3 - 3x - 5$$

looking for the point (or points) where the graph crosses the x-axis. These points correspond to the real solutions.

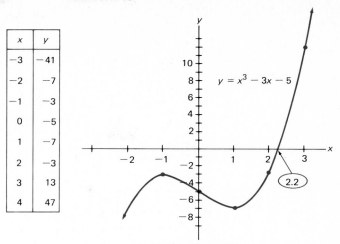

x	y
−3	−41
−2	−7
−1	−3
0	−5
1	−7
2	−3
3	13
4	47

$y = x^3 - 3x - 5$

First Approximation

Clearly there is only one real solution and it is between 2 and 3; a good first guess might be 2.2. Now we calculate y for values of x near 2.2 (for instance, 2.1, 2.2, and 2.3) until we find an interval of length .1 on which y changes sign. The interval is $2.2 \leq x \leq 2.3$. On this interval, we pretend the graph is a straight line. The point at which this line crosses the x-axis gives us our next approximation. It is about 2.28.

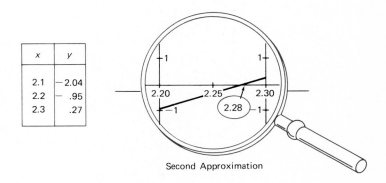

x	y
2.1	−2.04
2.2	− .95
2.3	.27

Second Approximation

Now we calculate y for values of x near 2.28 until we find an interval of length .01 on which y changes sign. This occurs on the interval $2.27 \leq x \leq 2.28$. Using a straight-line graph for this interval, we read our next approximation as 2.279 (see diagram at the top of the next page).

In effect, we are enlarging the graph (using a more and more powerful magnifying glass) at each stage, increasing the accuracy by one digit each time. We can continue this process as long as we have the patience to do the necessary calculations.

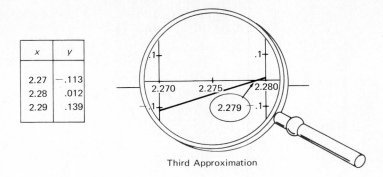

x	y
2.27	−.113
2.28	.012
2.29	.139

Third Approximation

NEWTON'S METHOD

Let $P(x) = 0$ be a polynomial equation with real coefficients. Suppose that by some means (perhaps graphing), we discover that it has a real solution r which we guess to be about x_1. Then, as the accompanying diagram suggests, a better approximation to r is x_2, the point at which the tangent line to the curve at x_1 crosses the x-axis.

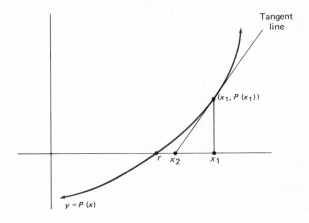

If the slope of the tangent line is m_1, then

$$m_1 = \frac{\text{rise}}{\text{run}} = \frac{P(x_1)}{x_1 - x_2}$$

Solving this for x_2 yields

$$x_2 = x_1 - \frac{P(x_1)}{m_1}$$

What has been done once can be repeated. A still better approximation to r would be x_3, obtained from x_2 as follows:

$$x_3 = x_2 - \frac{P(x_2)}{m_2}$$

where m_2 is the slope of the tangent line at $(x_2, P(x_2))$. In general, we find the $(k + 1)$st approximation from the kth by using Newton's formula

$$x_{k+1} = x_k - \frac{P(x_k)}{m_k}$$

where m_k is the slope of the tangent line at $(x_k, P(x_k))$.

The rub, of course, is that we do not know how to calculate m_k. That is precisely where Newton made his biggest contribution. He showed how to find the slope of the tangent line to any curve. This is done in every calculus course; we state one result without proof.

SLOPE THEOREM

If

$$P(x) = a_n x^n + a_{n-1} x^{n-1} + \cdots + a_2 x^2 + a_1 x + a_0$$

is a polynomial with real coefficients, then the slope of the tangent line to the graph of $y = P(x)$ at x is $P'(x)$, where $P'(x)$ is the polynomial

$$P'(x) = n a_n x^{n-1} + (n - 1) a_{n-1} x^{n-2} + \cdots + 2 a_2 x + a_1$$

For example, if

$$P(x) = 2x^4 + 4x^3 - 6x^2 + 2x + 15$$

then

$$P'(x) = 4 \cdot 2 \cdot x^3 + 3 \cdot 4 \cdot x^2 - 2 \cdot 6 \cdot x + 2$$
$$= 8x^3 + 12x^2 - 12 \cdot x + 2$$

In particular, the slope of the tangent line at $x = 2$ is $P'(2) = 8 \cdot 2^3 + 12 \cdot 2^2 - 12 \cdot 2 + 2 = 90$.

Taking the slope theorem for granted, we may write Newton's formula in the useful form

$$x_{k+1} = x_k - \frac{P(x_k)}{P'(x_k)}$$

USING NEWTON'S METHOD

Consider the equation

$$x^3 - 3x - 5 = 0$$

again. If we let

$$P(x) = x^3 - 3x - 5$$

then, by the slope theorem,

$$P'(x) = 3x^2 - 3$$

and Newton's formula becomes

$$x_{k+1} = x_k - \frac{x_k^3 - 3x_k - 5}{3x_k^2 - 3}$$

If we take $x_1 = 3$ as our initial guess, then

$$x_2 = x_1 - \frac{x_1^3 - 3x_1 - 5}{3x_1^2 - 3}$$

$$= 3 - \frac{3^3 - 3 \cdot 3 - 5}{3 \cdot 3^2 - 3}$$

$$\approx \boxed{2.5}$$

$$x_3 = x_2 - \frac{x_2^3 - 3x_2 - 5}{3x_2^2 - 3}$$

$$= 2.5 - \frac{(2.5)^3 - 3(2.5) - 5}{3(2.5)^2 - 3}$$

$$\approx \boxed{2.30}$$

$$x_4 = x_3 - \frac{x_3^3 - 3x_3 - 5}{3x_3^2 - 3}$$

$$= 2.30 - \frac{(2.30)^3 - 3(2.30) - 5}{3(2.30)^2 - 3}$$

$$\approx \boxed{2.279}$$

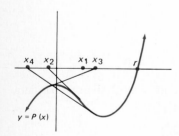

We can continue this repetitive process until we have the accuracy we desire.

When you use Newton's method, it is important to make your initial guess reasonably good. The sketch in the margin shows how badly the method can lead you astray if you choose x_1 too far off the mark.

Problem Set 7-4

In Problems 1–6, use the slope theorem to find the slope polynomial $P'(x)$ if $P(x)$ is the given polynomial. Then find the slope of the tangent line at $x = 1$.

1. $2x^2 - 5x + 6$ 2. $3x^2 + 4x - 9$
3. $2x^2 + x - 2$ 4. $x^3 - 5x + 8$
5. $2x^5 + x^4 - 2x^3 + 8x - 4$ 6. $x^6 - 3x^4 + 7x^3 - 4x^2 + 5x - 4$

Each of the equations in Problems 7–10 on the next page has exactly one real solution. By means of a graph, make an initial guess at the solution. Then use the method of successive enlargements to find a second and a third approximation.

7. $x^3 + 2x - 5 = 0$

8. $x^3 + x - 32 = 0$

9. $x^3 - 3x - 10 = 0$

10. $2x^3 - 6x - 15 = 0$

11. Use Newton's method to approximate the real solution of the equation $x^3 + 2x - 5 = 0$. Take as x_1 the initial guess you made in Problem 7, and then find x_2, x_3, and x_4.

12–14. Follow the instructions of Problem 11 for the equations in Problems 8–10.

EXAMPLE (Finding all real solutions by Newton's method)   Find all real solutions of the following equation by Newton's method.

$$P(x) = x^4 - 8x^3 + 22x^2 - 24x + 6 = 0$$

Solution.   First we sketch the graph of $P(x)$.

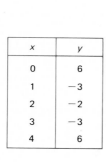

x	y
0	6
1	-3
2	-2
3	-3
4	6

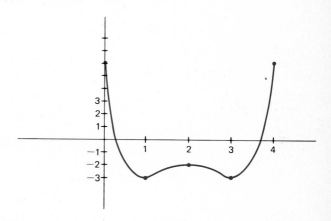

The graph crosses the x-axis at approximately .4 and 3.6. The slope polynomial is

$$P'(x) = 4x^3 - 24x^2 + 44x - 24$$

Take $x_1 = .4$.

$$x_2 = .4 - \frac{P(.4)}{P'(.4)}$$

$$= .4 - \frac{(-.57)}{(-9.98)} \approx .34$$

$$x_3 = .34 - \frac{P(.34)}{P'(.34)}$$

$$= .34 - \frac{.0821}{(-4.657)} \approx .347$$

Take $x_1 = 3.6$.

$$x_2 = 3.6 - \frac{P(3.6)}{P'(3.6)}$$

$$= 3.6 - \frac{(-.566)}{9.98} \approx 3.66$$

$$x_3 = 3.66 - \frac{P(3.66)}{P'(3.66)}$$

$$= 3.66 - \frac{.0821}{11.66} \approx 3.653$$

Each equation in Problems 15–18 has several real solutions. Draw a graph to get your first estimates. Then use Newton's method to find these solutions to three decimal places.

$\boxed{c}$ 15. $x^4 + x^3 - 3x^2 + 4x - 28 = 0$

$\boxed{c}$ 16. $x^4 + x^3 - 6x^2 - 7x - 7 = 0$

$\boxed{c}$ 17. $x^3 - 3x + 1 = 0$

$\boxed{c}$ 18. $x^3 - 12x + 1 = 0$

Miscellaneous Problems

$\boxed{c}$ 19. The line $y = x + 3$ intersects the curve $y = x^3 - 3x + 4$ at three points. Use Newton's method to find the x-coordinates of these points correct to 2 decimal places.

$\boxed{c}$ 20. A spherical shell has a thickness of 1 centimeter. What is the outer radius r of the shell if the volume of the shell is equal to the volume of the space inside? First find an equation for r and then solve it by Newton's method.

$\boxed{c}$ 21. Solve Problem 20 if the volume of the hollow space is twice the volume of the shell.

$\boxed{c}$ 22. The dimensions of a rectangular box are 6, 8, and 10 feet. Suppose that the volume of the box is increased by 300 cubic feet by equal elongations of the three dimensions. Find this elongation correct to two decimal places.

$\boxed{c}$ 23. What rate of interest compounded annually is implied in an offer to sell a house for $50,000 cash, or in annual installments of $20,000 each payable 1, 2, and 3 years from now? (*Hint:* The amount of $50,000 with interest for 3 years should equal the sum of the first payment accumulated for 2 years, the second accumulated for 1 year, and the third payment. Hence, if i is the interest rate, $50,000(1 + i)^3 = 20,000(1 + i)^2 + 20,000(1 + i) + 20,000$. Dividing by 10,000 and writing x for $1 + i$ gives the equation $5x^3 = 2x^2 + 2x + 2$.)

$\boxed{c}$ 24. Find the rate of interest implied if a house is offered for sale at $80,000 cash or in 4 annual installments of $23,000, the first payable now. (See Problem 23.)

25. Find the equation of the tangent line to the graph of the equation $y = 3x^2$ at the point (2, 12). (*Hint:* Remember that the line through (2, 12) with slope m has equation $y - 12 = m(x - 2)$. Find m by evaluating the slope polynomial at $x = 2$.)

26. Find the equation of the tangent line to the curve
(a) $y = x^2 + x$ at the point (2, 6);
(b) $y = 2x^3 - 4x + 5$ at the point $(-1, 7)$;
(c) $y = \frac{1}{5}x^5$ at the point $(2, \frac{32}{5})$.
(See the hint for Problem 25.)

27. Draw a careful graph of the equation $y = x^2$ for $-3 \le x \le 3$. Then draw tangent lines to the curve at $x = -2$, $x = 0$, $x = 1$, $x = 2$, and $x = 3$, and estimate their slopes. How well do your answers compare with the corre-

sponding values of the slope polynomial for x^2, that is, with the values of $2x$?

28. The point $P(2, 12)$ is on the graph of the equation $y = 3x^2$. Find the slope of the line segment PQ, where Q is also on the graph and
    (a) has x-coordinate 2.5;
    (b) has x-coordinate 2.1;
    (c) has x-coordinate $2 + h$.
    What number does your answer to part (c) approach as h approaches zero? How does this compare with the value of the slope polynomial of $3x^2$ for $x = 2$?

CHAPTER SUMMARY

The **division law for polynomials** asserts that if $P(x)$ and $D(x)$ are any given nonconstant polynomials, then there are unique polynomials $Q(x)$ and $R(x)$ such that

$$P(x) = D(x)Q(x) + R(x)$$

where $R(x)$ is either 0 or of lower degree than $D(x)$. In fact, we can find $Q(x)$ and $R(x)$ by the **division algorithm,** which is just a fancy name for ordinary long division. When $D(x)$ has the form $x - c$, $R(x)$ will have to be a constant R, since it is of lower degree than $D(x)$. The substitution $x = c$ then gives

$$P(c) = R$$

a result known as the **remainder theorem.** An immediate consequence is the **factor theorem,** which says that c is a zero of $P(x)$ if and only if $x - c$ is a factor of $P(x)$. Division of a polynomial by $x - c$ can be greatly simplified by use of **synthetic division.**

That every nonconstant polynomial has at least one zero is guaranteed by Gauss's **fundamental theorem of algebra.** But we can say much more than that. For any nonconstant polynomial

$$P(x) = a_n x^n + a_{n-1} x^{n-1} + \cdots + a_1 x + a_0$$

there are n numbers $c_1, c_2, \ldots, c_n$ (not necessarily all different) such that

$$P(x) = a_n(x - c_1)(x - c_2) \cdots (x - c_n)$$

We call the latter result the **complete factorization theorem.**

If the polynomial equation

$$P(x) = a_n x^n + a_{n-1} x^{n-1} + \cdots + a_1 x + a_0 = 0$$

has real coefficients, then its nonreal solutions (if any) must occur in conjugate pairs $a + bi$ and $a - bi$. If the coefficients are integers and if c/d is a rational solution in reduced form, then c divides a_0 and d divides a_n.

To find exact solutions to a polynomial equation may be very difficult; often we are more than happy to find good approximations. Two good methods for doing this are the **method of successive enlargements** and **Newton's method.** Both require plenty of calculating power.

CHAPTER REVIEW PROBLEM SET

1. Find the quotient and remainder if the first polynomial is divided by the second.
   (a) $2x^3 - x^2 + 4x - 5$; $x^2 + 2x - 3$
   (b) $x^4 - 8x^2 + 5$; $x^2 + 3x$

2. Use synthetic division to find the quotient and remainder if the first polynomial is divided by the second.
   (a) $x^3 - 2x^2 - 4x + 7$; $x - 2$
   (b) $2x^4 - 15x^2 + 4x - 3$; $x + 3$
   (c) $x^3 + (3 - 3i)x^2 - (9 + 15i)x - 3 - 3i$; $x - 2 - 3i$

3. Without dividing, find the remainder if $2x^4 - 6x^3 + 17$ is divided by $x - 2$; if it is divided by $x + 2$.

4. Find the zeros of the given polynomial and give their multiplicities.
   (a) $(x^2 - 1)^2(x^2 + 1)$
   (b) $x(x^2 - 2x + 4)(x + \pi)^3$

In Problems 5 and 6, use synthetic division to show that $x - c$ is a factor of $P(x)$. Then factor $P(x)$ completely into linear factors.

5. $P(x) = 2x^3 - x^2 - 18x + 9$; $c = 3$
6. $P(x) = x^3 + 4x^2 - 7x - 28$; $c = -4$

In Problems 7 and 8, find a polynomial $P(x)$ with integral coefficients that has the given zeros. Assume each zero to be simple unless otherwise indicated.

7. $3, -2, 4$ (multiplicity 2)
8. $3 + \sqrt{7}, 3 - \sqrt{7}, 2 - i, 2 + i$
9. Show that 1 is a zero of multiplicity 2 of the polynomial $x^4 - 4x^3 - 3x^2 + 14x - 8$ and find the remaining zeros.
10. Find the polynomial $P(x) = a_3x^3 + a_2x^2 + a_1x + a_0$ which has zeros $\frac{1}{2}$, $-\frac{1}{3}$, and 4 and for which $P(2) = -42$.
11. Find the value of k so that $\sqrt{2}$ is a zero of $x^3 + 3x^2 - 2x + k$.
12. Find a cubic polynomial with real coefficients that has $4 + 3i$ and -2 as two of its zeros.
13. Solve the equation $x^4 - 4x^3 + 24x^2 + 20x - 145 = 0$, given that $2 + 5i$ is one of its solutions.
14. The equation $2x^3 - 15x^2 + 20x - 3 = 0$ has a rational solution. Find it and then find the other solutions.
15. The equation $x^3 - x^2 - x - 7 = 0$ has at least one real solution. Why? Show that it has no rational solution.
16. The equation $x^3 - 6x + 6 = 0$ has exactly one real solution. By means of a graph, make an initial guess at the solution. Then use the method of successive enlargements to find a second and a third approximation.
17. Use Newton's method to approximate the real solution of the equation in Problem 16. Take as x_1 the initial guess you made in Problem 16, and find x_2 and x_3.

8

SYSTEMS OF EQUATIONS AND INEQUALITIES

Geometry may sometimes appear to take the lead over analysis but in fact precedes it only as a servant goes before the master to clear the path and light him on his way.

James Joseph Sylvester

Moving from System to System

On Algebraic
Wheels

I $\begin{cases} x - y = 1 \\ 2x - y = 4 \end{cases}$ II $\begin{cases} x - y = 1 \\ y = 2 \end{cases}$ III $\begin{cases} x = 3 \\ y = 2 \end{cases}$

On Geometric
Wheels

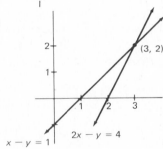

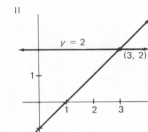

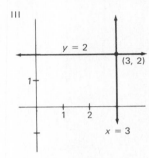

8-1
Equivalent Systems of Equations

In Section 3-3, you learned how to solve a system of two equations in two unknowns, but you probably did not think of the process as one of replacing the given system by another having the same solutions. It is this point of view that we now want to explore.

In the display above, system I is replaced by system II, which is simpler; system II is in turn replaced by system III, which is simpler yet. To go from system I to system II, we eliminated x from the second equation; we then substituted $y = 2$ in the first equation to get system III. What happened geometrically is shown in the bottom half of our display. Notice that the three pairs of lines have the same point of intersection $(3, 2)$.

Because the notion of changing from one system of equations to another having the same solutions is so important, we make a formal definition. We say that two systems of equations are **equivalent** if they have the same solutions.

OPERATIONS THAT LEAD TO EQUIVALENT SYSTEMS

Now we face a big question. What operations can we perform on a system without changing its solutions?

Operation 1 We can interchange the position of two equations.

Operation 2 We can multiply an equation by a nonzero constant, that is, we can replace an equation by a nonzero multiple of itself.

Operation 3 We can add a multiple of one equation to another, that is, we can replace an equation by the sum of that equation and a multiple of another.

Operation 3 is the workhorse of the set. We show how it is used in the example of the opening display.

$$\text{I} \quad \begin{cases} x - y = 1 \\ 2x - y = 4 \end{cases}$$

If we add -2 times the first equation to the second, we obtain

$$\text{II} \quad \begin{cases} x - y = 1 \\ y = 2 \end{cases}$$

We then add the second equation to the first. This gives

$$\text{III} \quad \begin{cases} x = 3 \\ y = 2 \end{cases}$$

This is one way to write the solution. Alternatively, we say that the solution is the ordered pair $(3, 2)$.

THE THREE POSSIBILITIES FOR A LINEAR SYSTEM

We are mainly interested in linear systems—that is, systems of linear equations—and we shall restrict our discussion to the case where there is the same number of equations as unknowns. There are three possibilities for the set of solutions: The set may be empty, it may have just one point, or it may have infinitely many points. These three cases are illustrated below.

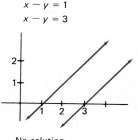

$x - y = 1$
$x - y = 3$

No solution

$x - y = 1$
$2x - y = 4$

One solution

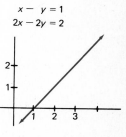

$x - y = 1$
$2x - 2y = 2$

Infinitely many solutions

Someone is sure to object and ask if we cannot have a linear system with exactly two solutions or exactly three solutions. The answer is no. If a linear system has two solutions, it has infinitely many. This is obvious in the case of two equations in two unknowns, since two points determine a line, but it is also true for n equations in n unknowns.

LINEAR SYSTEMS IN MORE THAN TWO UNKNOWNS

When we consider large systems, it is a good idea to be very systematic about our method of attack. Our method is to reduce the system to **triangular form** and then use **back substitution.** Let us explain. Consider

$$
\begin{aligned}
x - 2y + z &= -4 \\
5y - 3z &= 18 \\
2z &= -2
\end{aligned}
$$

which is already in triangular form. (The name arises from the fact that the terms within the dotted triangle have zero coefficients.) This system is easy to solve. Solve the third equation first ($z = -1$). Then substitute that value in the second equation

$$5y - 3(-1) = 18$$

which gives $y = 3$. Finally substitute these two values in the first equation

$$x - 2(3) + (-1) = -4$$

which gives $x = 3$. Thus the solution of the system is $(3, 3, -1)$. This process, called **back substitution**, works on any linear system that is in triangular form. Just start at the bottom and work your way up.

If the system is not in triangular form initially, we try to operate on it until it is. Suppose we start with

$$
\begin{aligned}
2x - 4y + 2z &= -8 \\
2x + y - z &= 10 \\
3x - y + 2z &= 4
\end{aligned}
$$

Begin by multiplying the first equation by $\frac{1}{2}$ so that its leading coefficient is 1.

$$
\begin{aligned}
x - 2y + z &= -4 \\
2x + y - z &= 10 \\
3x - y + 2z &= 4
\end{aligned}
$$

Next add -2 times the first equation to the second equation. Also add -3 times the first equation to the third.

$$
\begin{aligned}
x - 2y + z &= -4 \\
5y - 3z &= 18 \\
5y - z &= 16
\end{aligned}
$$

Finally, add -1 times the second equation to the third.

$$x - 2y + z = -4$$
$$5y - 3z = 18$$
$$2z = -2$$

The system is now in triangular form and can be solved by back substitution. It is, in fact, the triangular system we discussed earlier. The solution is $(3, 3, -1)$.

In Examples A and B of the problem set, we show how this process works when there are infinitely many solutions and when there are no solutions.

Problem Set 8-1

Solve each of the following systems of equations.

1. $2x - 3y = 7$
   $y = -1$

2. $5x - 3y = -25$
   $y = 5$

3. $x = -2$
   $2x + 7y = 24$

4. $x = 5$
   $3x + 4y = 3$

5. $x - 3y = 7$
   $4x + y = 2$

6. $5x + 6y = 27$
   $x - y = 1$

7. $2x - y + 3z = -6$
   $2y - z = 2$
   $z = -2$

8. $x + 2y - z = -4$
   $3y + z = 2$
   $z = 5$

9. $3x - 2y + 5z = -10$
   $y - 4z = 8$
   $2y + z = 7$

10. $4x + 5y - 6z = 31$
    $y - 2z = 7$
    $5y + z = 2$

11. $x + 2y + z = 8$
    $2x - y + 3z = 15$
    $-x + 3y - 3z = -11$

12. $x + y + z = 5$
    $-4x + 2y - 3z = -9$
    $2x - 3y + 2z = 5$

13. $x - 2y + 3z = 0$
    $2x - 3y - 4z = 0$
    $x + y - 4z = 0$

14. $x + 4y - z = 0$
    $-x - 3y + 5z = 0$
    $3x + y - 2z = 0$

15. $x + y + z + w = 10$
    $y + 3z - w = 7$
    $x + y + 2z = 11$
    $x - 3y + w = -14$

16. $2x + y + z = 3$
    $y + z + w = 5$
    $4x + z + w = 0$
    $3y - z + 2w = 0$

EXAMPLE A (Infinitely many solutions)   Solve

$$x - 2y + 3z = 10$$
$$2x - 3y - z = 8$$
$$4x - 7y + 5z = 28$$

Solution. Using Operation 3, we may eliminate x from the last two equations. First add -2 times the first equation to the second equation. Then add -4 times the first equation to the third equation. We obtain

$$x - 2y + 3z = 10$$
$$y - 7z = -12$$
$$y - 7z = -12$$

Next add -1 times the second equation to the third equation.

$$x - 2y + 3z = 10$$
$$y - 7z = -12$$
$$0 = 0$$

Finally, we solve the second equation for y in terms of z and substitute that result in the first equation.

$$y = 7z - 12$$
$$x = 2y - 3z + 10 = 2(7z - 12) - 3z + 10 = 11z - 14$$

Notice that there are infinitely many solutions; we can give any value we like to z, calculate the corresponding x and y values, and come up with a solution. Here is the format we use to list all the solutions to the system.

$$x = 11z - 14$$
$$y = 7z - 12$$
$$z \quad \text{arbitrary}$$

We could also say that the set of solutions consists of all ordered triples of the form $(11z - 14, 7z - 12, z)$. Thus if $z = 0$, we get the solution $(-14, -12, 0)$; if $z = 2$, we get $(8, 2, 2)$. Of course, it does not have to be z that is arbitrary; in our example, it could just as well be x or y. For example, if we had arranged things so y is treated as the arbitrary variable, we would have obtained.

$$x = \frac{11}{7}y + \frac{34}{7}$$

$$z = \frac{1}{7}y + \frac{12}{7}$$

$$y \quad \text{arbitrary}$$

Note that the solution corresponding to $y = 2$ is $(8, 2, 2)$, which agrees with one found above.

Solve each of the following systems. Some, but not all, have infinitely many solutions.

17. $x - 4y + z = 18$
     $2x - 7y - 2z = 4$
     $3x - 11y - z = 22$

18. $x + y - 3z = 10$
     $2x + 5y + z = 18$
     $5x + 8y - 8z = 48$

19. $\begin{aligned} x - 2y + 3z &= -2 \\ 3x - 6y + 9z &= -6 \\ -2x + 4y - 6z &= 4 \end{aligned}$

20. $\begin{aligned} -4x + y - z &= 5 \\ 4x - y + z &= -5 \\ -24x + 6y - 6z &= 30 \end{aligned}$

21. $\begin{aligned} 2x - y + 4z &= 0 \\ 3x + 2y - z &= 0 \\ 9x - y + 11z &= 0 \end{aligned}$

22. $\begin{aligned} x + 3y - 2z &= 0 \\ 2x + y + z &= 0 \\ y - z &= 0 \end{aligned}$

EXAMPLE B (No solution)   Solve

$$x - 2y + 3z = 10$$

$$2x - 3y - z = 8$$

$$5x - 9y + 8z = 20$$

Solution.   Using Operation 3, we eliminate x from the last two equations to obtain

$$x - 2y + 3z = 10$$

$$y - 7z = -12$$

$$y - 7z = -30$$

It is already apparent that the last two equations cannot get along with each other. Let us continue anyway, putting the system in triangular form by adding -1 times the second equation to the third.

$$x - 2y + 3z = 10$$

$$y - 7z = -12$$

$$0 = -18$$

This system has no solution; we say it is **inconsistent.**

Solve the following systems or show that they are inconsistent.

23. $\begin{aligned} x - 4y + z &= 18 \\ 2x - 7y - 2z &= 4 \\ 3x - 11y - z &= 10 \end{aligned}$

24. $\begin{aligned} x + y - 3z &= 10 \\ 2x + 5y + z &= 18 \\ 5x + 8y - 8z &= 50 \end{aligned}$

25. $\begin{aligned} x + 3y - 2z &= 10 \\ 2x + y + z &= 4 \\ 5y - 5z &= 16 \end{aligned}$

26. $\begin{aligned} x - 2y + 3z &= -2 \\ 3x - 6y + 9z &= -6 \\ -2x + 4y - 6z &= 0 \end{aligned}$

EXAMPLE C (Nonlinear systems)   Solve the following system of equations.

$$x^2 + y^2 = 25$$

$$x^2 + y^2 - 2x - 4y = 5$$

Solution.   We can use the same operations on this system as we did on linear systems. Adding (-1) times the first equation to the second, we get

$$x^2 + y^2 = 25$$
$$-2x - 4y = -20$$

We solve the second equation for x in terms of y, substitute in the first equation, and then solve the resulting quadratic equation in y.

$$x = 10 - 2y$$
$$(10 - 2y)^2 + y^2 = 25$$
$$5y^2 - 40y + 75 = 0$$
$$y^2 - 8y + 15 = 0$$
$$(y - 3)(y - 5) = 0$$

From this we get $y = 3$ or 5. Substituting these values in the equation $x = 10 - 2y$ yields two solutions to the original system, $(4, 3)$ and $(0, 5)$.

Solve each of the following systems.

27. $x + 2y = 10$
    $x^2 + y^2 - 10x = 0$

28. $x + y = 10$
    $x^2 + y^2 - 10x - 10y = 0$

29. $x^2 + y^2 - 4x + 6y = 12$
    $x^2 + y^2 + 10x + 4y = 96$

30. $x^2 + y^2 - 16y = 45$
    $x^2 + y^2 + 4x - 20y = 65$

31. $y = 4x^2 - 2$
    $y = x^2 + 1$

32. $x = 3y^2 - 5$
    $x = y^2 + 3$

Miscellaneous Problems

Solve each system in Problems 33–38 or show that it is inconsistent.

33. $3x + 4y = 3$
    $2x - y = 5$

34. $2x - y = 6$
    $-4x + 2y = 8$

35. $3x - y = 6$
    $-4x + 2y = -12$

36. $-4x + y - z = 5$
    $4x - y + z = -5$
    $-24x + 6y - 6z = 10$

37. $x - y + 3z = 1$
    $3x - 2y + 4z = 0$
    $4x + 2y - z = 3$

ⓒ 38. $.43x - .79y + 4.24z = .67$
    $3.61y - 9.74z = 2$
    $y + 1.22z = 1.67$

In Problems 39 and 40, find the values of a and b for which the system has infinitely many solutions.

39. $2x - 3y = 5$
    $ax + by = -15$

40. $x + 2y = 4$
    $ax + 3y = b$

41. Tommy claims that he has $4.10 in nickels, dimes, and quarters, that he has twice as many quarters as dimes, and that he has 28 coins in all. Is this possible? If so, find how many coins of each kind he has.

42. Suppose that Tommy makes the same claim as in Problem 41, except that instead of having twice as many quarters as dimes, he says that the number of dimes added to four times the number of quarters is 40. Is that possible? If so, find how many coins of each kind he has.

43. A three-digit number equals 25 times the sum of its digits. If the digits are reversed, the resulting number is greater than the given number by 198. The sum of the hundreds digit and the ones digit is 8. Find the number.

44. Find the dimensions of a rectangle whose diagonal and perimeter measure 25 and 62 inches, respectively.

45. A certain rectangle has an area of 130 square inches. Increasing the width by 3 inches and decreasing the length by 2 inches increases the area by 13 square inches. Find the dimensions of the original rectangle.

46. Suppose that the distances from the origin to the points (x, y) and $(x + 2, y + 16)$ are 5 and 13, respectively. Find x and y.

47. A chemist mixes three different nitric acid solutions with concentrations of 25 percent, 40 percent, and 50 percent to form 100 liters of a 32 percent solution. If she uses twice as much of the 25 percent solution as the 40 percent solution, how many liters of each kind does she use?

Arthur Cayley, lawyer, painter, mountaineer, Cambridge professor, but most of all creative mathematician, made his biggest contributions in the field of algebra. To him we owe the idea of replacing a linear system by its matrix.

$$\begin{array}{rl} 2x + 3y - z = 1 \\ x + 4y - z = 4 \\ 3x + y + 2z = 5 \end{array} \qquad \begin{bmatrix} 2 & 3 & -1 & 1 \\ 1 & 4 & -1 & 4 \\ 3 & 1 & 2 & 5 \end{bmatrix}$$

Arthur Cayley (1821–1895)

8-2
Matrix Methods

Contrary to what many people think, mathematicians do not enjoy long, involved calculations. What they do enjoy is looking for shortcuts, for labor-saving devices, and for elegant ways of doing things. Consider the problem of solving a system of linear equations, which as you know can become very complicated. Is there any way to simplify and systematize this process? There is. It is the method of matrices (plural of matrix).

A **matrix** is just a rectangular array of numbers. One example is shown in our opening panel. It has 3 rows and 4 columns and is referred to as a 3 × 4 matrix. We follow the standard practice of enclosing a matrix in brackets.

AN EXAMPLE WITH THREE EQUATIONS

Look at our opening display again. Notice how we obtained the matrix from the system of equations. We just suppressed all the unknowns, the plus signs, and the equal signs, and supplied some 1's. We call this matrix the **matrix of the system.** We are going to solve this system, keeping track of what happens to the matrix as we move from step to step.

$$2x + 3y - z = 1 \qquad \begin{bmatrix} 2 & 3 & -1 & 1 \\ 1 & 4 & -1 & 4 \\ 3 & 1 & 2 & 5 \end{bmatrix}$$
$$x + 4y - z = 4$$
$$3x + y + 2z = 5$$

Interchange the first and second equations.

$$x + 4y - z = 4 \qquad \begin{bmatrix} 1 & 4 & -1 & 4 \\ 2 & 3 & -1 & 1 \\ 3 & 1 & 2 & 5 \end{bmatrix}$$
$$2x + 3y - z = 1$$
$$3x + y + 2z = 5$$

Add -2 times the first equation to the second; then add -3 times the first equation to the third.

$$x + 4y - z = 4 \qquad \begin{bmatrix} 1 & 4 & -1 & 4 \\ 0 & -5 & 1 & -7 \\ 0 & -11 & 5 & -7 \end{bmatrix}$$
$$- 5y + z = -7$$
$$- 11y + 5z = -7$$

Multiply the second equation by $-\frac{1}{5}$.

$$x + 4y - z = 4 \qquad \begin{bmatrix} 1 & 4 & -1 & 4 \\ 0 & 1 & -\frac{1}{5} & \frac{7}{5} \\ 0 & -11 & 5 & -7 \end{bmatrix}$$
$$y - \tfrac{1}{5}z = \tfrac{7}{5}$$
$$- 11y + 5z = -7$$

Add 11 times the second equation to the third.

$$x + 4y - z = 4 \qquad \begin{bmatrix} 1 & 4 & -1 & 4 \\ 0 & 1 & -\frac{1}{5} & \frac{7}{5} \\ 0 & 0 & \frac{14}{5} & \frac{42}{5} \end{bmatrix}$$
$$y - \tfrac{1}{5}z = \tfrac{7}{5}$$
$$\tfrac{14}{5}z = \tfrac{42}{5}$$

Now the system is in triangular form and can be solved by backward substitution. The result is $z = 3$, $y = 2$, and $x = -1$; we say the solution is $(-1, 2, 3)$.

We make two points about what we have just done. First, the process is not unique. We happen to prefer having a leading coefficient of 1; that was the reason for our first step. One could have started by multiplying the first equation by $-\frac{1}{2}$ and adding to the second, then multiplying the first equation by $-\frac{3}{2}$ and adding to the third. Any process that ultimately puts the system in triangular form is fine.

The second and main point is this. It is unnecessary to carry along all the x's and y's. Why not work with just the numbers? Why not do all the operations on the matrix of the system? Well, why not?

EQUIVALENT MATRICES

Guided by our knowledge of systems of equations, we say that matrices **A** and **B** are **equivalent** if **B** can be obtained from **A** by applying the operations below (a finite number of times).

Operation 1 Interchanging two rows.
Operation 2 Multiplying a row by a nonzero number.
Operation 3 Replacing a row by the sum of that row and a multiple of another row.

When **A** and **B** are equivalent, we write **A** $\sim$ **B**. If **A** $\sim$ **B**, then **B** $\sim$ **A**. If **A** $\sim$ **B** and **B** $\sim$ **C**, then **A** $\sim$ **C**.

AN EXAMPLE WITH FOUR EQUATIONS
Consider

$$\begin{aligned}
x + 3y + z \quad\quad &= 1 \\
2x + 7y + z - w &= -1 \\
3x - 2y \quad\quad + 4w &= 8 \\
-x + y - 3z - w &= -6
\end{aligned}$$

To solve this system, we take its matrix and transform it to triangular form using the operations above. Here is one possible sequence of steps.

$$\begin{bmatrix}
1 & 3 & 1 & 0 & 1 \\
2 & 7 & 1 & -1 & -1 \\
3 & -2 & 0 & 4 & 8 \\
-1 & 1 & -3 & -1 & -6
\end{bmatrix}$$

Add -2 times the first row to the second; -3 times the first row to the third row; and 1 times the first row to the fourth row.

$$\begin{bmatrix} 1 & 3 & 1 & 0 & 1 \\ 0 & 1 & -1 & -1 & -3 \\ 0 & -11 & -3 & 4 & 5 \\ 0 & 4 & -2 & -1 & -5 \end{bmatrix}$$

Add 11 times the second row to the third and -4 times the second row to the fourth.

$$\begin{bmatrix} 1 & 3 & 1 & 0 & 1 \\ 0 & 1 & -1 & -1 & -3 \\ 0 & 0 & -14 & -7 & -28 \\ 0 & 0 & 2 & 3 & 7 \end{bmatrix}$$

Multiply the third row by $-\frac{1}{14}$.

$$\begin{bmatrix} 1 & 3 & 1 & 0 & 1 \\ 0 & 1 & -1 & -1 & -3 \\ 0 & 0 & 1 & \frac{1}{2} & 2 \\ 0 & 0 & 2 & 3 & 7 \end{bmatrix}$$

Add -2 times the third row to the fourth row.

$$\begin{bmatrix} 1 & 3 & 1 & 0 & 1 \\ 0 & 1 & -1 & -1 & -3 \\ 0 & 0 & 1 & \frac{1}{2} & 2 \\ 0 & 0 & 0 & 2 & 3 \end{bmatrix}$$

This last matrix represents the system

$$\begin{aligned} x + 3y + z \qquad\quad &= 1 \\ y - z - \quad w &= -3 \\ z + \tfrac{1}{2}w &= 2 \\ 2w &= 3 \end{aligned}$$

If we use back substitution, we get $w = \frac{3}{2}, z = \frac{5}{4}, y = -\frac{1}{4}$, and $x = \frac{1}{2}$. The solution is $(\frac{1}{2}, -\frac{1}{4}, \frac{5}{4}, \frac{3}{2})$.

THE CASES WITH MANY SOLUTIONS AND NO SOLUTION

A system of equations need not have a unique solution; it may have infinitely many solutions or none at all. We need to be able to analyze the latter two cases by our matrix method. Fortunately, this is easy to do. Consider Example A of Section 8-1 first. Here is how we handle it using matrices.

$$\begin{bmatrix} 1 & -2 & 3 & 10 \\ 2 & -3 & -1 & 8 \\ 4 & -7 & 5 & 28 \end{bmatrix} \sim \begin{bmatrix} 1 & -2 & 3 & 10 \\ 0 & 1 & -7 & -12 \\ 0 & 1 & -7 & -12 \end{bmatrix}$$

$$\sim \begin{bmatrix} 1 & -2 & 3 & 10 \\ 0 & 1 & -7 & -12 \\ 0 & 0 & 0 & 0 \end{bmatrix}$$

The appearance of a row of zeros tells us that we have infinitely many solutions. The set of solutions is obtained by considering the equations corresponding to the first two rows.

$$x - 2y + 3z = 10$$
$$y - 7z = -12$$

When we solve for y in the second equation and substitute in the first, we obtain

$$x = 11z - 14$$
$$y = 7z - 12$$
$$z \quad \text{arbitrary}$$

Next consider the inconsistent example treated in Section 8-1 (Example B). Here is what happens when the matrix method is applied to this example.

$$\begin{bmatrix} 1 & -2 & 3 & 10 \\ 2 & -3 & -1 & 8 \\ 5 & -9 & 8 & 20 \end{bmatrix} \sim \begin{bmatrix} 1 & -2 & 3 & 10 \\ 0 & 1 & -7 & -12 \\ 0 & 1 & -7 & -30 \end{bmatrix}$$

$$\sim \begin{bmatrix} 1 & -2 & 3 & 10 \\ 0 & 1 & -7 & -12 \\ 0 & 0 & 0 & -18 \end{bmatrix}$$

We are tipped off to the inconsistency of the system by the third row of the matrix. It corresponds to the equation

$$0x + 0y + 0z = -18$$

which has no solution. Consequently, the system as a whole has no solution.

We may summarize our discussion as follows. If the process of transforming the matrix of a system of n equations in n unknowns to triangular form leads to a row in which all elements but the last one are zero, then the system is inconsistent; that is, it has no solution. If the above does not occur and we are led to a matrix with one or more rows consisting entirely of zeros, then the system has infinitely many solutions.

Problem Set 8-2

Write the matrix of each system in Problems 1–8.

1. $2x - y = 4$
   $x - 3y = -2$

2. $x + 2y = 13$
   $11x - y = 0$

3. $x - 2y + z = 3$
   $2x + y = 5$
   $x + y + 3z = -4$

4. $x + 4z = 10$
   $2y - z = 0$
   $3x - y = 20$

5. $2x = 3y - 4$
   $3x + 2 = -y$

6. $x = 4y + 3$
   $y = -2x + 5$

7. $x = 5$
   $2y + x - z = 4$
   $3x - y + 13 = 5z$

8. $z = 2$
   $2x - z = -4$
   $x + 2y + 4z = -8$

Regard each matrix in Problems 9–18 as a matrix of a linear system of equations. Tell whether the system has a unique solution, infinitely many solutions, or no solution. You need not solve any of the systems.

9. $\begin{bmatrix} 1 & -2 & 3 \\ 0 & 1 & -4 \end{bmatrix}$

10. $\begin{bmatrix} 2 & 5 & 0 \\ 0 & -3 & 5 \end{bmatrix}$

11. $\begin{bmatrix} 1 & -3 & 5 \\ 2 & -6 & -10 \end{bmatrix}$

12. $\begin{bmatrix} 2 & 1 & -4 \\ -6 & -3 & 12 \end{bmatrix}$

13. $\begin{bmatrix} 1 & -2 & 4 & -2 \\ 0 & 3 & 1 & 4 \\ 0 & 0 & 1 & -3 \end{bmatrix}$

14. $\begin{bmatrix} 5 & 4 & 0 & -11 \\ 0 & 1 & -4 & 0 \\ 0 & 0 & 2 & -4 \end{bmatrix}$

15. $\begin{bmatrix} 2 & 1 & 5 & 4 \\ 0 & 3 & -2 & 10 \\ 0 & 3 & -2 & 10 \end{bmatrix}$

16. $\begin{bmatrix} 4 & 1 & -3 & 5 \\ 0 & 0 & 1 & -4 \\ 0 & 0 & 1 & -4 \end{bmatrix}$

17. $\begin{bmatrix} 3 & 2 & -1 & 0 \\ 0 & 1 & 0 & -4 \\ 0 & 1 & 0 & 5 \end{bmatrix}$

18. $\begin{bmatrix} -1 & 5 & 6 & -3 \\ 0 & 0 & 0 & 0 \\ 0 & 0 & 0 & 4 \end{bmatrix}$

In Problems 19–30, use matrices to solve each system or to show that it has no solution.

19. $x + 2y = 5$
    $2x - 5y = -8$

20. $2x + 4y = 16$
    $3x - y = 10$

21. $3x - 2y = 1$
    $-6x + 4y = -2$

22. $x + 3y = 12$
    $5x + 15y = 12$

23. $3x - 2y + 5z = -10$
    $y - 4z = 8$
    $2y + z = 7$

24. $4x + 5y + 2z = 25$
    $y - 2z = 7$
    $5y + z = 2$

25. $x + y - 3z = 10$
    $2x + 5y + z = 18$
    $5x + 8y - 8z = 48$

26. $x - 4y + z = 18$
    $2x - 7y - 2z = 4$
    $3x - 11y - z = 22$

27. $2x + 5y + 2z = 6$
    $x + 2y - z = 3$
    $3x - y + 2z = 9$

28. $x - 2y + 3z = -2$
    $3x - 6y + 9z = -6$
    $-2x + 4y - 6z = 0$

$\boxed{c}$ 29. $x + 1.2y - 2.3z = 8.1$
    $1.3x + .7y + .4z = 6.2$
    $.5x + 1.2y + .5z = 3.2$

30. $3x + 2y = 4$
    $3x - 4y + 6z = 16$
    $3x - y + z = 6$

Miscellaneous Problems

31. Write the matrix which represents the following system.

$$z = 2x - y + 1$$

$$y = 4x - 3$$

$$2x - y = 7$$

Regard each matrix in Problems 32–35 as the matrix of a linear system of equations. Without solving the system, tell whether it has a unique solution, infinitely many solutions, or no solution.

32. $\begin{bmatrix} 3 & -2 & 5 \\ 0 & 1 & -3 \end{bmatrix}$

33. $\begin{bmatrix} 2 & -1 & 5 \\ -4 & 2 & 8 \end{bmatrix}$

34. $\begin{bmatrix} 2 & -1 & 4 & 6 \\ 0 & 4 & -1 & 5 \\ 0 & 0 & 2 & 1 \end{bmatrix}$

35. $\begin{bmatrix} 3 & 3 & 0 & -4 \\ 0 & 1 & -3 & 2 \\ 0 & 0 & 0 & 0 \end{bmatrix}$

In Problems 36–38, use matrices to solve each system or to show that it has no solution.

36. $3x - 2y + 4z = 0$
    $x - y + 3z = 1$
    $4x + 2y - z = 3$

37. $-4x + y - z = 5$
    $4x - y + z = -5$
    $-24x + 6y - 6z = 10$

38. $2x + 4y - z = 8$
    $4x + 9y + 3z = 42$
    $8x + 17y + z = 58$

39. Find the values of a, b, and c for which the parabola $y = ax^2 + bx + c$ passes through the points $(-2, -32)$, $(1, 4)$, and $(3, -12)$.

40. Show that there is no parabola $y = ax^2 + bx + c$ which passes through the given set of points.
    (a) $(1, 2)$, $(4, 8)$, and $(5, 10)$
    (b) $(1, 2)$, $(4, 8)$, and $(1, -4)$

41. Find the values of D, E, and F for which the circle $x^2 + y^2 + Dx + Ey + F = 0$ passes through the points $(3, -3)$, $(8, 2)$, and $(6, 6)$.

$$\begin{bmatrix} a & b \\ c & d \end{bmatrix} + \begin{bmatrix} A & B \\ C & D \end{bmatrix} = \begin{bmatrix} a+A & b+B \\ c+C & d+D \end{bmatrix}$$

$$\begin{bmatrix} a & b \\ c & d \end{bmatrix} \cdot \begin{bmatrix} A & B \\ C & D \end{bmatrix} = \begin{bmatrix} aA+bC & aB+bD \\ cA+dC & cB+dD \end{bmatrix}$$

"Cayley is forging the weapons for future generations of physicists."

P. G. Tait

8-3
The Algebra of Matrices

When Arthur Cayley introduced matrices, he had much more in mind than the application described in the previous section. There, matrices served as a device to simplify solving systems of equations. Cayley saw that these number boxes could be studied independently of equations, that they could be thought of as a new type of mathematical object. He realized that if he could give appropriate definitions of addition and multiplication, he would create a mathematical system that might stand with the real numbers and the complex numbers as a potential model for many applications. Cayley did all of this in a major paper in 1858. Some of his contemporaries saw little of significance in this new abstraction. But one of them, P. G. Tait, uttered the prophetic words quoted in the opening box. Tait was right. During the 1920's, Werner Heisenberg found that matrices were just the tool he needed to formulate his quantum mechanics. And by 1950, it was generally recognized that matrix theory provides the best model for many problems in economics and the social sciences.

To simplify our discussion, we initially consider only 2×2 matrices, that is, matrices with two rows and two columns. Examples are

$$\begin{bmatrix} -1 & 3 \\ 4 & 0 \end{bmatrix} \qquad \begin{bmatrix} \log .1 & \frac{6}{2} \\ \frac{12}{3} & \log 1 \end{bmatrix} \qquad \begin{bmatrix} a & b \\ c & d \end{bmatrix}$$

The first two of these matrices are said to be equal. In fact, two matrices are **equal** if and only if the entries in corresponding positions are equal. Be sure to distinguish the notion of equality (written $=$) from that of equivalence (written $\sim$) introduced in Section 8-2. For example,

$$\begin{bmatrix} 2 & 1 \\ -3 & 4 \end{bmatrix} \quad \text{and} \quad \begin{bmatrix} -3 & 4 \\ 2 & 1 \end{bmatrix}$$

are equivalent matrices; however, they are not equal.

ADDITION AND SUBTRACTION

Cayley's definition of addition is straightforward. To add two matrices, add the entries in corresponding positions. Thus

$$\begin{bmatrix} 1 & 3 \\ -1 & 4 \end{bmatrix} + \begin{bmatrix} 6 & -2 \\ 5 & 1 \end{bmatrix} = \begin{bmatrix} 1+6 & 3+(-2) \\ -1+5 & 4+1 \end{bmatrix} = \begin{bmatrix} 7 & 1 \\ 4 & 5 \end{bmatrix}$$

and in general

$$\begin{bmatrix} a & b \\ c & d \end{bmatrix} + \begin{bmatrix} A & B \\ C & D \end{bmatrix} = \begin{bmatrix} a+A & b+B \\ c+C & d+D \end{bmatrix}$$

It is easy to check that the commutative and associative properties for addition are valid. Let **U**, **V**, and **W** be any three matrices.

1. (**Commutativity +**) $\mathbf{U} + \mathbf{V} = \mathbf{V} + \mathbf{U}$
2. (**Associativity +**) $\mathbf{U} + (\mathbf{V} + \mathbf{W}) = (\mathbf{U} + \mathbf{V}) + \mathbf{W}$

The matrix

$$\mathbf{O} = \begin{bmatrix} 0 & 0 \\ 0 & 0 \end{bmatrix}$$

behaves as the "zero" for matrices. And the additive inverse of the matrix

$$\mathbf{U} = \begin{bmatrix} a & b \\ c & d \end{bmatrix}$$

is given by

$$-\mathbf{U} = \begin{bmatrix} -a & -b \\ -c & -d \end{bmatrix}$$

We may summarize these statements as follows.

3. (**Neutral element +**)   There is a matrix **O** satisfying $\mathbf{O} + \mathbf{U} = \mathbf{U} + \mathbf{O} = \mathbf{U}$.
4. (**Additive inverses**)   For each matrix **U**, there is a matrix $-\mathbf{U}$ satisfying

$$\mathbf{U} + (-\mathbf{U}) = (-\mathbf{U}) + \mathbf{U} = \mathbf{O}$$

With the existence of an additive inverse settled, we can define subtraction by $\mathbf{U} - \mathbf{V} = \mathbf{U} + (-\mathbf{V})$. This amounts to subtracting the entries of **V** from the corresponding entires of **U**. Thus

$$\begin{bmatrix} 1 & 3 \\ -1 & 4 \end{bmatrix} - \begin{bmatrix} 6 & -2 \\ 5 & 1 \end{bmatrix} = \begin{bmatrix} -5 & 5 \\ -6 & 3 \end{bmatrix}$$

So far, all has been straightforward and nice. But with multiplication, Cayley hit a snag.

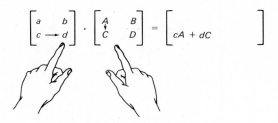

MULTIPLICATION

Cayley's definition of multiplication may seem odd at first glance. He was led to it by consideration of a special problem that we do not have time to describe. It is enough to say that Cayley's definition is the one that proves useful in modern applications (as you will see).

Here it is in symbols.

$$\begin{bmatrix} a & b \\ c & d \end{bmatrix} \cdot \begin{bmatrix} A & B \\ C & D \end{bmatrix} = \begin{bmatrix} aA + bC & aB + bD \\ cA + dC & cB + dD \end{bmatrix}$$

Stated in words, we multiply two matrices by multiplying the rows of the left matrix by the columns of the right matrix in pairwise entry fashion, adding the results. For example, the entry in the second row and first column of the product is obtained by multiplying the entries of the second row of the left matrix by the corresponding entries of the first column of the right matrix, adding the results. Until you get used to it, it may help to use your fingers as shown in the diagram below.

$$\begin{bmatrix} a & b \\ c \rightarrow d \end{bmatrix} \cdot \begin{bmatrix} A & B \\ C & D \end{bmatrix} = \begin{bmatrix} & \\ cA + dC & \end{bmatrix}$$

Here is an example worked out in detail.

$$\begin{bmatrix} 1 & 3 \\ -1 & 4 \end{bmatrix}\begin{bmatrix} 6 & -2 \\ 5 & 1 \end{bmatrix} = \begin{bmatrix} (1)(6) + (3)(5) & (1)(-2) + (3)(1) \\ (-1)(6) + (4)(5) & (-1)(-2) + (4)(1) \end{bmatrix}$$

$$= \begin{bmatrix} 21 & 1 \\ 14 & 6 \end{bmatrix}$$

Here is the same problem, but with the matrices multiplied in the opposite order.

$$\begin{bmatrix} 6 & -2 \\ 5 & 1 \end{bmatrix}\begin{bmatrix} 1 & 3 \\ -1 & 4 \end{bmatrix} = \begin{bmatrix} (6)(1) + (-2)(-1) & (6)(3) + (-2)(4) \\ (5)(1) + (1)(-1) & (5)(3) + (1)(4) \end{bmatrix}$$

$$= \begin{bmatrix} 8 & 10 \\ 4 & 19 \end{bmatrix}$$

Now you see the snag about which we warned you. The commutative property for multiplication fails. That is troublesome, but not fatal. We manage to get along in a world that is largely noncommutative (try removing your clothes and taking a shower in the opposite order). We just have to remember never to commute matrices under multiplication. Fortunately two other nice properties do hold.

5. (**Associativity ·**) $\mathbf{U} \cdot (\mathbf{V} \cdot \mathbf{W}) = (\mathbf{U} \cdot \mathbf{V}) \cdot \mathbf{W}$
6. (**Distributivity**) $\mathbf{U} \cdot (\mathbf{V} + \mathbf{W}) = \mathbf{U} \cdot \mathbf{V} + \mathbf{U} \cdot \mathbf{W}$
$(\mathbf{V} + \mathbf{W}) \cdot \mathbf{U} = \mathbf{V} \cdot \mathbf{U} + \mathbf{W} \cdot \mathbf{U}$

We have not said anything about multiplicative inverses; that comes in the next section. We do, however, want to mention a special operation called **scalar multiplication**—that is, multiplication of a matrix by a scalar (number). To multiply a matrix by a number, multiply each entry by that number. That is,

$$k \begin{bmatrix} a & b \\ c & d \end{bmatrix} = \begin{bmatrix} ka & kb \\ kc & kd \end{bmatrix}$$

Scalar multipliction satisfies the expected properties.

7. $k(\mathbf{U} + \mathbf{V}) = k\mathbf{U} + k\mathbf{V}$
8. $(k + m)\mathbf{U} = k\mathbf{U} + m\mathbf{U}$
9. $(km)\mathbf{U} = k(m\mathbf{U})$

LARGER MATRICES AND COMPATIBILITY

So far we have considered only 2×2 matrices. That is an unnecessary restriction; however, to perform operations on arbitrary matrices, we must make sure they are **compatible.** For addition, this simply means that the matrices must be of the same size. Thus

$$\begin{bmatrix} 1 & -1 & 3 \\ 4 & -5 & 2 \end{bmatrix} + \begin{bmatrix} 2 & 6 & 0 \\ -3 & 2 & 4 \end{bmatrix} = \begin{bmatrix} 3 & 5 & 3 \\ 1 & -3 & 6 \end{bmatrix}$$

but

$$\begin{bmatrix} 1 & -1 & 3 \\ 4 & -5 & 2 \end{bmatrix} + \begin{bmatrix} 2 & 6 \\ -3 & 2 \end{bmatrix}$$

makes no sense.

Two matrices are compatible for multiplication if the left matrix has the same number of columns as the right matrix has rows. For example,

$$\begin{bmatrix} 1 & -1 & 3 \\ 4 & -5 & 2 \end{bmatrix} \begin{bmatrix} 2 & 1 & 5 & 0 \\ -1 & 3 & 2 & 1 \\ 1 & -2 & 0 & 2 \end{bmatrix} = \begin{bmatrix} 6 & -8 & 3 & 5 \\ 15 & -15 & 10 & -1 \end{bmatrix}$$

The left matrix is 2×3, the right one is 3×4, and the result is 2×4. In general, we can multiply an $m \times n$ matrix by an $n \times p$ matrix, the result being an $m \times p$ matrix. All of the properties mentioned earlier are valid, provided we work with compatible matrices.

A BUSINESS APPLICATION

The ABC Company sells precut lumber for two types of summer cottages, standard and deluxe. The standard model requires 30,000 board feet of lumber and 100 worker-hours of cutting; the deluxe model takes 40,000 board feet of lumber and 110 worker-hours of cutting. This year, the ABC Company buys its lumber at $.20 per board foot and pays its laborers $9.00 per hour. Next year it expects these costs to be $.25 and $10.00, respectively. This information can be displayed in matrix form as follows.

	REQUIREMENTS A lumber	labor			UNITS COSTS B this year	next year
standard	30,000	100		lumber	$.20	$.25
deluxe	40,000	110		labor	$9.00	$10.00

Now we ask whether the product matrix **AB** has economic significance. It does: It gives the total dollar cost of standard and deluxe cottages both for this year and next. You can see this from the following calculation.

$$\mathbf{AB} = \begin{bmatrix} (30,000)(.20) + (100)(9) & (30,000)(.25) + (100)(10) \\ (40,000)(.20) + (110)(9) & (40,000)(.25) + (110)(10) \end{bmatrix}$$

$$= \begin{bmatrix} \text{this year} & \text{next year} \\ \$6900 & \$8500 \\ \$8990 & \$11,100 \end{bmatrix} \begin{matrix} \text{standard} \\ \text{deluxe} \end{matrix}$$

Problem Set 8-3

Calculate $\mathbf{A} + \mathbf{B}$, $\mathbf{A} - \mathbf{B}$, *and* $3\mathbf{A}$ *in Problems 1–4.*

1. $\mathbf{A} = \begin{bmatrix} 2 & -1 \\ 3 & 7 \end{bmatrix}$, $\mathbf{B} = \begin{bmatrix} 6 & 5 \\ -2 & 3 \end{bmatrix}$

2. $\mathbf{A} = \begin{bmatrix} -1 & 0 \\ 5 & 4 \end{bmatrix}$, $\mathbf{B} = \begin{bmatrix} 2 & -2 \\ 3 & 7 \end{bmatrix}$

3. $\mathbf{A} = \begin{bmatrix} 3 & -2 & 5 \\ 4 & 0 & -3 \end{bmatrix}$, $\mathbf{B} = \begin{bmatrix} 2 & 6 & -1 \\ 4 & 3 & -3 \end{bmatrix}$

4. $\mathbf{A} = \begin{bmatrix} 1 & 2 & 3 \\ 4 & 5 & 6 \\ 7 & 8 & 9 \end{bmatrix}$, $\mathbf{B} = \begin{bmatrix} -1 & -2 & -2 \\ -4 & -5 & -6 \\ -7 & -8 & -9 \end{bmatrix}$

Calculate **AB** *and* **BA** *if possible in Problems 5–12.*

5. $\mathbf{A} = \begin{bmatrix} 2 & -1 \\ 3 & 7 \end{bmatrix}$, $\mathbf{B} = \begin{bmatrix} 6 & 5 \\ -2 & 3 \end{bmatrix}$

6. $\mathbf{A} = \begin{bmatrix} -1 & 0 \\ 5 & 4 \end{bmatrix}$, $\mathbf{B} = \begin{bmatrix} 2 & -2 \\ 3 & 7 \end{bmatrix}$

7. $\mathbf{A} = \begin{bmatrix} 1 & -1 & 2 \\ 3 & 4 & -4 \\ 2 & 1 & 3 \end{bmatrix}$, $\mathbf{B} = \begin{bmatrix} 0 & 2 & -3 \\ 1 & 2 & 3 \\ -1 & -2 & 4 \end{bmatrix}$

8. $\mathbf{A} = \begin{bmatrix} -2 & 5 & 1 \\ 0 & -2 & 3 \\ 1 & 2 & -1 \end{bmatrix}$, $\mathbf{B} = \begin{bmatrix} -3 & 4 & 1 \\ 2 & 5 & 1 \\ 1 & 2 & 3 \end{bmatrix}$

9. $\mathbf{A} = \begin{bmatrix} 1 & -2 & 3 & 4 \\ 3 & 2 & -5 & 1 \end{bmatrix}$, $\mathbf{B} = \begin{bmatrix} 1 & 2 \\ 3 & 4 \end{bmatrix}$

10. $\mathbf{A} = \begin{bmatrix} -1 & 3 \\ 4 & 2 \\ 1 & 5 \end{bmatrix}$, $\mathbf{B} = \begin{bmatrix} -1 & 2 & 3 & 4 \\ 0 & -3 & 2 & 1 \end{bmatrix}$

11. $\mathbf{A} = \begin{bmatrix} 3 & 1 & -1 \\ 2 & 4 & 2 \\ -3 & 2 & -1 \end{bmatrix}$, $\mathbf{B} = \begin{bmatrix} 1 \\ 2 \\ 3 \end{bmatrix}$

12. $\mathbf{A} = \begin{bmatrix} 1 & 2 & -1 \end{bmatrix}$, $\mathbf{B} = \begin{bmatrix} 4 & 3 \\ 0 & 2 \\ -1 & 4 \end{bmatrix}$

13. Calculate **AB** and **BA** for

$$\mathbf{A} = \begin{bmatrix} 0 & 0 \\ 0 & 0 \end{bmatrix} \qquad \mathbf{B} = \begin{bmatrix} 2 & -1 \\ 3 & 4 \end{bmatrix}$$

14. State the general property illustrated by Problem 13.

15. Find **X** if

$$\begin{bmatrix} 2 & 1 & -3 \\ 1 & 5 & 0 \end{bmatrix} + \mathbf{X} = 2\begin{bmatrix} -1 & 4 & 3 \\ -2 & 0 & 4 \end{bmatrix}$$

16. Solve for **X**.

$$-3\mathbf{X} + 2\begin{bmatrix} 1 & -2 \\ 5 & 6 \end{bmatrix} = -\begin{bmatrix} 5 & -14 \\ 8 & 15 \end{bmatrix}$$

17. Calculate **A(B + C)** and **AB + AC** for

$$\mathbf{A} = \begin{bmatrix} 2 & -1 \\ 3 & 4 \end{bmatrix} \qquad \mathbf{B} = \begin{bmatrix} 2 & 4 \\ 6 & 1 \end{bmatrix} \qquad \mathbf{C} = \begin{bmatrix} -1 & -2 \\ 3 & 6 \end{bmatrix}$$

What property does this illustrate?

18. Calculate $(\mathbf{A} + \mathbf{B})(\mathbf{A} - \mathbf{B})$ and $\mathbf{A}^2 - \mathbf{B}^2$ for

$$\mathbf{A} = \begin{bmatrix} 3 & -2 \\ 1 & 4 \end{bmatrix} \qquad \mathbf{B} = \begin{bmatrix} 6 & -3 \\ 2 & 5 \end{bmatrix}$$

Why are your answers different?

© 19. Find the entry in the third row and second column of the product

$$\begin{bmatrix} 1.39 & 4.13 & -2.78 \\ 4.72 & -3.69 & 5.41 \\ 8.09 & -6.73 & 5.03 \end{bmatrix} \begin{bmatrix} 5.45 & 6.31 \\ 7.24 & -5.32 \\ 6.06 & 1.34 \end{bmatrix}$$

© 20. Find the entry in the second row and first column of the product in Problem 19.

Miscellaneous Problems

21. Compute $\mathbf{A} - 2\mathbf{B}$, $\mathbf{AB}$, and $\mathbf{A}^2$ for

$$\mathbf{A} = \begin{bmatrix} 4 & -1 & 3 \\ 2 & 5 & 3 \\ 6 & 2 & 1 \end{bmatrix} \qquad \mathbf{B} = \begin{bmatrix} 1 & -3 & 2 \\ 5 & 0 & 3 \\ -5 & 2 & 1 \end{bmatrix}$$

22. Let $\mathbf{A}$ and $\mathbf{B}$ be 3×4 matrices, $\mathbf{C}$ a 4×5 matrix, and $\mathbf{D}$ a 3×5 matrix. Which of the following satisfy the compatibility conditions?
    (a) $\mathbf{AB}$ (b) $\mathbf{AC}$ (c) $\mathbf{AC} + \mathbf{D}$
    (d) $(\mathbf{A} - \mathbf{B})\mathbf{C}$ (e) $(\mathbf{AC})\mathbf{D}$ (f) $\mathbf{A(CD)}$
    (g) $\mathbf{A}^2$

23. Expand $(\mathbf{A} + \mathbf{B})^2$. Does it equal $\mathbf{A}^2 + 2\mathbf{AB} + \mathbf{B}^2$?

24. Calculate

$$\begin{bmatrix} 1 & 5 \\ 3 & -4 \end{bmatrix} \begin{bmatrix} 3 & 0 \\ -2 & 0 \end{bmatrix}$$

25. If the second row of $\mathbf{A}$ consists of zeros, what can you say about the second row of $\mathbf{AB}$?

26. Calculate $\mathbf{AB}$ and $\mathbf{BA}$ for

$$\mathbf{A} = \begin{bmatrix} 1 & 2 & 3 & 4 \end{bmatrix} \qquad \mathbf{B} = \begin{bmatrix} 2 \\ 1 \\ -1 \\ -2 \end{bmatrix}$$

27. Show that if both $\mathbf{AB}$ and $\mathbf{BA}$ make sense ($\mathbf{B}$ and $\mathbf{A}$ are compatible for multiplication in either order), then $\mathbf{AB}$ and $\mathbf{BA}$ are both square matrices.

28. Show that if $\mathbf{AB}$ and $\mathbf{BC}$ both satisfy the compatibility condition, then $(\mathbf{AB})\mathbf{C}$ and $\mathbf{A}(\mathbf{BC})$ do also.

29. Art, Bob, and Curt work for a company that makes Flukes, Gizmos, and Horks. They are paid for their labor on a piecework basis, receiving $1 for each Fluke, $2 for each Gizmo, and $3 for each Hork. Below are matrices $\mathbf{U}$ and $\mathbf{V}$ representing their outputs on Monday and Tuesday. Matrix $\mathbf{X}$ is the wage/unit matrix.

	MONDAY'S OUTPUT U				TUESDAY'S OUTPUT V				WAGE/UNIT X

$$
\begin{array}{c}
\textit{MONDAY'S OUTPUT} \\
\mathbf{U} \\
\begin{array}{ccc} F & G & H \end{array}
\end{array}
\qquad
\begin{array}{c}
\textit{TUESDAY'S OUTPUT} \\
\mathbf{V} \\
\begin{array}{ccc} F & G & H \end{array}
\end{array}
\qquad
\begin{array}{c}
\textit{WAGE/UNIT} \\
\mathbf{X} \\
\end{array}
$$

$$
\begin{array}{c}
\text{Art} \\ \text{Bob} \\ \text{Curt}
\end{array}
\begin{bmatrix} 4 & 3 & 2 \\ 5 & 1 & 2 \\ 3 & 4 & 1 \end{bmatrix}
\qquad
\begin{array}{c}
\text{Art} \\ \text{Bob} \\ \text{Curt}
\end{array}
\begin{bmatrix} 3 & 6 & 1 \\ 4 & 2 & 2 \\ 5 & 1 & 3 \end{bmatrix}
\qquad
\begin{array}{c}
\text{F} \\ \text{G} \\ \text{H}
\end{array}
\begin{bmatrix} 1 \\ 2 \\ 3 \end{bmatrix}
$$

Compute the following matrices and decide what they represent.
(a) **UX** (b) **VX** (c) **U + V** (d) **(U + V)X**

30. Four friends A, B, C, and D have unlisted telephone numbers. Whether or not one person knows another's number is indicated by the matrix **U** below, where 1 indicates knowing and 0 indicates not knowing. For example, the 1 in row 3 and column 1 means that C knows A's number.

$$
\mathbf{U} =
\begin{array}{c}
A \\ B \\ C \\ D
\end{array}
\begin{array}{c}
\begin{array}{cccc} A & B & C & D \end{array} \\
\begin{bmatrix} 1 & 0 & 1 & 0 \\ 0 & 1 & 1 & 0 \\ 1 & 0 & 1 & 1 \\ 0 & 1 & 0 & 1 \end{bmatrix}
\end{array}
$$

(a) Calculate $\mathbf{U}^2$.
(b) Interpret $\mathbf{U}^2$ in terms of the possibility of each person being able to get a telephone message to another.
(c) Can D get a message to A via one other person?
(d) Interpret $\mathbf{U}^3$.

31. Let

$$
\mathbf{A} =
\begin{bmatrix} 3 & 0 & 0 \\ 0 & 4 & 0 \\ 0 & 0 & 5 \end{bmatrix}
$$

If **B** is any 3×3 matrix, what does multiplication on the left by **A** do to **B**? Multiplication on the right by **A**?

32. Calculate $\mathbf{A}^2$ and $\mathbf{A}^3$ for the matrix of Problem 31. State a general result for raising diagonal matrices to powers.

"What was that?" inquired Alice. "Reeling and Writhing, of course, to begin with," the mock turtle replied, "and then the different branches of Arithmetic–Ambition, Distraction, Uglification, and Derision."

from *Alice's Adventures*
in Wonderland
by Lewis Carroll

$$\frac{\begin{bmatrix} 2 & 3 \\ -4 & 1 \end{bmatrix}}{\begin{bmatrix} 6 & 7 \\ 1 & 2 \end{bmatrix}} = \begin{bmatrix} ? & ? \\ ? & ? \end{bmatrix}$$

8-4
Multiplicative Inverses

Even for ordinary numbers, the notion of division seems more difficult than that of addition, subtraction, and multiplication. Certainly this is true for division of matrices. Look at the example displayed above. It could tempt more than a mock turtle to derision. However, Arthur Cayley saw no need to sneer. He noted that in the case of numbers,

$$\frac{U}{V} = U \cdot \frac{1}{V} = U \cdot V^{-1}$$

What is needed is a concept of "one" for matrices; then we need the concept of multiplicative inverse. The first is easy.

THE MULTIPLICATIVE IDENTITY FOR MATRICES
Let

$$\mathbf{I} = \begin{bmatrix} 1 & 0 \\ 0 & 1 \end{bmatrix}$$

Then for any 2×2 matrix $\mathbf{U}$,

$$\mathbf{UI} = \mathbf{U} = \mathbf{IU}$$

This can be checked by noting that

$$\begin{bmatrix} a & b \\ c & d \end{bmatrix}\begin{bmatrix} 1 & 0 \\ 0 & 1 \end{bmatrix} = \begin{bmatrix} a & b \\ c & d \end{bmatrix} = \begin{bmatrix} 1 & 0 \\ 0 & 1 \end{bmatrix}\begin{bmatrix} a & b \\ c & d \end{bmatrix}$$

The symbol $\mathbf{I}$ is chosen because it is often called the **multiplicative identity.** In accordance with Section 1-4, it is also called the neutral element for multiplication.

For 3×3 matrices, the multiplicative identity has the form

$$\begin{bmatrix} 1 & 0 & 0 \\ 0 & 1 & 0 \\ 0 & 0 & 1 \end{bmatrix}$$

You should be able to guess its form for 4×4 and higher order matrices.

INVERSES OF 2 × 2 MATRICES

Suppose we want to find the multiplicative inverse of

$$\mathbf{V} = \begin{bmatrix} 6 & 7 \\ 1 & 2 \end{bmatrix}$$

We are looking for a matrix

$$\mathbf{W} = \begin{bmatrix} a & b \\ c & d \end{bmatrix}$$

that satisfies $\mathbf{VW} = \mathbf{I}$ and $\mathbf{WV} = \mathbf{I}$. Taking $\mathbf{VW} = \mathbf{I}$ first, we want

$$\begin{bmatrix} 6 & 7 \\ 1 & 2 \end{bmatrix} \begin{bmatrix} a & b \\ c & d \end{bmatrix} = \begin{bmatrix} 1 & 0 \\ 0 & 1 \end{bmatrix}$$

which means

$$\begin{bmatrix} 6a + 7c & 6b + 7d \\ a + 2c & b + 2d \end{bmatrix} = \begin{bmatrix} 1 & 0 \\ 0 & 1 \end{bmatrix}$$

or

$$6a + 7c = 1 \qquad 6b + 7d = 0$$
$$a + 2c = 0 \qquad b + 2d = 1$$

When these four equations are solved for a, b, c, d, we have

$$\mathbf{W} = \begin{bmatrix} \frac{2}{5} & -\frac{7}{5} \\ -\frac{1}{5} & \frac{6}{5} \end{bmatrix}$$

as a tentative solution to our problem. We say tentative, because so far we know only that $\mathbf{VW} = \mathbf{I}$. Happily, $\mathbf{W}$ works on the other side of $\mathbf{V}$ too, as we can check. (In this exceptional case, we do have commutativity.)

$$\mathbf{WV} = \begin{bmatrix} \frac{2}{5} & -\frac{7}{5} \\ -\frac{1}{5} & \frac{6}{5} \end{bmatrix} \begin{bmatrix} 6 & 7 \\ 1 & 2 \end{bmatrix} = \begin{bmatrix} 1 & 0 \\ 0 & 1 \end{bmatrix}$$

Success! $\mathbf{W}$ is the inverse of $\mathbf{V}$; we denote it by the symbol $\mathbf{V}^{-1}$.

The process just described can be carried out for any specific 2×2 matrix, or better, it can be carried out for a general 2×2 matrix. But before we give the result, we make an important comment. There is no reason to think that every 2×2 matrix has a multiplicative inverse. Remember that the number 0 does not have such an inverse; neither does the matrix $\mathbf{O}$. But here is a mild surpirse. Many other 2×2 matrices do not have inverses. The following theorem identifies in a very precise way those that do, and then gives a formula for their inverses.

THEOREM (MULTIPLICATIVE INVERSES)

The matrix

$$\mathbf{V} = \begin{bmatrix} a & b \\ c & d \end{bmatrix}$$

has a multiplicative inverse if and only if $D = ad - bc$ is nonzero. If $D \neq 0$, then

$$\mathbf{V}^{-1} = \begin{bmatrix} \dfrac{d}{D} & -\dfrac{b}{D} \\ -\dfrac{c}{D} & \dfrac{a}{D} \end{bmatrix}$$

Thus the number D determines whether a matrix has an inverse. This number, which we shall call a *determinant*, will be studied in detail in the next section. Each 2×2 matrix has such a number associated with it. Let us look at two examples.

$$\mathbf{X} = \begin{bmatrix} 2 & -3 \\ -4 & 6 \end{bmatrix} \qquad\qquad \mathbf{Y} = \begin{bmatrix} 5 & -3 \\ -4 & 3 \end{bmatrix}$$

$$D = (2)(6) - (-3)(-4) = 0 \qquad D = (5)(3) - (-3)(-4) = 3$$

$$\mathbf{X}^{-1} \text{ does not exist.} \qquad \mathbf{Y}^{-1} = \begin{bmatrix} \frac{3}{3} & \frac{3}{3} \\ \frac{4}{3} & \frac{5}{3} \end{bmatrix}$$

INVERSES FOR HIGHER-ORDER MATRICES

There is a theorem like the one above for square matrices of any size, which Cayley found in 1858. It is complicated and, rather than try to state it, we are going to illustrate a process which yields the inverse of a matrix whenever it exists. Briefly described, it is this. Take any square matrix $\mathbf{V}$ and write the corresponding identity matrix $\mathbf{I}$ next to it on the right. By using the three row operations of Section 8-2, attempt to reduce $\mathbf{V}$ to the identity matrix while simultaneously performing the same operations on $\mathbf{I}$. If you can reduce $\mathbf{V}$ to $\mathbf{I}$, you will simultaneously turn $\mathbf{I}$ into $\mathbf{V}^{-1}$. If you cannot reduce $\mathbf{V}$ to $\mathbf{I}$, $\mathbf{V}$ has no inverse.

Here is an illustration for the 2×2 matrix $\mathbf{V}$ that we used earlier.

$$\begin{bmatrix} 6 & 7 & \bigm| & 1 & 0 \\ 1 & 2 & \bigm| & 0 & 1 \end{bmatrix}$$

$$\sim \begin{bmatrix} 1 & 2 & \bigm| & 0 & 1 \\ 6 & 7 & \bigm| & 1 & 0 \end{bmatrix} \qquad \text{(interchange rows)}$$

$$\sim \begin{bmatrix} 1 & 2 & \bigm| & 0 & 1 \\ 0 & -5 & \bigm| & 1 & -6 \end{bmatrix} \qquad \text{(add } -6 \text{ times row 1 to row 2)}$$

$$\sim \begin{bmatrix} 1 & 2 & \bigg| & 0 & 1 \\ 0 & 1 & \bigg| & -\frac{1}{5} & \frac{6}{5} \end{bmatrix} \quad \text{(divide row 2 by } -5\text{)}$$

$$\sim \begin{bmatrix} 1 & 0 & \bigg| & \frac{2}{5} & -\frac{7}{5} \\ 0 & 1 & \bigg| & -\frac{1}{5} & \frac{6}{5} \end{bmatrix} \quad \text{(add } -2 \text{ times row 2 to row 1)}$$

Notice that the matrix $\mathbf{V}^{-1}$ that we obtained earlier appears on the right. We illustrate the same process for a 3×3 matrix in Example A of the problem set.

AN APPLICATION

Consider the system of equations

$$2x + 6y + 6z = 8$$
$$2x + 7y + 6z = 10$$
$$2x + 7y + 7z = 9$$

If we introduce matrices

$$\mathbf{A} = \begin{bmatrix} 2 & 6 & 6 \\ 2 & 7 & 6 \\ 2 & 7 & 7 \end{bmatrix} \quad \mathbf{X} = \begin{bmatrix} x \\ y \\ z \end{bmatrix} \quad \mathbf{B} = \begin{bmatrix} 8 \\ 10 \\ 9 \end{bmatrix}$$

this system can be written in the form

$$\mathbf{AX} = \mathbf{B}$$

Now divide both sides by $\mathbf{A}$, by which we mean, of course, multiply both sides by $\mathbf{A}^{-1}$. We must be more precise. Multiply both sides on the left by $\mathbf{A}^{-1}$ (do not forget the lack of commutativity).

$$\mathbf{A}^{-1}\mathbf{AX} = \mathbf{A}^{-1}\mathbf{B}$$
$$\mathbf{IX} = \mathbf{A}^{-1}\mathbf{B}$$
$$\mathbf{X} = \mathbf{A}^{-1}\mathbf{B}$$

In Example A on page 322, $\mathbf{A}^{-1}$ is found to be

$$\mathbf{A}^{-1} = \begin{bmatrix} \frac{7}{2} & 0 & -3 \\ -1 & 1 & 0 \\ 0 & -1 & 1 \end{bmatrix}$$

Thus

$$\mathbf{X} = \begin{bmatrix} \frac{7}{2} & 0 & -3 \\ -1 & 1 & 0 \\ 0 & -1 & 1 \end{bmatrix} \begin{bmatrix} 8 \\ 10 \\ 9 \end{bmatrix} = \begin{bmatrix} 1 \\ 2 \\ -1 \end{bmatrix}$$

and therefore $(1, 2, -1)$ is the solution to our system.

This method of solution is particularly useful when many systems with the same coefficient matrix $\mathbf{A}$ are under consideration. Once we have $\mathbf{A}^{-1}$, we can obtain any solution simply by doing an easy matrix multiplication. If only one system is being studied, the method of Section 8-2 is best.

Problem Set 8-4

Find the multiplicative inverse of each matrix. Use matrix multiplication as a check.

1. $\begin{bmatrix} 2 & 3 \\ -1 & -1 \end{bmatrix}$

2. $\begin{bmatrix} 4 & 3 \\ 1 & 2 \end{bmatrix}$

3. $\begin{bmatrix} 6 & -14 \\ 0 & 2 \end{bmatrix}$

4. $\begin{bmatrix} 0 & 3 \\ 2 & 4 \end{bmatrix}$

5. $\begin{bmatrix} 1 & 0 \\ 0 & 1 \end{bmatrix}$

6. $\begin{bmatrix} 4 & 0 \\ 0 & 5 \end{bmatrix}$

7. $\begin{bmatrix} a & 0 \\ 0 & b \end{bmatrix}$

8. $\begin{bmatrix} 3 & 0 & 0 \\ 0 & 4 & 0 \\ 0 & 0 & 5 \end{bmatrix}$

EXAMPLE A (Inverses of large matrices)   Find the multiplicative inverse of

$$\begin{bmatrix} 2 & 6 & 6 \\ 2 & 7 & 6 \\ 2 & 7 & 7 \end{bmatrix}$$

Solution.   We use the reduction method described in the text.

$$\left[\begin{array}{ccc|ccc} 2 & 6 & 6 & 1 & 0 & 0 \\ 2 & 7 & 6 & 0 & 1 & 0 \\ 2 & 7 & 7 & 0 & 0 & 1 \end{array}\right]$$

$$\sim \left[\begin{array}{ccc|ccc} 1 & 3 & 3 & \frac{1}{2} & 0 & 0 \\ 2 & 7 & 6 & 0 & 1 & 0 \\ 2 & 7 & 7 & 0 & 0 & 1 \end{array}\right] \quad \text{(divide row 1 by 2)}$$

$$\sim \left[\begin{array}{ccc|ccc} 1 & 3 & 3 & \frac{1}{2} & 0 & 0 \\ 0 & 1 & 0 & -1 & 1 & 0 \\ 0 & 1 & 1 & -1 & 0 & 1 \end{array}\right] \quad \begin{array}{l}\text{(add } -2 \text{ times row 1 to} \\ \text{row 2 and to row 3)}\end{array}$$

$$\sim \left[\begin{array}{ccc|ccc} 1 & 3 & 3 & \frac{1}{2} & 0 & 0 \\ 0 & 1 & 0 & -1 & 1 & 0 \\ 0 & 0 & 1 & 0 & -1 & 1 \end{array}\right] \quad \begin{array}{l}\text{(add } -1 \text{ times row 2 to} \\ \text{row 3)}\end{array}$$

$$\sim \left[\begin{array}{ccc|ccc} 1 & 0 & 3 & \frac{7}{2} & -3 & 0 \\ 0 & 1 & 0 & -1 & 1 & 0 \\ 0 & 0 & 1 & 0 & -1 & 1 \end{array}\right] \quad \begin{array}{l}\text{(add } -3 \text{ times row 2 to} \\ \text{row 1)}\end{array}$$

$$\sim \left[\begin{array}{ccc|ccc} 1 & 0 & 0 & \frac{7}{2} & 0 & -3 \\ 0 & 1 & 0 & -1 & 1 & 0 \\ 0 & 0 & 1 & 0 & -1 & 1 \end{array}\right] \quad \begin{array}{l}\text{(add } -3 \text{ times row 3 to} \\ \text{row 1)}\end{array}$$

Thus the desired inverse is

$$\begin{bmatrix} \frac{7}{2} & 0 & -3 \\ -1 & 1 & 0 \\ 0 & -1 & 1 \end{bmatrix}$$

Use the method illustrated above to find the multiplicative inverse of each of the following.

9. $\begin{bmatrix} 1 & 3 \\ 2 & 4 \end{bmatrix}$

10. $\begin{bmatrix} 2 & 6 \\ 3 & 1 \end{bmatrix}$

11. $\begin{bmatrix} 1 & 1 & 1 \\ 1 & -1 & 2 \\ 3 & 2 & 0 \end{bmatrix}$

12. $\begin{bmatrix} 2 & 1 & 1 \\ 1 & 3 & 1 \\ -1 & 4 & 0 \end{bmatrix}$

13. $\begin{bmatrix} 3 & 1 & 2 \\ 4 & 1 & -6 \\ 1 & 0 & 1 \end{bmatrix}$

14. $\begin{bmatrix} 2 & 4 & 6 \\ 3 & 2 & -5 \\ 2 & 3 & 1 \end{bmatrix}$

15. $\begin{bmatrix} 1 & 2 & 1 & 1 \\ 0 & 2 & 3 & 2 \\ 0 & 0 & 1 & 3 \\ 0 & 0 & 0 & 4 \end{bmatrix}$

16. $\begin{bmatrix} 1 & 1 & 1 & 1 \\ 1 & 1 & 1 & -1 \\ 1 & 1 & -1 & 1 \\ 1 & -1 & 1 & 1 \end{bmatrix}$

Solve the following systems by making use of the inverses you found in Problems 11–14. Begin by writing the system in the matrix form $AX = B$.

17. $\begin{aligned} x + y + z &= 2 \\ x - y + 2z &= -1 \\ 3x + 2y &= 5 \end{aligned}$

18. $\begin{aligned} 2x + y + z &= 4 \\ x + 3y + z &= 5 \\ -x + 4y &= 0 \end{aligned}$

19. $\begin{aligned} 3x + y + 2z &= 3 \\ 4x + y - 6z &= 2 \\ x \quad\ + z &= 6 \end{aligned}$

20. $\begin{aligned} 2x + 4y + 6z &= 9 \\ 3x + 2y - 5z &= 2 \\ 2x + 3y + z &= 4 \end{aligned}$

EXAMPLE B (Matrices without inverses) Try to find the multiplicative inverse of

$$\mathbf{U} = \begin{bmatrix} 1 & 4 & 2 \\ 0 & 2 & 4 \\ 0 & -3 & -6 \end{bmatrix}$$

Solution.

$$\left[\begin{array}{ccc|ccc} 1 & 4 & 2 & 1 & 0 & 0 \\ 0 & 2 & 4 & 0 & 1 & 0 \\ 0 & -3 & -6 & 0 & 0 & 1 \end{array}\right] \sim \left[\begin{array}{ccc|ccc} 1 & 4 & 2 & 1 & 0 & 0 \\ 0 & 1 & 2 & 0 & \frac{1}{2} & 0 \\ 0 & -3 & -6 & 0 & 0 & 1 \end{array}\right]$$

$$\sim \left[\begin{array}{ccc|ccc} 1 & 4 & 2 & 1 & 0 & 0 \\ 0 & 1 & 2 & 0 & \frac{1}{2} & 0 \\ 0 & 0 & 0 & 0 & \frac{3}{2} & 1 \end{array}\right]$$

Since we got a row of zeros in the left half above, we know we can never reduce it to the identity matrix **I**. The matrix **U** does not have an inverse.

Show that neither of the following matrices has a multiplicative inverse.

21. $\begin{bmatrix} 1 & 3 & 4 \\ 2 & 1 & -1 \\ 4 & 7 & 7 \end{bmatrix}$

22. $\begin{bmatrix} 2 & -2 & 4 \\ 5 & 3 & 2 \\ 3 & 5 & -2 \end{bmatrix}$

Miscellaneous Problems

In Problems 23–26, find the multiplicative inverse or indicate that it does not exist.

23. $\begin{bmatrix} -2 & 5 \\ 1 & -\frac{5}{2} \end{bmatrix}$

24. $\begin{bmatrix} 3 & -1 \\ 4 & 2 \end{bmatrix}$

25. $\begin{bmatrix} 1 & -3 & 4 \\ 2 & 3 & 5 \\ -1 & 4 & 2 \end{bmatrix}$

26. $\begin{bmatrix} -2 & 4 & 2 \\ 3 & 5 & 6 \\ 1 & 9 & 8 \end{bmatrix}$

27. Find the multiplicative inverse of

$$\begin{bmatrix} 2 & 0 & 0 \\ 0 & 3 & 0 \\ 0 & 0 & -4 \end{bmatrix}$$

28. Give a formula for $\mathbf{U}^{-1}$ if

$$\mathbf{U} = \begin{bmatrix} a & 0 & 0 \\ 0 & b & 0 \\ 0 & 0 & c \end{bmatrix}$$

When does the matrix **U** fail to have an inverse?

29. Use your result from Problem 25 to solve the system

$$x - 3y + 4z = a$$
$$2x + 3y + 5z = b$$
$$-x + 4y + 2z = c$$

30. Let **A** and **B** be 3×3 matrices with inverses $\mathbf{A}^{-1}$ and $\mathbf{B}^{-1}$. Show that **AB** has an inverse given by $\mathbf{B}^{-1}\mathbf{A}^{-1}$. (*Hint:* The product in either order must be **I**.)

31. Show that

$$\begin{bmatrix} 0 & 0 \\ 2 & 3 \end{bmatrix} \begin{bmatrix} -6 & 0 \\ 4 & 0 \end{bmatrix} = \begin{bmatrix} 0 & 0 \\ 0 & 0 \end{bmatrix}$$

Thus $\mathbf{AB} = \mathbf{O}$ but neither **A** nor **B** is **O**. This is another way in which matrices differ from ordinary numbers.

32. Suppose $\mathbf{AB} = \mathbf{O}$ and **A** has a multiplicative inverse. Show that $\mathbf{B} = \mathbf{O}$. (See Problem 31.)

Matrix	Determinant	Value of Determinant
$\begin{bmatrix} a & b \\ c & d \end{bmatrix}$	$\begin{vmatrix} a & b \\ c & d \end{vmatrix}$	$ad - bc$
$\begin{bmatrix} a_1 & b_1 & c_1 \\ a_2 & b_2 & c_2 \\ a_3 & b_3 & c_3 \end{bmatrix}$	$\begin{vmatrix} a_1 & b_1 & c_1 \\ a_2 & b_2 & c_2 \\ a_3 & b_3 & c_3 \end{vmatrix}$	$a_1b_2c_3 + a_2b_3c_1 + a_3b_1c_2$ $- a_1b_3c_2 - a_2b_1c_3 - a_3b_2c_1$

8-5
Second- and Third-Order Determinants

The notion of a determinant is usually attributed to the German mathematician Gottfried Wilhelm Leibniz (1646–1716), but it seems that Seki Kōwa of Japan had the idea somewhat earlier. It grew out of the study of systems of equations.

SECOND-ORDER DETERMINANTS

Consider the general system of two equations in two unknowns

$$ax + by = r$$
$$cx + dy = s$$

If we multiply the second equation by a and then add $-c$ times the first equation to it, we obtain the equivalent triangular system.

$$ax + \qquad by = r$$
$$(ad - bc)y = as - cr$$

If $ad - bc \neq 0$, we can solve by backward substitution.

$$x = \frac{rd - bs}{ad - bc}$$

$$y = \frac{as - rc}{ad - bc}$$

These formulas are hard to remember unless we associate special symbols with the numbers $ad - bc$, $rd - bs$, and $as - rc$. For the first of these, we propose

$$\begin{vmatrix} a & b \\ c & d \end{vmatrix} = ad - bc$$

The symbol on the left is called a **second-order determinant,** and we say that $ad - bc$ is its value. Thus

$$\begin{vmatrix} -2 & -1 \\ 5 & 6 \end{vmatrix} = (-2)(6) - (-1)(5) = -7$$

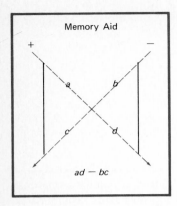

Memory Aid

$ad - bc$

The diagram in the margin may help you remember how to make the evaluation.

With this new symbol, we can write the solution to

$$ax + by = r$$
$$cx + dy = s$$

as

$$x = \frac{rd - bs}{ad - bc} = \frac{\begin{vmatrix} r & b \\ s & d \end{vmatrix}}{\begin{vmatrix} a & b \\ c & d \end{vmatrix}}$$

$$y = \frac{as - rc}{ad - bc} = \frac{\begin{vmatrix} a & r \\ c & s \end{vmatrix}}{\begin{vmatrix} a & b \\ c & d \end{vmatrix}}$$

These results are easy to remember when we notice that the denominator is the determinant of the coefficient matrix, and that the numerator is the same except that the coefficients of the unknown we are seeking are replaced by the constants from the right side of the system.

Here is an example.

$$3x - 2y = 7$$
$$4x + 5y = 2$$

$$x = \frac{\begin{vmatrix} 7 & -2 \\ 2 & 5 \end{vmatrix}}{\begin{vmatrix} 3 & -2 \\ 4 & 5 \end{vmatrix}} = \frac{(7)(5) - (-2)(2)}{(3)(5) - (-2)(4)} = \frac{39}{23}$$

$$y = \frac{\begin{vmatrix} 3 & 7 \\ 4 & 2 \end{vmatrix}}{\begin{vmatrix} 3 & -2 \\ 4 & 5 \end{vmatrix}} = \frac{(3)(2) - (7)(4)}{23} = -\frac{22}{23}$$

The choice of the name *determinant* is appropriate, for the determinants of a system completely *determine* its character.

1. If $ad - bc \neq 0$, the system has a unique solution, the one given at the top of this page.
2. If $ad - bc = 0$, $as - rc = 0$, and $rd - bs = 0$, then a, b, and r are

proportional to c, d, and s and the system has infinitely many solutions. Here is an example.

$$3x - 2y = 7 \qquad \frac{3}{6} = \frac{-2}{-4} = \frac{7}{14}$$
$$6x - 4y = 14$$

3. If $ad - bc = 0$ and $as - rc \neq 0$ or $rd - bs \neq 0$, then a and b are proportional to c and d, but this proportionality does not extend to r and s; the system has no solution. This is illustrated by the following.

$$3x - 2y = 7 \qquad \frac{3}{6} = \frac{-2}{-4} \neq \frac{7}{10}$$
$$6x - 4y = 10$$

THIRD-ORDER DETERMINANTS

When we consider the general system of three equations in three unknowns

$$a_1 x + b_1 y + c_1 z = d_1$$
$$a_2 x + b_2 y + c_2 z = d_2$$
$$a_3 x + b_3 y + c_3 z = d_3$$

things get more complicated, but the results are similar. The appropriate determinant symbol and its corresponding value are

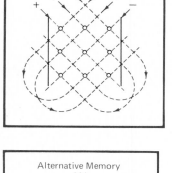

Memory Aid

$$\begin{vmatrix} a_1 & b_1 & c_1 \\ a_2 & b_2 & c_2 \\ a_3 & b_3 & c_3 \end{vmatrix} = a_1 b_2 c_3 + b_1 c_2 a_3 + c_1 b_3 a_2 - c_1 b_2 a_3 - b_1 a_2 c_3 - a_1 b_3 c_2$$

There are six terms in the sum on the right, three with a plus sign and three with a minus sign. The diagrams in the margin will help you remember the products that enter each term. Just follows the arrows. Here is an example.

$$\begin{vmatrix} 3 & 2 & 4 \\ 4 & -2 & 6 \\ 8 & 3 & 5 \end{vmatrix} = (3)(-2)(5) + (2)(6)(8) + (4)(3)(4)$$
$$-(4)(-2)(8) - (2)(4)(5) - (3)(3)(6)$$
$$= 84$$

CRAMER'S RULE

Alternative Memory Aid

We saw that the solutions for x and y in a second-order system could be written as the quotients of two determinants. That fact generalizes to the third-order case. We present it without proof. Consider

$$a_1 x + b_1 y + c_1 z = d_1$$
$$a_2 x + b_2 y + c_2 z = d_2$$
$$a_3 x + b_3 y + c_3 z = d_3$$

If

$$D = \begin{vmatrix} a_1 & b_1 & c_1 \\ a_2 & b_2 & c_2 \\ a_3 & b_3 & c_3 \end{vmatrix} \neq 0$$

then the system above has a unique solution given by

$$x = \frac{1}{D}\begin{vmatrix} d_1 & b_1 & c_1 \\ d_2 & b_2 & c_2 \\ d_3 & b_3 & c_3 \end{vmatrix} \qquad y = \frac{1}{D}\begin{vmatrix} a_1 & d_1 & c_1 \\ a_2 & d_2 & c_2 \\ a_3 & d_3 & c_3 \end{vmatrix} \qquad z = \frac{1}{D}\begin{vmatrix} a_1 & b_1 & d_1 \\ a_2 & b_2 & d_2 \\ a_3 & b_3 & d_3 \end{vmatrix}$$

The pattern is the same as in the second-order situation. The denominator D is the determinant of the coefficient matrix. The numerator in each case is obtained from D by replacing the coefficients of the unknown by the constants from the right side of the system.

This method of solving a system of equations is named after one of its discoverers, Gabriel Cramer (1704–1752). Historically, it has been a popular method. However, note that even for a system of three equations in three unknowns, it requires the evaluation of four determinants. The method of matrices (Section 8-2) is considerably more efficient, both for hand and computer calculation. Consequently, Cramer's rule is now primarily of theoretical rather than practical interest.

PROPERTIES OF DETERMINANTS

We are interested in how the matrix operations considered in Section 8-2 affect the values of the corresponding determinants.

1. *Interchanging two rows changes the sign of the determinant; for example,*

$$\begin{vmatrix} a & b \\ c & d \end{vmatrix} = -\begin{vmatrix} c & d \\ a & b \end{vmatrix}$$

2. *Multiplying a row by a constant k multiplies the value of the determinant by k; for example,*

$$\begin{vmatrix} ka & kb \\ c & d \end{vmatrix} = k\begin{vmatrix} a & b \\ c & d \end{vmatrix}$$

3. *Adding a multiple of one row to another does not affect the value of the determinant; for example,*

$$\begin{vmatrix} a & b \\ c + ka & d + kb \end{vmatrix} = \begin{vmatrix} a & b \\ c & d \end{vmatrix}$$

We mention also the effect of a new operation.

4. *Interchanging the rows and columns (pairwise) does not affect the value of the determinant; for example,*

$$\begin{vmatrix} a & c \\ b & d \end{vmatrix} = \begin{vmatrix} a & b \\ c & d \end{vmatrix}$$

Though they are harder to prove in the third-order case, we emphasize that all four properties hold for both second- and third-order determinants. We offer only one proof, a proof of Property 3 in the second-order case.

$$\begin{vmatrix} a & b \\ c + ka & d + kb \end{vmatrix} = a(d + kb) - b(c + ka)$$

$$= ad + akb - bc - bka$$

$$= ad - bc$$

$$= \begin{vmatrix} a & b \\ c & d \end{vmatrix}$$

Property 3 can be a great aid in evaluating a determinant. Using it, we can transform a matrix to triangular form without changing the value of its determinant. But the determinant of a triangular matrix is just the product of the elements on the main diagonal, since this is the only nonzero term in the determinant formula. Here is an example.

$$\begin{vmatrix} 1 & 3 & 4 \\ -1 & -2 & 3 \\ 2 & -6 & 11 \end{vmatrix} = \begin{vmatrix} 1 & 3 & 4 \\ 0 & 1 & 7 \\ 0 & -12 & 3 \end{vmatrix} = \begin{vmatrix} 1 & 3 & 4 \\ 0 & 1 & 7 \\ 0 & 0 & 87 \end{vmatrix} = 87$$

Problem Set 8-5

Evaluate each of the determinants in Problems 1–8 by inspection.

1. $\begin{vmatrix} 4 & 0 \\ 0 & -2 \end{vmatrix}$ 2. $\begin{vmatrix} 8 & 0 \\ 5 & 0 \end{vmatrix}$ 3. $\begin{vmatrix} 11 & 4 \\ 0 & 2 \end{vmatrix}$

4. $\begin{vmatrix} 2 & -1 & 5 \\ 0 & 4 & 2 \\ 0 & 0 & -1 \end{vmatrix}$ 5. $\begin{vmatrix} -1 & -7 & 9 \\ 0 & 5 & 4 \\ 0 & 0 & 10 \end{vmatrix}$ 6. $\begin{vmatrix} 3 & -2 & 1 \\ 0 & 0 & 0 \\ 1 & 5 & -8 \end{vmatrix}$

7. $\begin{vmatrix} 3 & 0 & 8 \\ 10 & 0 & 2 \\ -1 & 0 & -9 \end{vmatrix}$ 8. $\begin{vmatrix} 9 & 0 & 0 \\ 0 & 0 & -2 \\ 0 & 4 & 0 \end{vmatrix}$

9. If

$$\begin{vmatrix} a_1 & b_1 & c_1 \\ a_2 & b_2 & c_2 \\ a_3 & b_3 & c_3 \end{vmatrix} = 12$$

find the value of each of the following determinants.

(a) $\begin{vmatrix} a_1 & a_2 & a_3 \\ b_1 & b_2 & b_3 \\ c_1 & c_2 & c_3 \end{vmatrix}$ (b) $\begin{vmatrix} a_3 & b_3 & c_3 \\ a_2 & b_2 & c_2 \\ a_1 & b_1 & c_1 \end{vmatrix}$

(c) $\begin{vmatrix} a_1 & b_1 & c_1 \\ a_2 & b_2 & c_2 \\ 3a_3 & 3b_3 & 3c_3 \end{vmatrix}$ (d) $\begin{vmatrix} a_1 + 3a_3 & b_1 + 3b_3 & c_1 + 3c_3 \\ a_2 & b_2 & c_2 \\ a_3 & b_3 & c_3 \end{vmatrix}$

Evaluate each of the determinants in Problems 10–17.

10. $\begin{vmatrix} 3 & 2 \\ 5 & 6 \end{vmatrix}$ 11. $\begin{vmatrix} 5 & 3 \\ 5 & -3 \end{vmatrix}$

12. $\begin{vmatrix} 3 & 0 & 0 \\ -2 & 5 & 4 \\ 1 & 2 & -9 \end{vmatrix}$ 13. $\begin{vmatrix} 4 & 8 & -2 \\ 1 & -2 & 0 \\ 2 & 4 & 0 \end{vmatrix}$

14. $\begin{vmatrix} 3 & 2 & -4 \\ 1 & 0 & 5 \\ 4 & -2 & 3 \end{vmatrix}$ 15. $\begin{vmatrix} 2 & 4 & 1 \\ 1 & 3 & 6 \\ 2 & 3 & -1 \end{vmatrix}$

16. $\begin{vmatrix} 5.1 & -3.2 & 2.6 \\ 1.3 & 4.5 & 2.3 \\ 3.4 & -2.2 & 1.9 \end{vmatrix}$ 17. $\begin{vmatrix} 2.03 & 5.41 & -3.14 \\ 0 & 6.22 & 0 \\ -1.93 & 7.13 & 6.34 \end{vmatrix}$

Use Cramer's rule to solve the system of equations in Problems 18–21.

18. $2x - 3y = -11$
    $x + 2y = -2$

19. $5x + y = 7$
    $3x - 4y = 18$

20. $2x + 4y + z = 15$
    $x + 3y + 6z = 15$
    $2x + 3y - z = 11$

21. $5x - 3y + 2z = 18$
    $x + 4y + 2z = -4$
    $3x - 2y + z = 11$

Miscellaneous Problems

Evaluate the determinants in Problems 22–26.

22. $\begin{vmatrix} 11 & 4 \\ 0 & 2 \end{vmatrix}$ 23. $\begin{vmatrix} 3 & -2 & 5 \\ 0 & 0 & 0 \\ 1 & -4 & 6 \end{vmatrix}$

24. $\begin{vmatrix} -2 & 5 & 100 \\ 0 & 4 & 96 \\ 0 & 0 & -1 \end{vmatrix}$ 25. $\begin{vmatrix} 2 & 3 & -4 \\ 1 & 2 & 5 \\ 5 & 8 & -3 \end{vmatrix}$

26. $\begin{vmatrix} 4 & 3 & -2 \\ -1 & 12 & 13 \\ 1 & 3 & 5 \end{vmatrix}$

27. Solve by Cramer's rule.

$$3x - y + 2z = 7$$
$$2x \quad + z = 5$$
$$y - 2z = -4$$

28. Show that interchanging two rows of a second-order determinant changes the sign of the determinant; that is,

$$\begin{vmatrix} a & b \\ c & d \end{vmatrix} = - \begin{vmatrix} c & d \\ a & b \end{vmatrix}$$

29. Show that

$$\begin{vmatrix} ka & kb \\ c & d \end{vmatrix} = k \begin{vmatrix} a & b \\ c & d \end{vmatrix}$$

30. Show that

$$\begin{vmatrix} a & c \\ b & d \end{vmatrix} = \begin{vmatrix} a & b \\ c & d \end{vmatrix}$$

31. Let the determinant

$$\begin{vmatrix} a_1 & b_1 & c_1 \\ a_2 & b_2 & c_2 \\ a_1 & b_1 & c_1 \end{vmatrix}$$

have value D. Interchanging rows 1 and 3 leaves the determinant unchanged, so its value is still D. But Property 1 tells us that the value should now be $-D$. What must be true about D?

32. If we apply Property 3 to the determinant in Problem 31 we see that

$$D = \begin{vmatrix} a_1 & b_1 & c_1 \\ a_2 & b_2 & c_2 \\ a_1 & b_1 & c_1 \end{vmatrix} = \begin{vmatrix} 0 & 0 & 0 \\ a_2 & b_2 & c_2 \\ a_1 & b_1 & c_1 \end{vmatrix}$$

What is the value of the determinant on the right? Does your answer agree with what you got in Problem 31?

33. If you have worked Problems 31 and 32 you have probably reached the verdict that if two rows (or two columns) of a determinant are identical, the determinant has a value of zero. This is true also if two rows (or two columns) are proportional. Convince yourself of this by showing that

$$\begin{vmatrix} a_1 & b_1 & c_1 \\ a_2 & b_2 & c_2 \\ ka_1 & kb_1 & kc_1 \end{vmatrix} = 0$$

34. (a) Show that the determinant equation

$$\begin{vmatrix} x & y & 1 \\ 2 & 4 & 1 \\ -3 & 5 & 1 \end{vmatrix} = 0$$

is the equation of a line.

(b) How can you tell without expanding the determinant that the points (2, 4) and (−3, 5) are on the line?

(c) Write a determinant equation for the line that passes through the points (5, −1) and (4, 11).

James Joseph Sylvester
1814–1897

A Mathematician and a Poet

One of the most consistent workers on the theory of determinants over a period of 50 years was the Englishman, James Joseph Sylvester. Known as a poet, a wit, and a mathematician, he taught in England and in America. During his stay at Johns Hopkins University in Baltimore (1877–1883), he helped establish one of the first graduate programs in mathematics in America. Under his tutelage, mathematics began to flourish in the United States.

8-6
Higher-Order Determinants

Having defined determinants for 2×2 and 3×3 matrices, we expect to do it for 4×4 matrices, 5×5 matrices, and so on. Our problem is to do it in such a way that Cramer's rule and the determinant properties of Section 8-5 still hold. This will take some work.

MINORS

We begin by introducing the standard notation for a general $n \times n$ matrix.

$$\begin{bmatrix} a_{11} & a_{12} & a_{13} & \cdots & a_{1n} \\ a_{21} & a_{22} & a_{23} & \cdots & a_{2n} \\ a_{31} & a_{32} & a_{33} & \cdots & a_{3n} \\ \vdots & \vdots & \vdots & & \vdots \\ a_{n1} & a_{n2} & a_{n3} & \cdots & a_{nn} \end{bmatrix}$$

Note the use of the double subscript on each entry: the first subscript gives the row in which a_{ij} is and the second gives the column. For example, a_{32} is the entry in the third row and second column.

Associated with each entry a_{ij} in an $n \times n$ matrix is a determinant M_{ij} of order $n - 1$ called the **minor** of a_{ij}. It is obtained by taking the determinant of the submatrix that results when we blot out the row and column in which a_{ij} stands. For example, the minor M_{13} of a_{13} in the 4×4 matrix

$$\begin{bmatrix} a_{11} & a_{12} & a_{13} & a_{14} \\ a_{21} & a_{22} & a_{23} & a_{24} \\ a_{31} & a_{32} & a_{33} & a_{34} \\ a_{41} & a_{42} & a_{43} & a_{44} \end{bmatrix}$$

is the third-order determinant.

$$\begin{vmatrix} a_{21} & a_{22} & a_{24} \\ a_{31} & a_{32} & a_{34} \\ a_{41} & a_{42} & a_{44} \end{vmatrix}$$

THE GENERAL nth-ORDER DETERMINANT

Here is the definition to which we have been leading.

$$\begin{vmatrix} a_{11} & a_{12} & \cdots & a_{1n} \\ a_{21} & a_{22} & \cdots & a_{2n} \\ \vdots & \vdots & & \vdots \\ a_{n1} & a_{n2} & \cdots & a_{nn} \end{vmatrix} = a_{11}M_{11} - a_{12}M_{12} + a_{13}M_{13} \cdots + (-1)^{n+1}a_{1n}M_{1n}$$

There are three important questions to answer regarding this definition.

Does this definition really define? Only if the minors M_{ij} can be evaluated. They are themselves determinants, but here is the key point: They are of order $n - 1$, one less than the order of the determinant we started with. They can, in turn, be expressed in terms of determinants of order $n - 2$, and so on, using the same definition. Thus, for example, a fifth-order determinant can be expressed in terms of fourth-order determinants, and these fourth-order determinants can be expressed in terms of third-order determinants. But we know how to evaluate third-order determinants from Section 8-5.

Is this definition consistent with the earlier definition when applied to third-order determinants? Yes, for if we apply it to a general third-order determinant, we get

$$\begin{vmatrix} a_1 & b_1 & c_1 \\ a_2 & b_2 & c_2 \\ a_3 & b_3 & c_3 \end{vmatrix} = a_1 \begin{vmatrix} b_2 & c_2 \\ b_3 & c_3 \end{vmatrix} - b_1 \begin{vmatrix} a_2 & c_2 \\ a_3 & c_3 \end{vmatrix} + c_1 \begin{vmatrix} a_2 & b_2 \\ a_3 & b_3 \end{vmatrix}$$

$$= a_1 b_2 c_3 - a_1 c_2 b_3 - b_1 a_2 c_3 + b_1 c_2 a_3 + c_1 a_2 b_3 - c_1 b_2 a_3$$

This is the same value we gave in Section 8-5.

Does this definition preserve Cramer's rule and the properties of Section 8-5? Yes, it does. We shall not prove this because the proofs are lengthy and difficult.

EXPANSION ACCORDING TO ANY ROW OR COLUMN

Our definition expressed the value of a determinant in terms of the entries and minors of the first row; we call it an expansion according to the first row. It is a remarkable fact that we can expand a determinant according to any row or column (and always get the same answer).

Before we can show what we mean, we must explain a sign convention. We associate a plus or minus sign with every position in a matrix. To the ij-position, we assign a plus sign if $i + j$ is even and a minus sign otherwise. Thus for a 4×4 matrix, we have this pattern of signs.

$$\begin{bmatrix} + & - & + & - \\ - & + & - & + \\ + & - & + & - \\ - & + & - & + \end{bmatrix}$$

There is always a $+$ in the upper left position and then the signs alternate.

With this understanding about signs, we may expand according to any row or column. For example, to evaluate a fourth-order determinant, we can expand according to the second column if we wish. We multiply each entry in that column by its minor, prefixing each product with a plus or minus sign according to the pattern above. Then we add the results.

$$\begin{vmatrix} a_{11} & a_{12} & a_{13} & a_{14} \\ a_{21} & a_{22} & a_{23} & a_{24} \\ a_{31} & a_{32} & a_{33} & a_{34} \\ a_{41} & a_{42} & a_{43} & a_{44} \end{vmatrix} = -a_{12}M_{12} + a_{22}M_{22} - a_{32}M_{32} + a_{42}M_{42}$$

EXAMPLE

To evaluate

$$\begin{vmatrix} 6 & 0 & 4 & -1 \\ 2 & 0 & -1 & 4 \\ -2 & 4 & -2 & 3 \\ 4 & 0 & 5 & -4 \end{vmatrix}$$

it is obviously best to expand according to the second column, since three of the four resulting terms are zero. The single nonzero term is just $(-1)(4)$ times the minor M_{32}—that is,

$$-4 \begin{vmatrix} 6 & 4 & -1 \\ 2 & -1 & 4 \\ 4 & 5 & -4 \end{vmatrix}$$

We could now evaluate this third-order determinant as in Section 8-5. But having seen the usefulness of zeros, let us take a different tack. It is easy to get two zeros in the first column. Simply add -3 times the second row to the first row and -2 times the second row to the third. We get

$$-4\begin{vmatrix} 0 & 7 & -13 \\ 2 & -1 & 4 \\ 0 & 7 & -12 \end{vmatrix}$$

Finally, expand according to the first column.

$$(-4)(-1)(2)\begin{vmatrix} 7 & -13 \\ 7 & -12 \end{vmatrix} = 8(-84 + 91) = 56$$

The reason for the factor of -1 is that the entry 2 is in a minus position in the 3×3 pattern of signs.

Problem Set 8-6

Evaluate each of the determinants in Problems 1–6 according to a row or column of your choice. Make a good choice or suffer the consequences!

1. $\begin{vmatrix} 3 & -2 & 4 \\ 1 & 5 & 0 \\ 3 & 10 & 0 \end{vmatrix}$

2. $\begin{vmatrix} 4 & 0 & -6 \\ -2 & 3 & 5 \\ 1 & 0 & 8 \end{vmatrix}$

3. $\begin{vmatrix} 1 & 2 & 3 \\ 0 & 2 & 3 \\ 1 & 3 & 4 \end{vmatrix}$

4. $\begin{vmatrix} 2 & -1 & -1 \\ 3 & 4 & 2 \\ 0 & -1 & -1 \end{vmatrix}$

5. $\begin{vmatrix} 3 & 0 & 0 & 0 \\ -1 & 1 & 4 & 2 \\ 2 & 0 & 2 & -3 \\ -4 & 0 & 1 & 5 \end{vmatrix}$

6. $\begin{vmatrix} 0 & 5 & 0 & 0 \\ 1 & -3 & 0 & 2 \\ 4 & 1 & 2 & 8 \\ -3 & 2 & 0 & 5 \end{vmatrix}$

Evaluate each of the determinants in Problems 7–10 by first getting some zeros in a row or column and then expanding according to that row or column.

7. $\begin{vmatrix} 3 & 5 & -10 \\ 2 & 4 & 6 \\ -3 & -5 & 12 \end{vmatrix}$

8. $\begin{vmatrix} 2 & -1 & 2 \\ 4 & 3 & 4 \\ 7 & -5 & 10 \end{vmatrix}$

9. $\begin{vmatrix} 1 & -2 & 1 & 4 \\ -2 & 5 & -3 & 1 \\ 0 & 7 & -4 & 2 \\ 3 & -2 & 2 & 6 \end{vmatrix}$

10. $\begin{vmatrix} 1 & -2 & 0 & -4 \\ 3 & -4 & 3 & -10 \\ 2 & 1 & -2 & 1 \\ 4 & -5 & 1 & 4 \end{vmatrix}$

11. Solve the following system for x only.

$$\begin{aligned}
x - 2y + z + 4w &= 1 \\
-2x + 5y - 3z + w &= -2 \\
7y - 4z + 2w &= 3 \\
3x - 2y + 2z + 6w &= 6
\end{aligned}$$

(Make use of your answer to Problem 9.)

12. Solve the following system for z only.

$$\begin{aligned} x - 2y \quad\quad - 4w &= -14 \\ 3x - 4y + 3z - 10w &= -28 \\ 2x + y - 2z + w &= 0 \\ 4x - 5y + z + 4w &= 9 \end{aligned}$$

(Make use of your answer to Problem 10.)

Evaluate the determinants in Problems 13–18.

13. $\begin{vmatrix} 2 & -3 & 2 \\ 1 & 0 & -4 \\ -1 & 0 & 6 \end{vmatrix}$

14. $\begin{vmatrix} 3 & 1 & -5 \\ 2 & -2 & 7 \\ 1 & 0 & -1 \end{vmatrix}$

15. $\begin{vmatrix} 2 & -3 & 4 & 5 \\ 2 & -3 & 4 & 7 \\ 1 & 6 & 4 & 5 \\ 2 & 6 & 4 & -8 \end{vmatrix}$

16. $\begin{vmatrix} 2 & 2 & 3 & 7 \\ 1 & 2 & 3 & -2 \\ 4 & -3 & 9 & 6 \\ 1 & 2 & 3 & -1 \end{vmatrix}$

17. $\begin{vmatrix} 1 & 2 & -3 & 1 & 2 \\ -1 & 0 & 2 & 5 & -3 \\ 5 & 0 & 0 & -2 & 4 \\ 0 & 0 & 0 & 6 & 3 \\ 0 & 0 & 0 & 2 & -7 \end{vmatrix}$

18. $\begin{vmatrix} 1 & 2 & 3 & 4 & 5 \\ 2 & 1 & 1 & 1 & 1 \\ 3 & 1 & 1 & 1 & 1 \\ 4 & 1 & 1 & 1 & 1 \\ 5 & 1 & 1 & 1 & 1 \end{vmatrix}$

(*Hint:* Subtract row 2 from row 3.)

19. Evaluate the following determinant.

$$\begin{vmatrix} a & b & c & d \\ 0 & e & f & g \\ 0 & 0 & h & i \\ 0 & 0 & 0 & j \end{vmatrix}$$

Conjecture a general result about the determinant of a triangular matrix.

c 20. Use the result of Problem 19 to evaluate

$$\begin{vmatrix} 2.12 & 3.14 & -1.61 & 1.72 \\ 0 & -2.36 & 5.91 & 7.82 \\ 0 & 0 & 1.46 & 3.34 \\ 0 & 0 & 0 & 3.31 \end{vmatrix}$$

c 21. Evaluate by reducing to triangular form and using Problem 19.

$$\begin{vmatrix} 1 & 2 & 2.6 & 1.5 \\ 2.3 & 5.6 & -1.3 & 9.8 \\ 2.7 & 1.3 & 4.2 & -1.9 \\ 5.5 & 6.2 & 3.0 & 1.4 \end{vmatrix}$$

22. Show that

$$
\begin{vmatrix} a_1 + d_1 & b_1 & c_1 \\ a_2 + d_2 & b_2 & c_2 \\ a_3 + d_3 & b_3 & c_3 \end{vmatrix} = \begin{vmatrix} a_1 & b_1 & c_1 \\ a_2 & b_2 & c_2 \\ a_3 & b_3 & c_3 \end{vmatrix} + \begin{vmatrix} d_1 & b_1 & c_1 \\ d_2 & b_2 & c_2 \\ d_3 & b_3 & c_3 \end{vmatrix}
$$

23. Show that the determinant

$$
\begin{vmatrix} a_1 & b_1 & c_1 \\ a_2 & b_2 & c_2 \\ ra_1 + sa_2 & rb_1 + sb_2 & rc_1 + sc_2 \end{vmatrix}
$$

is zero for any values of r and s.

24. Show that for any value of n,

$$
\begin{vmatrix} n+1 & n+2 & n+3 \\ n+4 & n+5 & n+6 \\ n+7 & n+8 & n+9 \end{vmatrix} = 0
$$

25. Show that

$$
\begin{vmatrix} 1 & 1 & 1 \\ p & q & r \\ p^2 & q^2 & r^2 \end{vmatrix} = (p - q)(q - r)(r - p)
$$

(*Hint:* First get two zeros in the first row; then look for common factors in two of the columns.)

26. Show that if all the entries in a determinant are integers, its value is an integer.

27. From Problem 26 and Cramer's rule, conclude something about the kind of numbers that can arise as solutions to n linear equations in n unknowns if the coefficients are all integers.

Corn

Profit : $40 per acre

Labor :   2 hours per acre

Maximizing Profit

Farmer Brown has 480 acres of land on which he can grow either corn or wheat. He figures that he has 800 hours of labor available during the crucial summer season. Given the profit margins and labor requirements shown at the right, how many acres of each should he plant to maximize his profit? What is this maximum profit?

Wheat

Profit : $30 per acre

Labor :   1 hour per acre

8-7
Systems of
Inequalities

At first glance you might think that Farmer Brown should put all of his land into corn. Unfortunately, however, that requires 960 hours of labor and he has only 800 available. Well, maybe he should plant 400 acres of corn, using his allocated 800 hours of labor on them, and let the remaining 80 acres lie idle. Or would it be wise to at least plant enough wheat so all his land is in use? This problem is complicated enough so that no one is likely to find the best solution without a lot of work. And would not a method be better than blind experimenting? That is our subject—a method for handling Farmer Brown's problem and others of the same type.

Like all individuals and businesses, Farmer Brown must operate within certain limitations; we call them **constraints.** Suppose he plants x acres of corn and y acres of wheat. His constraints can be translated into inequalities.

Land constraint:	$x + y \leq 480$
Labor constraint:	$2x + y \leq 800$
Nonnegativity constraints:	$x \geq 0 \qquad y \geq 0$

His task is to maximize the profit $P = 40x + 30y$ subject to these constraints. Before we can solve his problem, we will need to know more about inequalities.

THE GRAPH OF A LINEAR INEQUALITY

The best way to visualize an inequality is by means of its graph. Consider, for example,

$$2x + y \leq 6$$

which can be rewritten as

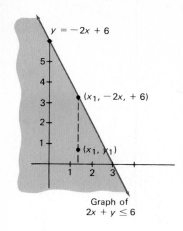

$y = -2x + 6$

$(x_1, -2x, + 6)$

(x_1, y_1)

Graph of
$2x + y \leq 6$

$2x - 3y < -6$

$$y \leq -2x + 6$$

The complete graph consists of those points which satisfy $y = -2x + 6$ (a line), together with those that satisfy $y < -2x + 6$ (the points below the line). To see that this description is correct, note that for any abscissa x_1, the point $(x_1, -2x_1 + 6)$ is on the line $y = -2x + 6$. The point (x_1, y_1) is directly below that point if and only if $y_1 < -2x_1 + 6$ (see the diagram in the margin). Thus the graph of $y \leq -2x + 6$ is the **closed half-plane** that we have shaded on the diagram. We refer to it as *closed* because the edge $y = -2x + 6$ is included. Correspondingly, the graph of $y < -2x + 6$ is called an **open half-plane.**

The graph of any linear inequality in x and y is a half-plane, open or closed. To sketch the graph, first draw the corresponding edge. Then determine the correct half-plane by taking a sample point, not on the edge, and checking to see if it satisfies the inequality.

To illustrate this procedure, consider

$$2x - 3y < -6$$

Its graph does not include the line $2x - 3y = -6$, although that line is crucial in determining the graph. We therefore show it as a dotted line. Since the sample point $(0, 0)$ does not satisfy the inequality, we choose the half-plane on the opposite side of the line from it. The complete graph is the shaded open half-plane shown in the margin.

GRAPHING A SYSTEM OF LINEAR INEQUALITIES

The graph of a system of inequalities like Farmer Brown's constraints

$$x + y \leq 480$$
$$2x + y \leq 800$$
$$x \geq 0 \qquad y \geq 0$$

is simply the intersection of the graphs of the individual inequalities. We can construct the graph in stages as we do at the top of the next page, though we are confident that you will quickly learn to do it in one operation.

The diagram on the right (page 340) is the one we want. All the points in the shaded region F satisfy the four inequalities simultaneously. The points $(0, 0)$, $(400, 0)$, $(320, 160)$, and $(0, 480)$ are called the **vertices** (or corner points) of F. Incidentally, the point $(320, 160)$ was obtained by solving the two equations $2x + y = 800$ and $x + y = 480$ simultaneously.

The region F has three important properties.

1. It is polygonal (its boundary consists of line segments).
2. It is convex (if points P and Q are in the region, then the line segment PQ lies entirely within the region).
3. It is bounded (it can be enclosed in a circle).

Polygonal, convex,
and bounded

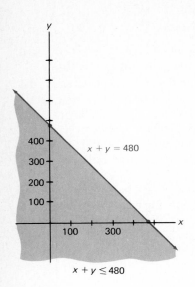

$x + y \leq 480$

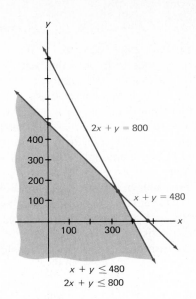

$x + y \leq 480$
$2x + y \leq 800$

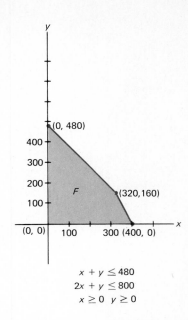

$x + y \leq 480$
$2x + y \leq 800$
$x \geq 0 \quad y \geq 0$

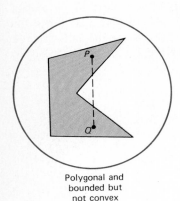

Polygonal and
bounded but
not convex

As a matter of fact, every region that arises as the solution set of a system of linear inequalities is polygonal and convex, though it need not be bounded. The shaded region in the lower diagram in the margin could not be the solution set for a system of linear inequalities because it is not convex.

LINEAR PROGRAMMING PROBLEMS

It is time that we solved Farmer Brown's problem.
   Maximize

$$P = 40x + 30y$$

subject to

$$\begin{cases} x + y \leq 480 \\ 2x + y \leq 800 \\ x \geq 0 \qquad y \geq 0 \end{cases}$$

Any problem that asks us to find the maximum (or minimum) of a linear function subject to linear inequality constraints is called a **linear programming problem.** Here is a method for solving such problems.

1. Graph the solution set corresponding to the inequality constraints.
2. Find the coordinates of the vertices of the solution set.
3. Evaluate the linear function that you want to maximize (or minimize) at each of these vertices. The largest of these gives the maximum, while the smallest gives the minimum.

To see why this method works, consider the diagram for Farmer Brown's problem below.

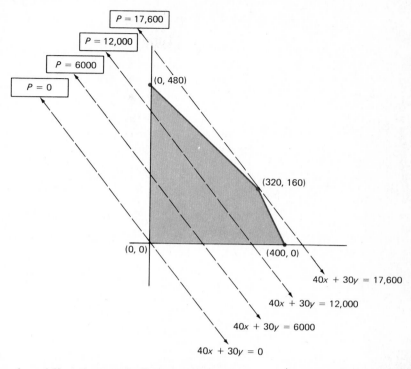

The dotted lines are profit lines, each with slope $-\frac{4}{3}$; they are the graphs of $40x + 30y = P$ for various values of P. All the points on a dotted line give the same total profit. Imagine a profit line moving from left to right across the shaded region with the slope constant, as indicated in the diagram. During this motion, the profit is zero at $(0, 0)$ and increases to its maximum of $17,600$ at $(320,160)$. It should be clear that such a moving line (no matter what its slope) will always enter the shaded set at a vertex and leave it at another vertex. The particular vertex depends upon the slope of the profit lines. In Farmer Brown's problem, the minimum profit of $0 occurs at $(0, 0)$; the maximum profit of $17,600 occurs at $(320, 160)$. In the margin, we show the total profit for each of the four vertices. Clearly Farmer Brown should plant 320 acres of corn and 160 acres of wheat.

Now suppose the price of corn goes up so that Farmer Brown can expect a profit of $80 per acre on corn but still only $30 per acre on wheat. Would this change his strategy? In the second table in the margin, we show his total profit $P = 80x + 30y$ at each of the four vertices. Evidently he should plant 400 acres of corn and no wheat to achieve maximum profit. Note that this means he should leave 80 acres of his land idle.

Finally, suppose that the profit per acre is $40 both for wheat and for

Vertex	$P = 40x + 30y$
(0, 0)	0
(0, 480)	14,400
(320, 160)	17,600
(400, 0)	16,000

Vertex	$P = 80x + 30y$
(0, 0)	0
(0, 480)	14,400
(320, 160)	30,400
(400, 0)	32,000

Vertex	$P = 40x + 40y$
(0, 0)	0
(0, 480)	19,200
(320, 160)	19,200
(400, 0)	16,000

corn. The table for this case shows the same total profit at the vertices (0, 480) and (320, 160). This means that the moving profit line leaves the shaded region along the side determined by those two vertices, so that every point on that side gives a maximum profit. It is still true, however, that the maximum profit occurs at a vertex.

The situation with an unbounded constraint set is slightly more complicated. It is discussed in Example A. Here we simply point out that in the unbounded case, there may not be a maximum (or minimum), but if there is one, it will still occur at a vertex.

Problem Set 8-7

In Problems 1–6, graph the solution set of each inequality in the xy-plane.

1. $4x + y \leq 8$
2. $2x + 5y \leq 20$
3. $x \leq 3$
4. $y \leq -2$
5. $4x - y \geq 8$
6. $2x - 5y \geq -20$

In Problems 7–10, graph the solution set of the given system. On the graph, label the coordinates of the vertices.

7. $4x + y \leq 8$
   $2x + 3y \leq 14$
   $x \geq 0 \quad y \geq 0$

8. $2x + 5y \leq 20$
   $4x + y \leq 22$
   $x \geq 0 \quad y \geq 0$

9. $4x + y \leq 8$
   $x - y \leq -2$
   $x \geq 0$

10. $2x + 5y \leq 20$
    $x - 2y \geq 1$
    $y \geq 0$

In Problems 11–14, find the maximum and minimum values of the given linear function P subject to the given inequalities.

11. $P = 2x + y$; the inequalities of Problem 7.
12. $P = 3x + 2y$; the inequalities of Problem 8.
13. $P = 2x - y$; the inequalities of Problem 9.
14. $P = 3x - 2y$; the inequalities of Problem 10.

EXAMPLE A (Unbounded region)   Find the maximum and minimum values of the function $3x + 4y$ subject to the constraints

$$\begin{cases} 3x + 2y \geq 13 \\ x + y \geq 5 \\ x \geq 1 \\ y \geq 0 \end{cases}$$

Solution.   We proceed to graph the solution set of our system, noting that the region must lie above the lines $3x + 2y = 13$ and $x + y = 5$ and to the right of the line $x = 1$. It is shown at the top of page 343. Notice that the region is unbounded and has (5, 0), (3, 2), and (1, 5) as its vertices. The point (1, 4) at which the lines $x + y = 5$ and $x = 1$ intersect is not

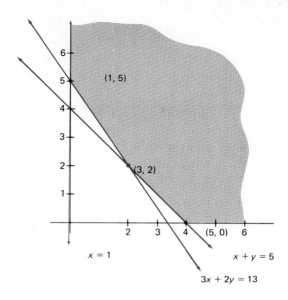

a vertex. It should be clear right away that $3x + 4y$ does not assume a maximum value in our region; its values can be made as large as we please by increasing x and y. To find the minimum value, we calculate $3x + 4y$ at the three vertices.

$$(5, 0): \quad 3 \cdot 5 + 4 \cdot 0 = 15$$
$$(3, 2): \quad 3 \cdot 3 + 4 \cdot 2 = 17$$
$$(1, 5): \quad 3 \cdot 1 + 4 \cdot 5 = 23$$

The minimum value of $3x + 4y$ is 15.

Solve each of the following problems.

15. Minimize $5x + 2y$ subject to

$$\begin{cases} x + y \geq 4 \\ \quad x \geq 2 \\ \quad y \geq 0 \end{cases}$$

16. Minimize $2x - y$ subject to

$$\begin{cases} x - 2y \geq 2 \\ \quad y \geq 2 \end{cases}$$

17. Minimize $2x + y$ subject to

$$\begin{aligned} 4x + \quad y &\geq 7 \\ 2x + 3y &\geq 6 \\ x &\geq 1 \\ y &\geq 0 \end{aligned}$$

18. Minimize $3x + 2y$ subject to

$$\begin{cases} x - 2y \leq 2 \\ x - 2y \geq -2 \\ 3x - 2y \geq 10 \end{cases}$$

EXAMPLE B (Systems witn nonlinear inequalities) Graph the solution set of the following system of inequalities.

$$\begin{cases} y \geq 2x^2 \\ y \leq 2x + 4 \end{cases}$$

Solution.   It helps to find the points at which the parabola $y = 2x^2$ intersects the line $y = 2x + 4$. Eliminating y between the two equations and then solving for x, we get

$$2x^2 = 2x + 4$$

$$x^2 - x - 2 = 0$$

$$(x - 2)(x + 1) = 0$$

$$x = 2 \qquad x = -1$$

The corresponding values of y are 8 and 2, respectively; so the points of intersection are $(-1, 2)$ and $(2, 8)$. Making use of these points, we draw the parabola and the line. Since the point $(0, 2)$ satisfies the inequality $y \geq 2x^2$, the desired region is above and including the parabola. The graph of the linear inequality $y \leq 2x + 4$ is to the right of and including the line. The shaded region in the diagram below is the graph we want.

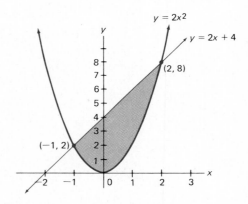

In Problems 19–22, graph the solution set of each system of inequalities.

19. $y \leq 4x - x^2$ 20. $y \geq 2^x$ 21. $y \leq \log_2 x$ 22. $x^2 + y^2 \geq 9$
    $y \leq x$ $y \leq 8$ $x \leq 8$ $0 \leq x \leq 3$
    $x \geq 0$ $y \geq 0$ $x \geq 0$ $y \geq 0$ $0 \leq y \leq 3$

Miscellaneous Problems

In Problems 23 and 24, graph the solution set and label the coordinates of the vertices.

23. $4x + y \leq 8$ 24. $2x + 5y \leq 20$
    $x - y \geq -2$ $x - 2y \leq 1$
    $x \geq 0$ $y \geq 0$ $x \geq 0$

25. Find the maximum and minimum values of $P = 2x - y$ subject to the inequalities of Problem 23.

26. Find the maximum and minimum values of $P = 3x - 2y$ subject to the inequalities of Problem 24.

27. Find the maximum and minimum values of $x + y$ (if they exist) subject to the following inequalities.

$$x - y \geq -1$$
$$x - 2y \leq 5$$
$$3x + y \leq 10$$
$$x \geq 0 \qquad y \geq 0$$

28. Find the maximum and minimum values of $x - 2y$ (if they exist) subject to the inequalities of Problem 27.

29. An oil refinery has a maximum production of 2000 barrels of oil per day. It produces two types of oil; type A, which is used for gasoline and type B, which is used for heating oil. There is a requirement that at least 300 barrels of type B be produced each day. If the profit is $3 a barrel for type A and $2 a barrel for type B, find the maximum profit per day.

30. A company makes a single product on two production lines, A and B. A labor force of 900 hours per week is available, and weekly running costs shall not exceed $1500. It takes 4 hours to produce one item on production line A and 3 hours on production line B. The cost per item is $5 on line A and $6 on line B. Find the largest number of items that can be produced in one week.

31. A shoemaker has a supply of 100 square feet of type A leather which is used for soles and 600 square feet of type B leather used for the rest of the shoe. The average shoe uses $\frac{1}{4}$ square feet of type A leather and 1 square foot of type B leather. The average boot uses $\frac{1}{4}$ square feet and 3 square feet of types A and B leather, respectively. If shoes sell at $40 a pair and boots at $60 a pair, find the maximum income.

32. A manufacturer of trailers wishes to determine how many camper units and how many house trailers she should produce in order to make the best possible use of her resources. She has 42 units of wood, 56 worker-weeks of labor and 16 units of aluminum. (Assume that all other needed resources are available and have no effect on her decision.) The amount of each resource needed to produce each camper and each trailer is given below.

	Wood	Worker-weeks	Aluminum
Per camper	3	7	3
Per trailer	6	7	1

If the manufacturer realizes a profit of $600 on a camper and $800 on a trailer, what should be her production in order to maximize her profit?

33. A grain farmer has 100 acres available for sowing oats and wheat. The seed oats costs $5 per acre and the seed wheat costs $8 per acre. The labor costs are $20 per acre for oats and $12 per acre for wheat. The farmer expects an income from oats of $220 per acre and from wheat of $250 per acre. How many acres of each crop should he sow to maximize his profit, if he does not wish to spend more than $620 for seed and $1800 for labor?

34. Suppose that the minimum monthly requirements for one person are 60 units of carbohydrates, 40 units of protein, and 35 units of fat. Two foods A and

B contain the following numbers of units of the three diet components per pound.

	Carbohydrates	Protein	Fat
A	5	3	5
B	2	2	1

If food A costs $3.00 a pound and food B costs $1.40 a pound, how many pounds of each should a person purchase per month to minimize the cost?

CHAPTER SUMMARY

Two systems of equations are **equivalent** if they have the same solutions. Three elementary operations (multiplying an equation by a nonzero constant, interchanging two equations, and adding a multiple of one equation to another) lead to equivalent systems. Use of these operations allows us to transform a system of linear equations to **triangular form** and then to solve the system by **back substitution** or to show it is **inconsistent.**

A **matrix** is a rectangular array of numbers. In solving a system of linear equations, it is efficient to work with just the **matrix of the system.** We solve the system by transforming its matrix to triangular form using the three operations mentioned above.

Addition and multiplication are defined for matrices with resulting algebraic rules. Matrices behave much like numbers, with the exception that the commutative law for multiplication fails. Even the notion of **multiplicative inverse** has meaning, though the process for finding such an inverse is lengthy.

Associated with every square matrix is a symbol called its **determinant.** For 2×2 and 3×3 matrices, the value of the determinant can be found by using certain arrow diagrams. For higher order cases, the value of a determinant is found by expanding it in terms of the elements of a row (or column) and their **minors** (determinants whose order is lower than the given determinant by 1). In doing this, it is helpful to know what happens to the determinant of a matrix when any of the three elementary operations are applied to it. **Cramer's rule** provides a direct way of solving a system of n equations in n unknowns using determinants.

The graph of a **linear inequality** in x and y is a **half-plane** (closed or open according as the inequality sign does or does not include the equal sign). The graph of a **system of linear inequalities** is the intersection of the half-planes corresponding to the separate inequalities. Such a graph is always **polygonal** and **convex** but may be **bounded** or **unbounded.** A **linear programming problem** asks us to find the maximum (or minimum) of a linear function (such as $2x + 5y$) subject to a system of linear inequalities called **constraints.** The maximum (or minimum) always occurs at a **vertex,** that is, at a corner point of the graph of the inequality constraints.

CHAPTER REVIEW PROBLEM SET

In Problems 1–5, solve each system or show that there is no solution.

1. $3x + y = 12$
   $2x - y = -2$

2. $x - 3y + 2z = 5$
   $2y - z = 1$
   $z = 3$

3. $2x - y + 3z = 10$
$\qquad y - 2z = 4$
$\qquad -2y + 4z = 8$

4. $\quad x - 3y + 2z = -5$
$\quad 4x + y - z = 4$
$\quad 5x + 11y - 8z = 20$

5. $x + 2y + 3z + 4w = 20$
$\qquad y - 4z + w = 6$
$\qquad z + 2w = 4$
$\qquad 2z + 4w = 8$

6. Evaluate each expression for

$$A = \begin{bmatrix} 1 & 1 & 1 \\ 3 & 1 & -1 \\ 2 & 2 & -1 \end{bmatrix} \quad \text{and} \quad B = \begin{bmatrix} -3 & 6 & -4 \\ 2 & -3 & 5 \\ 1 & 9 & 2 \end{bmatrix}$$

(a) $2A + B$ (b) $A - 2B$ (c) AB (d) BA

7. Find A^{-1} for the matrix A of Problem 6.

8. Consider the system

$$x + y + z = 4$$
$$3x + y - z = -4$$
$$2x + 2y - z = -1$$

Write this system in matrix form and then use the result of Problem 7 to solve it.

Evaluate the determinants in Problems 9–13.

9. $\begin{vmatrix} -2 & 5 \\ 2 & -6 \end{vmatrix}$

10. $\begin{vmatrix} 2 & 3 \\ -4 & -6 \end{vmatrix}$

11. $\begin{vmatrix} -2 & 1 & 4 \\ 0 & 5 & -1 \\ 4 & 0 & 3 \end{vmatrix}$

12. $\begin{vmatrix} 1 & -2 & 3 \\ 4 & 1 & 5 \\ 7 & -5 & 14 \end{vmatrix}$

13. $\begin{vmatrix} 6 & 0 & 0 & 0 \\ -1 & 3 & 1 & 0 \\ 3 & 2 & 3 & 2 \\ 4 & 5 & 1 & -4 \end{vmatrix}$

14. Use Cramer's rule to solve the following system.

$$2x + y - z = -4$$
$$x - 3y - 2z = -1$$
$$3x + 2y + 3z = 11$$

In Problems 15 and 16, graph the solution set of the given system. On the graph, label the coordinates of the vertices.

15. $x + y \leq 7$
$\quad 3x + y \leq 15$
$\quad x \geq 0 \qquad y \geq 0$

16. $x - 2y + 4 \geq 0$
$\quad x + y - 11 \geq 0$
$\quad x \geq 0 \qquad y \geq 0$

17. Find the maximum value of the function $P = x + 2y$ subject to the inequalities of Problem 15.

18. Find the minimum value of the function $P = x + 2y$ subject to the inequalities of Problem 16.

19. A certain company has 100 employees, some of whom get $4 an hour, others $5, and the rest $8. Half as many make $8 an hour as $5 an hour. If the total paid out in hourly wages is $544, find the number of employees who make $8 an hour.

20. A tailor has 110 yards of cotton material and 160 yards of woolen material. It takes $1\frac{1}{2}$ yards of cotton and 1 yard of wool to make a suit, while a dress requires 1 yard of cotton and 2 yards of wool. If a suit sells for $100 and a dress for $80, how many of each should the tailor make to maximize the total income?

9

SEQUENCES AND COUNTING PROBLEMS

Method consists entirely in properly ordering and arranging things to which we should pay attention.

René Descartes

What Comes Next?

(a) 1, 4, 7, 10, 13, 16, □, □, . . .

(c) 1, 4, 9, 16, 25, 36, □, □, . . .

(b) 2, 4, 6, 8, 10, 12, □, □, . . .

(d) 1, 4, 9, 16, 27, 40, □, □, . . .

(e) 1, 1, 2, 3, 5, 8, □, □, . . .

9-1
Number Sequences

Try filling in the boxes of our opening display. You will have little trouble with *a*, *b*, and *c*, but *d* and *e* may offer quite a challenge. We will give the answers we had in mind later. Right now, we merely point out that each sequence has a pattern; we used a definite rule in writing the first six terms of each of them.

The word *sequence* is commonly used in ordinary language. For example, your history teacher may talk about a sequence of events that led to World War II (for instance, the Versailles Treaty, world depression, Hitler's ascendancy, Munich Agreement). What characterizes this sequence is the notion of one event following another in a definite order. There is a first event, a second event, a third event, and so on. We might even give them labels.

E_1: Versailles Treaty
E_2: World depression
E_3: Hitler's ascendancy
E_4: Munich Agreement

We use a similar notation for number sequences. Thus

$$a_1, a_2, a_3, a_4, \ldots$$

could denote sequence (a) of our opening display. Then

$$a_1 = 1$$
$$a_2 = 4$$
$$a_3 = 7$$
$$a_4 = 10$$
$$\vdots$$

Note that a_3 stands for the 3rd term; a_{10} would represent the tenth term. The subscript indicates the position of the term in the sequence. For the general term, that is, the nth term, we use the symbol a_n. The three dots indicate that the sequence continues indefinitely.

There is another way to describe a number sequence. A **number sequence** is a function whose domain is the set of positive integers. That means it is a rule that associates with each positive integer n a definite number a_n. In conformity with Chapter 5, we could use the notation $a(n)$, but tradition dictates that we use a_n instead. We usually specify functions by giving formulas; this is true of sequences also.

EXPLICIT FORMULAS

Rather than give the first few terms of a sequence and hope that our readers see the pattern intended (different people sometimes see different patterns in the first few terms of a sequence), it is better to give a formula. Take sequence (a) for example. The formula

$$a_n = 3n - 2$$

tells all there is to know about that sequence. In particular,

$$a_1 = 3 \cdot 1 - 2 = 1$$
$$a_2 = 3 \cdot 2 - 2 = 4$$
$$a_3 = 3 \cdot 3 - 2 = 7$$
$$a_{100} = 3 \cdot 100 - 2 = 298$$

How does one find the formula for a sequence? Look at sequence (b). Suppose we let b_n stand for the nth term. Then $b_1 = 2$, $b_2 = 4$, $b_3 = 6$, $b_4 = 8$, and so on. Our job is to relate the value of b_n to the subscript n. Clearly, it is just twice the subscript—that is,

$$b_n = 2n$$

Knowing this formula, we can calculate the value of any term. For example,

$$b_{10} = 2 \cdot 10 = 20$$
$$b_{281} = 2 \cdot 281 = 562$$

If we follow the same procedure for sequence (c), we have

$$c_1 = 1 \qquad c_2 = 4 \qquad c_3 = 9 \qquad c_4 = 16$$

from which we infer the formula

$$c_n = n^2$$

Now look at sequence (d).

$$1, 4, 9, 16, 27, 40, \ldots$$

The fact that it starts out just like sequence (c) suggests that the pattern is subtle and incidentally warns us that we may have to look at many terms of a sequence before we can discover its rule of construction. Here, as in many sequences, it is a good idea to observe how each term relates to the previous one. Let us write the sequence again, indicating below it the numbers to be added as we progress from term to term.

$$1 \qquad 4 \qquad 9 \qquad 16 \qquad 27 \qquad 40$$
$$3 \qquad 5 \qquad 7 \qquad 11 \qquad 13$$

You may recognize the second row of numbers as consecutive primes (starting with 3). Thus the next two terms in sequence (d) are

$$40 + 17 = 57$$
$$57 + 19 = 76$$

But observing a pattern does not necessarily mean we can write an explicit formula. Though many have tried, no one has found a formula for the nth prime and, thus, no one is likely to find a formula for sequence (d).

Sequence (e) is a famous one. It was introduced by Leonardo Fibonacci around 1200 A.D. in connection with rabbit reproduction (see Problem 27). If you are a keen observer, you have noticed that any term (after the second) is the sum of the preceding two. It was not until 1724 that mathematician Daniel Bernoulli found the explicit formula for this sequence. You will agree that it is complicated, but at least you can check it for $n = 1, 2, 3$.

$$e_n = \frac{1}{\sqrt{5}} \left[\left(\frac{1 + \sqrt{5}}{2} \right)^n - \left(\frac{1 - \sqrt{5}}{2} \right)^n \right]$$

If it took 500 years to discover this formula, you should not be surprised when we say that explicit formulas are often difficult to find (the problem set will give more evidence). There is another type of formula that is usually easier to discover.

RECURSION FORMULA

An explicit formula relates the value of a_n to its subscript n (for example, $a_n = 3n - 2$). Often the pattern we first observe relates a term to the preceding term (or terms). If so, we may be able to describe this pattern by a recursion formula. Look at sequence (a) again. To get a term from the preceding one, we always add 3, that is,

$$a_n = a_{n-1} + 3$$

Or look at sequence (b). There we add 2 each time.

$$b_n = b_{n-1} + 2$$

Sequence (e) is more interesting. There we add together the two previous terms, that is,

$$e_n = e_{n-1} + e_{n-2}$$

We summarize our knowledge about the five sequences in the following chart.

TABLE 4

Sequence	Explicit Formula	Recursion Formula
(a) 1, 4, 7, 10, 13, 16, . . .	$a_n = 3n - 2$	$a_n = a_{n-1} + 3$
(b) 2, 4, 6, 8, 10, 12, . . .	$b_n = 2n$	$b_n = b_{n-1} + 2$
(c) 1, 4, 9, 16, 25, 36, . . .	$c_n = n^2$	$c_n = c_{n-1} + 2n - 1$
(d) 1, 4, 9, 16, 27, 40, . . .	?	$d_n = d_{n-1} + n\text{th prime}$
(e) 1, 1, 2, 3, 5, 8, . . .	$e_n = \dfrac{1}{\sqrt{5}}\left[\left(\dfrac{1 + \sqrt{5}}{2}\right)^n - \left(\dfrac{1 - \sqrt{5}}{2}\right)^n\right]$	$e_n = e_{n-1} + e_{n-2}$

Recursion formulas are themselves not quite enough to determine a sequence. For example, the recursion formula

$$f_n = 3f_{n-1}$$

does not determine a sequence until we specify the first term. But with the additional information that $f_1 = 2$, we can find any term. Thus

$$f_1 = 2$$
$$f_2 = 3f_1 = 3 \cdot 2 = 6$$
$$f_3 = 3f_2 = 3 \cdot 6 = 18$$
$$f_4 = 3f_3 = 3 \cdot 18 = 54$$
$$\vdots$$

The disadvantage of a recursion formula is apparent. To find the 100th term, we must first calculate the 99 previous terms. But if it is hard work, it is at least possible. Programmable calculators are particularly adept at calculating sequences by means of recursion formulas.

1. Discover a pattern and use it to fill in the boxes.
   (a) 1, 3, 5, 7, □, □, . . .
   (b) 17, 14, 11, 8, □, □, . . .
   (c) 1, $\frac{1}{2}$, $\frac{1}{4}$, $\frac{1}{8}$, □, □, . . .
   (d) 1, 9, 25, 49, □, □, . . .

2. Fill in the boxes.
   (a) 1, 3, 9, 27, □, □, . . .
   (b) 2, 2.5, 3, 3.5, □, □, . . .
   (c) 1, 8, 27, 64, □, □, . . .
   (d) $\frac{1}{2}$, $\frac{2}{3}$, $\frac{3}{4}$, $\frac{4}{5}$, □, □, . . .

3. In each case an explicit formula is given. Find the indicated terms.
   (a) $a_n = 2n + 3$; $a_4 =$ □; $a_{20} =$ □
   (b) $a_n = \dfrac{n}{n + 1}$; $a_5 =$ □; $a_9 =$ □
   (c) $a_n = (2n - 1)^2$; $a_4 =$ □; $a_5 =$ □
   (d) $a_n = (-3)^n$; $a_3 =$ □; $a_4 =$ □

4. Find the indicated terms.
   (a) $a_n = 2n - 5$; $a_4 =$ □; $a_{20} =$ □
   (b) $a_n = 1/n$; $a_5 =$ □; $a_{50} =$ □
   (c) $a_n = (2n)^2$; $a_5 =$ □; $a_{10} =$ □
   (d) $a_n = 4 - \frac{1}{2}n$; $a_5 =$ □; $a_{10} =$ □

5. Give an explicit formula for each sequence in Problem 1. Recall that you must relate the value of a term to its subscript (see the examples on page 351).

6. Give an explicit formula for each sequence in Problem 2.

7. In each case below, an initial term and a recursion formula are given. Find a_5. (*Hint:* First find a_2, a_3, and a_4, and then find a_5.)
   (a) $a_1 = 2$; $a_n = a_{n-1} + 3$
   (b) $a_1 = 2$; $a_n = 3a_{n-1}$
   (c) $a_1 = 8$; $a_n = \frac{1}{2}a_{n-1}$
   (d) $a_1 = 1$; $a_n = a_{n-1} + 8(n - 1)$

8. Find a_4 for each of the following sequences.
   (a) $a_1 = 2$; $a_n = 2a_{n-1} + 1$
   (b) $a_1 = 2$; $a_n = a_{n-1} + 3$
   (c) $a_1 = 1$; $a_n = a_{n-1} + 3n^2 - 3n + 1$
   (d) $a_1 = 3$; $a_n = a_{n-1} + .5$

9. Try to find a recursion formula for each of the sequences in Problem 1.

10. Try to find a recursion formula for each of the sequences in Problem 2.

EXAMPLE (Sum sequences)   Corresponding to a sequence $a_1, a_2, a_3, \ldots$, we introduce another sequence A_n, called the sum sequence, by

$$A_n = a_1 + a_2 + a_3 + \cdots + a_n$$

Thus

$$A_1 = a_1$$

$$A_2 = a_1 + a_2$$

$$A_3 = a_1 + a_2 + a_3$$

$$\vdots$$

Find A_5 for the sequence given by $a_n = 3n - 2$.

Solution. We begin by finding the first five terms of sequence a_n.

$$a_1 = 1$$

$$a_2 = 4$$

$$a_3 = 7$$

$$a_4 = 10$$

$$a_5 = 13$$

Then

$$A_5 = a_1 + a_2 + a_3 + a_4 + a_5$$

$$= 1 + 4 + 7 + 10 + 13$$

$$= 35$$

In each of the following problems, find A_6.

11. $a_n = 2n + 1$ 12. $a_n = 2^n$

13. $a_n = (-2)^n$ 14. $a_n = n^2$

15. $a_n = n^2 - 2$ 16. $a_n = 3n - 4$

17. $a_1 = 4; a_n = a_{n-1} + 3$ 18. $a_1 = 1; a_2 = 1; a_n = a_{n-1} + 2a_{n-2}$

Miscellaneous Problems

19. Find a_4 and a_{20} for each sequence.
    (a) $a_n = 3n - 1$ (b) $a_n = 2^n$

20. Find recursion formulas for the sequences of Problem 19.

21. Find a_4 for each of the following.
    (a) $a_1 = 8; a_n = \frac{3}{2}a_{n-1}$

    (b) $a_1 = 6; a_2 = 4; a_n = \dfrac{a_{n-1}}{a_{n-2}}$

22. Give an explicit formula for the first sequence of Problem 21.

23. For each sequence in Problem 19, find A_5, the sum of the first five terms.

24. Find a pattern in each of the following sequences and use it to fill in the boxes.
    (a) 2, 6, 18, 54, □, □, . . .
    (b) 2, 6, 10, 14, □, □, . . .
    (c) 2, 4, 8, 14, □, □, . . .
    (d) 2, 4, 6, 10, 16, □, □, . . .
    (e) 2, 1, $\frac{1}{2}$, $\frac{1}{4}$, □, □, . . .
    (f) 2, 5, 10, 17, □, □, . . .

25. Try to find an explicit formula for each sequence in Problem 24.

26. Find a recursion formula for each sequence in Problem 24.

27. Suppose that a pair of rabbits consisting of a male and a female matures so that is reproduces another male-female pair after two months and continues to do so each month thereafter. If each new male-female pair of rabbits has the same reproductive habits as its parents, how many rabbit pairs will there be after 1 month? 2 months? 3 months? 4 months? n months?

28. Let f_n denote the Fibonacci sequence determined by $f_1 = f_2 = 1$ and $f_n = f_{n-1} + f_{n-2}$. Let $F_n = f_1 + f_2 + \cdots + f_n$. Calculate $F_1, F_2, F_3, F_4, F_5,$ and F_6.

29. With regard to Problem 28, see if you can find a nice formula that connects F_n and f_{n+2}.

30. Let a_n be the remainder when n is divided by 5.
    (a) Find a_{11} and a_{400}.
    (b) If $a_m = 3$, find a_{m+4}.

31. Let a_n be the number of primes that are less than n. (The smallest prime is 2.)
    (a) Find a_{10} and a_{20}.
    (b) If m is a number such that $a_{m+20} = a_m$, what can you conclude?

32. Let a_n be the sum of the proper divisors of n. For example, $a_{12} = 1 + 2 + 3 + 4 + 6 = 16$ and $a_{25} = 1 + 5 = 6$
    (a) Find a_{10}, a_{16}, and a_{40}.
    (b) If k is a prime, what value does a_k have?

33. The Greeks were enchanted with sequences that arose in a geometric way (see the diagrams in the margin).
    (a) Find an explicit formula for s_n.
    (b) Find a recursion formula for s_n.
    (c) Find a recursion formula for t_n.
    (d) Find an explicit formula for t_n.

34. The numbers 1, 5, 12, 22, . . . are called *pentagonal numbers*. See if you can figure out why and then guess at an explicit formula for p_n, the nth pentagonal number. Use diagrams.

35. Let a_n be the nth digit in the decimal expansion of $\frac{1}{7} = .1428 \ldots$. Thus $a_1 = 1$, $a_2 = 4$, $a_3 = 2$, and so on. Find a pattern and use it to determine a_8, a_{27}, and a_{53}.

36. Suppose that January 1 occurs on Wednesday. Let a_n be the day of the week corresponding to the nth day of the year. Thus $a_1 = $ Wednesday, $a_2 = $ Thursday, and so forth. Find a_{39}, a_{57}, and a_{84}.

37. Try to find a pattern in the following array of numbers.:

$$3, 3, 5, 4, 4, 3, 5, 5, 4, 3, 6, 6, \ldots$$

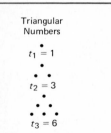

Square Numbers

$s_1 = 1$

$s_2 = 4$

$s_3 = 9$

Triangular Numbers

$t_1 = 1$

$t_2 = 3$

$t_3 = 6$

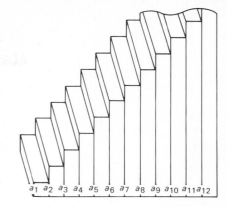

Jacob's Stairway

In the skyscraper that Jacob owns, there is a stairway going from ground level to the very top. The first step is 8 inches high. After that each step is 9 inches above the previous one. How high above the ground is the 800th step?

a_1 a_2 a_3 a_4 a_5 a_6 a_7 a_8 a_9 a_{10} a_{11} a_{12}

9-2
Arithmetic Sequences

We are going to answer the question above and others like it. Notice that if a_n denotes the height of the nth step, then

$$a_1 = 8$$
$$a_2 = 8 + 1(9) = 17$$
$$a_3 = 8 + 2(9) = 26$$
$$a_4 = 8 + 3(9) = 35$$
$$\vdots$$
$$a_{800} = 8 + 799(9) = 7199$$

The 800th step of Jacob's stairway is 7199 inches (almost 600 feet) above the ground.

FORMULAS

Now consider the following number sequences. When you see a pattern, fill in the boxes.

(a) 5, 9, 13, 17, $\square$, $\square$, . . .
(b) 2, 2.5, 3, 3.5, $\square$, $\square$, . . .
(c) 8, 5, 2, -1, $\square$, $\square$, . . .

What is it that these three sequences have in common? Simply this: In each case, you can get a term by adding a fixed number to the preceding term. In (a), you add 4 each time, in (b) you add 0.5, and in (c), you add -3. Such sequences are called **arithmetic sequences.** If we denote such a sequence by

$a_1, a_2, a_3, \ldots$, it satisfies the recursion formula

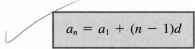

$$a_n = a_{n-1} + d$$

where d is a fixed number called the **common difference.**

Can we also obtain an explicit formula? Yes. The diagram in the margin should help. Notice that the number of d's to be added to a_1 is one less than the subscript. This means that

$$a_n = a_1 + (n - 1)d$$

Now we can give explicit formulas for each of the sequences (a), (b), and (c) above.

$$a_n = 5 + (n - 1)4 = 1 + 4n$$

$$b_n = 2 + (n - 1)(.5) = 1.5 + .5n$$

$$c_n = 8 + (n - 1)(-3) = 11 - 3n$$

ARITHMETIC SEQUENCES AND LINEAR FUNCTIONS

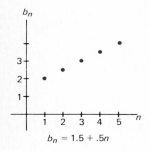

$b_n = 1.5 + .5n$

We have said that a sequence is a function whose domain is the set of positive integers. Functions are best visualized by drawing their graphs. Consider the sequence b_n discussed above; its explicit formula is

$$b_n = 1.5 + .5n$$

Its graph is shown in the margin.

Even a cursory look at this graph suggests that the points lie along a straight line. Consider now the function

$$b(x) = 1.5 + 0.5x = 0.5x + 1.5$$

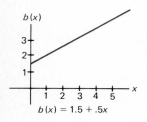

$b(x) = 1.5 + .5x$

where x is allowed to be any real number. This is a linear function, being of the form $mx + b$ (see Section 4-3). Its graph is a straight line (see the second graph in the margin), and its values at $x = 1, 2, 3, \ldots$ are equal to $b_1, b_2, b_3, \ldots$.

The relationship illustrated above between an arithmetic sequence and a linear function holds in general. An arithmetic sequence is just a linear function whose domain has been restricted to the positive integers.

SUMS OF ARITHMETIC SEQUENCES

There is an old story about Carl Gauss that aptly illustrates the next idea. We are not sure if the story is true, but if not, it should be.

When Gauss was about 10 years old, he was admitted to an arithmetic class. To keep the class busy, the teacher often assigned long addition problems. One day he asked his students to add the numbers from 1 to 100. Hardly had he made the assignment when young Gauss laid his slate on the teacher's desk with the answer 5050 written on it.

Here is how Gauss probably thought about the problem.

$$1 + 2 + \ldots + 49 + 50 + 51 + 52 + \ldots + 99 + 100$$

Each of the indicated pairs has 101 as its sum and there are 50 such pairs. Thus the answer is $50(101) = 5050$. For a 10-year-old boy, that is good thinking.

Gauss's trick works perfectly well on any arithmetic sequence where we want to add an even number of terms. And there is a slight modification that works whether the number of terms to be added is even or odd.

Suppose $a_1, a_2, a_3, \ldots$ is an arithmetic sequence and let

$$A_n = a_1 + a_2 + a_3 + \cdots + a_{n-1} + a_n$$

Write this sum twice, once forwards and once backwards, and then add.

$$
\begin{array}{ccccccccc}
A_n = & a_1 & + & a_2 & + \cdots + & a_{n-1} & + & a_n \\
A_n = & a_n & + & a_{n-1} & + \cdots + & a_2 & + & a_1 \\
\hline
2A_n = & (a_1 + a_n) & + & (a_2 + a_{n-1}) & + \cdots + & (a_2 + a_{n-1}) & + & (a_1 + a_n)
\end{array}
$$

Each group on the right has the same sum, namely, $a_1 + a_n$. For example,

$$a_2 + a_{n-1} = a_1 + d + a_n - d = a_1 + a_n$$

There are n such groups and so

$$2A_n = n(a_1 + a_n)$$

$$\boxed{A_n = \frac{n}{2}(a_1 + a_n)}$$

We call this the **sum formula** for an arithmetic sequence. You can remember this sum as being n times the average term $(a_1 + a_n)/2$.

Here is how we apply this formula to Gauss's problem. We want the sum of 100 terms of the sequence $1, 2, 3, \ldots$, that is, we want A_{100}. Here $n = 100$, $a_1 = 1$, and $a_n = 100$. Therefore

$$A_{100} = \frac{100}{2}(1 + 100) = 50(101) = 5050$$

For a second example, suppose we want to add the first 350 odd numbers, that is, the first 350 terms of the sequence $1, 3, 5, \ldots$. We can calculate the 350th odd number from the formula $a_n = a_1 + (n - 1)d$.

$$a_{350} = 1 + (349)2 = 699$$

Then we use the sum formula with $n = 350$.

$$A_{350} = 1 + 3 + 5 + \cdots + 699$$

$$= \frac{350}{2}(1 + 699) = 122,500$$

SIGMA NOTATION

There is a convenient shorthand that is frequently employed in connection with sums. The first letter of the word *sum* is *s*; the Greek letter for *S* is Σ(sigma). We use Σ in mathematics to stand for the operation of summation. In particular,

$$\sum_{i=1}^{n} a_i = a_1 + a_2 + a_3 + \cdots + a_n$$

The symbol $i = 1$ underneath the sigma tells where to start adding the terms a_i and the n at the top tells where to stop. Thus,

$$\sum_{i=1}^{4} a_i = a_1 + a_2 + a_3 + a_4$$

$$\sum_{i=3}^{7} b_i = b_3 + b_4 + b_5 + b_6 + b_7$$

$$\sum_{i=1}^{5} i^2 = 1^2 + 2^2 + 3^2 + 4^2 + 5^2$$

$$\sum_{i=1}^{30} 3i = 3 + 6 + 9 + \cdots + 90$$

If $a_1, a_2, a_3, \ldots$ is an *arithmetic sequence*, then the sum formula previously derived may be written

$$\sum_{i=1}^{n} a_i = \frac{n}{2}(a_1 + a_n)$$

Problem Set 9-2

1. Fill in the boxes below.
   (a) 1, 4, 7, 10, $\square$, $\square$, . . .
   (b) 2, 2.3, 2.6, 2.9, $\square$, $\square$, . . .
   (c) 28, 24, 20, 16, $\square$, $\square$, . . .

2. Fill in the boxes.
   (a) 4, 6, 8, 10, $\square$, $\square$, . . .
   (b) 4, 4.2, 4.4, 4.6, $\square$, $\square$, . . .
   (c) 4, 3.8, 3.6, 3.4, $\square$, $\square$, . . .

3. Determine d and the 30th term of each sequence in Problem 1.

4. Determine d and the 101st term of each sequence in Problem 2.

5. Determine $A_{30} = a_1 + a_2 + \cdots + a_{30}$ for the sequence in part (a) of Problem 1. Similarly, find B_{30} and C_{30} for the sequences in parts (b) and (c).

6. Determine A_{100}, B_{100}, and C_{100} for the sequences of Problem 2.

7. If $a_1 = 5$ and $a_{40} = 24.5$ in an arithmetic sequence, determine d.

8. If $b_1 = 6$ and $b_{30} = -52$ in an arithmetic sequence, determine d.

9. Calculate each sum.
   (a) $2 + 4 + 6 + \cdots + 200$
   (b) $1 + 3 + 5 + \cdots + 199$
   (c) $3 + 6 + 9 + \cdots + 198$
   (*Hint:* Before using the sum formula, you have to determine n. In part (a), n is 100 since we are adding the doubles of the integers from 1 to 100.)

10. Calculate each sum.
    (a) $4 + 8 + 12 + \cdots + 100$
    (b) $10 + 15 + 20 + \cdots + 200$
    (c) $6 + 9 + 12 + \cdots + 72$

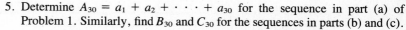

11. The bottom rung of a tapered ladder is 30 centimeters long and the top rung is 15 centimeters long. If there are 17 rungs, how many centimeters of rung material are needed to make the ladder, assuming no waste?

12. A clock strikes once at 1:00, twice at 2:00, and so on. How many times does it strike between 10:30 A.M. on Monday and 10:30 P.M. on Tuesday?

13. If $3, a, b, c, d, 7, \ldots$ is an arithmetic sequence, find a, b, c, and d.

14. If $8, a, b, c, 5$ is an arithmetic sequence, find a, b, and c.

15. How many multiples of 9 are there between 200 and 300? Find their sum.

16. If Ronnie is paid $10 on January 1, $20 on January 2, $30 on January 3, and so on, how much does he earn during January?

17. Calculate each sum.

    (a) $\displaystyle\sum_{i=2}^{6} i^2$ (b) $\displaystyle\sum_{i=1}^{4} \frac{2}{i}$

    (c) $\displaystyle\sum_{i=1}^{100} (3i + 2)$ (d) $\displaystyle\sum_{i=2}^{100} (2i - 3)$

18. Calculate each sum.

    (a) $\displaystyle\sum_{i=1}^{6} 2^i$ (b) $\displaystyle\sum_{i=1}^{5} (i^2 - 2i)$

    (c) $\displaystyle\sum_{i=1}^{101} (2i - 6)$ (d) $\displaystyle\sum_{i=3}^{102} (3i + 5)$

19. Write in sigma notation.
    (a) $b_3 + b_4 + \cdots + b_{20}$
    (b) $1^2 + 2^2 + \cdots + 19^2$
    (c) $1 + \dfrac{1}{2} + \dfrac{1}{3} + \cdots + \dfrac{1}{n}$

20. Write in sigma notation.
    (a) $a_6 + a_7 + a_8 + \cdots + a_{70}$
    (b) $2^3 + 3^3 + 4^3 + \cdots + 100^3$
    (c) $1 + 3 + 5 + 7 + \cdots + 99$

21. For the arithmetic sequence 5, 7.5, 10, 12.5, . . . ,
    (a) find d;
    (b) find the 51st term;
    (c) find the sum of the first 51 terms.

22. If $a_1 = 12$ and $a_{21} = 38$ in an arithmetic sequence,
    (a) find d;
    (b) find a_{56};
    (c) find m if $a_m = 61.4$.

23. Calculate the sum $4.25 + 4.5 + 4.75 + 5 + \cdots + 21.75$.

24. If $10, a, b, c, d, e, 30, \ldots$ is an arithmetic sequence, find $a, b, c, d,$ and e.

25. Find the sum of all multiples of 8 between 150 and 450.

26. Calculate each sum.

    (a) $\displaystyle\sum_{i=1}^{4} i^3$

    (b) $\displaystyle\sum_{i=1}^{100} \left(\frac{i}{i+1} - \frac{i-1}{i} \right)$

    (*Hint:* Write the first three or four terms of this sum without simplifying. You will see a pattern.)

27. Write in sigma notation.
    (a) $b_1 + b_2 + b_3 + \cdots + b_{112}$
    (b) $19 + 26 + 33 + \cdots + 719$

28. At a club meeting with 300 people present, everyone shook hands with every other person exactly once. How many handshakes were there? (*Hint:* Number the people from 1 to 300. Person 1 shakes hands with each of the other 299 people, person 2 shakes hands with 298 people (the handshake with person 1 should not be counted again), and so on.)

29. A pile of logs has 70 logs in the bottom layer, 69 in the second layer, 68 in the third layer, and so on. If there are 59 layers, how many logs are there in the pile?

30. Consider the sequence for which $a_n = n^2$, starting with $n = 0$.
    (a) If $b_n = a_n - a_{n-1}$, show that $b_1, b_2, b_3, \ldots$ is an arithmetic sequence.
    (b) Find b_{1001}.
    (c) Use your answer to part (b) to calculate $a_{1001} = 1001^2$.

31. Consider the sequence for which $a_n = n^3$, starting with $n = 0$.
    (a) If $b_n = a_n - a_{n-1}$, find $b_1, b_2, b_3, b_4,$ and b_5.
    (b) Let $c_n = b_n - b_{n-1}$, find $c_2, c_3, c_4,$ and c_5. The sequence $c_2, c_3, c_4, \ldots$ seems to be what kind of sequence?

Jacob's Golden Staircase

In his dreams, Jacob saw a golden staircase with angels walking up and down. The first step was 8 inches high but after that each step was 5/4 as high above the ground as the previous one. How high above the ground was the 800th step?

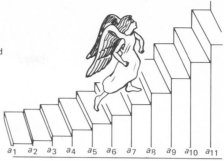

a_1 a_2 a_3 a_4 a_5 a_6 a_7 a_8 a_9 a_{10} a_{11}

9-3
Geometric Sequences

The staircase of Jacob's dream is most certainly one for angels, not for people. The 800th step actually stands 3.4×10^{73} miles high. By way of comparison, it is 9.3×10^6 miles to the sun and 2.5×10^{13} miles to Alpha Centauri, our nearest star beyond the sun. You might say the golden staircase reaches to heaven.

To see how to calculate the height of the 800th step, notice the pattern of heights for the first few steps and then generalize.

$$a_1 = 8$$

$$a_2 = 8\left(\frac{5}{4}\right)$$

$$a_3 = 8\left(\frac{5}{4}\right)^2$$

$$a_4 = 8\left(\frac{5}{4}\right)^3$$

$$\vdots$$

$$a_{800} = 8\left(\frac{5}{4}\right)^{799}$$

With a pocket calculator, it is easy to calculate $8(\frac{5}{4})^{799}$ and then change this number of inches to miles; the result is the figure given above.

FORMULAS

In the sequence above, each term was $\frac{5}{4}$ times the preceding one. You should be able to find a similar pattern in each of the following sequences. When you do, fill in the boxes.

    (a) 3, 6, 12, 24, $\square$, $\square$, . . .

    (b) 12, 4, $\frac{4}{3}$, $\frac{4}{9}$, $\square$, $\square$, . . .

    (c) .6, 6, 60, 600, $\square$, $\square$, . . .

SECTION 9-3 Geometric Sequences **363**

The common feature of these three sequences is that in each case, you can get a term by multiplying the preceding term by a fixed number. In sequence (a), you multiply by 2; in (b), by $\frac{1}{3}$; and in (c), by 10. We call such sequences **geometric sequences.** Thus a geometric sequence $a_1, a_2, a_3, \ldots$ satisfies the recursion formula

$$a_n = ra_{n-1}$$

where r is a fixed number called the **common ratio.**

To obtain the corresponding explicit formula, note that

$$a_2 = ra_1$$
$$a_3 = ra_2 = r(ra_1) = r^2 a_1$$
$$a_4 = ra_3 = r(r^2 a_1) = r^3 a_1$$

In each case, the exponent on r is one less than the subscript on a. Thus

$$a_n = a_1 r^{n-1}$$

From this, we can get explicit formulas for each of the sequences (a), (b), and (c) on page 363.

$$a_n = 3 \cdot 2^{n-1}$$
$$b_n = 12\left(\frac{1}{3}\right)^{n-1}$$
$$c_n = (.6)(10)^{n-1}$$

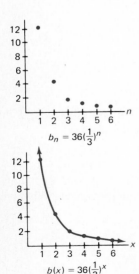

$$b_n = 36\left(\tfrac{1}{3}\right)^n$$

$$b(x) = 36\left(\tfrac{1}{3}\right)^x$$

GEOMETRIC SEQUENCES AND EXPONENTIAL FUNCTIONS

Let us consider sequence (b) once more; its explicit formula is

$$b_n = 12\left(\frac{1}{3}\right)^{n-1} = 36\left(\frac{1}{3}\right)^n$$

We have graphed this sequence and also the exponential function

$$b(x) = 36\left(\frac{1}{3}\right)^x$$

in the margin. (See Section 6-2 for a discussion of exponential functions.) It should be clear that the sequence b_n is the function $b(x)$ with its domain restricted to the positive integers.

What we have observed in this example is true in general. A geometric sequence is simply an exponential function with its domain restricted to the positive integers.

SUMS OF GEOMETRIC SEQUENCES

There is an old legend about geometric sequences and chessboards. When the king of Persia learned to play chess, he was so enchanted with the game that he determined to reward the inventor, a man named Sessa. Calling Sessa to the palace, the king promised to fulfill any request he might make. With an air of modesty, wily Sessa asked for one grain of wheat for the first square of the chessboard, two for the second, four for the third, and so on. The king was amused at such an odd request; nevertheless, he called a servant, told him to get a bag of wheat, and start counting. To the king's surprise, it soon became apparent that Sessa's request could never be fulfilled. The world's total production of wheat for a whole century would not be sufficient.

Sessa was really asking for

$$1 + 2 + 2^2 + 2^3 + \cdots + 2^{63}$$

grains of wheat, the sum of the first 64 terms of the geometric sequence 1, 2, 4, 8, ... We are going to develop a formula for this sum and for all others that arise from adding the terms of a geometric sequence.

Let $a_1, a_2, a_3, \ldots$ be a geometric sequence with ratio $r \neq 1$. As usual, let

$$A_n = a_1 + a_2 + a_3 + \cdots + a_n$$

which can be written

$$A_n = a_1 + a_1 r + a_1 r^2 + \cdots + a_1 r^{n-1}$$

Now multiply A_n by r, subtract the result from A_n, and use a little algebra to solve for A_n. We obtain

$$
\begin{aligned}
A_n &= a_1 + a_1 r + a_1 r^2 + \cdots + a_1 r^{n-1} \\
rA_n &= \qquad\quad a_1 r + a_1 r^2 + \cdots + a_1 r^{n-1} + a_1 r^n \\
\hline
A_n - rA_n &= a_1 + 0 + 0 + \cdots + 0 - a_1 r^n \\
A_n(1 - r) &= a_1(1 - r^n)
\end{aligned}
$$

$$\boxed{A_n = \frac{a_1(1 - r^n)}{1 - r} \qquad r \neq 1}$$

In sigma notation, this is

$$\sum_{i=1}^{n} a_i = \frac{a_1(1 - r^n)}{1 - r} \qquad r \neq 1$$

In the case where $r = 1$,

$$\sum_{i=1}^{n} a_i = a_1 + a_1 + \cdots + a_1 = na_1$$

Applying the sum formula to Sessa's problem (using $n = 64$, $a_1 = 1$, and $r = 2$), we get

$$A_{64} = \frac{1(1 - 2^{64})}{1 - 2} = 2^{64} - 1$$

Ignoring the -1 and using the approximation $2^{10} \approx 1000$ gives

$$A_{64} \approx 2^{64} \approx 2^4(1000)^6 = 1.6 \times 10^{19}$$

If you do this problem on a calculator, you will get $A_{64} \approx 1.845 \times 10^{19}$. Thus, if a bushel of wheat has one million grains, A_{64} grains would amount to more than 1.8×10^{13}, or 18 trillion, bushels. That exceeds the world's total production of wheat in 1 century.

THE SUM OF THE WHOLE SEQUENCE

Is it possible to add infinitely many numbers? Do the following sums make sense?

$$\frac{1}{2} + \frac{1}{4} + \frac{1}{8} + \frac{1}{16} + \cdots$$

$$1 + 3 + 9 + 27 + \cdots$$

Questions like this have intrigued great thinkers since Zeno first introduced his famous paradoxes of the infinite over 2000 years ago. We now show that we can make sense out of the first of these two sums but not the second.

Consider a string of length 1 kilometer. We may imagine cutting it into infinitely many pieces, as indicated in the figure below.

Since these pieces together make a string of length 1, it seems natural to say

$$\frac{1}{2} + \frac{1}{4} + \frac{1}{8} + \frac{1}{16} + \cdots = 1$$

Let us look at it another way. The sum of the first n terms of the geometric sequence $\frac{1}{2}, \frac{1}{4}, \frac{1}{8}, \frac{1}{16}, \ldots$ is given by

$$A_n = \frac{\frac{1}{2}\left[1 - \left(\frac{1}{2}\right)^n\right]}{1 - \frac{1}{2}} = 1 - \left(\frac{1}{2}\right)^n$$

As n gets larger and larger (tends to infinity), $\left(\frac{1}{2}\right)^n$ gets smaller and smaller (approaches 0). Thus A_n tends to 1 as n tends to infinity. We therefore say that 1 is the sum of *all* the terms of this sequence.

Now consider any geometric sequence with ratio r satisfying $|r| < 1$. We claim that when n gets large, r^n approaches 0. (As evidence, try calculating

$(0.99)^{100}$, $(0.99)^{1000}$, and $(0.99)^{10,000}$ on your calculator.) Thus, as n gets large,

$$A_n = \frac{a_1(1 - r^n)}{1 - r}$$

approaches $a_1/(1 - r)$. We write

$$\sum_{i=1}^{\infty} a_i = \frac{a_1}{1 - r} \qquad |r| < 1$$

For an important use of this formula in a familiar context, see the example in the problem set.

We emphasize that what we have just done is valid if $|r| < 1$. There is no way to make sense out of adding all the terms of a geometric sequence if $|r| \geq 1$.

Problem Set 9-3

1. Fill in the boxes.
   (a) $\frac{1}{2}$, 1, 2, 4, □, □, . . .
   (b) 8, 4, 2, 1, □, □, . . .
   (c) .3, 03, .003, .0003, □, □, . . .
2. Fill in the boxes.
   (a) 1, 3, 9, 27, □, □, . . .
   (b) 27, 9, 3, 1, □, □, . . .
   (c) .2, .02, .002, .0002, □, □, . . .
3. Determine r for each of the sequences in Problem 1 and write an explicit formula for the nth term.
4. Write a formula for the nth term of each sequence in Problem 2.
5. Evaluate the 30th term of each sequence in Problem 1.
6. Evaluate the 20th term of each sequence in Problem 2.
7. Use the sum formula to find the sum of the first five terms of each sequence in Problem 1.
8. Use the sum formula to find the sum of the first five terms of each sequence in Problem 2.
9. Find the sum of the first 30 terms of each sequence in Problem 1.
10. Find the sum of the first 20 terms of each sequence in Problem 2.
11. A certain culture of bacteria doubles every week. If there are 100 bacteria now, how many will there be after 10 full weeks?
12. A water lily grows so rapidly that each day it covers twice the area it covered the day before. At the end of 20 days, it completely covers a pond. If we start with two lilies, how long will it take to cover the same pond?
13. Johnny is paid $1 on January 1, $2 on January 2, $4 on January 3, and so on. Approximately how much will he earn during January?
14. If you were offered 1¢ today, 2¢ tomorrow, 4¢ the third day, and so on for 20 days or a lump sum of $10,000, which would you choose? Show why.

15. Calculate:

(a) $\displaystyle\sum_{i=1}^{\infty} \left(\frac{1}{3}\right)^i$ (b) $\displaystyle\sum_{i=2}^{\infty} \left(\frac{2}{5}\right)^i$

16. Calculate:

(a) $\displaystyle\sum_{i=1}^{\infty} \left(\frac{2}{3}\right)^i$ (b) $\displaystyle\sum_{i=3}^{\infty} \left(\frac{1}{6}\right)^i$

infinite

17. A ball is dropped from a height of 10 feet. At each bounce, it rises to a height of $\frac{1}{2}$ the previous height. How far will it travel altogether (up and down) by the time it comes to rest? (*Hint:* Think of the total distance as being the sum of the "down" distances $(10 + 5 + \frac{5}{2} + \cdot \cdot \cdot)$ and the "up" distances $(5 + \frac{5}{2} + \frac{5}{4} + \cdot \cdot \cdot).$)

18. Do Problem 17 assuming the ball rises to $\frac{2}{3}$ its previous height at each bounce.

EXAMPLE (Repeating decimals) Show that $.333\overline{3} \ldots$ and $.2323\overline{23} \ldots$ are rational numbers by using the methods of this section.

Solution.

$$.333\overline{3} = \frac{3}{10} + \frac{3}{100} + \frac{3}{1000} + \cdot \cdot \cdot$$

Thus we must add all the terms of an infinite geometric sequence with ratio $\frac{1}{10}$. Using the formula $a_1/(1 - r)$, we get

$$\frac{\dfrac{3}{10}}{1 - \dfrac{1}{10}} = \frac{\dfrac{3}{10}}{\dfrac{9}{10}} = \frac{1}{3}$$

Similarly,

$$.2323\overline{23} = \frac{23}{100} + \frac{23}{10000} + \frac{23}{1000000} + \cdot \cdot \cdot$$

$$= \frac{\dfrac{23}{100}}{1 - \dfrac{1}{100}} = \frac{\dfrac{23}{100}}{\dfrac{99}{100}} = \frac{23}{99}$$

Use this method to express each of the following as the ratio of two integers.

19. $.11\overline{1}$ 20. $77\overline{7}$ 21. $.2525\overline{25}$

22. $.99\overline{9}$ 23. $1.234\overline{34}$ 24. $.341\overline{41}$

Miscellaneous Problems

25. If $a_n = 625(0.2)^{n-1}$, find a_1, a_2, a_3, a_4, and a_5.

26. Which of the following sequences are geometric, which are arithmetic, and which are neither?
    (a) 120, 24, 4.8, 0.96, . . .
    (b) .1, .02, .003, .0004, . . .
    (c) 4π, 2π, 0, -2π, . . .
    (d) 120, 60, -30, -15, 7.5, . . .
    (e) $100(1.08)$, $100(1.08)^2$, $100(1.08)^3$, . . .
    (f) $100(1.08)$, $100(1.10)$, $100(1.12)$, $100(1.14)$, . . .

27. Write an explicit formula for each geometric or arithmetic sequence in Problem 26.

28. Express $0.441\overline{441}$ as a ratio of positive integers in reduced form.

$\boxed{c}$ 29. Calculate each sum to three decimal places.
    (a) $\pi + \pi^2 + \pi^3 + \cdots + \pi^7$

    (b) $\displaystyle\sum_{k=1}^{\infty} \left(\frac{1}{\pi}\right)^k$

    (c) $1 + .982 + (.982)^2 + \cdots + (.982)^{99}$

    (d) $\displaystyle\sum_{k=0}^{\infty} \left(\frac{\sqrt{2}}{\sqrt{2} + 1}\right)^k$

$\boxed{c}$ *Recall from Section 6-3 that if a sum of P dollars is invested today at a compound rate of i per conversion period, then the accumulated value after n periods is given by $P(1 + i)^n$. The sequence of accumulated values*

$$P(1 + i), P(1 + i)^2, P(1 + i)^3, P(1 + i)^4, \ldots$$

is geometric with ratio $1 + i$. In Problems 30–33, write a formula for the answer and then use a calculator to evaluate it.

30. If \$1 is put in the bank at 8 percent interest compounded annually, it will be worth $(1.08)^n$ dollars after n years. How much will \$100 be worth after 10 years? When will the amount first exceed \$250?

31. If \$1 is put in the bank at 8 percent interest compounded quarterly, it will be worth $(1.02)^n$ dollars after n quarters. How much will \$100 be worth after 10 years (40 quarters)? When will the amount first exceed \$250?

32. Suppose Karen puts \$100 in the bank today and \$100 at the beginning of each of the following 9 years. If this money earns interest at 8 percent compounded annually, what will it be worth at the end of 10 years?

33. José makes 40 deposits of \$25 each in a bank at intervals of three months, making the first deposit today. If money earns interest at 8 percent compounded quarterly, what will it all be worth at the end of 10 years (40 quarters)?

34. Suppose the government pumps an extra billion dollars into the economy. Assume that each business and individual saves 25 percent of its income and spends the rest, so that of the initial one billion dollars, 75 percent is re-spent by individuals and businesses. Of that amount, 75 percent is spent, and so

forth. What is the total increase in spending due to the government action? (This is called the *multiplier effect* in economics.)

35. Try to explain why it does not make sense to talk about the total sum of a geometric sequence when $r \geq 1$ or $r \leq -1$.

36. Show that $.9\overline{9} = 1$.

37. Assume square $ABCD$ has sides of length 1 and that E, F, G, and H are midpoints of their respective sides. If the indicated pattern is continued indefinitely, what will be the area of the painted region?

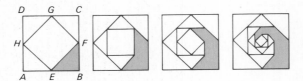

38. If the pattern shown below is continued indefinitely, what fraction of the original square will be painted?

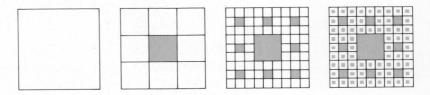

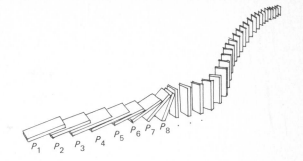

The Principle of Mathematical Induction

Let $P_1, P_2, P_3, \ldots$ be a sequence of statements with the following two properties:

1. P_1 is true.
2. The truth of P_k implies the truth of P_{k+1} $(P_k \Rightarrow P_{k+1})$.

Then the statement P_n is true for every positive integer n.

9-4 Mathematical Induction

The principle of mathematical induction deals with a sequence of statements. A **statement** is a sentence which is either true or false. In a sequence of statements, there is a statement corresponding to each positive integer. Here are four examples.

$$P_n: \quad \frac{1}{1 \cdot 2} + \frac{1}{2 \cdot 3} + \frac{1}{3 \cdot 4} + \cdots + \frac{1}{n(n+1)} = \frac{n}{n+1}$$

Q_n: $n^2 - n + 41$ is a prime number.

R_n: $(a + b)^n = a^n + b^n$

$$S_n: \quad 1 + 2 + 3 + \cdots + n = \frac{n^2 + n - 6}{2}$$

To be sure we understand the notation, let us write each of these statements for the case $n = 3$.

$$P_3: \quad \frac{1}{1 \cdot 2} + \frac{1}{2 \cdot 3} + \frac{1}{3 \cdot 4} = \frac{3}{4}$$

Q_3: $3^2 - 3 + 41$ is a prime number.

R_3: $(a + b)^3 = a^3 + b^3$

$$S_3: \quad 1 + 2 + 3 = \frac{3^2 + 3 - 6}{2}$$

n	$n^2 - n + 41$
1	41
2	43
3	47
4	53
5	61
6	71
7	83
.	.
.	.
.	.
40	1601
41	$1681 = 41^2$

Of these, P_3 and Q_3 are true, while R_3 and S_3 are false; you should verify this fact. A careful study of these four sequences will indicate the wide range of behavior that sequences of statements can display.

While it certainly is not obvious, we claim that P_n is true for every positive integer n; we are going to prove it soon. Q_n is a well-known sequence. It was thought by some to be true for all n and, in fact, it is true for $n = 1, 2, 3, \ldots,$ 40 (see the accompanying table). However, it fails for $n = 41$, a fact that allows us to make an important point. Establishing the truth of Q_n for a finite number of cases, no matter how many, does not prove its truth for *all* n.

Sequences R_n and S_n are rather hopeless cases, since R_n is true only for $n = 1$ and S_n is never true.

PROOF BY MATHEMATICAL INDUCTION

How does one prove that something is true for all n? The tool uniquely designed for this purpose is the **principle of mathematical induction;** it was stated in our opening display. Let us use mathematical induction to show that

$$P_n: \quad \frac{1}{1 \cdot 2} + \frac{1}{2 \cdot 3} + \frac{1}{3 \cdot 4} + \cdots + \frac{1}{(n-1)n} + \frac{1}{n(n+1)} = \frac{n}{n+1}$$

is true for every positive integer n. There are two steps to the proof. We must show that

1. P_1 is true;
2. $P_k \Rightarrow P_{k+1}$; that is, the truth of P_k implies the truth of P_{k+1}.

The first step is easy. P_1 is just the statement

$$\frac{1}{1 \cdot 2} = \frac{1}{1 + 1}$$

which is clearly true.

To handle the second step ($P_k \Rightarrow P_{k+1}$), it is a good idea to write down the statements corresponding to P_k and P_{k+1} (at least on scratch paper). We get them by substituting k and $k + 1$ for n in the statement for P_n.

$$P_k: \quad \frac{1}{1 \cdot 2} + \frac{1}{2 \cdot 3} + \cdots + \frac{1}{(k-1)k} + \frac{1}{k(k+1)} = \frac{k}{k+1}$$

$$P_{k+1}: \quad \frac{1}{1 \cdot 2} + \frac{1}{2 \cdot 3} + \cdots + \frac{1}{k(k+1)} + \frac{1}{(k+1)(k+2)} = \frac{k+1}{k+2}$$

Notice that the left side of P_{k+1} is the same as that of P_k except for the addition of one more term, $1/(k+1)(k+2)$.

Suppose for the moment that P_k is true and consider how this assumption allows us to simplify the left side of P_{k+1}.

$$\left[\frac{1}{1 \cdot 2} + \frac{1}{2 \cdot 3} + \cdots + \frac{1}{k(k+1)} \right] + \frac{1}{(k+1)(k+2)}$$

$$= \frac{k}{k+1} + \frac{1}{(k+1)(k+2)}$$

$$= \frac{k(k+2) + 1}{(k+1)(k+2)}$$

$$= \frac{(k+1)^2}{(k+1)(k+2)}$$

$$= \frac{k+1}{k+2}$$

If you read this chain of equalities from top to bottom, you will see that we have established the truth of P_{k+1}, but under the *assumption that P_k is true*. That is, we have established that the truth of P_k implies the truth of P_{k+1}.

SOME COMMENTS ABOUT MATHEMATICAL INDUCTION

Students never have any trouble with the verification step (showing that P_1 is true). The inductive step (showing that $P_k \Rightarrow P_{k+1}$) is harder and more subtle. In that step, we do *not* prove that P_k or P_{k+1} is true, but rather that the truth of P_k implies the truth of P_{k+1}. For a vivid illustration of the difference, we point out that in the fourth example of our opening paragraph, the truth of S_k does imply the truth of S_{k+1} ($S_k \Rightarrow S_{k+1}$) and yet not a single statement in that sequence is true (see Problems 31 and 32). To put it another way, what $S_k \Rightarrow S_{k+1}$ means is that *if* S_k were true, then S_{k+1} would be true also. It is like saying that if spinach were ice cream, then kids would want two helpings at every meal.

Perhaps the dominoes in the opening display can help illuminate the idea. For all the dominoes to fall it is sufficient that

1. the first domino is pushed over;
2. if any domino falls (say the kth one), it pushes over the next one (the $(k + 1)$st one).

The diagram on page 374 illustrates what happens to the dominoes in the four examples of our opening paragraph. Study them carefully.

ANOTHER EXAMPLE

Consider the statement

$$P_n: \quad 1^2 + 2^2 + 3^2 + \cdots + n^2 = \frac{n(n + 1)(2n + 1)}{6}$$

We are going to prove that P_n is true for all n by mathematical induction. For $n = 1$, k, and $k + 1$, the statements P_n are:

$$P_1: \quad 1^2 = \frac{1(2)(3)}{6}$$

$$P_k: \quad 1^2 + 2^2 + 3^2 + \cdots + k^2 = \frac{k(k + 1)(2k + 1)}{6}$$

$$P_{k+1}: \quad 1^2 + 2^2 + 3^2 + \cdots + k^2 + (k + 1)^2 = \frac{(k + 1)(k + 2)(2k + 3)}{6}$$

Clearly P_1 is true.

Assuming that P_k is true, we can write the left side of P_{k+1} as shown at the middle of the next page.

$P_n : \dfrac{1}{1 \cdot 2} + \dfrac{1}{2 \cdot 3} + \cdots + \dfrac{1}{n(n + 1)} = \dfrac{n}{n + 1}$ P_1 is true $P_k \Rightarrow P_{k + 1}$	$P_1P_2P_3P_4P_5P_6 \cdots$ First domino is pushed over. Each falling domino pushes over the next one.
$Q_n : n^2 - n + 41$ is prime. $Q_1, Q_2, \ldots, Q_{40}$ are true. $Q_k \nRightarrow Q_{k + 1}$	$Q_{35}\, Q_{36}\, Q_{37}\, Q_{38}\, Q_{39}\, Q_{40} \quad Q_{41} \quad Q_{42}$ First 40 dominoes are pushed over. 41st domino remains standing.
$R_n : (a + b)^n = a^n + b^n$ R_1 is true. $R_k \nRightarrow R_{k + 1}$	$R_1 \quad R_2 \quad R_3 \quad R_4 \quad R_5 \quad R_6 \quad R_7$ First domino is pushed over but dominoes are spaced too far apart to push each other over.
$S_n : 1 + 2 + 3 + \ldots + n = \dfrac{n^2 + n - 6}{2}$ S_1 is false. $S_k \Rightarrow S_{k + 1}$	$S_1S_2S_3S_4S_5$ Spacing is just right but no one can push over the first domino.

$$1^2 + 2^2 + 3^2 + \cdots + k^2 + (k + 1)^2 = \frac{k(k + 1)(2k + 1)}{6} + (k + 1)^2$$

$$= \frac{(k + 1)[k(2k + 1) + 6(k + 1)]}{6}$$

$$= \frac{(k + 1)(2k^2 + 7k + 6)}{6}$$

$$= \frac{(k + 1)(k + 2)(2k + 3)}{6}$$

Thus the truth of P_k does imply the truth of P_{k+1}. We conclude by mathematical induction that P_n is true for every positive integer n. Incidentally, the result just proved will be used in calculus.

Problem Set 9-4

In Problems 1–8 on the following page, prove by mathematical induction that P_n is true for every positive integer n.

1. P_n: $1 + 2 + 3 + \cdots + n = \dfrac{n(n+1)}{2}$

2. P_n: $1 + 3 + 5 + \cdots + (2n - 1) = n^2$

3. P_n: $3 + 7 + 11 + \cdots + (4n - 1) = n(2n + 1)$

4. P_n: $2 + 9 + 16 + \cdots + (7n - 5) = \dfrac{n(7n-3)}{2}$

5. P_n: $1 \cdot 2 + 2 \cdot 3 + 3 \cdot 4 + \cdots + n(n+1) = \frac{1}{3} n(n+1)(n+2)$

6. P_n: $\dfrac{1}{1 \cdot 3} + \dfrac{1}{3 \cdot 5} + \dfrac{1}{5 \cdot 7} + \cdots + \dfrac{1}{(2n-1)(2n+1)} = \dfrac{n}{2n+1}$

7. P_n: $2 + 2^2 + 2^3 + \cdots + 2^n = 2(2^n - 1)$

8. P_n: $1^2 + 3^2 + 5^2 + \cdots + (2n-1)^2 = \dfrac{n(2n-1)(2n+1)}{3}$

In Problems 9–18, tell what you can conclude from the information given about the sequence of statements. For example, if you are given that P_4 is true and that $P_k \Rightarrow P_{k+1}$ for any k, then you can conclude that P_n is true for every integer $n \geq 4$.

9. P_8 is true and $P_k \Rightarrow P_{k+1}$.

10. P_8 is not true and $P_k \Rightarrow P_{k+1}$.

11. P_1 is true but P_k does not imply P_{k+1}.

12. $P_1, P_2, \ldots, P_{1000}$ are all true.

13. P_1 is true and $P_k \Rightarrow P_{k+2}$.

14. P_{40} is true and $P_k \Rightarrow P_{k-1}$.

15. P_1 and P_2 are true; P_k and P_{k+1} together imply P_{k+2}.

16. P_1 and P_2 are true and $P_k \Rightarrow P_{k+2}$.

17. P_1 is true and $P_k \Rightarrow P_{4k}$.

18. P_1 is true, $P_k \Rightarrow P_{4k}$, and $P_k \Rightarrow P_{k-1}$.

EXAMPLE A (Mathematical induction applied to inequalities) Show that the following statement is true for every integer $n \geq 4$.

$$3^n > 2^n + 20$$

Solution. Let P_n represent the given statement. You might check that P_1, P_2, and P_3 are false. However, that does not matter to us. What we need to do is to show that P_4 is true and that $P_k \Rightarrow P_{k+1}$ for any $k \geq 4$.

$$P_4: \quad 3^4 > 2^4 + 20$$

$$P_k: \quad 3^k > 2^k + 20$$

$$P_{k+1}: \quad 3^{k+1} > 2^{k+1} + 20$$

Clearly P_4 is true (81 is greater than 36). Next we assume P_k to be true (for $k \geq 4$) and seek to show that this would force P_{k+1} to be true. Working with the left side of P_{k+1} and using the assumption that $3^k > 2^k + 20$, we get

$$3^{k+1} = 3 \cdot 3^k > 3(2^k + 20) > 2(2^k + 20) = 2^{k+1} + 40 > 2^{k+1} + 20$$

Therefore P_{k+1} is true, provided P_k is true. We conclude that P_n is true for every integer $n \geq 4$.

In Problems 19–24, find the smallest positive integer n for which the given statement is true. Then prove that the statement is true for all integers greater than that smallest value.

19. $n + 5 < 2^n$
20. $3n \leq 3^n$
21. $\log_{10} n < n$ (*Hint:* $k + 1 < 10k$.)
22. $n^2 \leq 2^n$ (*Hint:* $k^2 + 2k + 1 = k(k + 2 + 1/k) < k(k + k)$.)
23. $(1 + x)^n \geq 1 + nx$, where $x \geq -1$
24. $|\sin nx| \leq |\sin x| \cdot n$ for all x

EXAMPLE B (Mathematical induction and divisibility)   Prove that the statement

$$P_n: \quad x - y \text{ is a factor of } x^n - y^n$$

is true for every positive integer n.

Solution.   Trivially, $x - y$ is a factor of $x - y$ since $x - y = 1(x - y)$; so P_1 is true. Now suppose that P_k is true, that is, that

$$x - y \text{ is a factor of } x^k - y^k$$

This means that there is a polynomial $Q(x, y)$ such that

$$x^k - y^k = Q(x, y)(x - y)$$

Using this assumption, we may write

$$x^{k+1} - y^{k+1} = x^{k+1} - x^k y + x^k y - y^{k+1}$$
$$= x^k(x - y) + y(x^k - y^k)$$
$$= x^k(x - y) + yQ(x, y)(x - y)$$
$$= [x^k + yQ(x, y)](x - y)$$

Thus $x - y$ is a factor of $x^{k+1} - y^{k+1}$. We have shown that $P_k \Rightarrow P_{k+1}$ and that P_1 is true; we therefore conclude that P_n is true for all n.

Use mathematical induction to prove that each of the following is true for every positive integer n.

25. $x + y$ is a factor of $x^{2n} - y^{2n}$. (*Hint:* $x^{2k+2} - y^{2k+2} = x^{2k+2} - x^{2k}y^2 + x^{2k}y^2 - y^{2k+2}$.)
26. $x + y$ is a factor of $x^{2n-1} + y^{2n-1}$.
27. $n^2 - n$ is even (that is, has 2 as a factor).
28. $n^3 - n$ is divisible by 6.

Miscellaneous Problems

In Problems 29–30, use mathematical induction to prove that P_n is true for every positive integer n.

29. P_n:　$5 + 15 + 25 + \cdots + (10n - 5) = 5n^2$

30. P_n:　$(2n + 1)^2 - 1$ is divisible by 8.

31. Consider the statement

$$S_n:\quad 1 + 2 + 3 + \cdots + n = \frac{n^2 + n - 6}{2}$$

Show that $S_k \Rightarrow S_{k+1}$. Does this mean that S_n is true for any n? (See Problem 32.)

32. You know (Problem 1 or Section 9-2) that

$$1 + 2 + 3 + \cdots + n = \frac{n(n + 1)}{2}$$

Show that this means that S_n of Problem 31 is not true for any positive integer.

33. Use mathematical induction to prove

$$1^3 + 2^3 + 3^3 + \cdots + n^3 = \left[\frac{n(n + 1)}{2}\right]^2$$

34. Note that Problem 33 and the formula of Problem 32 imply that

$$1^3 + 2^3 + 3^3 + \cdots + n^3 = (1 + 2 + 3 + \cdots + n)^2$$

Check to see that this strange formula holds for $n = 1, 2, 3$, and 4.

35. Prove by mathematical induction that the sum formulas for arithmetic and geometric sequences are correct, that is, prove each of the following.

(a) $a + (a + d) + (a + 2d) + \cdots + (a + (n - 1)d)$
$= \frac{n}{2}[2a + (n - 1)d]$

(b) $a + ar + ar^2 + \cdots + ar^{n-1} = \frac{a(1 - r^n)}{1 - r}$　$(r \neq 1)$

36. Prove that the number of diagonals of an n-sided convex (that is, no holes or dents) polygon is $n(n - 3)/2$ for $n \geq 3$. The diagrams in the margin show the situation for $n = 4$ and $n = 5$.

37. Prove that the sum of the measures of the angles in an n-sided convex polygon is $(n - 2)180°$ for $n \geq 3$.

38. Prove that, for $n \geq 3$,

$$\frac{1}{n + 1} + \frac{1}{n + 2} + \cdots + \frac{1}{2n} > \frac{3}{5}$$

39. Prove that, for $n \geq 2$,

$$\left(1 - \frac{1}{4}\right)\left(1 - \frac{1}{9}\right)\left(1 - \frac{1}{16}\right)\cdots\left(1 - \frac{1}{n^2}\right) = \frac{n + 1}{2n}$$

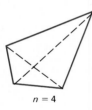

$n = 4$

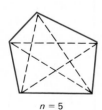

$n = 5$

40. Let $f_1 = 1$, $f_2 = 1$, and $f_n = f_{n-1} + f_{n-2}$ for $n \geq 2$ (this is the Fibonacci sequence of Section 9-1). Show by mathematical induction that

$$f_n = \frac{1}{\sqrt{5}}\left[\left(\frac{1+\sqrt{5}}{2}\right)^n - \left(\frac{1-\sqrt{5}}{2}\right)^n\right]$$

(*Hint:* In the inductive step, show that P_k and P_{k+1} together imply P_{k+2}.)

41. Let $F_n = f_1 + f_2 + \cdots + f_n$, where f_n is as in Problem 40. Show by mathematical induction that $F_n = f_{n+2} - 1$.

42. Let $a_0 = 0$, $a_1 = 1$, and $a_{n+2} = 2a_{n+1} - a_n$ for $n \geq 0$. Prove that $a_n = n$ for $n \geq 0$. (See the hint of Problem 40.)

43. Let $a_0 = 0$; $a_1 = 1$, and $a_{n+2} = (a_{n+1} + a_n)/2$ for $n \geq 0$. Prove that

$$a_n = \frac{2}{3}\left[1 - \left(-\frac{1}{2}\right)^n\right]$$

for $n \geq 0$.

44. What is wrong with the following argument?

Theorem. All horses in the world have the same color.

Proof. Let P_n be the statement: All the horses in any set of n horses are identically colored. Certainly P_1 is true. Suppose that P_k is true, that is, that all the horses in any set of k horses are identically colored. Let W be any set of $k + 1$ horses. Now we may think of W as the union of two overlapping sets X and Y, each with k horses. (The situation for $k = 4$ is shown in the margin.) By assumption, the horses in X are identically colored and the horses in Y are identically colored. Since X and Y overlap, all the horses in $X \cup Y$ must be identically colored. We conclude that P_n is true for all n. Thus the set of all horses in the world (some finite number) have the same color.

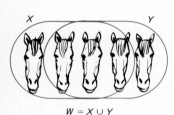

$W = X \cup Y$

9-5
Counting Ordered Arrangements

The Senior Birdwatchers' Club, consisting of 4 women and 2 men, is about to hold its annual meeting. In addition to having a group picture taken, they plan to elect a president, a vice president, and a secretary. Here are some questions that they (and we) might consider.

1. In how many ways can they line up for their group picture?
2. In how many ways can they elect their three officers if there are no restrictions as to sex?
3. In how many ways can they elect their three officers if the president is required to be female and the vice president male?
4. In how many ways can they elect their three officers if the president is to be of one sex and the vice president and secretary of the other?

In order to answer these questions and others of a similar nature, we need two counting principles. One involves multiplication; the other involves addition.

MULTIPLICATION PRINCIPLE IN COUNTING

Suppose that there are three roads a, b, and c leading from Clearwater to Longview and two roads m and n from Longview to Sun City. How many different routes can you choose from Clearwater to Sun City going through Longview? The diagram in the margin clarifies the situation. For each of the 3 choices from Clearwater to Longview, you have 2 choices from Longview to Sun City. Thus, you have $3 \cdot 2$ routes from Clearwater to Sun City. Here is the general principle.

Clearwater

a b c

Longview

m n

Sun City

Possible Routes

am	an
bm	bn
cm	cn

MULTIPLICATION PRINCIPLE

*Suppose that an event H can occur in h ways and, after it has occurred, event K can occur in k ways. Then the number of ways in which both H **and** K can occur is hk.*

This principle extends in an obvious way to three or more events.

Consider now the Birdwatchers' third question. It involves three events.

P: Elect a female president.
V: Elect a male vice-president.
S: Elect a secretary of either sex.

We understand that the election will take place in the order indicated and that no person can fill more than one position. Thus

P can occur in 4 ways   (there are 4 women);
V can occur in 2 ways   (there are 2 men);
S can occur in 4 ways   (after *P* and *V* occur, there are 4 people left from whom to select a secretary).

The entire selection process can be accomplished in $4 \cdot 2 \cdot 4 = 32$ ways.

PERMUTATIONS

To permute a set of objects is to rearrange them. Thus a **permutation** of a set of objects is an ordered arrangement of those objects. Take the set of letters in the word *FACTOR* as an example. Imagine that these 6 letters are printed on small cards so they can be arranged at will. Then we may form words like *COTARF, TRAFOC,* and *FRACTO,* none of which are in a dictionary but all of which are perfectly good words from our point of view. Let us call them code words. How many 6-letter code words can be made from the letters of the word *FACTOR;* that is, how many permutations of 6 objects are there?

Think of this as the problem of filling 6 slots.

We may fill the first slot in 6 ways. Having done that, we may fill the second slot in 5 ways, the third in 4 ways, and so on. By the multiplication principle, we can fill all six slots in

$$6 \cdot 5 \cdot 4 \cdot 3 \cdot 2 \cdot 1 = 720$$

ways.

Do you see that this is also the answer to the first question about the Birdwatchers, which asked in how many ways the 6 members could be ar-

---

Permutations of *ART*
ART
ATR
RAT
RTA
TAR
TRA

ranged for a group picture? Let us identify each person by a letter; the letters of *FACTOR* will do just fine. Then to arrange the Birdwatchers is to make a 6-letter code word out of *FACTOR*. It can be done in 720 ways.

What if we want to make 3-letter code words from the letters of the word *FACTOR,* words like *ACT, COF,* and *TAC*? How many such words can be made? This is the problem of filling 3 slots with 6 letters available. We can fill the first slot in 6 ways, then the second in 5 ways, and then the third in 4 ways. Therefore we can make $6 \cdot 5 \cdot 4 = 120$ 3-letter code words from the word *FACTOR*.

The number 120 is also the answer to the second question about the Birdwatchers. If there are no restrictions as to sex, they can elect a president, vice president, and secretary in $6 \cdot 5 \cdot 4 = 120$ ways.

Consider the corresponding general problem. Suppose that from n distinguishable objects, we select r of them and arrange them in a row. The resulting arrangement is called a **permutation of n things taken r at a time.** The number of such permutations is denoted by the symbol $_nP_r$. Thus

$$_6P_3 = 6 \cdot 5 \cdot 4 = 120$$

$$_6P_6 = 6 \cdot 5 \cdot 4 \cdot 3 \cdot 2 \cdot 1 = 720$$

$$_8P_2 = 8 \cdot 7 = 56$$

and in general

$$_nP_r = n(n - 1)(n - 2) \cdots (n - r + 2)(n - r + 1)$$

Notice that $_nP_r$ is the product of r consecutive positive integers starting with n and going down. In particular, $_nP_n$ is the product of n positive integers starting with n and going all the way down to 1, that is,

$$_nP_n = n(n - 1)(n - 2) \cdots 3 \cdot 2 \cdot 1$$

The symbol $n!$ (read **n factorial**) is also used for this product. Thus

$$_5P_5 = 5! = 5 \cdot 4 \cdot 3 \cdot 2 \cdot 1 = 120$$

$$_4P_4 = 4! = 4 \cdot 3 \cdot 2 \cdot 1 = 24$$

ADDITION PRINCIPLE IN COUNTING

We still have not answered the fourth Birdwatchers' question. In how many ways can they elect their three officers if the president is to be of one sex and the vice president and secretary of the other? This means that the president should be female and the other two officers male, *or* the president should be male and the other two female. To answer a question like this we need another principle.

ADDITION PRINCIPLE

*Let H and K be disjoint events, that is, events that cannot happen simultaneously. If H can occur in h ways and K in k ways, then H **or** K can occur in h + k ways.*

This principle generalizes to three or more disjoint events.

Applying this principle to the question at hand, we define H and K as follows.

> H: Elect a female president, male vice president, and male secretary.
> K: Elect a male president, female vice president, and female secretary.

Clearly H and K are disjoint. From the multiplication principle,

> H can occur in $4 \cdot 2 \cdot 1 = 8$ ways;
> K can occur in $2 \cdot 4 \cdot 3 = 24$ ways.

Then by the addition principle, H or K can occur in $8 + 24 = 32$ ways.

Here is another question that requires the addition principle. Consider again the letters of *FACTOR*, which we supposed were printed on 6 cards. How many code words of any length can we make using these 6 letters? We immediately translate this into 6 disjoint events: make 6-letter words, or 5-letter words, or 4-letter words, or 3-letter words, or 2-letter words, or 1-letter words. We can do this in the following number of ways.

$$_6P_6 + {_6P_5} + {_6P_4} + {_6P_3} + {_6P_2} + {_6P_1}$$

$$= 6 \cdot 5 \cdot 4 \cdot 3 \cdot 2 \cdot 1 + 6 \cdot 5 \cdot 4 \cdot 3 \cdot 2 + 6 \cdot 5 \cdot 4 \cdot 3 + 6 \cdot 5 \cdot 4 + 6 \cdot 5 + 6$$

$$= \quad 720 \quad + \quad 720 \quad + \quad 360 \quad + 120 + 30 + 6$$

$$= 1956$$

Students sometimes find it hard to decide whether to multiply or to add in a counting problem. Notice that the words **and** and **or** are in boldface type in the statements of the multiplication principle and of the addition principle. They are the key words; **and** goes with multiplication; **or** goes with addition.

Problem Set 9-5

1. Calculate.
   (a) 3!   (b) (3!)(2!)   (c) 10!/8!
2. Calculate.
   (a) 7!   (b) 7! + 5!   (c) 12!/9!
3. Calculate.
   (a) $_5P_2$ (b) $_9P_4$ (c) $_{10}P_3$
4. Calculate.
   (a) $_4P_3$ (b) $_8P_4$ (c) $_{20}P_3$

5. In how many ways can a president and a secretary be chosen from a group of 6 people?

6. Suppose that a club consists of 3 women and 2 men. In how many ways can a president and a secretary be chosen if
   (a) the president is to be female and the secretary male;
   (b) the president is to be male and the secretary female;
   (c) the president and secretary are to be of opposite sex?

7. A box contains 12 cards numbered 1 through 12. Suppose one card is drawn from the box. Find the number of ways each of the following can occur.
   (a) The number drawn is even.
   (b) The number is greater than 9 or less than 3.

8. Suppose that two cards are drawn in succession from the box in Problem 7. Assume that the first card is not replaced before the second one is drawn. In how many ways can each of the following occur?
   (a) Both numbers are even.
   (b) The two numbers are both even or both odd?
   (c) The first number is greater than 9 and the second one less than 3.

9. Do Problem 8 with the assumption that the first card is replaced before the second one is drawn.

10. In how many ways can a president, a vice-president, and a secretary be chosen from a group of 10 people?

11. How many 4-letter code words can be made from the letters of the word *EQUATION*? (Letters are not to be repeated.)

12. How many 3-letter code words can be made from the letters of the word *PROBLEM* if
    (a) letters cannot be repeated;
    (b) letters can be repeated?

13. Five roads connect Cheer City and Glumville. Starting at Cheer City, how many different ways can Smith drive to Glumville and return, that is, how many different round trips can he make? How many different round trips can he make if he wishes to return by a different road than he took to Glumville?

14. Filipe has 4 ties, 6 shirts, and 3 pairs of trousers. How many different outfits can he wear? Assume that he wears one of each kind of article.

15. Papa's Pizza Place offers 3 choices of salad, 20 kinds of pizza, and 4 different desserts. How many different 3-course meals can one order?

16. Minnesota license plate numbers consist of 3 letters followed by 3 digits (for example, AFF033). How many different plates could be issued? (You need not multiply out your answer.)

17. The letters of the word *CREAM* are printed on 5 cards. How many 3-, 4-, or 5-letter code words can be formed?

EXAMPLE A (Arrangements with side conditions)   Suppose that the letters of the word *COMPLEX* are printed on 7 cards. How many 3-letter code words can be formed from these letters if
(a) the first and last letters must be consonants (that is, C, M, P, L, or X);
(b) all vowels used (if any) must occur in the right-hand portion of a word (that is, a vowel cannot be followed by a consonant)?

Solution.

(a) Let *c* denote consonant, *v* vowel, and *a* any letter. We must fill the three slots below.

c   a   c

We begin by filling the two restricted slots, which can be done in $5 \cdot 4 = 20$ ways. Then we fill the unrestricted slot using one of the 5 remaining letters. It can be done in 5 ways. There are $20 \cdot 5 = 100$ code words of the required type. The diagram below summarizes the procedure.

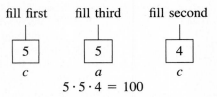

$$5 \cdot 5 \cdot 4 = 100$$

(b) We want to count words of the form *cvv*, *ccv*, or *ccc*. Note the use of the addition principle (as well as the multiplication principle) in the following solution.

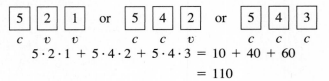

$$5 \cdot 2 \cdot 1 + 5 \cdot 4 \cdot 2 + 5 \cdot 4 \cdot 3 = 10 + 40 + 60$$
$$= 110$$

18. Using the letters of the word *FACTOR* (without repetition), how many 4-letter code words can be formed
    (a) starting with *R*;
    (b) with vowels in the two middle positions;
    (c) with only consonants;
    (d) with vowels and consonants alternating;
    (e) with all the vowels (if any) in the left-hand portion of a word (that is, a vowel cannot be preceded by a consonant)?

19. Using the letters of the word *EQUATION* (without repetition), how many 4-letter code words can be formed
    (a) starting with *T* and ending with *N*;
    (b) starting and ending with a consonant;
    (c) with vowels only;
    (d) with three consonants;
    (e) with all the vowels (if any) in the right-hand portion of the word?

20. Three brothers and 3 sisters are lining up to be photographed. How many arrangements are there
    (a) altogether;
    (b) with brothers and sisters in alternating positions;
    (c) with the 3 sisters standing together?

21. A baseball team is to be formed from a squad of 12 people. Two teams made up of the same 9 people are different if at least some of the people are assigned different positions. In how many ways can a team be formed if
    (a) there are no restrictions;
    (b) only 2 of the people can pitch and these 2 cannot play any other position;
    (c) only 2 of the people can pitch but they can also play any other position?

EXAMPLE B (Permutations with some indistinguishable objects)   Lucy has 3 identical red flags, 1 white flag, and 1 blue flag. How many different 5-flag signals could she display from the flagpole of her small boat?

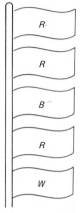

Solution.   If the 3 red flags were distinguishable, the answer would be $_5P_5 = 5! = 120$. Pretending they are distinguishable leads to counting an arrangement such as *RRBRW* six times, corresponding to the 3! ways of arranging the 3 red flags (see margin). For this reason, we must divide by 3!. Thus the number of signals Lucy can make is

$$\frac{5!}{3!} = \frac{5 \cdot 4 \cdot \not{3} \cdot \not{2} \cdot \not{1}}{\not{3} \cdot \not{2} \cdot \not{1}} = 20$$

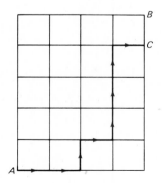

This result can be generalized. For example, given a set of *n* objects in which *j* are of one kind, *k* of a second kind, and *m* of a third kind, then the number of distinguishable permutations is

$$\frac{n!}{j! \; k! \; m!}$$

22. How many different signals consisting of 8 flags can be made using 4 white flags, 3 red flags, and 1 blue flag?
23. How many different signals consisting of 7 flags can be made using 3 white, 2 red, and 2 blue flags?
24. How many different 5-letter code words can be made from the 5 letters of the word *MIAMI*?
25. How many different 11-letter code words can be made from the 11 letters of the word *MISSISSIPPI*?
26. In how many different ways can a^4b^6 be written without using exponents? (*Hint:* One way is *aaaabbbbbb*.)
27. In how many different ways can a^3bc^6 be written without using exponents?
28. Consider the part of a city map shown in the margin. How many different shortest routes (no backtracking, no cutting across blocks) are there from *A* to *C*? Note that the route shown might be given the designation *EENENNNE*, with *E* denoting *East* and *N* denoting *North*.
29. How many different shortest routes are there from *A* to *B* (see Problem 28)?

Miscellaneous Problems

30. Calculate.
    • (a) $\dfrac{7!}{4!}$ (b) $\dfrac{7!}{4! \, 3!}$ (c) $7! - 4!$

31. The permutation symbol $_7P_3$ can be expressed in factorial notation as follows.

$$_7P_3 = \frac{_7P_3 \cdot 4!}{4!} = \frac{(7 \cdot 6 \cdot 5)(4 \cdot 3 \cdot 2 \cdot 1)}{4!} = \frac{7!}{4!}$$

In the same manner, express each of the following in terms of factorials.
   (a) $_5P_3$ (b) $_{10}P_8$ (c) $_8P_2$

32. Write $_nP_r$ in factorial notation. Assume $r < n$. (See Problem 31.)

33. Six horses run in a race.
   (a) How many different orders of finishing are there?
   (b) How many possibilities are there for the first three places?

34. In how many ways can first and second prizes be awarded in a baking competition in which there are 8 entries?

35. (a) In how many different ways can the letters of the word *CYCLIC* be arranged?
   (b) In how many of these arrangements are the three *C*'s in consecutive positions?

36. In how many different ways can Manuel select a dinner roll, a salad, a main entrée, and a dessert if there are 2 kinds of dinner rolls, 3 salads, 4 main entrées, and 3 desserts?

37. In how many different ways can a true-false test of 10 questions be answered?

38. The letters of the word *WRONG* are written on 5 cards. How many code words can be made? (Count 1-, 2-, 3-, 4-, and 5-letter words.)

39. How many 7-digit phone numbers are possible, assuming the first digit cannot be 0?

40. How many 4-digit numbers are there that use only the digits 1, 2, 3, 4, 5? How many of these are even? (*Note:* The question is worded in a way that allows repetition of digits.)

41. How many 3-digit numbers are there that use only the digits 0, 1, 2, 3, 4? (Be careful; a number cannot start with 0.)

42. In how many ways can 6 people be seated at a round table? (*Note:* Moving each person one place to the right (or left) does not constitute a new arrangement.)

43. A husband and wife invite 4 couples to dinner. The dinner table is rectangular. They decide on a seating arrangement in which the hostess will sit at the end nearest the kitchen, the host at the opposite end, and 4 guests on each side. Furthermore, no man shall sit next to another man, nor shall he sit next to his own wife. In how many ways can this be done?

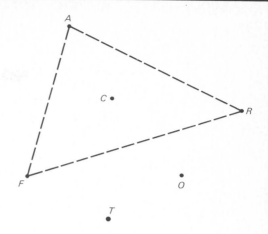

How Many Triangles?

Consider six points in the plane, no three on the same line. Label them *F, A, C, T, O, R*. How many triangles can be drawn using these points as vertices?

9-6
Counting
Unordered
Collections

For each choice of three points in the opening display, we can draw a triangle. But notice that the order in which we choose the three points does not matter. For example, *FAR*, *FRA*, *AFR*, *ARF*, *RAF*, and *RFA* all determine the same triangle, namely, the one that is shown by dotted lines in the picture. The question about triangles is very different from the question about 3-letter code words raised in the last section; yet there is a connection.

We learned that we can make

$$_6P_3 = 6 \cdot 5 \cdot 4 = 120$$

3-letter code words out of the letters of *FACTOR*. However, every triangle can be labeled with $3! = 6$ different code words. To find the number of triangles, we should therefore divide the number of code words by $3!$. We conclude that the number of triangles that can be drawn is

$$\frac{_6P_3}{3!} = \frac{6 \cdot 5 \cdot 4}{3 \cdot 2 \cdot 1} = 20$$

COMBINATIONS

An unordered collection of objects is called a **combination** of those objects. If we select *r* objects from a set of *n* distinguishable objects, the resulting subset is called a **combination of *n* things taken *r* at a time.** The number of such combinations is denoted by $_nC_r$. Thus $_6C_3$ is the number of combinations of 6 things taken 3 at a time. We calculated it in connection with the triangle problem.

$$_6C_3 = \frac{_6P_3}{3!} = \frac{6 \cdot 5 \cdot 4}{3 \cdot 2 \cdot 1} = 20$$

More generally, if $1 \leq r \leq n$, the combination symbol $_nC_r$ is given by

$$_nC_r = \frac{_nP_r}{r!} = \frac{n(n-1)(n-2) \cdots (n-r+2)(n-r+1)}{r(r-1)(r-2) \cdots 2 \cdot 1}$$

A good way to remember this is that you want r factors in both numerator and denominator. In the numerator, you start with n and go down; in the denominator you start with r and go down. Incidentally, the answer must be an integer. That means the denominator has to divide the numerator. Here are some more examples.

$$_5C_2 = \frac{5 \cdot \overset{2}{\cancel{4}}}{\cancel{2} \cdot 1} = 10$$

$$_{12}C_4 = \frac{\cancel{12} \cdot 11 \cdot \overset{5}{\cancel{10}} \cdot 9}{\cancel{4} \cdot \cancel{3} \cdot \cancel{2} \cdot 1} = 495$$

$$_{10}C_8 = \frac{\overset{5}{\cancel{10}} \cdot 9 \cdot \cancel{8} \cdot \cancel{7} \cdot \cancel{6} \cdot \cancel{5} \cdot \cancel{4} \cdot \cancel{3}}{\cancel{8} \cdot \cancel{7} \cdot \cancel{6} \cdot \cancel{5} \cdot \cancel{4} \cdot \cancel{3} \cdot \cancel{2} \cdot 1} = 45$$

$$_{10}C_2 = \frac{\overset{5}{\cancel{10}} \cdot 9}{\cancel{2} \cdot 1} = 45$$

Notice that $_{10}C_8 = {}_{10}C_2$. This is not surprising, since in selecting a subset of 8 objects out of 10, you automatically select 2 to leave behind (you might call it selection by omission). The general fact that follows by the same reasoning is

$$_nC_r = {}_nC_{n-r}$$

We can express $_nC_r$ entirely in terms of factorials. Recall that

$$_nC_r = \frac{n(n-1)(n-2) \cdots (n-r+1)}{r!}$$

If we multiply both numerator and denominator by $(n-r)!$, we get

$$_nC_r = \frac{n!}{r! \, (n-r)!}$$

For this formula to hold when $r = 0$ and $r = n$, it is necessary to define $_nC_0 = 1$ and $0! = 1$. Even in mathematics, there is truth to that old proverb "Necessity is the mother of invention."

COMBINATIONS VERSUS PERMUTATIONS

Whenever we are faced with the problem of counting the number of ways of selecting r objects from n objects, we are faced with this question. Is the notion of order significant? If the answer is yes, it is a permutation problem; if no, it is a combination problem.

Consider the Birdwatchers' Club of Section 9-5 again. Suppose the club members want to select a president, a vice president, and a secretary. Is order significant? Yes. The selection can be done in

$$_6P_3 = 6 \cdot 5 \cdot 4 = 120$$

ways.

But suppose they decide simply to choose an executive committee consisting of 3 members. Is order relevant? No. A committee consisting of Filipe, Celia, and Amanda is the same as a committee consisting of Celia, Amanda, and Filipe. A 3-member committee can be chosen from 6 people in

$$_6C_3 = \frac{6 \cdot 5 \cdot 4}{3 \cdot 2 \cdot 1} = 20$$

ways.

The words *arrangement, lineup,* and *signal* all suggest order. The words *set, committee, group,* and *collection* do not.

Problem Set 9-6

1. Calculate each of the following.
   (a) $_{10}P_3$ (b) $_{10}C_3$
   (c) $_5P_5$ (d) $_5C_5$
   (e) $_6P_1$ (f) $_6C_1$

2. Calculate each of the following.
   (a) $_{12}P_2$ (b) $_{12}C_2$
   (c) $_4P_4$ (d) $_4C_4$
   (e) $_{10}P_1$ (f) $_{10}C_1$

3. Use the fact that $_nC_r = {_nC_{n-r}}$ to calculate each of the following.
   (a) $_{20}C_{17}$ (b) $_{100}C_{97}$

4. Calculate each of the following.
   (a) $_{41}C_{39}$ (b) $_{1000}C_{998}$
   (Hint: See Problem 3.)

5. In how many ways can a committee of 3 be selected from a class of 8 students?

6. In how many ways can a committee of 5 be selected from a class of 8 students?

7. A political science professor must select 4 students from her class of 12 students for a field trip to the state legislature. In how many ways can she do it?

8. The professor of Problem 7 was asked to rank the top 4 students in her class of 12. In how many ways could that be done?

9. A police chief needs to assign officers from the 10 available to control traffic at junctions A, B, and C. In how many ways can he do it?

10. If 12 horses are entered in a race, in how many ways can the first 3 places (win, place, show) be taken?

11. From a class of 6 members, in how many ways can a committee of any size be selected (including a committee of one)?

12. From a penny, a nickel, a dime, a quarter, and a half dollar, how many different sums can be made?

EXAMPLE A (More on committees)   A committee of 4 is to be selected from a group of 3 seniors, 4 juniors, and 5 sophomores. In how many ways can it be done if
(a) there are no restrictions on the selection;
(b) the committee must have 2 sophomores, 1 junior, and 1 senior;
(c) the committee must have at least 3 sophomores;
(d) the committee must have at least 1 senior?

Solution.

(a) $_{12}C_4 = \dfrac{\overset{5}{\cancel{12}} \cdot 11 \cdot \cancel{10} \cdot 9}{\cancel{4} \cdot \cancel{3} \cdot \cancel{2} \cdot 1} = 495$

(b) Two sophomores can be chosen in $_5C_2$ ways, 1 junior in $_4C_1$ ways, and 1 senior in $_3C_1$ ways. By the multiplication principle of counting, the committee can be chosen in

$$_5C_2 \cdot {_4C_1} \cdot {_3C_1} = 10 \cdot 4 \cdot 3 = 120$$

ways. We used the multiplication principle because we choose 2 sophomores *and* 1 junior *and* 1 senior.

(c) At least 3 sophomores means 3 sophomores and 1 nonsophomore *or* 4 sophomores. The word *or* tells us to use the addition principle of counting. We get

$$_5C_3 \cdot {_7C_1} + {_5C_4} = 10 \cdot 7 + 5 = 75$$

(d) Let x be the number of selections with at least one senior and let y be the number of selections with no seniors. Then $x + y$ is the total number of selections, that is, $x + y = 495$ (see part (a)). We calculate y rather than x because it is easier.

$$y = {_9C_4} = \dfrac{9 \cdot 8 \cdot 7 \cdot 6}{4 \cdot 3 \cdot 2 \cdot 1} = 126$$

$$x = 495 - 126 = 369$$

13. An investment club has a membership of 4 women and 6 men. A research committee of 3 is to be formed. In how many ways can this be done if
(a) there are to be 2 women and 1 man on the committee;
(b) there is to be at least 1 woman on the committee;
(c) all 3 are to be of the same sex?

14. A senate committee of 4 is to be formed from a group consisting of 5 Republicans and 6 Democrats. In how many ways can this be done if
    (a) there are to be 2 Republicans and 2 Democrats on the committee;
    (b) there are to be no Republicans on the committee;
    (c) there is to be at most one Republican on the committee?

15. Suppose that a bag contains 4 black and 7 white balls. In how many ways can a group of 3 balls be drawn from the bag consisting of
    (a) 1 black and 2 white balls;
    (b) balls of just one color;
    (c) at least 1 black ball?
    (*Note:* Assume the balls are distinguishable; for example, they may be numbered.)

16. John is going on a vacation trip and wants to take 5 books with him from his personal library, which consists of 6 science books and 10 novels. In how many ways can he make his selection if we wants to take
    (a) 2 science books and 3 novels;
    (b) at least 1 science book;
    (c) 1 book of one kind and 4 books of the other kind?

EXAMPLE B (Bridge card problems)   a standard deck consists of 52 cards. There are 4 suits (spades, clubs, hearts, diamonds), each with 13 cards $(2, 3, 4, \ldots, 10$, jack, queen, king, ace). A bridge hand consists of 13 cards.
(a) How many different possible bridge hands are there?
(b) How many of them have exactly 3 aces?
(c) How many of them have no aces?
(d) How many of them have cards from just 3 suits?

Solution.

(a) The order of the cards in a hand is irrelevant; it is a combination problem. We can select 13 cards out of 52 in $_{52}C_{13}$ ways, a number so large we will not bother to calculate it.

(b) The three aces can be selected in $_4C_3$ ways, the 10 remaining cards in $_{48}C_{10}$ ways. The answer (using the multiplication principle) is $_4C_3 \cdot {}_{48}C_{10}$.

(c) From 48 non-aces, we select 13 cards; the answer is $_{48}C_{13}$.

(d) We think of this as no clubs, or no spades, or no hearts, or no diamonds and use the addition principle.

$$_{39}C_{13} + {}_{39}C_{13} + {}_{39}C_{13} + {}_{39}C_{13} = 4 \cdot {}_{39}C_{13}$$

Problems 17–22 deal with bridge hands. Leave your answers in terms of combination symbols.

17. How many of the hands have only red cards? (*Note:* Half of the cards are red.)

18. How many of the hands have only honor cards (aces, kings, queens, and jacks)?

19. How many of the hands have one card of each kind (one ace, one king, one queen, and so on)?
20. How many of the hands have exactly 2 kings?
21. How many of the hands have 2 or more kings?
22. How many of the hands have exactly 2 aces and 2 kings?

Problems 23–26 deal with poker hands, which consist of 5 cards.

23. How many different poker hands are possible?
24. How many of them have exactly 2 hearts and 2 diamonds?
25. How many have 2 pairs of different kinds (for example, 2 aces and 2 fives)?
26. How many are 5-card straights (for example, 7, 8, 9, 10, jack)? An ace may count either as the highest or the lowest card, that is, as 1 or 13.

Miscellaneous Problems

27. From a committee of 17 members, in how many ways can a subcommittee of 3 be chosen?
28. If 5 distinguishable coins are tossed, in how many ways can they fall?
29. In how many ways can a student select 4 college courses from a set of 9 courses?
30. A man drives from town A to town B and then returns by a different route. If there are 5 routes between the two towns, how many choices of round trips does he have?
31. From a cent, a nickel, a dime, a quarter, and a half dollar, how many sums can be formed consisting of
    (a) 3 coins each;
    (b) 2 coins each;
    (c) at least 3 coins each?
32. From a group of 5 representatives of labor, 4 representatives of business, and 7 representatives of the general public, how many mediation committees can be formed consisting of 2 people from each group?
33. In how many ways can 9 presents be distributed to 3 children, if each is to receive 3 presents?
34. A committee of 6 is to be formed from a group of 5 freshmen, 6 sophomores, and 8 seniors. In how many ways can this be done if
    (a) there shall be 2 members of each class on the committee;
    (b) there shall be no freshmen on the committee;
    (c) there shall be exactly 4 seniors on the committee;
    (d) there shall be 3 members of each of two classes on the committee?
35. Calculate each of the following.
    (a) $_2C_0 + {_2C_1} + {_2C_2}$
    (b) $_3C_0 + {_3C_1} + {_3C_2} + {_3C_3}$
    (c) $_4C_0 + {_4C_1} + {_4C_2} + {_4C_3} + {_4C_4}$
    Now conjecture a formula for
    $$_nC_0 + {_nC_1} + {_nC_2} + \cdots + {_nC_n}$$

36. Use the factorial formula for $_nC_r$ to prove that

$$_{n+1}C_r = {}_nC_{r-1} + {}_nC_r$$

37. A class of 10 will elect a president, a secretary, and a social committee of 3 with no overlapping of positions. In how many ways can it be done?

38. A test consists of 10 true-false items.
    (a) How many different sets of answers are possible?
    (b) How many of them have exactly 4 answers right?

39. An ice cream parlor has 10 different flavors. How many different double dip cones can be made if
    (a) the two dips must be of different flavors but the order of putting them on the cone does not matter;
    (b) the two dips must be different and order does matter;
    (c) the two dips need not be different but order matters;
    (d) the two dips need not be different and order does not matter?

40. How many positive integers less than 1,000,000 involve only the digits 1 and 2?

41. Two distinguishable dice are tossed.
    (a) In how many ways can they fall?
    (b) How many of them give a total of 7?
    (c) How many give a total less than 7?

42. Answer the questions of Problem 41 if 3 distinguishable dice are tossed.

43. In how many ways can a class of 6 girls and 6 boys be seated in a room with 12 chairs if the boys must take the odd-numbered seats?

Raising a Binomial to a Power

$$(x + y)(x + y)(x + y)(x + y) = xxxx + xxxy + xxyy + xyyy + yyyy$$
$$+ xxyx + xyxy + yxyy$$
$$+ xyxx + \boxed{xyyx} + yyxy$$
$$+ yxxx + yxxy + yyyx$$
$$+ yxyx$$
$$+ yyxx$$

$$(x + y)^4 = x^4 + 4x^3y + 6x^2y^2 + 4xy^3 + y^4$$
$$(x + y)^4 = {}_4C_0x^4y^0 + {}_4C_1x^3y^1 + {}_4C_2x^2y^2 + {}_4C_3x^1y^3 + {}_4C_4x^0y^4$$

9-7
The Binomial Formula

In the opening box, we have shown how to expand $(x + y)^4$. Admittedly, it looks complicated; however, it leads to the remarkable formula at the bottom of the display. That formula generalizes to handle $(x + y)^n$, where n is any positive integer. It is worth a careful investigation.

To produce any given term in the expansion of $(x + y)^4$, each of the four factors $x + y$ contributes either an x or a y. There are $2 \cdot 2 \cdot 2 \cdot 2 = 16$ ways in which they can make this contribution, hence the 16 terms in the long expanded form. But many of these terms are alike; in fact, only five different types occur, namely, x^4, x^3y, x^2y^2, xy^3, and y^4. The number of times each occurs is ${}_4C_0$, ${}_4C_1$, ${}_4C_2$, ${}_4C_3$, and ${}_4C_4$, respectively. (Remember we defined ${}_4C_0$ to be 1.)

Why do the combination symbols arise in this expansion? For example, why is ${}_4C_2$ the coefficient of x^2y^2? If you follow the arrows in the opening display, you see how the term $xyyx$ comes about. It gets its two y's from the second and third $x + y$ factors (the x's then must come from the first and fourth $x + y$ factors). Thus, the number of terms of the form x^2y^2 is the number of ways of selecting two factors out of four from which to take y's (the x's must come from the remaining two factors). We can select two objects out of four in ${}_4C_2$ ways; hence the coefficient of x^2y^2 is ${}_4C_2$.

THE BINOMIAL FORMULA

What we have just done for $(x + y)^4$ can be carried out for $(x + y)^n$, where n is any positive integer. The result is called the **binomial formula.**

$$(x + y)^n = {}_nC_0x^ny^0 + {}_nC_1x^{n-1}y^1 + \cdots + {}_nC_{n-1}x^1y^{n-1} + {}_nC_nx^0y^n$$

$$= \sum_{k=o}^{n} {}_nC_kx^{n-k}y^k$$

Notice that the k in ${}_nC_k$ is the exponent on y and that the two exponents in each term sum to n.

Let us apply the binomial formula with $n = 6$.

$$(x + y)^6 = {}_6C_0x^6 + {}_6C_1x^5y + {}_6C_2x^4y^2 + \cdots + {}_6C_6y^6$$
$$= x^6 + 6x^5y + 15x^4y^2 + 20x^3y^3 + 15x^2y^4 + 6xy^5 + y^6$$

This same result applies to the expansion of $(2a - b^2)^6$. We simply think of $2a$ as x and $-b^2$ as y. Thus

$$[2a + (-b^2)]^6 = (2a)^6 + 6(2a)^5(-b^2) + 15(2a)^4(-b^2)^2 + 20(2a)^3(-b^2)^3$$
$$+ 15(2a)^2(-b^2)^4 + 6(2a)(-b^2)^5 + (-b^2)^6$$
$$= 64a^6 - 192a^5b^2 + 240a^4b^4 - 160a^3b^6$$
$$+ 60a^2b^8 - 12ab^{10} + b^{12}$$

THE BINOMIAL COEFFICIENTS

The combination symbols ${}_nC_k$ are often called **binomial coefficients,** for reasons that should be obvious. Their remarkable properties have been studied for hundreds of years. Let us see if we can discover some of them. We begin by expanding $(x + y)^k$ for increasing values of k, listing only the coefficients.

The triangle of numbers composed of the binomial coefficients is called **Pascal's triangle** after the gifted French philosopher and mathematician, Blaise Pascal (1622–1662). Notice its symmetry. If folded across a vertical

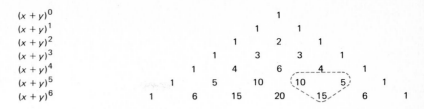

$(x + y)^0$
$(x + y)^1$
$(x + y)^2$
$(x + y)^3$
$(x + y)^4$
$(x + y)^5$
$(x + y)^6$

center line, the numbers match. This corresponds to an algebraic fact you learned earlier.

$$_nC_k = {}_nC_{n-k}$$

Next notice that any number in the body of the triangle is the sum of the two numbers closest to it in the line above the number. For example, $15 = 10 + 5$, as the dotted triangle

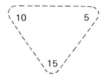

was meant to suggest. In symbols

$$_{n+1}C_k = {}_nC_{k-1} + {}_nC_k$$

a fact that can be proved rigorously by using the factorial formula for $_nC_k$ (see Problem 36 of Section 9-6).

Now add the numbers in each row of Pascal's triangle.

$$1 = 1 = 2^0$$

$$1 + 1 = 2 = 2^1$$

$$1 + 2 + 1 = 4 = 2^2$$

$$1 + 3 + 3 + 1 = 8 = 2^3$$

This suggests the formula

$$_nC_0 + {}_nC_1 + {}_nC_2 + \cdots + {}_nC_n = 2^n$$

Its truth can be demonstrated by substituting $x = 1$ and $y = 1$ in the binomial formula. It has an important interpretation. Take a set with n elements. This set has $_nC_n$ subsets of size n, $_nC_{n-1}$ subsets of size $n - 1$, and so on. The left side of the boxed formula is the total number of subsets (including the empty set) of a set with n elements; remarkably, it is just 2^n. For example, $\{a, b, c\}$ has $2^3 = 8$ subsets.

$$\{a, b, c\} \quad \{a, b\} \quad \{a, c\} \quad \{b, c\} \quad \{a\} \quad \{b\} \quad \{c\} \quad \emptyset$$

A related formula is

$$_nC_0 - {_nC_1} + {_nC_2} - \cdots + (-1)^n {_nC_n} = 0$$

which can be obtained by setting $x = 1$ and $y = -1$ in the binomial formula. When rewritten as

$$_nC_0 + {_nC_2} + {_nC_4} + \cdots = {_nC_1} + {_nC_3} + {_nC_5} + \cdots$$

it says that the number of subsets with an even number of elements is equal to the number of subsets with an odd number of elements.

Problem Set 9-7

In Problems 1–6, expand and simplify.

1. $(x + y)^3$
2. $(x - y)^3$
3. $(x - 2y)^3$
4. $(3x + b)^3$
5. $(c^2 - 3d^3)^4$
6. $(xy - 2z^2)^4$

Write the first three terms of each expansion in Problems 7–10 in simplified form.

7. $(x + y)^{20}$
8. $(x + y)^{30}$
9. $\left(x + \dfrac{1}{x^5}\right)^{20}$
10. $\left(xy^2 + \dfrac{1}{y}\right)^{14}$

11. Find the number of subsets of each of the following sets.
    (a) $\{a, b, c, d\}$
    (b) $\{1, 2, 3, 4, 5\}$
    (c) $\{x_1, x_2, x_3, x_4, x_5, x_6\}$

EXAMPLE A (Finding a specific term of a binomial expansion)   Find the term in the expansion of $(2x + y^2)^{10}$ that involves y^{12}.

Solution.   This term will arise from raising y^2 to the 6th power. It is therefore

$$_{10}C_6(2x)^4(y^2)^6 = 210 \cdot 16x^4y^{12} = 3360x^4y^{12}$$

12. Find the term in the expansion of $(y^2 - z^3)^{10}$ that involves z^9.
13. Find the term in the expansion of $(3x - y^3)^{10}$ that involves y^{24}.
14. Find the term in the expansion of $(2a - b)^{12}$ that involves a^3.
15. Find the term in the expansion of $(x^2 - 2/x)^5$ that involves x^4.

EXAMPLE B (An application to compound interest)   If \$100 is invested at 12 percent compounded monthly, it will accumulate to $100(1.01)^{12}$ dollars by the end of one year. Use the binomial formula to approximate this amount.

Solution.

$100(1.01)^{12} = 100(1 + .01)^{12}$

$$= 100\left[1 + 12(.01) + \frac{12 \cdot 11}{2}(.01)^2 + \frac{12 \cdot 11 \cdot 10}{6}(.01)^3 + \cdots \right]$$

$$= 100[1 + .12 + .0066 + .00022 + \cdots]$$

$$\approx 100(1.12682) \approx 112.86$$

This answer of $112.68 is accurate to the nearest penny since the last nine terms of the expansion do not add up to as much as a penny.

In Problems 16–19 use the first three terms of a binomial expansion to find an approximate value of the given expression.

16. $20(1.02)^8$ 17. $100(1.002)^{20}$ 18. $500(1.005)^{20}$ 19. $200(1.04)^{10}$

20. Bacteria multiply in a certain medium so that by the end of k hours their number N is $N = 100(1.02)^k$. Approximate the number of bacteria after 20 hours.

21. Do Problem 20 assuming $N = 1000(1.01)^k$.

Miscellaneous Problems

In Problems 22–23, expand and simplify.

22. $(x + h)^4$

23. $\left(2x^3 - \dfrac{1}{x}\right)^5$

In Problems 24 and 25, write and simplify the first three terms of the expansion.

24. $(a^2 + 2b)^{15}$

25. $(b^3 + 3c)^{12}$

26. Simplify the following expression.

$$[2(x + h)^4 - 3(x + h)^2] - [2x^4 - 3x^2]$$

27. Let $A = \{a_1, a_2, \ldots, a_n\}$ be a set with n elements. To make a subset B is to decide for each element a_k whether or not it goes into B. Show why this approach also leads to the conclusion that A has 2^n subsets.

28. Calculate $(0.99)^{10}$ accurate to six decimal places using the binomial formula.

29. Show that $(1.0003)^{10} > 1.003004$.

30. How many committees consisting of 2 or more members can be selected from a group of 10 people? (*Hint:* Think of subsets.)

31. How many selections of 1 or more books can be made from a set of 8 books?

32. Use the binomial formula to expand $(x^2 + x + 1)^3$. (*Hint:* Write $x^2 + x + 1$ as $x^2 + (x + 1)$.)

33. In the expansion of the trinomial $(x + y + z)^n$, the coefficient of $x^r y^s z^t$, where $r + s + t = n$, is

$$\frac{n!}{r! \, s! \, t!}$$

What is the coefficient of the term $x^2 y^4 z$ in the expansion of $(x + y + z)^7$?

34. Using the information given in Problem 33, write the term of the expansion of $(x + y + z)^{10}$ that involves y^2z^3.

CHAPTER SUMMARY

A **number sequence** $a_1, a_2, a_3, \ldots$ is a function that associates with each positive integer n a number a_n. Such a sequence may be described by an **explicit formula** (for instance, $a_n = 2n + 1$), by a **recursion formula** (for instance, $a_n = 3a_{n-1}$), or by giving enough terms so a pattern is evident (for instance, 1, 11, 21, 31, 41, . . .).

If any term in a sequence can be obtained by adding a fixed number d to the preceding term, we call it an **arithmetic sequence.** There are three key formulas associated with this type of sequence.

$$\text{Recursion formula:} \quad a_n = a_{n-1} + d$$

$$\text{Explicit formula:} \quad a_n = a_1 + (n - 1)d$$

$$\text{Sum formula:} \quad A_n = \frac{n}{2}(a_1 + a_n)$$

In the last formula, A_n represents

$$A_n = a_1 + a_2 + \cdots + a_n = \sum_{i=1}^{n} a_i$$

A **geometric sequence** is one in which any term results from multiplying the previous term by a fixed number r. The corresponding key formulas are

$$\text{Recursion formula:} \quad a_n = ra_{n-1}$$

$$\text{Explicit formula:} \quad a_n = a_1 r^{n-1}$$

$$\text{Sum formula:} \quad A_n = \frac{a_1(1 - r^n)}{1 - r}, \, r \neq 1$$

In the last formula, we may ask what happens as n grows larger and larger. If $|r| < 1$, the value of A_n gets closer and closer to $a_1/(1 - r)$, which we regard as the sum of *all* the terms of the sequence.

Often in mathematics, we wish to demonstrate that a whole **sequence of statements** P_n is true. For this, a powerful tool is the **principle of mathematical induction,** which asserts that if P_1 is true and if the truth of P_k implies the truth of P_{k+1}, then all the statements of the sequence are true.

Given enough time, anyone can count the number of elements in a set. But if the set consists of arrangements of objects (for example, the letters in a word), the work can be greatly simplified by using two principles, the **multiplication principle** and the **addition principle.** Of special interest are the number of **permutations** (ordered arrangements) of n things taken r at a time and the number of **combinations** (unordered collections) of n things taken r at a time. They can be calculated from the formulas

$$_nP_r = n(n - 1)(n - 2) \cdots (n - r + 1)$$

$$_nC_r = \frac{_nP_r}{r!} = \frac{n(n - 1)(n - 2) \cdots (n - r + 1)}{r(r - 1)(r - 2) \cdots 1}$$

One important use of the symbol $_nC_r$ is in the **binomial formula**

$$(x + y)^n = {_nC_0}x^ny^0 + {_nC_1}x^{n-1}y^1 + \cdots + {_nC_{n-1}}x^1y^{n-1} + {_nC_n}x^0y^n$$

CHAPTER REVIEW PROBLEM SET

Problems 1–6 refer to the sequences below.

(a) 2, 5, 8, 11, 14, . . .
(b) 2, 6, 18, 54, . . .
(c) 2, 1.5, 1, 0.5, 0, . . .
(d) 2, 4, 6, 10, 16, . . .
(e) 2, $\frac{2}{3}$, $\frac{2}{9}$, $\frac{2}{27}$, $\frac{2}{81}$, . . .

1. Which of these sequences are arithmetic? Which are geometric?
2. Give a recursion formula for each of the sequences (a)–(e).
3. Give an explicit formula for sequences (a) and (b).
4. Find the sum of the first 67 terms of sequence (a).
5. Write a formula for the sum of the first 100 terms of sequence (b).
6. Find the sum of *all* the terms of sequence (e).
7. If $a_n = 3a_{n-1} - a_{n-2}$, $a_1 = 1$, and $a_2 = 2$, find a_6.
8. If $a_n = n^2 - n$, find $A_5 = \Sigma_{n=1}^{5} a_n$.
9. Calculate $2 + 4 + 6 + 8 + \cdots + 1000$.
10. Write $.55\overline{55}$ as a ratio of two integers.
11. If \$100 is put in the bank today earning 8 percent interest compounded quarterly, write a formula for its value at the end of 12 years.
12. Show by mathematical induction that
    (a) $5 + 9 + \cdots + (4n + 1) = 2n^2 + 3n$;
    (b) $n! > 2^n$ when $n \geq 4$.
13. Suppose P_3 is true and $P_k \Rightarrow P_{k+3}$. What can we conclude about the sequence P_n?
14. John has 4 sport coats, 3 pairs of trousers, and 5 shirts. How many different outfits could he wear?
15. How many code words of all lengths can be made from the letters of the word *SNOW*, assuming letters cannot be repeated?
16. Evaluate.
    (a) $_{10}P_4$ (b) $_{10}C_4$ (c) $6!/4!$
    (d) $_{50}C_{48}$ (e) $8!/(2!\ 6!)$ (f) $10!/(2!\ 3!\ 5!)$
17. In how many ways can a class of 5 girls and 4 boys select
    (a) a president, vice president, and secretary;
    (b) a president, vice president, and secretary if the secretary must be a boy;
    (c) a social committee of 3 people;
    (d) a social committee of 3 people consisting of 2 girls and 1 boy?
18. The letters of the word *BARBARIAN* are written on 9 cards. How many 9-letter code words can be formed?

19. Consider 8 points in the plane, no 3 on the same line. How many triangles can be formed using these points as vertices?

20. How many subsets does a set of 8 elements have?

21. Find the first 4 terms in simplified form in the expansion of $(x + 2y)^{10}$.

22. Find the term involving a^3b^6 in the expansion of $(a - b^2)^6$.

23. Find $(1.002)^{20}$ accurate to 4 decimal places.

10

PROBABILITY AND EXPECTATION

It is remarkable that a science which began with the consideration of games of chance should have become the most important object of human knowledge . . . . The most important questions of life are, for the most part, really only problems of probability.

Pierre-Simon de Laplace

Here's how we'll decide.
Heads, we go to the game.
Tails, we go to the movie.
On edge, we study.

Probability is the very guide of life.

Cicero

10-1
Equally Likely Outcomes

If we can be sure of anything, we can be sure that our two college students are not going to study. There may be three outcomes of this experiment, but a coin that balances on edge, when weighed on the scales of justice, is likely to be found wanting. A fair coin landing on a flat surface will show either heads or tails.

More is true. A fair coin is just as likely to show heads as tails. That is, if a fair coin were tossed over and over again, say millions of times, it would show heads approximately one-half of the time and tails about one-half of the time. How do we know that this is true for a particular coin? We do not; but it is an assumption that most of us are willing to make, particularly if the coin appears to be perfectly round and unworn.

There are many situations in which the assumption that certain events are equally likely seems plausible. Toss a standard die. Is there any reason to think that one side has an edge in the battle to show its face? We think not. Draw a card from a well-shuffled deck. Nature is perfectly democratic. It gives each card the same chance of being drawn.

But one should be careful. If a stockbroker tries to convince you that the market can do only one of three things—go up, stay the same, or go down—he or she is, of course, right. But if he or she further insists that these three possibilities are equally likely and therefore that the odds are two to one against losing, hang onto your wallet.

DEFINITION OF PROBABILITY

Now we are ready to introduce the main notion of this chapter, the probability of an event. Consider the experiment of tossing a die, which can result in six equally likely outcomes. Of these six outcomes, three show an even number of spots. We therefore say that the probability that an even number will show is $\frac{3}{6}$.

Or consider the experiment of picking a card from a well-shuffled deck. Thirteen of the 52 cards are diamonds. Thus we say that the probability of getting a diamond is $\frac{13}{52}$.

These examples serve to illustrate the general situation. If an experiment can result in any one of n equally likely outcomes and if exactly m of them result in the event E, we say that the **probability of E** is m/n. We write this as

$$P(E) = \frac{m}{n}$$

PROPERTIES OF PROBABILITY

We use two dice, one colored and the other white, to illustrate the major properties of probabilities. Since one die can fall in 6 ways, two dice can fall in $6 \times 6 = 36$ ways. The 36 equally likely outcomes are shown below. As you read the next several paragraphs, you will want to refer back to this diagram.

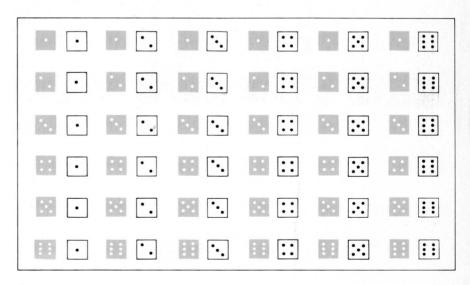

Six ways to get 7.

P (getting 7) $= \dfrac{6}{36}$

Two ways to get 11.

P (getting 11) $= \dfrac{2}{36}$

Most questions about a pair of dice have to do with the total number of spots showing after a toss. For example, what is the probability of getting a total of 7? Six of the 36 outcomes result in this event (see the margin). Therefore the required probability is $\frac{6}{36} = \frac{1}{6}$. Similarly,

$$P(\text{getting a total of } 11) = \frac{2}{36} = \frac{1}{18}$$

$$P(\text{getting a total of } 12) = \frac{1}{36}$$

$$P(\text{getting a total greater than } 7) = \frac{15}{36} = \frac{5}{12}$$

$$P(\text{getting a total less than } 13) = \frac{36}{36} = 1$$

$$P(\text{getting a total of } 13) = \frac{0}{36} = 0$$

The last two events are worthy of comment. The event "getting a total less than 13" is certain to occur; we call it a **sure event.** The probability of a sure event is always 1. However, the event "getting a total of 13" cannot occur; it is called an **impossible event.** The probability of an impossible event is zero.

Next, consider the event "getting 7 or 11," which is important in the dice game called craps. From the diagram, we see that 8 of the 36 outcomes give a total of 7 or 11, so the probability of this event is $\frac{8}{36}$. But note that we could have calculated this probability by adding the probability of getting 7 and the probability of getting 11.

$$P(7 \textit{ or } 11) = P(7) + P(11)$$

$$\frac{8}{36} = \frac{6}{36} + \frac{2}{36}$$

It would appear that we have found a very useful property of probability. When an event is described by using the conjunction *or,* we may find its probability by adding the probabilities of the two parts of the conjunction. However, a different example makes us take a second look. Note that

$$P(\text{odd } \textit{or} \text{ over } 7) \neq P(\text{odd}) + P(\text{over } 7)$$

$$\frac{27}{36} \qquad \neq \qquad \frac{18}{36} \quad + \quad \frac{15}{36}$$

Why is it that we can add probabilities in one case but not in the other? The reason is a simple one. The events "getting 7" and "getting 11" are *disjoint* (they cannot both happen). But the events "getting an odd total" and "getting over 7" overlap (a number such as 9 satisfies both conditions).

Considerations like these lead us to the main properties of probability.

> FOUR PROPERTIES OF PROBABILITY
>
> 1. $P(\text{impossible event}) = 0$; $P(\text{sure event}) = 1$
> 2. $0 \leq P(A) \leq 1$ for any event A.
> 3. $P(A \text{ or } B) = P(A) + P(B)$, provided A and B are disjoint (that is, cannot both happen at the same time).
> 4. $P(A) = 1 - P(\text{not } A)$

Property 4 deserves comment since it follows directly from Properties 1 and 3. The events "A" and "not A" are certainly disjoint. Thus, from Property 3,

$$P(A \text{ or not } A) = P(A) + P(\text{not } A)$$

The event "A or not A" must occur; it is a sure event. Hence from Property 1,

$$1 = P(A) + P(\text{not } A)$$

which is equivalent to Property 4.

Let us illustrate this fourth property. Suppose we want to know the probability of "getting a total less than 11" when tossing two dice. If we call this event "A" then "not A" is the event "getting a total of at least 11." The probability of "not A" is exceptionally easy to calculate since it involves just 3 outcomes. Thus

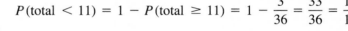

$$P(\text{total} < 11) = 1 - P(\text{total} \geq 11) = 1 - \frac{3}{36} = \frac{33}{36} = \frac{11}{12}$$

RELATION TO SET LANGUAGE

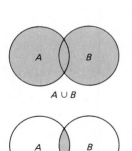

$A \cup B$

$A \cap B$

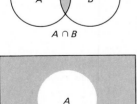

A'

Most American students are introduced to the language of sets in the early grades. Even so, a brief review may be helpful. In everyday language, we talk of a bunch of grapes, a class of students, a herd of cattle, a flock of birds, or perhaps even a team of toads, a passel of possums, or a gaggle of geese. Why are there so many words to express the same idea? We do not know, but we do know that mathematicians prefer to use one word—**set.** The objects that make up a set are called **elements,** or **members,** of the set.

Sets can be put together in various ways. We have $A \cup B$ (read "A **union** B"), which consists of the elements in A or B. The set $A \cap B$ (read "A **intersection** B") is made up of the elements that are in both A and B. We have the notion of an **empty set** ϕ and of a **universe** S (the set of all elements under discussion). Finally, we have A' (read "A **complement**"), which is composed of all elements in the universe that are not in A.

The language of sets and the language used in probability are very closely related, as the list below suggests.

PROBABILITY LANGUAGE	SET LANGUAGE
Outcome	Element
Event	Set
Or	Union
And	Intersection
Not	Complement
Impossible event	Empty set
Sure event	Universe

If we borrow the notation of set theory, we can state the laws of probability in a very succinct form.

1. $P(\phi) = 0; \qquad P(S) = 1$
2. $0 \le P(A) \le 1$
3. $P(A \cup B) = P(A) + P(B)$, provided $A \cap B = \phi$
4. $P(A) = 1 - P(A')$.

Problem Set 10-1

1. An ordinary die is tossed. What is the probability that the number of spots on the upper face will be
   (a) three;   (b) greater than 3;   (c) less than 3;
   (d) an even number;   (e) an odd number?

2. Nine balls, numbered 1, 2, . . . , 9, are in a bag. If one is drawn at random, what is the probability that its number is
   (a) 9;   (b) greater than 5;   (c) less than 6;
   (d) even;   (e) odd?

3. A penny, a nickel, and a dime are tossed. List the 8 possible outcomes of this experiment. What is the probability of
   (a) 3 heads;   (b) exactly 2 heads;
   (c) more than 1 head?

4. A coin and a die are tossed. Suppose that one side of the coin has 1 on it and the other a 2. List the 12 possible outcomes of this experiment, for example, (1, 1), (1, 2), (1, 3). What is the probability of
   (a) a total of 4;   (b) an even total;   (c) an odd total?

5. Two ordinary dice are tossed. What is the probability of
   (a) a double (both showing the same number);
   (b) the number on one of the dice being twice that on the other;
   (c) the numbers on the two dice differing by at least 2?

6. A letter is chosen at random from the word *PROBABILITY*. What is the probability that it will be
   (a) *P;*   (b) *B;*   (c) *M;*   (d) a vowel?   (Treat *Y* as a vowel.)

7. Two regular tetrahedra (tetrahedra have four identical equilateral triangles for faces) have faces numbered 1, 2, 3, and 4. Suppose they are tossed and

we keep track of the outcome by listing the numerals on the bottom faces, for example, (1, 1), (1, 2).
(a) How many outcomes are there?
(b) What is the probability of a sum of 7?
(c) What is the probability of a sum less than 7?

8. Two regular octahedra (polyhedra having eight identical faces) have faces numbered 1, 2, . . . , 8. Suppose they are tossed and we record the outcomes by listing the numerals on the bottom faces.
(a) How many outcomes are there?
(b) What is the probability of a sum of 7?
(c) What is the probability of a sum less than 7?

9. What is wrong with each of the following statements?
(a) Since there are 50 states, the probability of being born in Wyoming is $\frac{1}{50}$.
(b) The probability that a person smokes is .45, and that he or she drinks, .54; therefore the probability that he or she smokes or drinks is .54 + .45 = .99.
(c) The probability that a certain candidate for president of the United States will win is $\frac{3}{5}$, and that he or she will lose, $\frac{1}{4}$.
(d) Two football teams A and B are evenly matched; therefore the probability that A will win is $\frac{1}{2}$.

10. During the past 30 years, Professor Witquick has given only 100 A's and 200 B's in Math 13 to the 1200 students who registered for the class. Based on these data, what is the probability that a student who registers next year
(a) will get an A or a B;
(b) will not get either an A or a B?

11. A poll was taken at Podunk University on the question of coeducational dormitories, with the following results.

	Administrators	Faculty	Students	Total
For	4	16	100	120
Against	3	32	100	135
No opinion	3	2	40	45
Total	10	50	240	300

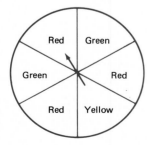

On the basis of this poll, what is the probability that
(a) a randomly chosen faculty member will favor coed dorms;
(b) a randomly chosen student will be against coed dorms;
(c) a person selected at random at Podunk University will favor coed dorms;
(d) a person selected at random at Podunk University will be a faculty member who is against coed dorms?

12. The well-balanced spinner shown in the margin is spun. What is the probability that the pointer will stop at
(a) red;   (b) green;
(c) red or green;   (d) not green?

13. Four balls numbered 1, 2, 3, and 4 are placed in a bag, mixed; and drawn

out, one at a time. What is the probability that they will be drawn in the order 1, 2, 3, 4?

14. A five-volume set of books is placed on a shelf at random. What is the probability they will be in the right order?

EXAMPLE A (Odds)   Gamblers often use the language of odds rather than of probability. For example, Las Vegas may report that the odds in favor of the Yankees winning the seventh game of a World Series are 3 to 2. Translate this statement into probability language.

Solution.   To say that the odds are 3 to 2 in favor of the Yankees means that if the game were played many times, the Yankees would win 3 times to every 2 times their opponent won. Thus the Yankees would win $\frac{3}{5}$ of the time, which means the probability of the Yankees winning is $\frac{3}{5}$. In general, if the odds in favor of event E are m to n, then the probability of E is $m/(m + n)$. Conversely, if the probability of E is j/k, then the odds in favor of E are j to $k - j$.

15. Suppose that the odds in favor of the Vikings winning the Super Bowl game are given as 2 to 9. What is the probability they will lose the Super Bowl game?

16. What is the probability that David will marry Jane if he says the odds in favor are 1 to 8?

17. What are the odds in favor of getting two heads when tossing two coins?

18. What are the odds in favor of 7 or 11 in tossing two dice?

EXAMPLE B (Card problems)   A standard deck consists of 52 cards. There are 4 suits (spades, clubs, hearts, and diamonds), each with 13 cards (2, 3, . . . , 10, jack, queen, king, ace). A poker hand consists of 5 cards. If a poker hand is to be dealt from a standard deck, what is the probability it will be (a) a diamond flush (that is, all diamonds);   (b) a flush?

Solution.
(a) Recall our study of combinations from Section 9-6. There are $_{52}C_5$ possible 5-card hands, of which $_{13}C_5$ consists of all diamonds. Thus

$$P(\text{diamond flush}) = \frac{_{13}C_5}{_{52}C_5} = \frac{\dfrac{13 \cdot 12 \cdot 11 \cdot 10 \cdot 9}{5 \cdot 4 \cdot 3 \cdot 2 \cdot 1}}{\dfrac{52 \cdot 51 \cdot 50 \cdot 49 \cdot 48}{5 \cdot 4 \cdot 3 \cdot 2 \cdot 1}} \approx .0005.$$

(b) A flush is a diamond flush or a heart flush or a club flush or a spade flush. Thus

$$P(\text{flush}) = P(\text{diamond flush}) + P(\text{heart flush})$$
$$+ P(\text{club flush}) + P(\text{spade flush})$$
$$\approx .0005 + .0005 + .0005 + .0005$$
$$= .002$$

19. From a standard deck, one card is drawn. What is the probability that it will be
    (a) red;   (b) a spade;   (c) an ace?
    (*Note:* Two of the suits are red and two are black.)

20. Two cards are drawn from a standard deck (there are $_{52}C_2$ ways of doing it). What is the probability both will be
    (a) red;   (b) of the same color;   (c) aces?

21. Three cards are drawn from a standard deck. What is the probability that
    (a) all will be red;   (b) all will be diamonds;
    (b) exactly one will be a queen;   (d) all will be queens?

22. A poker hand is drawn from a standard deck. What is the probability of at least one ace? (*Hint:* Look at the complementary event consisting of no aces.)

23. What is the probability of a poker hand consisting of all kings and queens?

24. What is the probability of a full-house poker hand (two cards of one value, three of another)?

Miscellaneous Problems

25. If a coin is tossed three times, what is the probability of getting exactly three heads? Of getting at least one head?

26. If two dice are tossed, what is the probability of getting a total greater than 3?

27. A poker hand is drawn from a standard deck. What is the probability that all the cards will be red? That all will be the same color?

28. What are the odds against getting an ace in a single draw from a standard deck?

29. What is the probability of getting at least one king or queen if three cards are drawn from a standard deck?

30. If three men and two women are seated at random in a row, what is the probability that
    (a) men and women will alternate;
    (b) the men will be together;
    (c) the women will be together?

31. A die has been loaded so that the probabilities of getting 1, 2, 3, 4, 5, and 6 are $\frac{1}{3}, \frac{1}{4}, \frac{1}{6}, \frac{1}{12}, \frac{1}{12}$, and $\frac{1}{12}$, respectively. Assuming that the properties of probability are still valid in this situation (where the outcomes are not equally likely), find the probability of rolling
    (a) an even number;
    (b) a number less than 5;
    (c) an even number or a number less than 5.

32. If a whole number between 1 and 1000 (inclusive) is picked at random, what is the probability that it will have no zeros in its decimal representation?

33. The third property of probability is $P(A \cup B) = P(A) + P(B)$, provided $A \cap B = \emptyset$. Show that the following extension always holds.

$$P(A \cup B) = P(A) + P(B) - P(A \cap B)$$

34. Three dice are tossed. What is the probability that 1, 2, and 3 all will appear?

35. In poker, what is the probability that a player will be dealt a straight flush (five consecutive cards in the same suit; an ace counts both high and low)? A royal flush (five highest cards in a suit)?

36. A careless secretary typed four letters and four envelopes and then inserted the letters at random into the envelopes. Find the probability that
    (a) none went into the correct envelope;
    (b) at least one went into the correct envelope;
    (c) all went into the correct envelopes;
    (d) exactly two went into the correct envelopes.

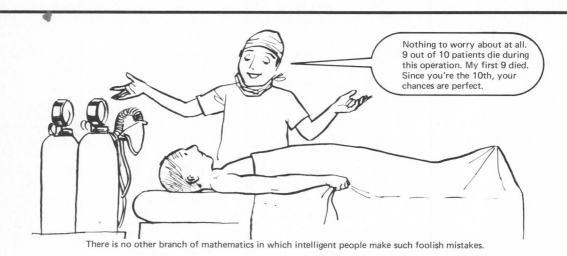

There is no other branch of mathematics in which intelligent people make such foolish mistakes.

10-2
Independent Events

A perfectly balanced coin has shown nine tails in a row. What is the probability that it will show a tail on the tenth flip? Some people argue that a mystical law of averages makes the appearance of a head practically certain. It is as if the coin had a memory and a conscience; it must atone for falling on its face nine times in a row. Such thinking is pure, unadulterated nonsense. The probability of showing a tail on the tenth flip is $\frac{1}{2}$, just as it was on each of the previous nine flips. The outcome of any flip is independent of what happened on previous flips.

Here is a different question, not to be confused with the one just answered. If one plans to flip a coin 10 times, what is the probability of getting all tails? To answer, we reason that there are 2 possibilities on the first flip, 2 on the second, and so forth. By the multiplication principle for counting (see Chapter 9), there are $2 \cdot 2 \cdot 2 \cdot 2 \cdot 2 \cdot 2 \cdot 2 \cdot 2 \cdot 2 \cdot 2$ or 1024 possible outcomes of this experiment, only one of which consists of all tails. The probability of 10 tails is $\frac{1}{1024}$, a very unlikely event indeed.

Here are the same questions stated for a game with higher stakes. A couple already has 9 girls. What is the probability that its tenth child will also be a girl? Assuming that boys and girls are equally likely (which is not quite true), the answer is $\frac{1}{2}$. But suppose a newly married couple decides, with great passion and little thought, to have 10 children. Assuming that nature assents to their decision, what is the probability that they will have all girls? It is $\frac{1}{1024}$.

Consider now the worried patient in the opening cartoon. Most of us would not (and should not) find the doctor's logic very comforting. In fact, if we make the assumption that the outcome of the tenth operation is completely independent of the first nine, the probability of a tenth failure is $\frac{9}{10}$. The assumption of independence may be questioned. Perhaps doctors improve with experience, but this particular patient had better hope for a miracle.

DEPENDENCE VERSUS INDEPENDENCE

Perhaps we can make the distinction between dependence and independence clear by describing an experiment in which we again toss two dice—one colored and the other white. Let A, B, and C designate the following events.

A: colored die shows 6
B: white die shows 5
C: total on the two dice is greater than 7

Only 6 outcomes if we know that colored die shows 6

Consider first the relationship between B and A. It seems quite clear that the chance of B occurring is not affected by our knowledge that A has occurred. In fact, if we let $P(B \mid A)$ denote the probability of B given that A has occurred, then (see illustration in margin)

$$P(B \mid A) = \frac{1}{6}$$

But this is equivalent to the answer we get if we calculate $P(B)$ without any knowledge of A. For then we look at all 36 outcomes for two dice and note that 6 of them have the white die showing 5; that is,

$$P(B) = \frac{6}{36} = \frac{1}{6}$$

We conclude that $P(B \mid A) = P(B)$, just as we expected.

The relation between C and A is very different; C's chances are greatly improved if we know that A has occurred. From the illustration in the margin, we see that

$$P(C \mid A) = \frac{5}{6}$$

However, if we have no knowledge of A and calculate $P(C)$ by looking at the 36 outcomes (page 403) for two dice, we find

$$P(C) = \frac{15}{36} = \frac{5}{12}$$

Clearly $P(C \mid A) \neq P(C)$.

This discussion has prepared the way for a formal definition. If $P(B \mid A) = P(B)$, we say that A and B are **independent** events. If $P(B \mid A) \neq P(B)$, A and B are dependent events.

And now, recalling that $A \cap B$ means A *and* B, we can state the **multiplication rule for probabilities.**

$$P(A \cap B) = P(A)P(B \mid A) = P(B)P(A \mid B)$$

In words, the probability of both A and B occurring is equal to the probability that A will occur multiplied by the probability that B will occur given that A has already occurred. In the case of independence, this takes a particularly elegant form.

$$P(A \cap B) = P(A)P(B)$$

To illustrate these rules, first consider the problem of drawing two cards one after another from a well-shuffled deck. What is the probability that both will be spades? Based on previous knowledge (that is, Section 10-1), we respond:

$$P(\text{two spades}) = \frac{_{13}C_2}{_{52}C_2} = \frac{\dfrac{13 \cdot 12}{2 \cdot 1}}{\dfrac{52 \cdot 51}{2 \cdot 1}} = \frac{13 \cdot 12}{52 \cdot 51}$$

But here is another approach. Consider the events

A: getting a spade on the first draw
B: getting a spade on the second draw

Our interest is in $P(A \cap B)$. According to the rule above, it is given by

$$P(A \cap B) = P(A)P(B \mid A) = \frac{13}{52} \cdot \frac{12}{51}$$

which naturally agrees with our earlier answer.

Here is a different but related problem: Suppose we draw a card, replace it, shuffle the deck, and draw a second card. What is the probability that both cards are spades? Now A and B are independent.

$$P(A \cap B) = P(A)P(B) = \frac{13}{52} \cdot \frac{13}{52}$$

URNS AND BALLS

For reasons not entirely clear, teachers have always illustrated the central ideas of probability by talking about urns (vases) containing colored balls. Most of

6 red and
4 green balls.

18 red and
2 green balls.

us have never seen an urn containing colored balls, but it won't hurt to use a little imagination,

Consider two urns labeled A and B, A containing 6 red balls and 4 green balls, and B containing 18 red balls and 2 green balls. If a ball is drawn from each urn, what is the probability that both will be red?

$$P(\text{red from A } and \text{ red from B}) = P(\text{red from A}) \cdot P(\text{red from B})$$

$$= \frac{6}{10} \cdot \frac{18}{20}$$

$$= \frac{6}{10} \cdot \frac{9}{10} = \frac{54}{100} = \frac{27}{50}$$

As a second problem, suppose we choose an urn at random and then draw one ball. What is the probability that it will be red? In other words, what is the probability of the event

"choose A *and* draw red"   or   "choose B *and* draw red"

The events in quotation marks are disjoint, so their probabilities can be added. We obtain the answer

$$\frac{1}{2} \cdot \frac{6}{10} + \frac{1}{2} \cdot \frac{9}{10} = \frac{15}{20} = \frac{3}{4}$$

You should note the procedure we use. We describe the event using the words *and* and *or*. When we determine probabilities, *and* translates into *times*, and *or* into *plus*.

A HISTORICAL EXAMPLE

Two French mathematicians, Pierre de Fermat (1601–1665) and Blaise Pascal (1623–1662), are usually given credit for originating the theory of probability. This is how it happened: The famous gambler, Chevalier de Méré, was fond of a dice game in which he would bet that a 6 would appear at least once in four throws of a die. He won more often than he lost for, though he probably did not know it,

$$P(\text{at least one 6}) = 1 - P(\text{no 6's})$$

$$= 1 - \frac{5}{6} \cdot \frac{5}{6} \cdot \frac{5}{6} \cdot \frac{5}{6} \approx 1 - .48 = .52$$

Growing tired of this game, he introduced a new one played with two dice. Méré then bet that at least one double 6 would appear in 24 throws of two dice. Somehow (perhaps he noted that $\frac{4}{6} = \frac{24}{36}$) he thought he should do just as well as before. But he lost more often than he won. Mystified, he proposed it as a problem to Pascal, who in turn wrote to Fermat. Together they produced the following explanation.

$$P(\text{at least one double 6}) = 1 - P(\text{no double 6's})$$

$$= 1 - \left(\frac{35}{36}\right)^{24} \approx 1 - .51 = .49$$

From this humble, slightly disreputable origin grew the science of probability.

Problem Set 10-2

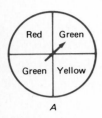

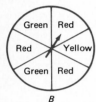

A

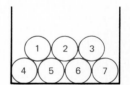

B

1. Toss a balanced die three times in succession. What is the probability of all 1's?

2. Toss a fair coin four times in succession. What is the probability of all heads?

3. Spin the two spinners pictured in the margin. What is the probability that
   (a) both will show red (that is, *A* shows red *and B* shows red);
   (b) neither will show red (that is, *A* shows not red *and B* shows not red);
   (c) spinner *A* will show red and spinner *B* not red;
   (d) spinner *A* will show red and spinner *B* red or green;
   (e) just one of the spinners will show green?

4. The two boxes shown below are shaken thoroughly and a ball is drawn from each. What is the probability that
   (a) both will be 1's;
   (b) exactly one of them will be a 2 (2 from first box and not 2 from second box, or not 2 from first box and 2 from second box);
   (c) both will be even;
   (d) exactly one of them will be even;
   (e) at least one of them will be even?

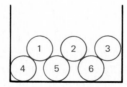

5. In each case, indicate whether or not the two events seem independent to you. Explain.
   (a) Getting an *A* in math and getting an *A* in physics.
   (b) Getting an *A* in math and winning a tennis match.
   (c) Getting a new shirt for your birthday and stubbing your toe the next day.
   (d) In tossing two dice, getting an odd total and getting a 5 on one of the dice.
   (e) Being a woman and being a doctor.
   (f) Walking under a ladder and having an accident the next day.

6. In each case indicate whether or not the two events are necessarily disjoint.
   (a) Getting an *A* in math and getting an *A* in physics.
   (b) Getting an *A* in Math 101 and getting a *B* in Math 101.
   (c) In tossing two dice, getting an odd total and getting the same number on both dice.
   (d) In tossing two dice, getting an odd total and getting a 5 on one of the dice.

(e) The sun shining on Tuesday and being rainy on Tuesday.

(f) Not losing the college football game and not winning it.

7. A machine produces bolts which are put in boxes. It is known that 1 box in 10 will have at least one defective bolt in it. Assuming that the boxes are independent of each other, what is the probability that a customer who ordered 3 boxes will get all good bolts?

[c] 8. Suppose that the probability of being hospitalized during a year is .152. Assuming that family members are hospitalized independently of each other, what is the probability that no one in a family of five will be hospitalized this year? Do you think the assumption of independence is reasonable?

9. Consider two urns, one with 3 red balls and 7 white balls, the other with 6 red balls and 6 white balls. If an urn is chosen at random and then a ball drawn, what is the probabilty it will be red? (*Hint:* See the urn example of this section).

10. Suppose that 4 percent of males are colorblind, that 1 percent of females are colorblind, and that males and females each make up 50 percent of the population. If a person is chosen at random, what is the probability that this person wll be colorblind? (This is like the urn problem in Problem 9.)

EXAMPLE A (Replacement or nonreplacement) An urn contains 2 red, 3 green, and 5 black balls. Two balls are drawn at random. What is the probability both are green (a) if we replace the first ball before the second is drawn; (b) if we do not replace the first ball before the second is drawn?

Solution.

(a) Here we have independence.

$$P(\text{green } and \text{ green}) = P(\text{green}) \cdot P(\text{green})$$

$$= \frac{3}{10} \cdot \frac{3}{10} = \frac{9}{100}$$

(b) Now we have dependence. The outcome of the first draw does affect what happens on the second draw.

$$P(\text{green } and \text{ green})$$

$$= P(\text{first ball green}) \cdot P(\text{second ball green} \mid \text{first ball green})$$

$$= \frac{3}{10} \cdot \frac{2}{9} = \frac{6}{90} = \frac{1}{15}$$

Problems 11–16 refer to the urn of Example A.

11. If 2 balls are drawn, what is the probability both are black assuming
    (a) replacement between draws;
    (b) nonreplacement?

12. If 2 balls are drawn without replacement between draws, what is the probability of getting
    (a) red on the first draw and green on the second;
    (b) a red and a green ball in either order;

(c) 2 red balls;

(d) 2 nonred balls?

13. If 3 balls are drawn with replacement between draws, what is the probability of getting
    (a) 3 green balls;
    (b) 3 red balls;
    (c) 3 balls of the same color;
    (d) 3 balls of all different colors?

14. Answer the questions of Problem 13 if there is no replacement between draws.

15. Two balls are drawn. What is the probability that the second one drawn is red if
    (a) the first is replaced before the second is drawn;
    (b) the first is not replaced before the second is drawn?

16. If 4 balls are drawn without replacement between draws, what is the probability of getting
    (a) 4 balls of the same color;
    (b) at least two different colors;
    (c) 2 balls of one color and 2 of another?

EXAMPLE B (Using the laws of probability) Recall two general laws of probability.

$$P(A \cup B) = P(A) + P(B) - P(A \cap B)$$

$$P(A \cap B) = P(A)P(B \mid A)$$

If $P(A) = .7$, $P(B) = .3$, and $P(B \mid A) = .4$, calculate
(a) $P(A \cap B)$; (b) $P(A \cup B)$; (c) $P(A \mid B)$.

Solution.
(a) $P(A \cap B) = P(A)P(B \mid A) = (.7)(.4) = .28$
(b) $P(A \cup B) = P(A) + P(B) - P(A \cap B) = .7 + .3 - .28 = .72$
(c) $P(A \mid B) = P(A \cap B)/P(B) = .28/.3 \approx .93$

17. Given that $P(A) = .8$, $P(B) = .5$, and $P(A \cap B) = .4$, find
    (a) $P(A \cup B)$;   (b) $P(B \mid A)$;   (c) $P(A \mid B)$.

18. Given that $P(A) = .8$, $P(B) = .4$, and $P(B \mid A) = .3$, find
    (a) $P(A \cap B)$;   (b) $P(A \cup B)$;   (c) $P(A \mid B)$.

19. In Sudsville, 30 percent of the people are Catholics, 55 percent of the people are Democrats, and 90 percent of the Catholics are Democrats. Find the probability that a person chosen at random from Sudsville is
    (a) both Catholic and Democrat;
    (b) Catholic or Democrat;
    (c) Catholic given that he or she is Democrat.

20. At Podunk U, $\frac{1}{4}$ of the applicants fail the entrance examination in mathematics, $\frac{1}{5}$ fail the one in English, and $\frac{1}{9}$ fail both mathematics and English. What is the probability that an applicant will
    (a) fail mathematics or English;

(b) fail mathematics, given that he or she failed English;

(c) fail English, given that he or she failed mathematics?

Miscellaneous Problems

21. An urn contains 5 red balls and 7 black balls. Two balls are drawn in succession. What is the probability of drawing 2 red balls if the first ball
    (a) is replaced before the second is drawn;
    (b) is not replaced before the second is drawn?

22. Three balls are drawn from the urn of Problem 21 without replacement between draws. What is the probability that
    (a) all are red;
    (b) at least one is red?

23. An urn contains 5 red, 6 black, and 7 green balls. If 2 balls are drawn at random and the first ball is replaced before the second is drawn, what is the probability that
    (a) both are red;
    (b) both are of the same color;
    (c) they are of different colors?

24. Consider the urns of Problems 21 and 23. If one of these urns is picked at random and then a ball drawn, what is the probability that the ball is
    (a) red;   (b) black;   (c) green?

25. A committee of 3 people is chosen at random from among a group of 6 boys and 4 girls. What is the probability that it will consist of all boys?

ⓒ 26. A coin, loaded so that the probability of a head is .6, is tossed eight times. What is the probability of getting at least one head?

27. Let us suppose that, in a World Series, the probability that team A will win any given game is $\frac{2}{3}$. Then the probability that team B will win a given game is $\frac{1}{3}$. What is the probability that
    (a) A will win the series in four games;
    (b) B will win in four games;
    (c) the series will end in four games;
    (d) A will win in five games;
    (e) the series will end in five games?

ⓒ 28. (Birthday problem)   In a class of 23 students, what is the probability that at least 2 people have the same birthday? Make a guess first. Then make a computation based on the following assumptions: (1) There are 365 days in a year; (2) One day is as likely as another for a birthday; (3) The 23 students were born (chose their birthdays) independently of each other. *Suggestion:* Look at the complementary event. For example, if there were only 3 people in the class, the probability that no two have the same birthday is

$$1 \cdot \frac{364}{364} \cdot \frac{363}{365}$$

In the game Yahtzee, which involves rolling five dice, it is very desirable to get several 6's. What is the probability of getting exactly three 6's on the first roll? Four 6's?

10-3
The Binomial Distribution

The manufacturer of Yahtzee chooses to make all five dice of the same color, usually white. This is a matter of convenience; surely you will agree that a little paint on each of the dice does not affect the way they roll. Let's imagine that the five dice have been painted black, green, purple, red, and white, respectively. This will allow us to keep track of each die separately. Now call getting a 6 a success (S) and getting anything else a failure (F). Thus, for each die,

$$P(S) = \frac{1}{6} \qquad P(F) = \frac{5}{6}$$

If we roll five dice, there are 10 ways of getting exactly three 6's (see the chart in the margin). Because of the independence of the five dice, we can calculate the probability of each of these events by multiplication. For example,

Color of Die				
B	**G**	**P**	**R**	**W**
S	S	S	F	F
S	S	F	S	F
S	S	F	F	S
S	F	S	S	F
S	F	F	S	S
S	F	S	F	S
F	S	S	F	S
F	S	F	S	S
F	S	S	S	F
F	F	S	S	S

$$P(SSSFF) = \frac{1}{6} \cdot \frac{1}{6} \cdot \frac{1}{6} \cdot \frac{5}{6} \cdot \frac{5}{6} = \left(\frac{1}{6}\right)^3 \left(\frac{5}{6}\right)^2$$

and

$$P(SSFSF) = \frac{1}{6} \cdot \frac{1}{6} \cdot \frac{5}{6} \cdot \frac{1}{6} \cdot \frac{5}{6} = \left(\frac{1}{6}\right)^3 \left(\frac{5}{6}\right)^2$$

In fact, each of the 10 disjoint events has this same probability. Consequently,

$$P(\text{three 6's in rolling five dice}) = 10\left(\frac{1}{6}\right)^3 \left(\frac{5}{6}\right)^2 \approx .03$$

We need to know why the number 10 appears in this problem. Look at the chart again. A row is determined as soon as we decide where to put the three

S's, that is, as soon as we select from the set $\{B, G, P, R, W\}$ three dice to classify as S's. We can choose three objects from five in $_5C_3$ ways; and recall that

$$_5C_3 = \frac{5 \cdot 4 \cdot 3}{3 \cdot 2 \cdot 1} = 10$$

There is another way to view this problem. We are really asking for the number of five-letter words that can be made using three S's and two F's. From Example B of Section 9-5 we know that there are $5!/3!\ 2!$ such words. But

$$\frac{5!}{3!\ 2!} = {}_5C_3$$

No matter how we look at it, there are $_5C_3$ ways of getting three 6's in a roll of five dice. Thus

$$P(\text{three 6's in rolling five dice}) = {}_5C_3\left(\frac{1}{6}\right)^3\left(\frac{5}{6}\right)^2$$

and, by similar reasoning,

$$P(\text{four 6's in rolling five dice}) = {}_5C_4\left(\frac{1}{6}\right)^4\left(\frac{5}{6}\right)^1$$

We make one final remark about rolling five dice. Whether we roll five dice at once (as in Yahtzee) or roll one die five times is of no significance in probability questions. From now on we adopt the second point of view.

THE BINOMIAL DISTRIBUTION

The general situation we have in mind is this. Suppose that the outcomes of an experiment fall into two categories. One we call a success (S) and the other a failure (F). The probabilities of S and F are presumed to be known, say

$$P(S) = p \qquad P(F) = q$$

For example, if p were .3, then q would be .7. The experiment is repeated n times. What is the probability of getting exactly k success? The answer, based on the same reasoning as in the five-dice problem, is

$$\boxed{P(k \text{ successes in } n \text{ trials}) = {}_nC_k p^k q^{n-k}}$$

This result is closely related to the binomial formula. Recall, for example, that

$$(p + q)^5 = (q + p)^5$$
$$= {}_5C_0 p^0 q^5 + {}_5C_1 p^1 q^4 + {}_5C_2 p^2 q^3 + {}_5C_3 p^3 q^2 + {}_5C_4 p^4 q^1 + {}_5C_5 p^5 q^0$$
$$= q^5 + 5pq^4 + 10p^2q^3 + 10p^3q^2 + 5p^4q + p^5$$

The terms in the last row are, respectively, the probabilities of zero, one, two, three, four, and five successes in an experiment repeated five times (for example, the Yahtzee problem). For this reason, the number of successes in n trials of an experiment is said to exhibit a binomial distribution.

The case $p = q = \frac{1}{2}$ is particularly interesting. Suppose that a perfectly balanced coin is tossed nine times. What is the probability of getting exactly one head? Two heads? Three heads?

$$P(\text{one head in nine tosses}) \quad = {}_9C_1\left(\frac{1}{2}\right)^1\left(\frac{1}{2}\right)^8 = 9 \cdot \frac{1}{512} \approx .02$$

$$P(\text{two heads in nine tosses}) \quad = {}_9C_2\left(\frac{1}{2}\right)^2\left(\frac{1}{2}\right)^7 = 36 \cdot \frac{1}{512} \approx .07$$

$$P(\text{three heads in nine tosses}) = {}_9C_3\left(\frac{1}{2}\right)^3\left(\frac{1}{2}\right)^6 = 84 \cdot \frac{1}{512} \approx .16$$

The bar graph below shows the probability for any number of heads from zero to nine. We have superimposed on it the famous normal curve, which is often used to approximate the binomial distribution.

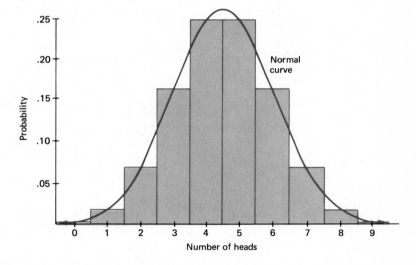

If you toss a coin 9 times, you may of course get anything from zero to nine heads. But if you repeat this experiment 100 or 1000 times and record the proportion of times you get zero, one, two, . . . , nine heads, you should expect a distribution something like that in the chart.

TWO PRACTICAL APPLICATIONS

A certain rare disease has been studied for over 50 years. It is known that 30 percent of the people afflicted with it will eventually recover without treatment while the rest will die. A drug company has discovered what it claims is a miracle cure, citing as evidence the fact that 8 of the 10 people on whom it was tested recovered. Of course, this might have happened by chance even if the drug is absolutely worthless. We would like to know the probability that 8 or more of the 10 patients would have recovered without treatment.

$$P(8 \text{ would have recovered}) = {}_{10}C_8(.3)^8(.7)^2 \approx .0014$$

$$P(9 \text{ would have recovered}) = {}_{10}C_9(.3)^9(.7)^1 \approx .0001$$

$$P(10 \text{ would have recovered}) = {}_{10}C_{10}(.3)^{10}(.7)^0 \approx .0000$$

Therefore the probability of 8 or more recovering naturally is .0014 + .0001 = .0015. Either we have observed a very rare event or the drug is useful. The latter conclusion seems likely; thus the drug merits further experimentation.

A company that manufactures $\frac{1}{2}$-inch steel bolts advertises that at most 1 percent of its bolts will break under a stress of 10,000 pounds. To maintain quality control, it tests a random sample of 100 bolts from each day's output and keeps track of the number of defective ones. A typical record for 10 days is: 0, 2, 0, 1, 0, 3, 0, 0, 0, 1. The question the company faces is to decide when its manufacturing process is out of control. For example, does getting 3 defective bolts in the sixth day establish that something is wrong? Not necessarily. For by chance, one may get 3 or 4 or even 10 defective bolts in a random sample of 100 even if the day's total production is only 1 percent defective.

Based on many years' experience, the company has established the following rule. When samples from two consecutive days yield 3 or more defective bolts each, the manufacturing process is shut down and the equipment carefully checked. For if the process were under control (that is, 1 percent or fewer defective bolts),

$$P(0 \text{ defective}) = {}_{100}C_0(.01)^0(.99)^{100} \approx .37$$

$$P(1 \text{ defective}) = {}_{100}C_1(.01)^1(.99)^{99} \approx .37$$

$$P(2 \text{ defective}) = {}_{100}C_2(.01)^2(.99)^{98} \approx .18$$

Thus

$$P(3 \text{ or more defective}) = 1 - P(\text{less than 3 defective})$$

$$\approx 1 - (.37 + .37 + .18) = .08$$

The probability of 3 or more defective bolts being produced on 2 consecutive days is $(.08)(.08) = .0064$. An event with this small a probability is so rare that is is unlikely to have occurred by chance; it seems likely that something is wrong with the process.

Problem Set 10-3

1. Calculate.
   (a) ${}_6C_1$ (b) ${}_6C_2$ (c) ${}_6C_3$
2. Calculate.
   (a) ${}_8C_1$ (b) ${}_8C_2$ (c) ${}_8C_3$

3. A balanced coin is tossed six times. Calculate the probability of getting
   (a) no heads;
   (b) exactly one head;
   (c) exactly two heads;
   (d) exactly three heads;
   (e) more than three heads.

4. A fair coin is tossed eight times. Calculate the probability of getting
   (a) no tails;
   (b) exactly one tail;
   (c) exactly two tails;
   (d) exactly three tails;
   (e) at most three tails.

5. Experiments indicate that, for an ordinary thumbtack, the probability of falling head down is $\frac{1}{3}$ and head up, $\frac{2}{3}$. Write an expression (but do not evaluate) for the probability in 12 tosses of its falling
   (a) head up exactly four times;
   (b) head up exactly six times.

6. On a true-false test of 20 items, Homer estimates the probability of his getting any one item right is $\frac{3}{4}$. Write an expression for the probability of his getting
   (a) exactly 19 right;
   (b) at least 19 right.

7. A certain machine will work if each of five parts operates successfully. If the probability of failure during a day is $\frac{1}{10}$ for each part, what is the probability the machine will work for a whole day without a breakdown? If only four of the five parts must operate successfully, what would the answer be?

[c] 8. A small airline has learned that 10 percent of those who make reservations for a given flight will fail to use their reservations. There are 22 reservations for a plane that has 20 seats. What is the probability that all who show will get seats? (*Hint:* Look at the complementary event, namely, 21 or 22 using their reservations.)

Prob. $= \frac{1}{3}$

Prob. $= \frac{2}{3}$

TABLE OF $_{10}C_k p^k q^{10-k}$		
k	p = .25	p = .35	p = .50
0	.0563	.0135	.0010
1	.1877	.0725	.0098
2	.2816	.1757	.0439
3	.2503	.2522	.1172
4	.1460	.2377	.2051
5	.0584	.1536	.2461
6	.0162	.0689	.2051
7	.0031	.0212	.1172
8	.0004	.0043	.0439
9	.0000	.0005	.0098
10	.0000	.0000	.0010

EXAMPLE (Using tables of binomial probabilities)   Binomial probabilities are messy to calculate, although they can be computed with the aid of a scientific calculator. To make the job simpler, extensive tables have been developed; a small sample is shown in the margin. Use this table to solve the following problem: Veterinarians know that the probability that German shepherd pups will die from a certain disease is .25. Assuming independence between pups, what is the probability that in a litter of 10 pups
(a) exactly two will die;
(b) at most two will die?

Solution.
   (a) $_{10}C_2(.25)^2(.75)^8 = .2816$
   (b) At most two will die means that 0, 1, or 2 will die. Thus the probability is

$$_{10}C_0(.25)^0(.75)^{10} + {}_{10}C_1(.25)^1(.75)^9 + {}_{10}C_2(.25)^2(.75)^8$$

$$= .0563 + .1877 + .2816 = .5256$$

Use the table in the margin to do Problems 9–18.

9. Calculate
   (a) $_{10}C_6(.25)^6(.75)^4$;   (b) $_{10}C_3(.35)^3(.65)^7$.

10. If a fair coin is tossed 10 times, what is the probability of getting at least 8 heads? Less than 8 heads?

11. Major Electronics sells transistors to the United States Government in lots of 1000. The government takes a random sample of 10 from each lot and accepts the lot if no more than 3 of the 10 break down under severe testing. Major Electronics feels certain that at least $\frac{3}{4}$ of its transistors will pass these tests. If they are correct, what is the probability that any given lot will be accepted by the government?

12. Assuming that men and women are equally likely to enroll in a chemistry class that has 10 stations, which are always filled, what is the probability that Professor Snodgrass will have a class of 9 women and 1 man?

13. Slugger Brown has a batting average of .350. In a 3-game series, he expects to have 10 official times at bat. What is the probability that he will get 3 or more hits? What assumptions did you make in arriving at your answer?

14. A multiple-choice test has 10 questions, each with 4 alternative answers. If Gertrude guesses on all the questions, what is the probability she will get at least 5 right?

15. In a certain city, 35 percent of all days are rainy. Assuming independence between days, what is the probability that the Kuipers family will have at most 2 rainy days during their 10-day vacation in that city?

16. What percentage of all families with exactly 10 children, would you expect to have an equal number of boys and girls?

17. A coin is loaded so that the probability of a head is .25. If this coin is tossed 10 times, what is the most likely number of heads?

18. One card is drawn from a standard deck 10 times with replacement after each draw. What is the probability of getting at least one spade?

Miscellaneous Problems

19. A fair coin is tossed 20 times. Write an expression for the probability of getting exactly 10 heads.

ⓒ 20. Calculate the answer to Problem 19 correct to 3 decimal places.

21. A fair die is tossed 20 times. Write an expression for the probability of getting at least two 6's.

ⓒ 22. Calculate the answer to Problem 21 correct to 3 decimal places.

23. Experience shows that about $\frac{1}{3}$ of those who receive a mail questionnaire respond to it. If Susan Thatcher plans to send out 100 questionaires, write an expression for the probability that she will get at least 25 responses. Use sigma notation.

24. Each item on a multiple-choice test has 5 responses. If Curt guesses on every item on a 30-item test, use sigma notation to write an expression for the probability that he will get at least 15 right.

25. A doctor knows that 25 percent of the patients to whom he administers a certain drug will have undesirable side effects. Use the table that goes with the example of this problem set to calculate the probability that among 10 patients
   (a) none will have undesirable side effects;
   (b) at most two will have undesirable side effects.

26. A packer of mixed nuts claims that at least 65 percent of all 8-ounce cans it sells will contain three or more pecans. To check on this, a consumer decides to take a random sample of 10 cans and accept the packer's claim if at least 6 of them pass the test. Assuming the packer is making an honest claim, what is the probability the sample will pass the test? Suppose that only 50 percent of the cans contain three or more pecans. What is the probability the sample will still pass the test?

27. In the game of Yahtzee, 5 dice are thrown at once. What is the probability on one throw of getting
   (a) 5 of a kind;
   (b) 4 of a kind;
   (c) 3 of a kind;
   (d) a full house (two of one kind and three of another);
   (e) a large straight (1, 2, 3, 4, and 5 or 2, 3, 4, 5, and 6)?

c 28. A player in Yahtzee has reached her last turn to play and needs at least three 6's to achieve a top score. She has three chances to get it. On the first throw, she tosses all 5 dice, but on the second and third throws, she tosses only those that do not already show 6's. What is the probability that she will succeed?

SUPER TRIPLE

SWEEPSTAKES

1st Prize $50,000
2nd Prize $30,000
3rd Prize $10,000

To enter simply send your name and address to: Reader's Repeat, Somewhere, USA

Is it Worth the Cost of a Stamp?

To publicize their products, many companies run contests which may be entered by sending in one's name. Winners are determined by a random drawing. If one million people enter the contest described at the left, are the expected winnings for any individual large enough to cover the cost of the stamp it takes to enter?

10-4
Mathematical
Expectation

People gamble for various reasons. Some simply enjoy the psychological thrill associated with the unpredictable. Others honestly believe that they can make money at games of chance, even those conducted at gaming houses. If you are in this latter group, we hope you will study this section. You may find the results discouraging.

A SIMPLE GAME

Before we answer the question of our opening display, let us consider a simpler example. Jack's rich uncle has offered to pay him 10¢ every time he flips a coin when it shows heads and 5¢ when it shows tails. To make a good thing even better, Jack has loaded a coin so that it shows heads with probability .6. How much, on the average, can Jack expect to receive every time he flips the loaded coin?

Look at the problem this way. If Jack flipped his coin 1000 times, he could expect it to come up heads about 600 times and tails about 400 times. Thus his total payoff would be approximately

$$5(400) + 10(600) = 8000¢$$

That is an average of 8¢ per flip. Notice that the expected payoff per flip (which we will call the *expected value* of the game) can be written as

$$E = \frac{5(400) + 10(600)}{1000}$$

$$= 5 \cdot \frac{400}{1000} + 10 \cdot \frac{600}{1000}$$

$$= 5(.4) + 10(.6)$$

$$= a_1 p_1 + a_2 p_2$$

Here a_1 and a_2 are the payoffs, while p_1 and p_2 are their corresponding probabilities.

This example suggests a definition. If a game (experiment) can result in outcomes with payoffs $a_1, a_2, \ldots, a_k$ occurring with probabilities $p_1, p_2, \ldots, p_k$, respectively, then the **expected value** E of the game is

$$E = a_1 p_1 + a_2 p_2 + \cdots + a_k p_k$$

You can think of the expected value of a game as the amount you can afford to pay to play it if you are satisfied with coming out even over the long run. In Jack's case, he can actually afford to pay his uncle 8¢ a game to play, though he probably would not relish it as much in this case. If he were to pay anything less than 8¢ a game, he could count on coming out ahead in the long run.

THE SWEEPSTAKES CONTEST

Here is the answer to the question of our opening display. The probability of winning any one of the three prizes is approximately $1/1,000,000$. Thus the expected winnings of any individual are

$$E = (50{,}000)\frac{1}{1{,}000{,}000} + (30{,}000)\frac{1}{1{,}000{,}000} + (10{,}000)\frac{1}{1{,}000{,}000}$$

$$= \frac{90{,}000}{1{,}000{,}000} = .09$$

That is only 9¢. In strictly mathematical terms, the game is not worth the cost of the stamp it takes to enter it. However, we would guess that the contest sponsors need not worry about getting people to participate; most people do not look at such contests mathematically. In fact, if the rules allow, some will enter the contest many times and thereby most likely waste a whole roll of stamps.

Incidentally, you will notice that in calculating E, we ignored one of the outcomes, namely, failure to win any prize, even though its probability is very high (near 1). We could ignore this outcome because it has a zero payoff and therefore does not affect the value of E.

ROULETTE AT MONTE CARLO

Next consider a typical game that one finds at a gaming house in Monte Carlo. There a roulette wheel has slots numbered $0, 1, 2, \ldots, 36$. Each slot is equally likely as a place for a ball to stop after the wheel is spun. A player may stake an amount of money on any of the slots numbered 1 to 36 (but not 0); if the ball stops in that slot, the player receives 36 times the stake. For example, if the player stakes \$1 on number 17 and the ball stops in that slot he or she receives back the \$1 stake plus \$35. If the ball stops in any other slot, the player loses the stake (that is wins $-\$1$). Thus if the player stakes \$1 on 17, the expected winnings are

$$E = (35)\frac{1}{37} + (-1)\frac{36}{37}$$

$$= -\frac{1}{37} = -.027$$

This means that a Monte Carlo roulette player can expect to lose 2.7¢ (on the average) every time he or she stakes a dollar on a number on the spinning wheel.

There are other games that can be played on the roulette wheel, but you can be sure that all of them offer you negative expected winnings. Incidentally, American gaming houses offer even poorer expected winnings than do those in Monte Carlo, as you will see in the problem set.

A BUSINESS EXAMPLE

In analyzing whether she should purchase the Tooterville Toads baseball franchise, Susan Sharp consulted the local weather bureau. She learned that during the baseball season, the probabilities of rainy nights, cold clear nights, hot muggy nights, and ideal weather nights are .12, .11, .31, and .46, respectively.

With good promotion and a good team, she estimates that she could attract crowds of 3000, 9000, 7000, and 15,000, respectively, on such nights. What size crowd can she expect on the average?

Here the payoffs are in numbers of people rather than amounts of money, but that is fine. The expected crowd is

$$E = 3000(.12) + 9000(.11) + 7000(.31) + 15,000(.46)$$
$$= 10,420$$

Problem Set 10-4

1. Maurine is to receive $2 if a die shows an odd number and $.50 if it shows an even number. What are her expected winnings?

2. Maria will receive a number of dollars equal to the number of spots showing after a die is tossed. What are her expected winnings?

3. Roberto will receive a number of dollars equal to the number of spots showing if two dice show 7 or 11 but will lose a dollar if any other number shows. What are his expected winnings?

4. Jerry will toss 3 coins and receive $1 for each head and $.50 for each tail. What are his expected winnings?

5. I have 1 dime and 5 pennies in my hand. You may draw two of these coins at random and keep them. What are your expected winnings?

6. Karen is to draw one card from a standard deck. If it is an ace, she will receive $6; if it is a jack, queen, or king, she will receive $2; otherwise, she will lose $1. What are her expected winnings?

7. The prizes for first and second place in a golf tournament are $30,000 and $20,000, respectively. Mary and Martha are playing in the championship round. What are Mary's expected winnings if
   (a) the two players are evenly matched;
   (b) the odds favor Mary 3 to 2?

8. A bag contains 4 white and 6 black balls. Peter may draw out as many balls as he wishes (without replacing the balls between draws). He will receive $1 for each white ball and pay $.60 for each black ball. What are his expected winnings if he draws
   (a) just 1 ball;
   (b) 2 balls;
   (c) all 10 balls?

9. Sylvia Stockton has been offered a job at $12,500 per year and must make a decision today on whether to accept it. She has made applications for two other positions, one at $15,000 and the other at $16,000. She estimates her probabilities of job offers for these two positions at $\frac{1}{2}$ and $\frac{1}{4}$, respectively. If she decides purely on the basis of mathematical expectation, will she accept the $12,500 job? Why? (Assume the last two offers come together.)

10. Ruth Brown has applied for two jobs, one at $14,000 per year and the other at $20,000 per year. She has a firm offer on the first, but will not hear about the second job for a month. However, she must make a decision on the first job before then. If she chooses to accept the first job, what can we say about the probability she assigns to being offered the $20,000 job?

11. Players A and B independently choose a whole number from 1 to 3. The payoffs are computed by adding the two numbers chosen. If the sum is even, player B pays A that number of dollars. If it is odd, then A pays B that number of dollars. Assuming that each player chooses his or her number at random, what are A's expected winnings? (*Hint:* There are nine outcomes, which may be represented as (1, 1), (1, 2), . . . , (3, 3).)

12. In Problem 11, suppose that A continues to pick his or her number at random, but that B always picks 2. Now what are A's expected winnings?

13. Valerie is to draw 10 cards from a standard deck, replacing each card and reshuffling before the next one is drawn. She will receive $5 if she gets 3 spades, $10 if she gets 4 spades, and $15 if she gets more than 4 spades. Otherwise, she gets nothing. What are her expected winnings? (*Hint:* Use the binomial table in Problem Set 10-3.)

c 14. Andy will toss a die eight times and will receive $20 if he gets more than three 6's and pay $2 if he gets three or fewer 6's. What are his expected winnings?

EXAMPLE (More on Monte Carlo roulette)   One game that can be played on the roulette wheel was described in the text. Here is another: Slots 1 to 36 are evenly divided between red and black (slot 0 is neither red nor black). If a player bets on red and red turns up, he or she receives twice the original stake (the original stake plus an equal amount). If black turns up, he or she loses the stake. If 0 turns up, the wheel is spun until a number different from 0 appears. If this is black, the player loses the stake and if it is red, he or she receives the original stake (not twice it). What are a player's expected winnings if he or she stakes $1 on red?

Solution.   A player receives twice the stake (wins $1) with probability $\frac{18}{37}$, loses the stake with probability $\frac{18}{37} + \frac{1}{37} \cdot \frac{1}{2}$, and breaks even with probability $\frac{1}{37} \cdot \frac{1}{2}$. Thus, the player's expected winnings are

$$E = \frac{18}{37}(1) + \left(\frac{18}{37} + \frac{1}{37} \cdot \frac{1}{2}\right)(-1) + \frac{1}{37} \cdot \frac{1}{2}(0)$$

$$= -\frac{1}{74} = -.0135$$

15. American roulette is just like Monte Carlo roulette, except that there are two slots numbered 0 (actually one is 00 and the other 0). If the rules are as in the example, what are the expected winnings of a player who stakes $1 on red?

16. If a player stakes $1 on number 17 in American roulette, what are his or her expected winnings? (See Problem 15 and also the discussion in the text.)

Miscellaneous Problems

17. A hat contains 1 penny, 4 nickels, 2 dimes, and 1 quarter. How much should a person be willing to pay for the chance to draw one coin from the hat, assuming he or she keeps it?

18. Kira draws one card from a standard deck. If it is a red ace, she receives $100; if it is a king she receives $50. Otherwise, she must pay $4. What are her expected winnings?

19. In the game of chuck-a-luck, a player chooses a number from 1 to 6 and then throws three dice. If the player's number appears on just 1 die, he or she receives $1; if it appears on just 2 dice, the player receives $2; if it appears on all 3 dice, the player receives $3. What are the player's expected winnings?

20. Suppose that the probability that a 40-year-old man will die by age 45 is .05. If such a man wants to purchase $10,000 of term insurance for 5 years by paying a single premium now, what should he expect to have to pay? Keep in mind that an insurance company would have to pay $10,000 to his beneficiaries if he died within the 5-year period, but nothing if he lived through it. (*Note:* In making your calculation, ignore the fact that the insurance company has overhead and also that it can invest the premium.)

21. The probability that a 48-year-old man will live to at least 65 is approximately .70. If such a man purchases a pure endowment of $10,000 by payment of a single premium now, what should he expect to pay? Keep in mind that for a pure endowment, an insurance company must pay $10,000 at age 65 if the man lives till then, but nothing if he dies before then. (*Note:* Neglect insurance company overhead and income on premiums in making your calculation.)

22. If a teacher wants to grade a true-false test in such a way that the expected score for a person who simply guesses is 0, he or she might give 1 point for a correct answer and −1 for a wrong answer. How should the teacher score a multiple choice test where each item has 4 responses to achieve a similar result?

23. Find the expected number of games in a baseball World Series in which the two teams are evenly matched. (*Note:* The first team to win 4 games wins the series.)

24. A coin is tossed until a head appears or until 5 tails in a row occur. What is the expected number of tosses required?

CHAPTER SUMMARY

If an experiment can result in n **equally likely outcomes,** of which m result in the event E, then the **probability** of E is m/n. There are four main properties of probability. Stated in set language, they are

1. $P(\phi) = 0; P(S) = 1;$
2. $0 \leq P(A) \leq 1;$
3. $P(A \cup B) = P(A) + P(B),$ provided $A \cap B = \phi;$
4. $P(A) = 1 - P(A').$

If the probability of event A is unaffected by whether B has occurred or not, we say that A and B are **independent.** In this case, $P(A \cap B) = P(A) \cdot P(B)$. But if A and

B are **dependent,** then

$$P(A \cap B) = P(B)P(A \mid B)$$

We call this the **multiplication rule** for probabilities.

Suppose that an experiment can result in two types of outcomes, success or failure. If this experiment is performed n times (independently of each other) then

$$P(k \text{ successes}) = {}_nC_k p^k q^{n-k}$$

where p and q are the probabilities of success and failure, respectively, on each trial. The number of successes is said to have a **Binomial Distribution.**

If an experiment can result in k outcomes, which produce corresponding payoffs $a_1, a_2, a_3, \ldots, a_k$ with probabilities $p_1, p_2, \ldots, p_k$, respectively, then the **expected value** of the experiment is

$$E = a_1 p_1 + a_2 p_2 + \cdots + a_k p_k$$

CHAPTER REVIEW PROBLEM SET

1. The letters *ALGEBRA* are written on seven cards and put in a bag. If a letter is drawn at random, what is the probability it is an *A*? A consonant?

2. Three dice are tossed. What is the probability of getting a total of 3? Of 4? Of more than 4?

3. A whole number between 1 and 100 (inclusive) is picked at random. What is the probability that its decimal representation has no 9's?

4. Find the probability that at least one 5 will appear in three throws of a die.

5. If the probability that Oscar will marry is .8, that he will graduate from college is .5, and that he will do one or the other is .95, what is the probability that he will do both?

6. An urn contains 3 red and 7 green balls. If two balls are drawn at random, what is the probability both are red if
   (a) the first ball is not replaced before the second is drawn;
   (a) the first ball is replaced before the second is drawn?

7. Consider three urns. Each of the first two have 3 red and 7 green balls, but the third has 4 red and 4 green balls. If an urn is chosen at random and then a ball drawn, what is the probability it is red?

8. At Wescott College, both English 11 and History 13 are required courses. On the first attempt 30 percent of the students fail English 11, 20 percent fail History 13, and 8 percent fail both. Find
   (a) the probability that a student will fail one or the other;
   (b) the probability that a student will fail English 11, given that he or she has already failed History 13.

9. To decide who among 4 friends pays for lunch each day, one of them tosses two coins. If both coins come up heads, John pays. What is the probability that John will have to pay
   (a) at least once in the next 4 days;
   (b) exactly twice in the next 4 days?

10. A die is tossed 100 times. Write an expression for the probability of getting

(a) exactly thirty 6's;

(b) fourteen or fewer 6's.

11. I have been offered a payoff in dollars equal to the square of the number of spots showing when I toss a die, provided I pay a certain fee. How much can I afford to pay per toss if I expect to play many times?

12. Suppose that you are to toss a die and will receive $1 if you get a 5 or a 6, pay 50¢ if you get a 4, and pay 30¢ if you get anything else. What are your expected winnings?

13. Ken will toss a fair coin 10 times. He will receive $1 if he gets more than 7 heads and pay 5¢ if he gets 7 or less heads. What are his expected winnings?

11

MATHEMATICS
OF FINANCE

*Then you ought to have put
my money on deposit, and
on my return I should have
got it back with interest.*

*Parable of the talents
Matthew 25:27*

11-1
Simple and Compound Interest

In two earlier problem sets, we took a quick look at the subject of compound interest. In this section, we want to be more systematic and thereby set the stage for consideration of several problems from the world of finance.

A basic expectation in American society is that one's money should earn more money. Some people try for the fast dollar at the gambling table, while others play the stock market. But the surest and simplest way of having one's money make more money is by putting it in a savings account at a bank. The father in our opening display is hoping that a $5000 deposit now will earn enough additional money (called **interest**) so that he will have enough to pay for his daughter's college education some years from now. The amount of interest earned will depend on the rate of interest, the time involved, and how interest is computed.

SIMPLE INTEREST

If the interest is always based on the original investment, called the **principal,** then we refer to the interest as **simple interest.** It is usually described as a certain percent per year. Suppose that we invest $2000 at 6 percent interest per year. The interest for one year is then

$$2000(.06) = \$120$$

For 8 years, it would be

$$2000(.06)(8) = \$960$$

To generalize, let P stand for the principal, r for the rate of interest per year, and t for the time in years. Then the simple interest is given by

$$I = Prt$$

The **final amount** F is the principal plus the interest.

$$F = P + I = P(1 + rt)$$

If the man in our opening display could invest his $5000 at 8 percent simple interest for 18 years, he would have a final amount of

$$5000[1 + (.08)(18)] = \$12,200$$

hardly enough for 4 years of college. But wait! Simple interest is for simple-minded people. A clever man would demand compound interest. That is our next subject.

COMPOUND OF INTEREST

If, at regular intervals of time, interest is converted to principal, we say that money is earning **compound interest.** As an example, suppose that $2000 is invested at 7 percent compounded (that is, converted to principal) annually. At the end of 1 year, the interest earned ($2000(.07) = $140) is added to the principal, giving

$$2000 + 2000(.07) = 2000(1.07) = \$2140$$

This new principal earns interest during the second year. Thus, at the end of the second year, there is an amount of

$$2140 + 2140(.07) = 2140(1.07) = 2000(1.07)^2 = \$2289.80$$

which then becomes the principal for the third year. At the end of 3 years, there will be

$$2289.90(1.07) = 2000(1.07)^3 = \$2450.09$$

Notice that each year the principal is multiplied by 1.07, so that after n years it accumulates to $2000(1.07)^n$.

Consider again the concerned father of our opening display. If he invested his $5000 at 8 percent compounded annually, it would grow to

$5000(1.08)^{18}$ by the end of 18 years. We can compute this on a scientific calculator (with algebraic logic) by pressing the keys

$$5000 \;\boxed{\times}\; 1.08 \;\boxed{y^x}\; 18 \;\boxed{=}$$

Note that the calculator automatically does the exponential calculation before the multiplication, so no parentheses are needed. The result ($19,980.10) exceeds the corresponding simple interest amount ($12,200) by more than 60 percent. It's still not enough to pay for 4 years of college, but it's a good start.

Incidentally, older books offered tables for calculating numbers like $(1.08)^{18}$. We do not do this, partly because tables greatly restrict the variety of interest rates that can be used (try finding an interest rate of 8.231 percent in a table), but also because interest problems are extremely easy to do on a calculator. And, of course, that is the way banks now calculate interest, although they use a computer rather than a pocket calculator.

There is no law that says interest should be compounded annually. Until a few years ago, most banks compounded interest quarterly. Today many compound the interest daily and a few try to get advertising mileage out of compounding instantly.

To see what compounding more frequently means, consider again an investment of $5000 at 8 percent interest, but now compounded quarterly. Interest rates are always stated as annual rates, so 8 percent compounded quarterly really means $8/4 = 2$ percent per quarter. In 18 years, there are $4(18) = 72$ quarters. Consequently, an investment of $5000 at 8 percent compounded quarterly will grow to

$$5000(1.02)^{72} \approx \$20{,}805.70$$

after 18 years. At 8 percent compounded monthly ($.08/12 \approx .00666667$ per month), it will grow to

$$5000(1.00666667)^{216} \approx \$21{,}003$$

At 8 percent compounded daily ($.08/365 \approx .00021918$) for 18 years (or $18 \cdot 365 = 6570$ days), $5000 will accumulate to

$$5000(1.00021918)^{6570} \approx \$21{,}100$$

It seems clear that it is to the saver's advantage to have interest (at a given rate) compounded as often as possible.

Incidentally, you may have trouble reproducing our answers to the last two problems on your calculator. With such large exponents, errors due to rounding become significant and these errors may vary with the calculator model and may be affected by the order in which operations are done. To make the last calculation above, we wrote

$$5000\left(1 + \frac{.08}{365}\right)^{6570}$$

and pressed the following keys.

$$5000 \;\boxed{\times}\; \boxed{(}\; 1 \;\boxed{+}\; .08 \;\boxed{\div}\; 365 \;\boxed{)}\; \boxed{y^x}\; 6570 \;\boxed{=}$$

The general formulas for compound interest are obtained in a manner analogous to what we have just done. If an original principal P is invested at an annual rate r compounded m times a year for a total of n conversion periods, the final amount F is

$$F = P\left(1 + \frac{r}{m}\right)^n$$

The interest earned is

$$I = F - P = P\left(1 + \frac{r}{m}\right)^n - P = P\left[\left(1 + \frac{r}{m}\right)^n - 1\right]$$

Notice that r/m is the interest rate per conversion period and that n is the number of these periods (not the number of years).

EFFECTIVE RATES

Suppose that we want to find out which gives better returns, 7.1 percent compounded monthly or 7.2 percent compounded semiannually. To answer, we use the concept of effective rate. For a given compound rate, the **effective rate** is that rate which compounded annually yields the same return as the given rate. Here is a simpler way to say the same thing: The effective rate is just the interest earned on $1 during 1 year at the given rate. Thus, for a given annual rate r compounded m times per year, the effective rate r_e is

$$r_e = \left(1 + \frac{r}{m}\right)^m - 1$$

For $r = .071$ and $m = 12$

$$r_e = \left(1 + \frac{.071}{12}\right)^{12} - 1 \approx .07335664$$

and for $r = .072$ with $m = 2$

$$r_e = \left(1 + \frac{.072}{2}\right)^2 - 1 = .073296$$

We see that 7.1 percent compounded monthly gives a better return.

ⓒ Problem Set 11-1

Find the simple interest and the final amount for the given data.

1. $P = \$235$; $r = 7$ percent; $t = 5$.
2. $P = \$450$; $r = 8$ percent; $t = 4$.
3. $P = \$1200$; $r = .075$; $t = 10$.

4. $P = \$1850$; $r = .065$; $t = 12$.

5. $P = \$2250$; $r = 8.25$ percent; $t = 14$.

6. $P = \$1775$; $r = 7.75$ percent; $t = 9.5$.

In Problems 7–12, find the final amount and the compound interest for the given principal, compound rate, and time period.

7. $2000; 7 percent compounded annually; 11 years.

8. $3500; 8 percent compounded semiannually; 7 years.

9. $1800; 9 percent compounded quarterly; 8.5 years.

10. $1500; 8.5 percent compounded quarterly; 14 years.

11. $2250; 9.5 percent compounded monthly; 6 years 8 months.

12. $1780; 8.5 percent compounded daily; 4 years.

13. Find the effective rate corresponding to an annual rate of 8.5 percent compounded quarterly.

14. Find the effective rate corresponding to an annual rate of 7 percent compounded daily.

15. Which gives a better return, 7.8 percent compounded quarterly or 7.5 percent compounded daily?

16. Which gives a better return, 9 percent compounded annually or 8 percent compounded daily?

EXAMPLE A (Present value)   If we multiply both sides of the compound interest formula

$$F = P\left(1 + \frac{r}{m}\right)^{n}$$

by $\left(1 + \dfrac{r}{m}\right)^{-n}$, we get

$$P = F\left(1 + \frac{r}{m}\right)^{-n}$$

This gives the principal that will accumulate to F after n periods. It is called the **present value,** or *discounted value,* of F. Use this formula to find out how much should be invested at 9 percent compounded quarterly in order to have an amount of $12,000 after 12 years.

Solution.   We have $F = 12{,}000$, $r/m = .0225$, and $n = 4 \cdot 12 = 48$. Therefore

$$P = 12{,}000(1.0225)^{-48} \approx \$4124.22$$

On a typical scientific calculator, you can calculate this value by pressing

$$12000 \;\boxed{\times}\; 1.0225 \;\boxed{y^x}\; 48 \;\boxed{+/-}\;\boxed{=}$$

17. Find the present value of $10,000 due in 9.5 years at 8 percent compounded semiannually.

18. Find the present value of $8500 due 12 years from now at 10 percent compounded monthly.

19. How much should Sally invest at age 30 at 8 percent compounded quarterly in order to have a lump sum of $80,000 at her retirement age of 65?

20. How much should David invest in a savings account now if he expects to need $8000 for a car 4 years from now? Interest is at 7 percent compounded daily.

EXAMPLE B (Unknown rates) It is desired to have $3000 accumulate to $7410 by the end of 14 years.

(a) What simple interest rate is necessary to achieve this?

(b) What rate compounded monthly is necessary to do it?

Solution.

(a) Since $I = Prt$, $r = I/Pt$. Applying this with $I = 4410$, $P = 3000$, and $t = 14$, we get

$$r = \frac{4410}{(3000)(14)} = .105 \quad \text{or} \quad 10.5 \text{ percent}$$

(b) If $F = P\left(1 + \dfrac{r}{m}\right)^n$, then

$$\frac{F}{P} = \left(1 + \frac{r}{m}\right)^n$$

$$\left(\frac{F}{P}\right)^{1/n} = 1 + \frac{r}{m}$$

$$r = m\left[\left(\frac{F}{P}\right)^{1/n} - 1\right] = m\left[\sqrt[n]{\frac{F}{P}} - 1\right]$$

With $F = 7410$, $P = 3000$, $m = 12$, and $n = 168$, this gives

$$r = 12\left[\sqrt[168]{\frac{741}{300}} - 1\right] \approx .0648$$

We got this answer by pressing

$$12 \; \boxed{\times} \; \boxed{(} \; \boxed{(} \; 741 \; \boxed{\div} \; 300 \; \boxed{)} \; \boxed{\text{INV}} \; \boxed{y^x} \; 168 \; \boxed{-} \; 1 \; \boxed{)} \; \boxed{=}$$

21. At what simple interest rate will $2000 accumulate to $3480 in 8 years?

22. At what simple interest rate will $1480 accumulate to $3034 in 12 years?

23. At what rate compounded semiannually will $2000 accumulate to $3480 in 8 years?

24. At what rate compounded monthly will $1480 accumulate to $3034 in 12 years?

25. What is the rate if money doubles in value in 12 years and the interest is
    (a) simple;
    (b) compounded annually;
    (c) compounded quarterly?

26. Do Problem 25 with the word *doubles* replaced by *triples*.

EXAMPLE C (Unknown times) If $1500 is invested today, how long will it take to accumulate to $3400

(a) at 9 percent simple interest;

(b) at 8.5 percent compounded semiannually?

Solution.

(a) Solving the formula $I = Prt$ for t yields $t = I/Pr$. Thus, for our problem,

$$t = \frac{1900}{(1500)(.09)} \approx 14.1 \text{ years}$$

(b) We substitute our values in the formula $P(1 + r/m)^n = F$ and then solve for n by taking the logarithm of each side.

$$1500(1.0425)^n = 3400$$

$$(1.0425)^n = \frac{34}{15}$$

$$n \log 1.0425 = \log 34 - \log 15$$

$$n = \frac{\log 34 - \log 15}{\log 1.0425} \approx 19.66$$

Now 19.66 periods is 9.83 years. Ignoring the technicality that interest for the 20th period is not converted to principal until the end of that period, we conclude that the time required is approximately 9.83 years.

27. If $1400 is invested at $8\frac{1}{2}$ percent simple interest, when will the amount reach $3200?

28. If $2500 is invested at $9\frac{1}{2}$ percent simple interest, when will the amount reach $4000?

29. If $1400 is invested at $8\frac{1}{2}$ percent compounded quarterly, when will the amount be $3200?

30. If $2500 is invested at $9\frac{1}{2}$ percent compounded monthly, when will the amount be $4000?

31. How long does it take money to double its value if it is invested at

(a) 9 percent simple interest;

(b) 9 percent compounded monthly;

(c) 9 percent compounded daily?

32. Work Problem 31 with the word *double* replaced by *triple*.

Miscellaneous Problems

33. If $3000 is invested at 9.6 percent simple interest, what is the final amount at the end of 11 years?

34. What is the amount at the end of 11 years if $3000 is invested at 9.6 percent interest compounded quarterly?

35. What amount invested today will accumulate to $3400 by the end of 8 years if it is invested at 8 percent compounded semiannually?

36. What rate compounded quarterly is necessary so $3250 will accumulate to $12,000 in $15\frac{1}{2}$ years?

37. If $6000 is deposited in a bank that pays 8 percent compounded quarterly, when will its value reach $10,000?

38. If I borrow money at 9.75 percent interest compounded quarterly, what is the effective rate that I am paying?

39. What annual rate compounded monthly is equivalent to 9.5 percent compounded semiannually? (*Hint:* Use effective rates.)

40. Nine years ago, Helen Smith borrowed $6400 at 8.5 percent compounded annually. What is the amount of her debt today, assuming that she has not made any payment at all?

41. Suppose that Ms. Smith (see Problem 40) made a payment of $2000 exactly 3 years ago. How much does she owe today? (*Hint:* First determine how much she owed right after that payment.)

42. Suppose, by some stroke of fortune, you are able to get 100 percent interest. What will a dollar be worth at the end of a year if interest is compounded
    (a) quarterly;
    (b) monthly;
    (c) daily;
    (d) hourly?
    Look at the numbers you obtained above and make a conjecture about the number that is approached as interest is compounded more and more frequently (see Section 6-5).

43. If $1 is put in the bank today at 8 percent interest compounded n times per year, it will grow to

$$\left(1 + \frac{.08}{n}\right)^n$$

at the end of a year. If n is allowed to grow without bound, we obtain what is called **instant compounding** or **continuous compounding** (see Section 6.5). If this is done, to what value will the dollar grow by the end of one year? (*Note:* To complete this calculation, you need to know that $(1 + h)^{1/h}$ approaches $e = 2.71828 \ldots$ as h approaches 0.)

Question Two

If at the end of each month, I put $100 into a savings account earning 7 percent compounded monthly, how much would that account be worth at the end of 18 years?

11-2
Annuities

Any sequence of equal payments occurring at regular intervals of time constitutes an **annuity.** The monthly deposits contemplated by the man above constitute an annuity of 216 payments. Most people pay off the mortgage on their house in the form of an annuity, part of each payment covering the interest, with the balance used to reduce the principal.

One of the things we want to know about an annuity is its total value at the time of the last payment. That is what our concerned father wants to know. We call this value the amount of an annuity.

AMOUNT OF AN ANNUITY

Consider an annuity consisting of n payments of R dollars each, with interest rate r compounded m times a year. Suppose, as we almost always will, that the intervals between payments coincide with the interest conversion periods, and that the first payment occurs at the end of the first period. A line diagram offers a convenient way to picture this.

Notice that the first payment will draw interest for $n - 1$ periods, the second for $n - 2$ periods, and so on. The last payment will not earn any interest. The **amount of the annuity** is simply the sum of the values of each of these payments at the end of the line, that is, at the time of the last payment. Because it is the *final* value of a *sum*, we denote it by S_f. If $i = r/m$ is the interest rate per period, then S_f is given by

$$S_f = R(1 + i)^{n-1} + R(1 + i)^{n-2} + \cdots + R(1 + i) + R$$

$$= R + R(1 + i) + \cdots + R(1 + i)^{n-2} + R(1 + i)^{n-1}$$

The terms in this sum form a geometric sequence (see Section 9-3) with common ratio $1 + i$. Recalling that

$$a_1 + a_1 r + \cdots + a_1 r^{n-1} = a_1 \frac{r^n - 1}{r - 1}, \qquad r \neq 1$$

we see that

$$S_f = R \frac{(1 + i)^n - 1}{i}$$

Now we can answer the question in our opening example. There $R = 100$, $i = .07/12$, and $n = 216$, so

$$S_f = 100 \frac{[(1 + .07/12)^{216} - 1]}{.07/12} = \frac{1200}{.07}\left[\left(1 + \frac{.07}{12}\right)^{216} - 1\right] \approx 43{,}072.11$$

That might be enough for a college education, even if tuition continues to inflate over the next 18 years.

PRESENT VALUE OF AN ANNUITY

Suppose that a young couple in the market for a new house is wondering how large a mortgage loan they can afford. They expect to secure the loan at 9 percent interest compounded monthly for a period of 30 years, and they are prepared to make monthly payments as high as $420. What they really want to know is the value today of an annuity of 360 monthly payments of $420. This is called the **present value of an annuity** and is denoted by S_p.

A formula for the present value S_p is easily derived from the formula for the final value S_f displayed earlier. We need only discount the final value S_f back to the present to get S_p. Or, to put it another way, we need to find the amount of money S_p which, if accumulated for n periods, would give S_f. Clearly

$$S_f = S_p(1 + i)^n$$

and so

$$S_p = S_f(1 + i)^{-n} = R \frac{(1 + i)^n - 1}{i}(1 + i)^{-n}$$

which simplifies to

$$S_p = R \frac{1 - (1 + i)^{-n}}{i}$$

To make the computation for our young couple, we use $R = 420$, $i = .09/12 = .0075$, and $n = 360$. The result is

$$S_p = 420\frac{1 - (1.0075)^{-360}}{.0075} = \$52{,}198.38$$

They can afford a loan of \$52,198.38.

Now consider some shocking facts. To pay off this loan, they will make 360 payments of \$420, for a total of \$151,200. That means that almost \$100,000 will be paid in interest. but even more sobering is the fact that if, instead of buying a house, they were to put these \$420 payments in the bank at 9 percent interest compounded monthly, these payments would accumulate to a whopping

$$S_f = R\frac{(1 + i)^n - 1}{i} = 420\frac{(1.0075)^{360} - 1}{.0075} \approx \$768{,}912.34$$

at the end of 30 years.

What is the moral of this story? Should young couples avoid buying houses? No, since houses, at least in recent years, have tended to inflate in value at rates comparable to interest rates. But we do suggest that people should beware of another version of the same phenomenon, namely installment buying. In addition to the exceptionally high interest rates (often 18 percent) usually associated with installment buying, such buying is almost always used to purchase products that deflate in value with time. We shall discuss installment buying in Example B.

▣ Problem Set 11-2

In Problems 1–4, calculate the present value S_p and the amount (final value) S_f of annuities satisfying the given data.

1. $R = \$250$; $i = .038$; $n = 40$.
2. $R = \$400$; $i = .043$; $n = 64$.
3. $R = \$360$; $r = 9$ percent; $m = 4$; $n = 50$.
4. $R = \$520$; $r = 8.5$ percent; $m = 12$; $n = 300$.
5. Jeff Chow is taking out a 20-year mortgage loan on a house, with an interest charge of 8 percent compounded monthly. If his monthly payments, starting one month from now, are \$245, find the value of the loan.
6. Roberta Rivard has purchased a summer home by putting down \$5000 cash, with the balance due in monthly payments of \$150 spread over 10 years. The first payment is due 1 month after purchase. If interest is at 9 percent compounded monthly, what is the actual purchase price?
7. Helen Bergquist is going to make 24 semiannual deposits of \$850 in a savings account, starting 6 months from now. If the account earns interest at 6 percent compounded semiannually, how much will she have in her account after her last deposit?
8. Maria Valdez is going to make 48 quarterly deposits of \$425 in a savings

account starting 3 months from now. How much will she have after her last deposit if interest is at 6 percent compounded quarterly?

9. John Jones has decided to prepare for retirement by putting $50 in a savings account at the end of each month for the next 40 years. If interest is at 7.5 percent compounded monthly, how much will he have at retirement time?

10. Suppose that John Jones (see Problem 9) decided to puchase a piece of property with his monthly payments instead of putting them in a savings account. If the payments and interest are the same as in Problem 9, how much can he afford to pay for the property?

EXAMPLE A (Unknown payments) How much must I deposit in a savings account at the end of each month in order to have $5000 in the account at the end of 10 years if interest is at 8 percent compounded monthly?

Solution. This is an annuity whose final amount S_f is known. If we solve

$$S_f = R\frac{(1 + i)^n - 1}{i}$$

for R we get

$$R = \frac{iS_f}{(1 + i)^n - 1}$$

Substituting $S_f = 5000$, $i = .08/12$, and $n = 120$ yields

$$R = \frac{(.08/12)(5000)}{(1 + .08/12)^{120} - 1} \approx 27.33$$

Note that if S_p had been given rather than S_f, we would have solved the present value formula for R, obtaining

$$R = \frac{iS_p}{1 - (1 + i)^{-n}}$$

In Problems 11–14, find the value of R for the given data.

11. $S_f = \$12,400$; $i = .035$; $n = 30$.
12. $S_f = \$9500$; $i = .045$; $n = 40$.
13. $S_p = \$12,000$; $r = 9.2$ percent; $m = 4$; $n = 50$.
14. $S_p = \$20,000$; $r = 8.8$ percent; $m = 12$; $n = 48$.
15. Clara Henley wants to have $12,500 in 10 years. To accumulate this amount she will make 20 equal semiannual payments starting 6 months from now. If interest is at 8.6 percent compounded semiannually, how large will her payments be?

16. John Kim expects to buy a new car 5 years from now. In order to have $10,000 available by that time, he plans to make 20 equal quarterly deposits in a fund that accumulates at 7.6 percent compounded quarterly, with the first deposit 3 months from now. Find the size of these deposits.

17. Mr. Newburg owes $3800 to Ms. Kennedy who insists on charging 10 percent interest compounded monthly. Newburg agrees to pay off his debt by making 36 equal monthly payments, starting one month from now. How much must he pay each month?

18. If John Kim (see Problem 16) chose to buy a new car today, paying it off with the payments of that problem, how expensive a car could he afford?

EXAMPLE B (Installment buying)   Sylvia Carr bought a watch marked $200 and has agreed to pay for it by making monthly payments of $10 for 24 months, starting 1 month from now. The clerk pointed out that this results in interest of $40 over 2 years, or $20 per year, only 10 percent per year. What is wrong with the clerk's reasoning and what is the actual interest rate?

Solution.   The clerk is wrong because the principal does not stay at $200; it will steadily be reduced. Now we do know the present value S_p, the payment R, and the number of periods n for an annuity. From the formula

$$S_p = R\frac{1 - (1 + i)^{-n}}{i}$$

we obtain

$$200 = 10\frac{1 - (1 + i)^{-24}}{i}$$

Even with all our algebraic expertise, this equation is too hard to solve for i. But being forwarned that installment rates typically run about 18 percent per year (1.5 percent per month), let us try substituting $i = .015$ on the right side of our last equation and see what it gives.

$$10\frac{1 - (1.015)^{-24}}{.015} = 200.30$$

We conclude that the rate is very close to 18 percent compounded monthly. Incidentally, that corresponds to an effective rate

$$r_e = (1.015)^{12} - 1 = .196$$

which is almost 20 percent.

19. If an item marked $163.50 can be purchased by paying 20 monthly installments of $10 starting one month from now, find the annual interest rate, assuming it is compounded monthly. (You will have to experiment; an answer to the nearest percent is acceptable.)

20. Find to the nearest percent the annual interest rate that is being charged if a used car costing $2062 is to be purchased in 24 monthly installments of $100, with the first payment one month from now. Assume that interest is compounded monthly.

21. Find the present value and final amount of an annuity with payments of $400 at the end of each 6 months for the next 17 years if interest is at 9.4 percent compounded semiannually.

22. Find the payment R if the final amount S_f is $2800 with $i = .049$ and $n = 20$.

23. Find R if $S_p = $24,000$ $i = .052$, and $n = 60$.

24. For the last 10 years, Mary has put $40 a month into a savings account paying interest at 6 percent compounded monthly. Today she made her last (120th) deposit. How much does she have in her account?

25. Tony has $18,000 in a savings account that pays 8 percent interest compounded quarterly. He plans to make eight equal quarterly withdrawals starting 3 months from now. How much can he withdraw each time?

26. After Tony (see Problem 25) has made 2 withdrawals, how much was in his account? (*Hint:* This can be done as a present value problem.)

27. How expensive a car did Helen purchase if she is paying for it in $100 monthly installements spread over 4 years and interest is at 12 percent compounded monthly?

28. Linda owes her father $4500, which she will repay by making semiannual payments of $250, starting 6 months from now, until a last, smaller payment will settle the debt. How many regular payments will be required? Assume interest at 8 percent compounded semiannually.

29. So far we have always assumed that the payments in an annuity occur at the ends of periods, with the first payment at the end of the first period. Develop formulas for S_p and S_f assuming that the payments come at the beginning of each period with the first one right now.

30. What is the value of an annuity 10 years from now if it consists of 10 payments of $100 at the beginning of each year starting right now? Interest is at 8 percent compounded annually.

31. By using the concept of annuities, decide which of the following numbers is larger.

$$\frac{1 - (1.04)^{-80}}{.04} \quad \text{or} \quad \frac{1 - (1.05)^{-80}}{.05}$$

Now check that you are right by using your calculator.

FIRST NATIONAL BANK

Question Three

This is my 50th payment. How much do I still owe on my house?

11-3
Amortization and Sinking Funds

Most problems associated with saving or borrowing money are just compound interest problems related to an annuity. But here, as in other fields, special words are used to describe special situations. When a debt is retired by making equal payments at equal intervals of time with part of each payment covering the interest due and the rest being applied to the principal, the debt is said to be **amortized**. It is, of course, being paid off by an annuity.

The woman above, call her Ella Flynn, has just made her 50th house payment. She wants to know how much she still owes. She might also ask how much of her next payment will be interest and how much of it will lower the principal. Such questions are usually answered by referring to an **amortization schedule**.

AMORTIZATION SCHEDULES

Let us suppose that Ella's original mortgage loan was for $30,000, to be paid off in equal monthly payments for 20 years with interest at 9 percent compounded monthly. That information determines her monthly payment R (see Example A of Section 11-2).

$$R = \frac{iS_p}{1 - (1 + i)^{-n}} = \frac{(.0075)(30,000)}{1 - (1.0075)^{-240}} \approx \$269.92$$

The first portion of Ella's amortization schedule should look like Table 5 at the top of the next page. To see how these numbers are obtained, note that the interest for the first month is $(.0075)(30,000) = 225$ and so the principal repaid is $\$269.92 - \$225.00 = \$44.92$. This leaves a new principal of $\$30,000.00 - \$44.92 = \$29,955.08$. The same procedure can be continued to

TABLE 5 Amortization Schedule

Payment Number	Principal Before Payment	Interest Due	Payment	Amount of Principal Repaid
1	30,000.00	225.00	269.92	44.92
2	29,955.08	224.66	269.92	45.26
3	29,909.82	224.32	269.92	45.60
4	29,864.22	223.98	269.92	45.94

obtain the remaining numbers in the schedule. The complete schedule would tell Ella anything she wanted to know about her debt.

But suppose she and the bank had both lost their copies of the schedule or that they never had such copies. Could the banker still tell her how much she owes right after the 50th payment? Yes. If the banker has a pocket calculator and knows the formulas of this chapter, all he or she needs to do is to calculate the present value of the annuity formed by the 190 (that is, $240 - 50$) remaining payments. And that is

$$S_p = R\frac{1 - (1 + i)^{-n}}{i} = 269.92\frac{1 - (1.0075)^{-190}}{.0075} \approx \$27{,}287.45$$

Knowing this, the banker can calculate what part of the next payment will be interest, namely,

$$(27{,}287.45)(.0075) = \$204.66$$

This means that $\$269.92 - \$204.66 = \$65.26$ will apply to the principal.

SINKING FUNDS

Suppose that the ABC Company anticipates a capital expenditure of $40,000 for a new piece of equipment 5 years from now. In order to prepare for this expense, the manager decides to make equal annual deposits at the end of each of the next 5 years in a savings account that draws 8 percent interest compounded annually. An account such as this, whose sole purpose is to pay an obligation which will come due at a future date, is called a **sinking fund.** (It is, of course, just an annuity and we think it might better be called a *rising fund* since it grows with time.)

The first problem is to determine the size of the annual payment. This time we know the final value S_f, namely $40,000, and need the payment R (see Example A of Section 11-2).

$$R = \frac{iS_f}{(1 + i)^n - 1} = \frac{(.08)(40{,}000)}{(1.08)^5 - 1} \approx \$6818.26$$

For income tax purposes or for its report to stockholders, the ABC Company may want to know the status of the sinking fund at the end of each year. A **sinking fund schedule** is shown below. Notice that the amount in the fund at the end of each year exceeds the amount at the beginning of that year by the interest earned that year plus one deposit.

TABLE 6   Sinking Fund Schedule

Deposit Number	Amount at Beginning of Year	Interest Earned	Deposit	Amount at End of Year
1	—	—	6818.26	6818.26
2	6818.26	545.46	6818.26	14,181.98
3	14,181.98	1134.56	6818.26	22,134.80
4	22,134.80	1770.78	6818.26	30,723.84
5	30,723.84	2457.91	6818.26	40,000.01

Problem Set 11-3

1. A debt of $12,000 is being retired by means of equal payments at the end of each 6 months over a period of 15 years. The interest rate is 8.6 percent compounded semiannually. Find
   (a) the semiannual payment;
   (b) the amount still owed right after the 20th payment;
   (c) the interest portion of the 21st payment.

2. Paul McPherson borrowed $8000 from the Acme Loan Association and agreed to repay it by making equal payments at the end of each month for 5 years with interest at 10.4 percent compounded monthly. Find
   (a) his monthly payment;
   (b) the amount still owed right after the 40th payment;
   (c) the part of the 41st payment that will apply to the principal.

3. Construct the first four lines of the amortization schedule for Problem 1.

4. Construct the first four lines of the amortization schedule for Problem 2.

5. Hazel McNabb plans to buy a house 8 years from now and wants to accumulate $8400 for a down payment. To provide for this, she will make equal deposits at the end of each month in a sinking fund that earns interest at 9.6 percent compounded monthly. Find
   (a) her monthly deposit;
   (b) the amount in her fund right after the 40th deposit;
   (c) the amount by which the fund increases at the end of the 41st month.

6. To provide for replacement of equipment at a cost of $100,000 in 10 years, the XYZ Company decides to make equal deposits at the end of each of these years in a sinking fund that earns 9.5 percent interest compounded annually. Find

(a) the size of the ar.nual deposit;

(b) the amount in the fund right after the 6th deposit;

(c) the amount of increase in the fund at the end of the 7th year.

7. Construct the first four lines of a sinking fund schedule for Problem 5.

8. Construct the complete schedule for the sinking fund of Problem 6.

9. Mrs. King, who bought a car on the installment plan by agreeing to pay $100 at the end of each month for 36 months, inherited a large sum of money right after making her 20th payment. If the finance company will allow her to pay her remaining debt in a lump sum, what is that sum if interest is at 15 percent compounded monthly?

10. James Johnson is paying for his house by making payments of $300 at the end of each month for 25 years, with interest at 10 percent compounded monthly. After making his 200th payment, he decides to pay his remaining debt in one lump sum. Find this sum.

Miscellaneous Problems

11. Robert Hillman borrowed $24,000 from the Mayberry Savings and Loan Association with interest at 8.64 percent compounded semiannually. He has agreed to pay off his debt by making equal payments at the end of each 6- month period over the next 10 years. Find

(a) the amount of each payment;

(b) the total amount of interest he will pay over the 10 years.

12. Construct the first two lines of the amortization schedule for Problem 11.

13. Stewart Crosby knows he will need to buy a new combine for his farm in 4 years at a cost of $32,000. To provide for this, he decides to make 4 equal annual deposits, starting one year from now, in a fund that earns 9 percent interest compounded annually. Find

(a) the amount of each annual payment;

(b) the total amount of interest the fund will earn.

14. Construct the complete sinking fund schedule for Problem 13.

15. Suppose that in Problem 13 the interest was at 9 percent compounded daily. How much would Stewart's four payments be? (*Hint:* Find the effective rate—that is, the equivalent annual rate.)

16. Lucy Gaines is paying off her mortgage loan by making $200 monthly payments at the end of each month over a period of 25 years (300 payments). After making her 295th payment, she became ill and asked that the remaining debt be discharged by 12 equal monthly payments. If interest is at 9 percent compounded monthly, what will each of these remaining payments be?

17. Henrietta Smith has agreed to pay her cousin $300 at the end of each quarter for the next 5 years and then $200 at the end of each quarter for the 3 years after that. If interest is at 9 percent compounded quarterly, what does Henrietta actually owe today? (*Hint:* Think in terms of two annuities, one starting now.)

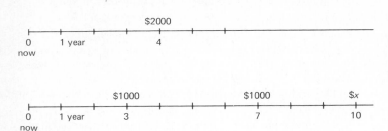

Question Four

What must x be so that the three payments on the lower line are equivalent to the single payment on the upper line if interest is at 8 percent compounded annually?

11-4
Equations of Value

If 1 plus 1 is not always 2 in money problems, it is for the same reason that the $1 I have today is more valuable than the $1 I will receive a year from now. Today's dollar can be invested and will be worth more than a dollar a year later.

The question posed above introduces the problem of replacing one set of obligations by another. Of course, the two sets are to be equivalent in value. This equivalence presupposes given interest rates and can then be expressed by an equation called an **equation of value.** Such an equation states that on a certain date, called the **comparison date**, two sets of obligations have the same value.

Before writing any equations, we need to recall how to accumulate a single payment R and how to discount it, given an interest rate of i per period. The line diagram below tells the story.

To accumulate R (bring it forward) k periods, multiply by $(1 + i)^k$; to discount it (bring it back) j periods, multiply by $(1 + i)^{-j}$.

AN EQUATION OF VALUE

We are ready to write an equation of value for our opening example. First we select a comparison date. We might as well take the present, though any other time would work. Thus we equate the present values of the obligations represented on the two lines

$$2000(1.08)^{-4} = 1000(1.08)^{-3} + 1000(1.08)^{-7} + x(1.08)^{-10}$$

Instead of solving this equation for x, let us write the corresponding equation using the end of 10 years as the comparison date. Now each payment must be accumulated to the 10-year mark.

$$2000(1.08)^6 = 1000(1.08)^7 + 1000(1.08)^3 + x$$

This latter equation is slightly easier to solve since the coefficient of x is 1, but note that the equations are really equivalent. If we multiply both sides of the first equation by $(1.08)^{10}$, we get the second one. When we solve for x, we obtain

$$x = 2000(1.08)^6 - 1000(1.08)^7 - 1000(1.08)^3 \approx \$200.21$$

ANOTHER EXAMPLE

Elmer Boyd has an obligation of $5000 due $3\frac{1}{2}$ years from now. He is negotiating with his creditor to make two equal payments instead, the first one 3 years from now and the other 4 years after that. What should these payments be, assuming that interest is at 7 percent compounded semiannually?

First we draw two lines on which each interval represents 6 months and x is the size of the unknown payments.

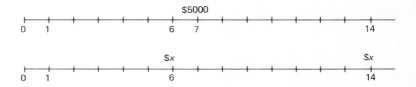

No choice of comparison date is especially good; we choose the 7-year mark, which corresponds to 14 on the two lines. Our equation is then

$$5000(1.035)^7 = x(1.035)^8 + x$$

or

$$5000(1.035)^7 = x[(1.035)^8 + 1]$$

This yields

$$x = \frac{5000(1.035)^7}{(1.035)^8 + 1} \approx \$2745.76$$

Thus each of the payments should be $2745.76.

DEFERRED ANNUITIES

A **deferred annuity** differs from an ordinary annuity in that payments start later than the end of the first conversion period. If the first payment occurs after $k + 1$ periods (rather than after 1 period), we say it is deferred k periods.

Suppose that a debt carrying an interest charge of 8 percent compounded monthly is to be repaid in 10 monthly installments of $250, the first one occurring after 4 months (that is, deferred 3 periods). What is the present value of the debt?

Here again, we draw a line diagram and use it to write an equation of value.

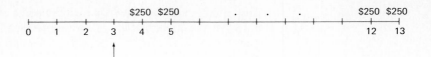

If we let x represent the present value of the annuity but use the end of the third period as the comparison date, we get

$$x\left(1 + \frac{.08}{12}\right)^3 = 250\frac{1 - (1 + .08/12)^{-10}}{.08/12}$$

Notice that on the left we accumulated x over 3 periods, whereas on the right we used the present value formula for an ordinary annuity. Solving for x gives

$$x = \frac{(250)(12)}{.08}\left[\left(1 + \frac{.08}{12}\right)^{-3} - \left(1 + \frac{.08}{12}\right)^{-13}\right] \approx \$2363.15$$

If you enjoy elegant tricks, you will appreciate another method of doing this problem. It involves introducing 3 fictitious payments, as shown below, finding the present value of the 13-payment ordinary annuity, and, finally, subtracting the present value of the fictitious payments.

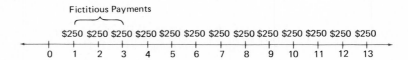

Thus, if x denotes the present value of the debt, we have

$$x = 250\frac{1 - (1 + .08/12)^{-13}}{.08/12} - 250\frac{1 - (1 + .08/12)^{-3}}{.08/12}$$

$$= \frac{(250)(12)}{.08}\left[\left(1 + \frac{.08}{12}\right)^{-3} - \left(1 + \frac{.08}{12}\right)^{-13}\right] \approx \$2363.15$$

© **Problem Set 11-4**

1. Assuming that money is worth 8 percent compounded annually, what single payment 6 years from now can replace the following two payments: $2500 due 3 years from now and $4000 due 8 years from now?

2. At 7 percent compounded semiannually, what single payment made $4\frac{1}{2}$ years from now can replace the following two payments: $1800 due now and $3200 due 8 years from now?

3. Suppose that the given two payments of Problem 1 are to be replaced by two equal payments, one at the end of 5 years and the other at the end of 10 years. How large must these payments be?

4. Suppose that the given two payments of Problem 2 are to be replaced by a payment of $2500 3 years from now and another payment 9 years from now. Find this last payment.

5. If money is worth 9 percent compounded monthly, find the present value of the deferred annuity of eleven $50 payments shown below.

6. If money is worth 10 percent compounded quarterly, find the present value of a 25-payment annuity, assuming that payments of $100 occur at 3-month intervals with the first one 18 months from now.

7. Assuming that money is worth 8.5 percent compounded annually, find the value of x such that the two annuities below are equivalent in value.

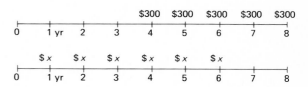

8. If interest is 9 percent compounded annually and the two annuities below are equivalent in value, find x.

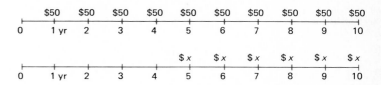

EXAMPLE A (Comparing values) At 7 percent compounded annually, which of the following two sets of payments is worth more?

(a)

(b)

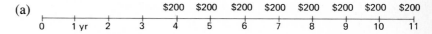

Solution. We must choose a comparison date. It does not make any difference what date we choose, since, with the same interest rate, if one set of payments is worth more than the other at one time, it is worth more at any other time. Computing both values at the 3-year mark, gives

(a) $$200\frac{1 - (1.07)^{-8}}{.07} \approx \$1194.26$$

(b) $$400 + 1100(1.07)^{-5} \approx \$1184.28$$

Thus at the end of 3 years, the first set of payments is worth about $10 more than the second. This difference in value would grow with time.

9. At 8 percent compounded annually, which of the following sets of payments is worth more?

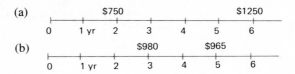

(a)

(b)

10. At 9 percent compounded annually, which of the two sets of payments below has the greater value?

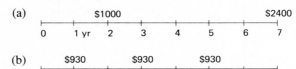

(a)

(b)

11. At 7 percent compounded annually, which of the annuities below is more valuable?

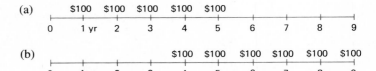

(a)

(b)

12. Suppose that each of the six payments of the second set in Problem 11 occurs 1 year earlier. Then which set is more valuable?

13. Solve Problem 11 using an interest rate of 4 percent compounded annually.

14. Suppose we want to alter the last payment of the second set in Problem 11 to make the two sets equivalent in value. What should the altered payment be if the interest rate is 7 percent compounded annually? Four percent compounded annually?

EXAMPLE B (Variable annuities) Just recently, banks have begun offering what they call **graduated payment loans.** The Smiths took out such a loan to pay for the $50,000 mortgage on their new home. During the first 5 years, they will make equal monthly payments of a certain amount. For the next 20 years, their payments will be twice as large. If interest is at 9 percent compounded monthly, how much are their initial payments?

Solution. We really have two annuities involved, one an ordinary annuity with payments of x dollars each and the other a deferred annuity with payments of $2x$ dollars each.

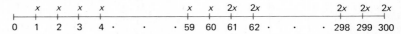

We might add the present values of these annuities and equate this sum to 50,000. We choose instead to take the end of 60 periods as the comparison date, accumulating the first annuity and discounting the second.

$$x\frac{(1.0075)^{60} - 1}{.0075} + 2x\frac{1 - (1.0075)^{-240}}{.0075} = 50,000(1.0075)^{60}$$

or

$$x[(1.0075)^{60} - 2(1.0075)^{-240} + 1] = (.0075)(50,000)(1.0075)^{60}$$

$$x \approx \$262.95$$

The method of fictitious payments (add in 60 payments of size x) also works well in this problem.

15. Robert Cooke took out a graduated payment loan with equal payments at the end of each of the first 48 months and double those payments for the next 120 months. If interest is at 9.6 percent compounded monthly and his initial loan was for $20,000, what is the size of his initial payments?

16. Mary Tompkins signed a mortgage for $5000 to be paid on a graduated schedule of equal monthly payments for the first year and monthly payments $1\frac{1}{2}$ times as large for the next 4 years. If interest is at 6 percent compounded monthly, what are her initial payments?

17. What is the present value of a (variable) annuity that consists of annual payments of $800 at the ends of the first 4 years and then increases to $1000 for each of the following 6 years? Assume that interest is at 9 percent compounded annually.

18. What is the present value of a (variable) annuity that consists of payments of $400 at the ends of the first 3 years, reduces to $300 for the following 3 years, and, finally, reduces to $200 for the last 3 years? Interest is at 8.37 percent compounded annually.

19. Jean Dubois plans to put $100 in the bank at the end of each month for the next year and increase this deposit to $150 a month for the following year. How much will he have at the end of that time if interest is at 7.5 percent compounded monthly?

20. Sylvia Kerr took out a graduated loan consisting of payments of $100 at the end of each month for 4 years, followed by payments of $150 at the end of each month for the next 6 years. At the end of 3 years, she decides to pay her loan off in one lump sum. How large is that payment if interest is at 8.5 percent compounded monthly?

21. Ian Worth plans to deposit $500 in a savings account at the end of each year for the next 6 years, then make five equal withdrawals at the ends of the following 5 years. Assuming the account is then depleted, how large will these withdrawals be if interest is at 6 percent compounded annually?

22. Jane Pompidou wishes to accumulate $6000 in a savings account by making semiannual deposits of $350 each until a smaller deposit 6 months after the last $350 deposit brings the account to exactly $6000. Assuming the account earns 7 percent compounded semiannually, find
    (a) the number of regular payments;
    (b) the last, smaller payment.

23. How much should Maria Ricardo deposit in a fund today so that she will be able to make six annual withdrawals of $1500 each starting 8 years from now, if the fund accumulates at 7.5 percent compounded annually?

24. Isaac Boldt will deposit $2000 in a fund at the end of each year for the next 12 years. If the fund accumulates at 7 percent compounded annually, how much will there be in the fund 20 years from now?

25. Clarence Dick's debt on his farm will be completely paid off after he has paid $12,000 5 years from now and $8000 9 years from now. He has made arrangements instead to pay $10,000 in 4 years and a final amount in 12 years. Find this last payment if interest on the debt is 9.25 percent compounded annually.

26. The Alliotos owed $13,250.87 on their house on January 1, 1980. Their monthly payment is $245.45 and the interest rate on their loan is 9 percent compounded monthly. Find the date of the last regular payment they will make and the amount of the smaller payment they will make one month later.

27. Suppose that Ed McNurty is to receive from Adam Stein a payment of $2000 after 2 years, $3000 after 4 years, and finally $4000 after 6 years. Stein offers instead to pay $1240 at the end of each year for the next 7 years. Should McNurty accept this offer if he demands 8 percent interest compounded annually? Explain.

28. To repay a debt, John Hall agrees to make payments of $500 at the end of each of the next 7 years followed by payments of $800 each at the ends of the 5 years after that. If the interest rate is 8.25 percent compounded annually, what is his present debt?

29. Ruth Nyhus took out a 20-year graduated loan on her house for $30,000. She will repay it by making equal payments at the end of each month during the first 6 years and will double those payments during the last 14 years. What are her initial payments if interest is at 9.25 percent compounded monthly?

30. A Boston bank in September 1978 made the following offer. Deposit $100 each month for 12 years and then we will pay you $100 a month forever. If the interest is i per month, write the equation that i must satisfy and solve. The bank claimed that this corresponded to 5.75 percent interest compounded monthly. Were they right?

CHAPTER SUMMARY

Savers earn interest, while borrowers pay it. If the interest is always based on the original **principal**, it is **simple interest**. But if the interest is periodically used to increase the principal, it is **compound interest**. While a compound interest rate is practically always stated as an annual rate r, the interest may be compounded (converted to principal) several times (say m times) per year. Then the interest rate per period is $i = r/m$. The key formulas are those for the amount (final value) F of a principal P invested for t years (n periods) at rate r.

Simple interest: $F = P(1 + rt)$

Compound interest: $F = P(1 + i)^n = P\left(1 + \dfrac{r}{m}\right)^n$

The rate r_e compounded annually which is equivalent to the rate r compounded m times a year is called the **effective rate.**

Annuities (that is, ordinary annuities) are just sequences of equal payments made at equal intervals of time. Here the key ideas are the **present value** S_p and the **amount** (final value) S_f. The corresponding formulas for payments of size R are

$$S_p = R\frac{1 - (1 + i)^{-n}}{i} \qquad S_f = R\frac{(1 + i)^n - 1}{i}$$

When an annuity is used to pay off a debt, it is an **amortization** of that debt. When it is used to prepare for an expected future expense, it is a **sinking fund.** Two relatives of an ordinary annuity are the **deferred annuity** and the **variable annuity.**

An important idea is that of replacing one set of obligations by another, which leads to **equations of value.** Setting up and solving such equations is, of course, what algebra is all about.

▣ CHAPTER REVIEW PROBLEM SET

1. Find the amount F if $800 is invested for 10 years at
   (a) 8 percent simple interest;
   (b) 8 percent compounded annually;
   (c) 8 percent compounded monthly;
   (d) 8 percent compounded daily.

2. What effective rate corresponds to 8 percent compounded daily?

3. Find the present value of a fund that will be worth $1000 in 10 years if interest is at 9 percent compounded quarterly.

4. At what rate, compounded monthly, will $750 grow to $1000 in 6 years?

5. How long does it take money to double in value if interest is at 7.5 percent compounded daily?

6. Find the amount S_f and the present value S_p of an annuity consisting of payments of $100 a month at the end of each month for 6 years if interest is at 8 percent compounded monthly.

7. Sally Jones has bought a refrigerator on the installment plan, paying $20 at the end of each month for the next 2 years. If interest is at 18 percent compounded monthly, how much would she have paid if she had paid cash?

8. Clara Knutson is amortizing her house loan of $30,000 by making equal payments at the end of each month for the next 20 years with interest at 9.1 percent compounded monthly. What are her payments?

9. Sam Kingland anticipates that he will need to buy a new truck costing $10,000 three years from now. To prepare for this purchase, he wants to establish a sinking fund, into which he will make equal payments at the end of each month until the time of purchase. If he can get 8.5 percent interest compounded monthly, what will his payments be?

10. For 10 long years, Audrey Saufferer put $1000 in a savings account at the end of each year. Now 8 years after her last payment, she will use this money for a down payment on a house. How much does she have if her money drew 6 percent during the first 10 years and 8 percent during the last 8 years, both rates compounded annually?

11. Bud Herganhahn expects to receive a lump sum of $10,000 from his publisher 3 years from now, but prefers to take it in three equal payments now, 2 years from now, and 4 years from now. What will these payments be if interest is at 7.5 percent compounded annually?

12. Tom Washington has negotiated a graduated payment plan for his new car in which he will pay $50 a month for 6 months and then $100 a month for 36 months after that. If interest is at 11 percent compounded monthly, what is the present value of his new car?

13. Rather than pay $40 per month over the next 3 years, Randy Carpenter has offered to pay a lump sum 1 year from now. What should he pay if interest is at 7.75 percent compounded monthly?

APPENDIX

Does the pursuit of truth give you as much pleasure as before? Surely it is not the knowing but the learning, not the possessing but the acquiring, not the being-there but the getting-there, that afford the greatest satisfaction. If I have clarified and exhausted something, I leave it in order to go again into the dark. Thus is that insatiable man so strange: when he has completed a structure it is not in order to dwell in it comfortably, but to start another.

Carl Friedrich Gauss

Table A. Natural logarithms (see Section 6-5).

Table B. Common logarithms (see Sections 6-6 and 6-7).

To find values between the values given in any of these tables, we suggest a process called **linear interpolation.** If we know $f(a)$ and $f(b)$ and want $f(c)$, where c is between a and b, we may write

$$f(c) \approx f(a) + d$$

where d is obtained by pretending that the graph of $y = f(x)$ is a straight line on the interval $a \le x \le b$. A complete description is given in Section 6-6; we show the process with two examples.

EXAMPLE A (Natural logarithms)   Find ln 2.133.

Solution.

$$.010 \left[.003 \left[\begin{array}{l} \ln 2.130 = .7561 \\ \ln 2.133 = \quad ? \\ \ln 2.240 = .7608 \end{array} \right] d \right] .0047$$

$$\frac{d}{.0047} = \frac{.003}{.010} = .3$$

$$d = .3(.0047) \approx .0014$$

$$\ln 2.133 \approx \ln 2.130 + d = .7561 + .0014 = .7575$$

EXAMPLE B (Common logarithms)   Find log 63.26.

Solution.

$$.10 \left[.06 \left[\begin{array}{l} \log 63.20 = 1.8007 \\ \log 63.26 = \quad ? \\ \log 63.30 = 1.8014 \end{array} \right] d \right] .0007$$

$$\frac{d}{.0007} = \frac{.06}{.10} = .6$$

$$d = .6(.0007) \approx .0004$$

$$\log 63.26 \approx \log 63.20 + d = 1.8007 + .0004 = 1.8011$$

	.00	.01	.02	.03	.04	.05	.06	.07	.08	.09
1.0	0.0000	0.0100	0.0198	0.0296	0.0392	0.0488	0.0583	0.0677	0.0770	0.0862
1.1	0.0953	0.1044	0.1133	0.1222	0.1310	0.1398	0.1484	0.1570	0.1655	0.1740
1.2	0.1823	0.1906	0.1989	0.2070	0.2151	0.2231	0.2311	0.2390	0.2469	0.2546
1.3	0.2624	0.2700	0.2776	0.2852	0.2927	0.3001	0.3075	0.3148	0.3221	0.3293
1.4	0.3365	0.3436	0.3507	0.3577	0.3646	0.3716	0.3784	0.3853	0.3920	0.3988
1.5	0.4055	0.4121	0.4187	0.4253	0.4318	0.4383	0.4447	0.4511	0.4574	0.4637
1.6	0.4700	0.4762	0.4824	0.4886	0.4947	0.5008	0.5068	0.5128	0.5188	0.5247
1.7	0.5306	0.5365	0.5423	0.5481	0.5539	0.5596	0.5653	0.5710	0.5766	0.5822
1.8	0.5878	0.5933	0.5988	0.6043	0.6098	0.6152	0.6206	0.6259	0.6313	0.6366
1.9	0.6419	0.6471	0.6523	0.6575	0.6627	0.6678	0.6729	0.6780	0.6831	0.6881
2.0	0.6931	0.6981	0.7031	0.7080	0.7130	0.7178	0.7227	0.7275	0.7324	0.7372
2.1	0.7419	0.7467	0.7514	0.7561	0.7608	0.7655	0.7701	0.7747	0.7793	0.7839
2.2	0.7885	0.7930	0.7975	0.8020	0.8065	0.8109	0.8154	0.8198	0.8242	0.8286
2.3	0.8329	0.8372	0.8416	0.8459	0.8502	0.8544	0.8587	0.8629	0.8671	0.8713
2.4	0.8755	0.8796	0.8838	0.8879	0.8920	0.8961	0.9002	0.9042	0.9083	0.9123
2.5	0.9163	0.9203	0.9243	0.9282	0.9322	0.9361	0.9400	0.9439	0.9478	0.9517
2.6	0.9555	0.9594	0.9632	0.9670	0.9708	0.9746	0.9783	0.9821	0.9858	0.9895
2.7	0.9933	0.9969	1.0006	1.0043	1.0080	1.0116	1.0152	1.0188	1.0225	1.0260
2.8	1.0296	1.0332	1.0367	1.0403	1.0438	1.0473	1.0508	1.0543	1.0578	1.0613
2.9	1.0647	1.0682	1.0716	1.0750	1.0784	1.0818	1.0852	1.0886	1.0919	1.0953
3.0	1.0986	1.1019	1.1053	1.1086	1.1119	1.1151	1.1184	1.1217	1.1249	1.1282
3.1	1.1314	1.1346	1.1378	1.1410	1.1442	1.1474	1.1506	1.1537	1.1569	1.1600
3.2	1.1632	1.1663	1.1694	1.1725	1.1756	1.1787	1.1817	1.1848	1.1878	1.1909
3.3	1.1939	1.1970	1.2000	1.2030	1.2060	1.2090	1.2119	1.2149	1.2179	1.2208
3.4	1.2238	1.2267	1.2296	1.2326	1.2355	1.2384	1.2413	1.2442	1.2470	1.2499
3.5	1.2528	1.2556	1.2585	1.2613	1.2641	1.2669	1.2698	1.2726	1.2754	1.2782
3.6	1.2809	1.2837	1.2865	1.2892	1.2920	1.2947	1.2975	1.3002	1.3029	1.3056
3.7	1.3083	1.3110	1.3137	1.3164	1.3191	1.3218	1.3244	1.3271	1.3297	1.3324
3.8	1.3350	1.3376	1.3403	1.3429	1.3455	1.3481	1.3507	1.3533	1.3558	1.3584
3.9	1.3610	1.3635	1.3661	1.3686	1.3712	1.3737	1.3762	1.3788	1.3813	1.3838
4.0	1.3863	1.3888	1.3913	1.3938	1.3962	1.3987	1.4012	1.4036	1.4061	1.4085
4.1	1.4110	1.4134	1.4159	1.4183	1.4207	1.4231	1.4255	1.4279	1.4303	1.4327
4.2	1.4351	1.4375	1.4398	1.4422	1.4446	1.4469	1.4493	1.4516	1.4540	1.4563
4.3	1.4586	1.4609	1.4633	1.4656	1.4679	1.4702	1.4725	1.4748	1.4770	1.4793
4.4	1.4816	1.4839	1.4861	1.4884	1.4907	1.4929	1.4952	1.4974	1.4996	1.5019
4.5	1.5041	1.5063	1.5085	1.5107	1.5129	1.5151	1.5173	1.5195	1.5217	1.5239
4.6	1.5261	1.5282	1.5304	1.5326	1.5347	1.5369	1.5390	1.5412	1.5433	1.5454
4.7	1.5476	1.5497	1.5518	1.5539	1.5560	1.5581	1.5602	1.5623	1.5644	1.5665
4.8	1.5686	1.5707	1.5728	1.5748	1.5769	1.5790	1.5810	1.5831	1.5851	1.5872
4.9	1.5892	1.5913	1.5933	1.5953	1.5974	1.5994	1.6014	1.6034	1.6054	1.6074
5.0	1.6094	1.6114	1.6134	1.6154	1.6174	1.6194	1.6214	1.6233	1.6253	1.6273
5.1	1.6292	1.6312	1.6332	1.6351	1.6371	1.6390	1.6409	1.6429	1.6448	1.6467
5.2	1.6487	1.6506	1.6525	1.6544	1.6563	1.6582	1.6601	1.6620	1.6639	1.6658
5.3	1.6677	1.6696	1.6715	1.6734	1.6753	1.6771	1.6790	1.6808	1.6827	1.6845
5.4	1.6864	1.6882	1.6901	1.6919	1.6938	1.6956	1.6974	1.6993	1.7011	1.7029

$$\ln(N \cdot 10^m) = \ln N + m \ln 10, \qquad \ln 10 = 2.3026$$

TABLE A. Natural Logarithms

	.00	.01	.02	.03	.04	.05	.06	.07	.08	.09
5.5	1.7047	1.7066	1.7084	1.7102	1.7120	1.7138	1.7156	1.7174	1.7192	1.7210
5.6	1.7228	1.7246	1.7263	1.7281	1.7299	1.7317	1.7334	1.7352	1.7370	1.7387
5.7	1.7405	1.7422	1.7440	1.7457	1.7475	1.7492	1.7509	1.7527	1.7544	1.7561
5.8	1.7579	1.7596	1.7613	1.7630	1.7647	1.7664	1.7682	1.7699	1.7716	1.7733
5.9	1.7750	1.7766	1.7783	1.7800	1.7817	1.7834	1.7851	1.7867	1.7884	1.7901
6.0	1.7918	1.7934	1.7951	1.7967	1.7984	1.8001	1.8017	1.8034	1.8050	1.8066
6.1	1.8083	1.8099	1.8116	1.8132	1.8148	1.8165	1.8181	1.8197	1.8213	1.8229
6.2	1.8245	1.8262	1.8278	1.8294	1.8310	1.8326	1.8342	1.8358	1.8374	1.8390
6.3	1.8406	1.8421	1.8437	1.8453	1.8469	1.8485	1.8500	1.8516	1.8532	1.8547
6.4	1.8563	1.8579	1.8594	1.8610	1.8625	1.8641	1.8656	1.8672	1.8687	1.8703
6.5	1.8718	1.8733	1.8749	1.8764	1.8779	1.8795	1.8810	1.8825	1.8840	1.8856
6.6	1.8871	1.8886	1.8901	1.8916	1.8931	1.8946	1.8961	1.8976	1.8991	1.9006
6.7	1.9021	1.9036	1.9051	1.9066	1.9081	1.9095	1.9110	1.9125	1.9140	1.9155
6.8	1.9169	1.9184	1.9199	1.9213	1.9228	1.9242	1.9257	1.9272	1.9286	1.9301
6.9	1.9315	1.9330	1.9344	1.9359	1.9373	1.9387	1.9402	1.9416	1.9430	1.9445
7.0	1.9459	1.9473	1.9488	1.9502	1.9516	1.9530	1.9544	1.9559	1.9573	1.9587
7.1	1.9601	1.9615	1.9629	1.9643	1.9657	1.9671	1.9685	1.9699	1.9713	1.9727
7.2	1.9741	1.9755	1.9769	1.9782	1.9796	1.9810	1.9824	1.9838	1.9851	1.9865
7.3	1.9879	1.9892	1.9906	1.9920	1.9933	1.9947	1.9961	1.9974	1.9988	2.0001
7.4	2.0015	2.0028	2.0042	2.0055	2.0069	2.0082	2.0096	2.0109	2.0122	2.0136
7.5	2.0149	2.0162	2.0176	2.0189	2.0202	2.0215	2.0229	2.0242	2.0255	2.0268
7.6	2.0282	2.0295	2.0308	2.0321	2.0334	2.0347	2.0360	2.0373	2.0386	2.0399
7.7	2.0412	2.0425	2.0438	2.0451	2.0464	2.0477	2.0490	2.0503	2.0516	2.0528
7.8	2.0541	2.0554	2.0567	2.0580	2.0592	2.0605	2.0618	2.0631	2.0643	2.0656
7.9	2.0669	2.0681	2.0694	2.0707	2.0719	2.0732	2.0744	2.0757	2.0769	2.0782
8.0	2.0794	2.0807	2.0819	2.0832	2.0844	2.0857	2.0869	2.0882	2.0894	2.0906
8.1	2.0919	2.0931	2.0943	2.0956	2.0968	2.0980	2.0992	2.1005	2.1017	2.1029
8.2	2.1041	2.1054	2.1066	2.1078	2.1090	2.1102	2.1114	2.1126	2.1138	2.1150
8.3	2.1163	2.1175	2.1187	2.1199	2.1211	2.1223	2.1235	2.1247	2.1258	2.1270
8.4	2.1282	2.1294	2.1306	2.1318	2.1330	2.1342	2.1353	2.1365	2.1377	2.1389
8.5	2.1401	2.1412	2.1424	2.1436	2.1448	2.1459	2.1471	2.1483	2.1494	2.1506
8.6	2.1518	2.1529	2.1541	2.1552	2.1564	2.1576	2.1587	2.1599	2.1610	2.1622
8.7	2.1633	2.1645	2.1656	2.1668	2.1679	2.1691	2.1702	2.1713	2.1725	2.1736
8.8	2.1748	2.1759	2.1770	2.1782	2.1793	2.1804	2.1815	2.1827	2.1838	2.1849
8.9	2.1861	2.1872	2.1883	2.1894	2.1905	2.1917	2.1928	2.1939	2.1950	2.1961
9.0	2.1972	2.1983	2.1994	2.2006	2.2017	2.2028	2.2039	2.2050	2.2061	2.2072
9.1	2.2083	2.2094	2.2105	2.2116	2.2127	2.2138	2.2148	2.2159	2.2170	2.2181
9.2	2.2192	2.2203	2.2214	2.2225	2.2235	2.2246	2.2257	2.2268	2.2279	2.2289
9.3	2.2300	2.2311	2.2322	2.2332	2.2343	2.2354	2.2364	2.2375	2.2386	2.2396
9.4	2.2407	2.2418	2.2428	2.2439	2.2450	2.2460	2.2471	2.2481	2.2492	2.2502
9.5	2.2513	2.2523	2.2534	2.2544	2.2555	2.2565	2.2576	2.2586	2.2597	2.2607
9.6	2.2618	2.2628	2.2638	2.2649	2.2659	2.2670	2.2680	2.2690	2.2701	2.2711
9.7	2.2721	2.2732	2.2742	2.2752	2.2762	2.2773	2.2783	2.2793	2.2803	2.2814
9.8	2.2824	2.2834	2.2844	2.2854	2.2865	2.2875	2.2885	2.2895	2.2905	2.2915
9.9	2.2925	2.2935	2.2946	2.2956	2.2966	2.2976	2.2986	2.2996	2.3006	2.3016

TABLE B.　Common Logarithms

n	0	1	2	3	4	5	6	7	8	9
1.0	.0000	.0043	.0086	.0128	.0170	.0212	.0253	.0294	.0334	.0374
1.1	.0414	.0453	.0492	.0531	.0569	.0607	.0645	.0682	.0719	.0755
1.2	.0792	.0828	.0864	.0899	.0934	.0969	.1004	.1038	.1072	.1106
1.3	.1139	.1173	.1206	.1239	.1271	.1303	.1335	.1367	.1399	.1430
1.4	.1461	.1492	.1523	.1553	.1584	.1614	.1644	.1673	.1703	.1732
1.5	.1761	.1790	.1818	.1847	.1875	.1903	.1931	.1959	.1987	.2014
1.6	.2041	.2068	.2095	.2122	.2148	.2175	.2201	.2227	.2253	.2279
1.7	.2304	.2330	.2355	.2380	.2405	.2430	.2455	.2480	.2504	.2529
1.8	.2553	.2577	.2601	.2625	.2648	.2672	.2695	.2718	.2742	.2765
1.9	.2788	.2810	.2833	.2856	.2878	.2900	.2923	.2945	.2967	.2989
2.0	.3010	.3032	.3054	.3075	.3096	.3118	.3139	.3160	.3181	.3201
2.1	.3222	.3243	.3263	.3284	.3304	.3324	.3345	.3365	.3385	.3404
2.2	.3424	.3444	.3464	.3483	.3502	.3522	.3541	.3560	.3579	.3598
2.3	.3617	.3636	.3655	.3674	.3692	.3711	.3729	.3747	.3766	.3784
2.4	.3802	.3820	.3838	.3856	.3874	.3892	.3909	.3927	.3945	.3962
2.5	.3979	.3997	.4014	.4031	.4048	.4065	.4082	.4099	.4116	.4133
2.6	.4150	.4166	.4183	.4200	.4216	.4232	.4249	.4265	.4281	.4298
2.7	.4314	.4330	.4346	.4362	.4378	.4393	.4409	.4425	.4440	.4456
2.8	.4472	.4487	.4502	.4518	.4533	.4548	.4564	.4579	.4594	.4609
2.9	.4624	.4639	.4654	.4669	.4683	.4698	.4713	.4728	.4742	.4757
3.0	.4771	.4786	.4800	.4814	.4829	.4843	.4857	.4871	.4886	.4900
3.1	.4914	.4928	.4942	.4955	.4969	.4983	.4997	.5011	.5024	.5038
3.2	.5051	.5065	.5079	.5092	.5105	.5119	.5132	.5145	.5159	.5172
3.3	.5185	.5198	.5211	.5224	.5237	.5250	.5263	.5276	.5289	.5302
3.4	.5315	.5328	.5340	.5353	.5366	.5378	.5391	.5403	.5416	.5428
3.5	.5441	.5453	.5465	.5478	.5490	.5502	.5514	.5527	.5539	.5551
3.6	.5563	.5575	.5587	.5599	.5611	.5623	.5635	.5647	.5658	.5670
3.7	.5682	.5694	.5705	.5717	.5729	.5740	.5752	.5763	.5775	.5786
3.8	.5798	.5809	.5821	.5832	.5843	.5855	.5866	.5877	.5888	.5899
3.9	.5911	.5922	.5933	.5944	.5955	.5966	.5977	.5988	.5999	.6010
4.0	.6021	.6031	.6042	.6053	.6064	.6075	.6085	.6096	.6107	.6117
4.1	.6128	.6138	.6149	.6160	.6170	.6180	.6191	.6201	.6212	.6222
4.2	.6232	.6243	.6253	.6263	.6274	.6284	.6294	.6304	.6314	.6325
4.3	.6335	.6345	.6355	.6365	.6375	.6385	.6395	.6405	.6415	.6425
4.4	.6435	.6444	.6454	.6464	.6474	.6484	.6493	.6503	.6513	.6522
4.5	.6532	.6542	.6551	.6561	.6571	.6580	.6590	.6599	.6609	.6618
4.6	.6628	.6637	.6646	.6656	.6665	.6675	.6684	.6693	.6702	.6712
4.7	.6721	.6730	.6739	.6749	.6758	.6767	.6776	.6785	.6794	.6803
4.8	.6812	.6821	.6830	.6839	.6848	.6857	.6866	.6875	.6884	.6893
4.9	.6902	.6911	.6920	.6928	.6937	.6946	.6955	.6964	.6972	.6981
5.0	.6990	.6998	.7007	.7016	.7024	.7033	.7042	.7050	.7059	.7067
5.1	.7076	.7084	.7093	.7101	.7110	.7118	.7126	.7135	.7143	.7152
5.2	.7160	.7168	.7177	.7185	.7193	.7202	.7210	.7218	.7226	.7235
5.3	.7243	.7251	.7259	.7267	.7275	.7284	.7292	.7300	.7308	.7316
5.4	.7324	.7332	.7340	.7348	.7356	.7364	.7372	.7380	.7388	.7396

n	0	1	2	3	4	5	6	7	8	9
5.5	.7404	.7412	.7419	.7427	.7435	.7443	.7451	.7459	.7466	.7474
5.6	.7482	.7490	.7497	.7505	.7513	.7520	.7528	.7536	.7543	.7551
5.7	.7559	.7566	.7574	.7582	.7589	.7597	.7604	.7612	.7619	.7627
5.8	.7634	.7642	.7649	.7657	.7664	.7672	.7679	.7686	.7694	.7701
5.9	.7709	.7716	.7723	.7731	.7738	.7745	.7752	.7760	.7767	.7774
6.0	.7782	.7789	.7796	.7803	.7810	.7818	.7825	.7832	.7839	.7846
6.1	.7853	.7860	.7868	.7875	.7882	.7889	.7896	.7903	.7910	.7917
6.2	.7924	.7931	.7938	.7945	.7952	.7959	.7966	.7973	.7980	.7987
6.3	.7993	.8000	.8007	.8014	.8021	.8028	.8035	.8041	.8048	.8055
6.4	.8062	.8069	.8075	.8082	.8089	.8096	.8102	.8109	.8116	.8122
6.5	.8129	.8136	.8142	.8149	.8156	.8162	.8169	.8176	.8182	.8189
6.6	.8195	.8202	.8209	.8215	.8222	.8228	.8235	.8241	.8248	.8254
6.7	.8261	.8267	.8274	.8280	.8287	.8293	.8299	.8306	.8312	.8319
6.8	.8325	.8331	.8338	.8344	.8351	.8357	.8363	.8370	.8376	.8382
6.9	.8388	.8395	.8401	.8407	.8414	.8420	.8426	.8432	.8439	.8445
7.0	.8451	.8457	.8463	.8470	.8476	.8482	.8488	.8494	.8500	.8506
7.1	.8513	.8519	.8525	.8531	.8537	.8543	.8549	.8555	.8561	.8567
7.2	.8573	.8579	.8585	.8591	.8597	.8603	.8609	.8615	.8621	.8627
7.3	.8633	.8639	.8645	.8651	.8657	.8663	.8669	.8675	.8681	.8686
7.4	.8692	.8698	.8704	.8710	.8716	.8722	.8727	.8733	.8739	.8745
7.5	.8751	.8756	.8762	.8768	.8774	.8779	.8785	.8791	.8797	.8802
7.6	.8808	.8814	.8820	.8825	.8831	.8837	.8842	.8848	.8854	.8859
7.7	.8865	.8871	.8876	.8882	.8887	.8893	.8899	.8904	.8910	.8915
7.8	.8921	.8927	.8932	.8938	.8943	.8949	.8954	.8960	.8965	.8971
7.9	.8976	.8982	.8987	.8993	.8998	.9004	.9009	.9015	.9020	.9025
8.0	.9031	.9036	.9042	.9047	.9053	.9058	.9063	.9069	.9074	.9079
8.1	.9085	.9090	.9096	.9101	.9106	.9112	.9117	.9122	.9128	.9133
8.2	.9138	.9143	.9149	.9154	.9159	.9165	.9170	.9175	.9180	.9186
8.3	.9191	.9196	.9201	.9206	.9212	.9217	.9222	.9227	.9232	.9238
8.4	.9243	.9248	.9253	.9258	.9263	.9269	.9274	.9279	.9284	.9289
8.5	.9294	.9299	.9304	.9309	.9315	.9320	.9325	.9330	.9335	.9340
8.6	.9345	.9350	.9355	.9360	.9365	.9370	.9375	.9380	.9385	.9390
8.7	.9395	.9400	.9405	.9410	.9415	.9420	.9425	.9430	.9435	.9440
8.8	.9445	.9450	.9455	.9460	.9465	.9469	.9474	.9479	.9484	.9489
8.9	.9494	.9499	.9504	.9509	.9513	.9518	.9523	.9528	.9533	.9538
9.0	.9542	.9547	.9552	.9557	.9562	.9566	.9571	.9576	.9581	.9586
9.1	.9590	.9595	.9600	.9605	.9609	.9614	.9619	.9624	.9628	.9633
9.2	.9638	.9643	.9647	.9652	.9657	.9661	.9666	.9671	.9675	.9680
9.3	.9685	.9689	.9694	.9699	.9703	.9708	.9713	.9717	.9722	.9727
9.4	.9731	.9736	.9741	.9745	.9750	.9754	.9759	.9763	.9768	.9773
9.5	.9777	.9782	.9786	.9791	.9795	.9800	.9805	.9809	.9814	.9818
9.6	.9823	.9827	.9832	.9836	.9841	.9845	.9850	.9854	.9859	.9863
9.7	.9868	.9872	.9877	.9881	.9886	.9890	.9894	.9899	.9903	.9908
9.8	.9912	.9917	.9921	.9926	.9930	.9934	.9939	.9943	.9948	.9952
9.9	.9956	.9961	.9965	.9969	.9974	.9978	.9983	.9987	.9991	.9996

ANSWERS
TO ODD-NUMBERED
PROBLEMS

PROBLEM SET 1-1 (Page 4)

1. Let x be one number and y the other; $x + \frac{1}{3}y$. **3.** Let x be one number and y the other; $2x/3y$.
5. Let x be the number; $0.10x + x$, or $1.10x$. **7.** Let x be one side and y the other; $x^2 + y^2$. **9.** xy **11.** y/x
13. $(30/x) + (30/y)$ **15.** x^2 **17.** $6x^2$ **19.** $4\pi(x/2)^2 = \pi x^2$ **21.** $10x^2$ **23.** $\frac{4}{3}\pi(x/2)^3 = \pi x^3/6$
25. $x^3 - 4\pi x$ **27.** $A = x^2 + \frac{1}{4}\pi x^2$; $P = 2x + \pi x$ **29.** Let x be the number; $x + \frac{1}{2}x = 45$; $x = 30$.
31. Let x be the smaller odd number; $x + x + 2 = 168$; $x = 83$. **33.** $\frac{9}{2}x = 252$; $x = 56$ **35.** 10.5 meters
37. $V = 64$ cubic inches; $S = 96$ square inches **39.** Volume is 8 times as large; surface area is 4 times as large.
41. $(D/100) + (D/200) = 4$; $D = 266\frac{2}{3}$ miles **43.** $400/[(200/50) + (200/70)] \approx 58.3$ kilometers per hour

PROBLEM SET 1-2 (Page 11)

1. 12 **3.** -31 **5.** -67 **7.** $-60 + 9x$ **9.** $14t$ **11.** $\frac{8}{9}$ **13.** $-\frac{3}{4}$ **15.** $(1 - 3x)/2$
17. $(-2x + 3)/2$ **19.** $\frac{21}{12} = \frac{7}{4}$ **21.** $\frac{19}{20}$ **23.** $\frac{23}{36}$ **25.** $\frac{106}{108} = \frac{53}{54}$ **27.** $\frac{1}{2}$ **29.** $\frac{3}{4}$ **31.** $\frac{5}{4}$ **33.** $\frac{3}{8}$
35. $\frac{54}{7}$ **37.** $\frac{17}{7}$ **39.** -17 **41.** $\frac{9}{17}$ **43.** $\frac{25}{18}$ **45.** $-\frac{7}{9}$ **47.** $\frac{1}{24}$ **49.** $\frac{22}{189}$ **51.** $\frac{4}{11}$ **53.** $-\frac{5}{16}$
55. $\frac{15}{4}$ **57.** $\frac{3}{11}$

PROBLEM SET 1-3 (Page 18)

1. $2 \cdot 5 \cdot 5 \cdot 5$ **3.** $2 \cdot 2 \cdot 2 \cdot 5 \cdot 5$ **5.** $2 \cdot 2 \cdot 3 \cdot 5 \cdot 5 \cdot 7$ **7.** $2 \cdot 2 \cdot 2 \cdot 5 \cdot 5 \cdot 5 = 1000$
9. $2 \cdot 2 \cdot 3 \cdot 5 \cdot 5 \cdot 5 \cdot 7 = 10{,}500$ **11.** $2 \cdot 2 \cdot 2 \cdot 3 \cdot 5 \cdot 5 \cdot 5 \cdot 7 = 21{,}000$ **13.** $97/1000$ **15.** $289/10{,}500$
17. $-1423/21{,}000$ **19.** $.\overline{6}$ **21.** $.625\overline{0}$ **23.** $.\overline{461538}$ **25.** $7/9$ **27.** $235/999$ **29.** $13/40$
31. $318/990 = 53/165$ **33.** $10{,}500$ **35.** (a) $.3125\overline{0}$ (b) $.\overline{27}$ **37.** $.\overline{637728}$
39. No; for example, $\sqrt{2} + (-\sqrt{2}) = 0$. **41.** Follow hint; also, see solution to Problem 43.
43. Suppose $\frac{3}{4}\sqrt{2} = r$, where r is rational. Then $\sqrt{2} = \frac{4}{3}r$; however, $\frac{4}{3}r$ is rational, which is a contradiction since $\sqrt{2}$ is
irrational. **45.** a; b; d; e; g **47.** $.000000001$; no. **49.** Suppose $\sqrt{3} = m/n$, where m and n are positive integers
greater than 1. Then $3n^2 = m^2$. Both n^2 and m^2 must have an even number of 3's as factors. This contradicts $3n^2 = m^2$, since $3n^2$
must have an odd number of 3's. **51.** It is nonrepeating. **53.** Negative.

PROBLEM SET 1-4 (Page 25)

1. 1431 **3.** 1700 **5.** 61 **7.** 3 **9.** 2 **11.** Associative and commutative properties of multiplication.
13. Commutative property of addition.
15. Associative property of addition; additive inverse; zero is neutral element for addition. **17.** Distributive property.
19. Distributive property. **21.** True; $a - (b - c) = a + (-1)[b + (-c)] = a + (-1)b + (-1)(-c) = a - b + c$.
23. False; $1 \div (1 + 1) = \frac{1}{2}$, but $(1 \div 1) + (1 \div 1) = 2$. **25.** True; $ab(a^{-1} + b^{-1}) = aba^{-1} + abb^{-1} = b + a$.
27. False; $(1 + 2)(1^{-1} + 2^{-1}) = 3(1 + \frac{1}{2}) = 3(3/2) = \frac{9}{2}$. **29.** False; $(1 + 2)(1 + 2) = 3 \cdot 3 = 9$, but $1^2 + 2^2 = 5$.
31. False; $1 \div (2 \div 3) = 1/\frac{2}{3} = \frac{3}{2}$ but $(1 \div 2) \div 3 = \frac{1}{2}/3 = \frac{1}{6}$. **33.** Additive inverse; zero is neutral element for addition;
distributive property; associative property of addition; additive inverse; zero is neutral element for addition. **35.** No; no.

PROBLEM SET 1-5 (Page 30)

1. $>$ **3.** $>$ **5.** $>$ **7.** $>$ **9.** $<$ **11.** $=$ **13.** $-\frac{3}{2}\sqrt{2}; -2; -\frac{\pi}{2}; \frac{3}{4}; \sqrt{2}; \frac{43}{24}$

15. **17.**

19. **21.**

23. **25.** $-2 \le x \le 3$ **27.** $x \ge 2$ **29.** $-2 < x \le 3$

31. $-1 < x < 2$ **33.** $-4 \le x \le 4$

35. $1 < x < 5$

37. $-4 \le x \le 2$

39. $x < 0$ or $x > 10$

41.

43.

45. No solutions.

47.

49. No; $-2 < -1$ but $(-2)^2 > (-1)^2$. **51.** $1/a > 1/b$ **53.** (a) $|x - 6| < 4$ (b) $|x + 1| \le 5$ (c) $|x - 4.5| < 1.5$
(d) $|x - 1.5| < 4.5$ **55.** $a^2 = |a^2| = |a||a| = |a|^2$ **57.** $|(a + b) + c| \le |a + b| + |c| \le |a| + |b| + |c|$
59. If $ab \ge 0$.

PROBLEM SET 1-6 (Page 37)

1. $-2 + 8i$ **3.** $-4 - i$ **5.** $0 + 6i$ **7.** $-4 + 7i$ **9.** $11 + 4i$ **11.** $14 + 22i$ **13.** $16 + 30i$
15. $61 + 0i$ **17.** $\frac{3}{2} + \frac{7}{2}i$ **19.** $2 - 5i$ **21.** $\frac{11}{2} + \frac{3}{2}i$ **23.** $\frac{2}{5} + \frac{1}{5}i$ **25.** $\sqrt{3}/4 - i/4$ **27.** $\frac{1}{5} + \frac{8}{5}i$
29. $(\sqrt{3} - 1/4) + (-\sqrt{3}/4 - 1)i$ **31.** -1 **33.** $-i$ **35.** $-729i$ **37.** $-2 + 2i$
39. $i^4 = (i^2)(i^2) = (-1)(-1) = 1$. The four 4th roots of 1 are 1, -1, i, and $-i$.
41. $(1 - i)^4 = (1 - i)^2(1 - i)^2 = (-2i)(-2i) = 4i^2 = -4$ **43.** (a) $4 - 12i$ (b) $14 + 10i$ (c) $\frac{7}{25} - \frac{51}{25}i$
45. $(1 + \sqrt{3}i)(1 + \sqrt{3}i)(1 + \sqrt{3}i) = (-2 + 2\sqrt{3}i)(1 + \sqrt{3}i) = -8$ **47.** $3 - \frac{1}{3}i$

CHAPTER 1. REVIEW PROBLEM SET (Page 39)

1. $V = x^2y$; $S = x^2 + 4xy$ **2.** $100/(x + y) + 100/(x - y)$ **3.** (a) $\frac{15}{24} = \frac{5}{8}$ (b) $\frac{1}{2}$ (c) $\frac{1}{9}$ (d) $\frac{1}{5}$ (e) $\frac{25}{7}$ (f) $\frac{27}{10}$
4. $2 \cdot 2 \cdot 2 \cdot 5 \cdot 5 \cdot 5$; $2 \cdot 2 \cdot 3 \cdot 3 \cdot 5$ **5.** $2 \cdot 2 \cdot 2 \cdot 3 \cdot 3 \cdot 5 \cdot 5 \cdot 5 = 9000$ **6.** $.\overline{384615}$; $1.\overline{571428}$ **7.** $257/999$, $122/99$
8. Yes; no. **9.** $a(bc) = (ab)c$; $a + b = b + a$ **10.** 1.4; $\sqrt{2}$; $1.\overline{4}$; $\frac{29}{20}$, $\frac{13}{8}$

11. $-8 < x < 4$

$-1 \le x \le 2$

12. No. If $x < 0$, then $|-x| = -x$. For example, $|-(-2)| = -(-2)$.
13. (a) $-6 + 7i$ (b) $17 - 8i$ (c) $\frac{29}{13} + \frac{28}{13}i$ (d) $2 + 11i$ (e) $\frac{5}{34} + \frac{3}{34}i$
14. Suppose that $5 + \sqrt{5} = r$, where r is rational. Then $\sqrt{5} = r - 5$, which is rational, a contradiction. Handle $5\sqrt{5}$ similarly.
15. 320 square centimeters

PROBLEM SET 2-1 (Page 47)

1. $3^3 = 27$ **3.** $2^2 = 4$ **5.** 8 **7.** $1/5^2 = 1/25$ **9.** $-1/5^2 = -1/25$ **11.** $1/(-2)^5 = -1/32$
13. $-27/8$ **15.** $27/4$ **17.** $81/16$ **19.** $3/64$ **21.** $1/72$ **23.** $81x^4$ **25.** x^6y^{12} **27.** $16x^8y^4/w^{12}$
29. $27y^6/(x^3z^6)$ **31.** $25x^8$ **33.** $1/(16y^6)$ **35.** $a/(5b^2x^2)$ **37.** $2z^3/(x^6y^2)$ **39.** $-x^3z^2/(2y^7)$
41. a^4b^3 **43.** $d^{40}/(32b^{15})$ **45.** $a^3/(a + 1)$ **47.** 0 **49.** $3x^4/y^4$ **51.** $3y^4/x^3$
53. $5184/(x^8z^4)$ **55.** $(x^4 + 1)/x^2$ **57.** $x/(x^2 + 1)$ **59.** $x + 1$ **61.** (a) 2^{-5} (b) 2^0 (c) 2^{-10}
63. (a) $2^{19} = \frac{1}{2}(2^{20}) \approx \frac{1}{2}(1000)^2 = 500,000\cent = \$5000$ (b) $2^{49} = \frac{1}{2}(2^{50}) \approx \frac{1}{2}(1000)^5 = \frac{1}{2}(10)^{15}\cent = \$(5 \times 10^{12})$
65. About February 7.

PROBLEM SET 2-2 (Page 55)

1. 3.41×10^8 **3.** 5.13×10^{-8} **5.** 1.245×10^{-10} **7.** 8.4×10^{-4} **9.** 7.2×10^{-4} **11.** 1.08×10^{10}
13. 4.132×10^4 **15.** 4×10^9 **17.** 9.144×10^2
(*Note:* Answers to Problems 19–43 may vary depending on the calculator used.)
19. 48.35 **21.** -2441.7393 **23.** 303.27778 **25.** 2.7721×10^{15} **27.** 1.286×10^{10} **29.** -13.138859
31. 1.7891883 **33.** 1.067068 **35.** $.90569641$ **37.** $.00000081$ **39.** About 1.28 seconds.
41. About 4.068×10^{16}. **43.** About 9.3×10^{32}. **45.** 5.4857873×10^8 **47.** 3.59694

PROBLEM SET 2-3 (Page 63)

1. Polynomial of degree 2. 3. Polynomial of degree 5. 5. Polynomial of degree 0. 7. Not a polynomial.
9. Not a polynomial. 11. $-2x + 1$ 13. $4x^2$ 15. $-2x^2 + 5x + 1$ 17. $6x - 15$ 19. $-10x + 12$
21. $35x^2 - 55x + 19$ 23. $t^2 + 16t + 55$ 25. $x^2 - x - 90$ 27. $2t^2 + 13t - 7$ 29. $y^2 + 2y - 8$
31. $7.9794x^2 + 0.1155x - 4.4781$ 33. $x^2 + 20x + 100$ 35. $x^2 - 64$ 37. $4t^2 - 20t + 25$
39. $4x^8 - 25x^2$ 41. $(t + 2)^2 + 2(t + 2)t^3 + t^6 = t^6 + 2t^4 + 4t^3 + t^2 + 4t + 4$
43. $(t + 2)^2 - (t^3)^2 = -t^6 + t^2 + 4t + 4$ 45. $5.29x^2 - 6.44x + 1.96$ 47. $x^3 + 6x^2 + 12x + 8$
49. $8t^3 - 36t^2 + 54t - 27$ 51. $8t^3 + 12t^4 + 6t^5 + t^6$ 53. $t^6 + 6t^5 + 15t^4 + 20t^3 + 15t^2 + 6t + 1$
55. $x^2 - 6xy + 9y^2$ 57. $9x^2 - 4y^2$ 59. $12x^2 + 11xy - 5y^2$ 61. $2x^4y^2 - x^2yz - z^2$
63. $t^2 + 2t + 1 - s^2$ 65. $8t^3 - 36t^2s + 54ts^2 - 27s^3$ 67. $-3x^3 - 19x^2 + 2x + 5$ 69. $2y^2 + y - 15$
71. $4y^2 + 20y + 25$ 73. $4z^2 - 49$ 75. $x^4 + x^2 - 12$ 77. $6x^3 + 7x^2 - 13x - 10$
79. $x^2 - 2$ 81. $x^2 - 4\sqrt{7}x + 28$ 83. $8x$ 85. $8x^3 + 12x^2 + 6x + 1$
87. $5.7121x^2 + 15.5636x + 16.9744$ 89. $25x^2 - y^2$ 91. $25x^2 - 10xy + 2y^2$ 93. $8x^2 - 10xy - 3y^2$
95. $x^3 + y^3$ 97. $x^3 + 6x^2y + 12xy^2 + 8y^3$ 99. $x^4 - y^4$ 101. $x^4 - x^2y^2 - 6y^4$ 103. $6x^2y^2 - 7xyz - 3z^2$
105. $x^4 + 3x^3y + 2x^2y^2 + xy^3 - y^4$

PROBLEM SET 2-4 (Page 70)

1. $x(x + 5)$ 3. $(x + 6)(x - 1)$ 5. $y^3(y - 6)$ 7. $(y + 6)(y - 2)$ 9. $(y + 4)^2$ 11. $(2x - 3y)^2$
13. $(y - 8)(y + 8)$ 15. $(1 - 5b)(1 + 5b)$ 17. $(2z - 3)(2z + 1)$ 19. $(5x + 2y)(4x - y)$
21. $(x + 3)(x^2 - 3x + 9)$ 23. $(a - 2b)(a^2 + 2ab + 4b^2)$ 25. $x^3(1 - y^3) = x^3(1 - y)(1 + y + y^2)$
27. Does not factor over integers. 29. Does not factor over integers. 31. $(y - \sqrt{5})(y + \sqrt{5})$
33. $(\sqrt{5}z - 2)(\sqrt{5}z + 2)$ 35. $t^2(t - \sqrt{2})(t + \sqrt{2})$ 37. $(y - \sqrt{3})^2$ 39. Does not factor over real numbers.
41. $(x + 3i)(x - 3i)$ 43. $(x^3 + 7)(x^3 + 2)$ 45. $(2x - 1)(2x + 1)(x - 3)(x + 3)$ 47. $(x + 4y + 3)^2$
49. $(x^2 - 3y^2)(x^2 + 2y^2)$ 51. $(x - 2)(x^2 + 2x + 4)(x + 2)(x^2 - 2x + 4)$ 53. $x^4(x - y)(x + y)(x^2 + y^2)$
55. $(x^2 + y^2)(x^4 - x^2y^2 + y^4)$ 57. $(x^2 + 1)(x - 4)$ 59. $(2x - 1 - y)(2x - 1 + y)$ 61. $(3 - x)(x + y)$
63. $(x + 3y)(x + 3y + 2)$ 65. $(x + y + 2)(x + y + 1)$ 67. $(x^2 - 4x + 8)(x^2 + 4x + 8)$
69. $(x^2 - x + 1)(x^2 + x + 1)$ 71. $5x(x - 2)$ 73. $(2x + 1)^2$ 75. $(2y - 1)(2y + 1)$ 77. $(2y - 5)(y - 1)$
79. $4(x^2 - 2)(x^2 + 2)$ 81. $(a + b - 5)(a + b + 5)$ 83. $(a + b)(a + b - 2)$ 85. $(-2x)(y - 3)(y^2 + 3y + 9)$
87. $(x^2 - y + 2)(x^2 + y - 2)$ 89. $(a + 2b - 3)(a + 2b + 2)$ 91. Does not factor over integers.
93. $(x^2 - x + 1)(x^2 + x + 1)$ 95. $(x + 3)(x - 7)^2(x^2 - 4x - 20)$ 97. $(x - 1.7674841)(x + 1.7674841)$
99. $(x + 1.69965391)(x^2 - 1.69965391x + 2.8888234)$ 101. $(2x + 3i)(2x - 3i)$
103. $(x - 2i)(x^2 + 2ix - 4) = (x - 2i)(x + i + \sqrt{3})(x + i - \sqrt{3})$

PROBLEM SET 2-5 (Page 76)

1. $1/(x - 6)$ 3. $y/5$ 5. $(x + 2)^2/(x - 2)$ 7. $z(x + 2y)/(x + y)$ 9. $(9x + 2)/(x - 2)(x + 2)$
11. $2(4x - 3)/(x - 2)(x + 2)$ 13. $(2xy + 3x - 1)/x^2y^2$ 15. $(x^2 + 10x - 3)/(x - 2)^2(x + 5)$
17. $(4 - x)/(2x - 1)$ 19. $(6y^2 + 9y + 2)/(3y - 1)(3y + 1)$ 21. $(3m^2 + m - 1)/3(m - 1)^2$
23. $5x/(2x - 1)(x + 1)$ 25. $1/(x - 3)(x - 2)$ 27. $y^2(x^3 - y^3)$ 29. $x/(x - 4)$ 31. $5(x + 1)/x(2x - 1)$
33. $x/(x - 2)$ 35. $(x - a)(x^2 + a^3)/(x + 2a)(x^2 + ax + a^2)$ 37. $(y - 1)/2y$ 39. $-2/(2x + 2h + 3)(2x + 3)$
41. $(-2x - h)/(x + h)^2x^2$ 43. $(y - 1)(y + 2)/(7y + 19)$ 45. $(a^2 + b^2)/ab$ 47. 1 49. $(y - 1)/y$
51. $(2x^2 - y^2)/(x - y)$ 53. $(-3x^2 - 26x + 2)/2(x + 10)(x - 6)$ 55. $(2x - 3y)(x^2 + xy + y^2)$
57. $(13x + 15)/(2x + 3)(2x - 3)$ 59. $(x - 2)/3x^2$

CHAPTER 2. REVIEW PROBLEM SET (Page 81)

1. $16/9$ **2.** $36/25$ **3.** $36/49$ **4.** $y^5/2x^3$ **5.** $x^6/27y^9$ **6.** $b^{11}/2a^8$ **7.** 1.382×10^6
8. 6.82×10^{-3} **9.** 5×10^7 **10.** Polynomial of degree 2.   **11.** Not a polynomial.   **12.** Not a polynomial.
13. Not a polynomial.   **14.** $2x^2 + x + 6$ **15.** $x^3 - x^2 - x + 7$ **16.** $2x^2 + 7x - 15$ **17.** $-4x + 1$
18. $z^6 - 16$ **19.** $2x^4 + 5x^2w - 12w^2$ **20.** $x^3 + 8a^3$ **21.** $6y^3 - 19y^2 + 27y - 18$
22. $9t^4 - 6t^3 + 7t^2 - 2t + 1$ **23.** $a^2 - b^2c^2d^2$ **24.** $x^2(2x^2 - x + 11)$ **25.** $(y - 4)(y - 3)$
26. $(3z - 1)(2z + 1)$ **27.** $(7a - 5)(7a + 5)$ **29.** $(a - 3)(a^2 + 3a + 9)$ **30.** $(2ab + 1)(4a^2b^2 - 2ab + 1)$
31. $x^4(x - y)(x + y)(x^2 + y^2)$ **32.** $(x + y - z^2)(x + y + z^2)$ **33.** $(2c + d)(2c - d - 3)$
34. $(3x - \sqrt{11})(3x + \sqrt{11})$ **35.** $(x - 4i)(x + 4i)$ **36.** $(x^2 + 2x + 4)/2$ **37.** $-2/(x - 1)$ **38.** $2/x$
39. $(x - 2)/(x + 1)$ **40.** $(x + 1)/10$

PROBLEM SET 3-1 (Page 88)

1. Conditional equation.   **3.** Identity.   **5.** Conditional equation.   **7.** Conditional equation.   **9.** Identity.
11. 2   **13.** $\frac{2}{5}$ **15.** $\frac{9}{2}$ **17.** $2/\sqrt{3}$ **19.** 7.57   **21.** -8.71×10^1 **23.** -24 **25.** $-\frac{1}{4}$ **27.** $\frac{22}{5}$
29. 3   **31.** 6   **33.** No solution (2 is extraneous).   **35.** $-\frac{3}{4}$ **37.** $-\frac{13}{2}$ **39.** -10 **41.** -21
43. $P = A/(1 + rt)$ **45.** $r = (nE - IR)/nI$ **47.** $h = (A - 2\pi r^2)/2\pi r$ **49.** $R_1 = RR_2/(R_2 - R)$ **51.** 8
53. 60   **55.** -1.732 **57.** $\frac{19}{3}$ **59.** $-\frac{1}{3}$ **61.** No solution (-3 is extraneous).
63. (a) $F = \frac{9}{5}C + 32$ (b) 95 (c) -40 (d) 72.5 (e) $\frac{80}{3} = 26.7$ **65.** (a) $2\frac{1}{2}$ seconds (b) $3\frac{3}{4}$ seconds

PROBLEM SET 3-2 (Page 95)

1. 9   **3.** 12   **5.** 23   **7.** 31 centimeters   **9.** 89   **11.** 7   **13.** 12   **15.** 12:40 A.M.   **17.** 4:20 P.M.
19. 9 miles per hour   **21.** 15   **23.** 6000   **25.** 142.86 liters   **27.** $145   **29.** 9 feet   **31.** 7.5 days
33. 4.6 feet from fulcrum   **35.** 125   **37.** 44.44

PROBLEM SET 3-3 (Page 101)

1. $x = -13; y = 13$ **3.** $u = \frac{3}{2}; v = -4$ **5.** $x = -1; y = 3$ **7.** $x = 4; y = 3$ **9.** $x = 6; y = -8$
11. $s = 1; t = 4$ **13.** $x = 3; y = -1$ **15.** $x = 4; y = 7$ **17.** $x = 9; y = -2$ **19.** $x = 16; y = -5$
21. $x = \frac{1}{2}; y = \frac{1}{3}$ **23.** $x = 4; y = 9$ **25.** $x = 1; y = 2; z = 0$ **27.** $x = \frac{4}{3}, y = -\frac{2}{3}; z = -\frac{5}{3}$ **29.** 7; 11
31. $7200; $5800   **33.** $28,500 (certificates); $19,000 (bonds)   **35.** 32,500 ($8 tickets); 12,500 ($10 tickets)
37. 20 nickels; 3 dimes   **39.** 5 hours; 11 hours   **41.** 62.79 pounds ($1.69 coffee); 37.21 pounds ($1.26 coffee)
43. 28 coats; 50 dresses

PROBLEM SET 3-4 (Page 109)

1. $5\sqrt{2}$ **3.** $\frac{1}{2}$ **5.** $\frac{3}{2}$ **7.** 22   **9.** $(5 + 6\sqrt{2})/5$ **11.** $(6 + i)/2$ **13.** ± 5 **15.** $-1; 7$
17. $-\frac{15}{2}; \frac{5}{2}$ **19.** $\pm 3i$ **21.** $0; 3$ **23.** ± 3 **25.** $\pm .12$ **27.** $-2; 5$ **29.** $-2; \frac{1}{3}$ **31.** $-\frac{4}{3}; \frac{7}{2}$
33. $-9; 1$ **35.** $-\frac{1}{2}; \frac{3}{2}$ **37.** $-2 \pm \sqrt{5}i$ **39.** $-6; -2$ **41.** $(-5 \pm \sqrt{13})/2$ **43.** $(3 \pm \sqrt{42})/3$
45. $(-5 \pm \sqrt{5})/2$ **47.** $(3 \pm \sqrt{13}i)/2$ **49.** $-.2714; 1.8422$ **51.** $-1.6537; .8302$ **53.** $y = 2 \pm 2x$
55. $y = -6x$ or $y = 0$ **57.** $y = -2x + 3$ or $y = -2x + 5$ **59.** ± 12 **61.** $-\frac{3}{2}; 2$ **63.** $-3; \frac{1}{2}$ **65.** $-\frac{5}{4}, 0$
67. $-1 \pm \sqrt{6}$ **69.** $2 \pm i$ **71.** 13 feet by 7 feet   **73.** About 89 feet square.
75. $2 + 8\sqrt{2} \approx 13.31$ inches square   **77.** (a) After 8 seconds. (b) After 2 seconds and after 6 seconds. (c) Never.
79. $x = 3, y = 4; x = 4.8, y = -1.4$ **81.** $x = 2, y = -1; x = -2, y = -3$

PROBLEM SET 3-5 (Page 117)

1. Conditional   **3.** Unconditional.   **5.** Conditional.   **7.** Conditional   **9.** Conditional   **11.** Unconditional

13. $\{x : x < -6\}$

15. $\{x : x > -24\}$

17. $\{x : x < \frac{30}{7}\}$

19. $\{x : -5 \le x \le 2\}$

21. $\{x : x < -3 \text{ or } x > \frac{1}{2}\}$

23. $\{x : x \le 1 \text{ or } x \ge 4\}$

25. $\{x : \frac{1}{2} < x < 3\}$

27. $\{x : -\frac{5}{2} < x < -\frac{1}{2}\}$

29. $\{x : -1 \le x \le 0\}$

31. $\{x : -4 \le x \le 0 \text{ or } x \ge 3\}$

33. $\{x : x < 5 \text{ and } x \ne 2\}$

35. $\{x : -2 < x \le 5\}$

37. $\{x : -2 < x < 0 \text{ or } x > 5\}$

39. $\{x : -2 < x < 2 \text{ or } x > 3\}$

41. $|x - 3| < 3$ **43.** $|x - 3| \le 4$ **45.** $|x - 6.5| < 4.5$

47. $\{x : -\sqrt{7} < x < \sqrt{7}\}$

49. $\{x : x \le 2 - \sqrt{2} \text{ or } x \ge 2 + \sqrt{2}\}$

51. $\{x : x < -5.71 \text{ or } x > -.61\}$

53. 4   **55.** 100

57. $\{x : x > -\frac{9}{2}\}$

59. $\{x : x < -4 \text{ or } x > 2\}$

61. $\{x : -\frac{1}{3} \le x \le \frac{5}{3}\}$

63. $\{x : x < 0 \text{ or } 3 < x \le 5\}$

65. $k \le \frac{9}{4}$ **67.** All real numbers.   **69.** (a) A score higher than 57.   (b) A score between 132 and 192.
71. (a) 144 feet above the ground.   (b) $0 < t < 4$ (c) After 5 seconds.   (d) $4.55 < t \le 5$ **73.** At most 100 miles.
75. $\{x : x \ge 3\}$ **77.** $\{x : x > -2\}$

PROBLEM SET 3-6 (Page 123)

1. About 18,333 feet.   **3.** 525 miles   **5.** About 21.82 minutes.   **7.** 4 hours and 48 minutes
9. About 4.15 hours after takeoff.   **11.** 12 standard; 20 deluxe   **13.** 60 miles per hour   **15.** 500 meters
17. 20 feet   **19.** About 8.17 feet.   **21.** $\frac{5}{6}$ kiloliters   **23.** About 3.29 milligrams.
25. .298 grams sodium chloride; .202 grams sodium bromide   **27.** 36 shares
29. 14 pounds walnuts; 11 pounds cashews   **31.** 23,500   **33.** \$785,714.29
35. 2340 undergraduate students; 864 graduate students   **37.** 2 feet   **39.** 5; 12; 13   **41.** $a = 8,\ b = 5,\ c = 10$
43. About 13.34 feet.   **45.** Two solutions: longer piece is 28 inches or $\frac{236}{7}$ inches.

CHAPTER 3. REVIEW PROBLEM SET (Page 127)

1. (a) Identity. (b) Conditional equation. (c) Conditional equation. (d) Identity.
2. (a) $-\frac{11}{12}$ (b) 7 (c) $\frac{5}{2}$ (d) No solution. **3.** $v_0 = (s - \frac{1}{2}at^2)/t$ **4.** (a) $-24.62°C$ (b) Above $-40°F$.
5. (a) $x = 2; y = -1$ (b) $x = 1; y = -2$ **6.** (a) $(2 + \sqrt{2})/4$ (b) $5 - 5\sqrt{3}$ (c) $(-1 + \sqrt{2})/3$ **7.** ± 7
8. $-5; 2$ **9.** -3 **10.** $-3; 7$ **11.** $-5; 4$ **12.** $4; 5$ **13.** $0; 4$ **14.** $-1 \pm \sqrt{5}$
15. $(-1 \pm \sqrt{13})/6$ **16.** $(-m \pm \sqrt{m^2 - 8n})/2$ **17.** $y = 2x \pm 2$ **18.** $y = (3 - 3x)/2$ or $y = (1 - 3x)/2$
19. $\{x : x \le \frac{12}{7}\}$ **20.** $\{x : x \le -6 \text{ or } x \ge 1\}$ **21.** $\{x : (-1 - \sqrt{13})/2 < x < (-1 + \sqrt{13})/2\}$
22. $\{x : x < -1 \text{ or } x > 4\}$ **23.** \$3500 in the bank; \$6500 in the credit union. **24.** 50 miles per hour
25. 2 miles per hour

PROBLEM SET 4-1 (Page 133)

1. 5 **3.** 4 **5.** 1.7 **7.** $2\pi - 5$ **9.** $3; 7$ **11.** $-6; 2$ **13.** $\frac{7}{4}; \frac{13}{4}$ **15.** Rectangle.
17. Parallelogram. **19.** $5; (\frac{7}{2}, 1)$ **21.** $2\sqrt{2}; (3, 3)$ **23.** $3; (\sqrt{3}/2, \sqrt{6}/2)$ **25.** 14.54; (3.974, 1.605)
27. (a) $\sqrt{10}; \sqrt{5}; \sqrt{10}; \sqrt{5}$ (b) Each midpoint has coordinates $(5/2, 5)$. (c) Opposite sides have the same length; the two diagonals bisect each other. **29.** (a) $(-2, 3); (4, 0)$ (b) $(2, 7); (8, -1)$
31. If the points are labeled A, B, and C, respectively, then $d(A, B) = \sqrt{20}$, $d(B, C) = \sqrt{20}$, and $d(A, C) = \sqrt{40}$. Then note that $(\sqrt{40})^2 = (\sqrt{20})^2 + (\sqrt{20})^2$.
33. (a) The three distances are 5, 10, and 15, and $15 = 5 + 10$. (b) The distances are 5, 10, and 15.
35. $(9, 0)$ **37.** $(17, 35)$ **39.** $-\frac{3}{2}; 1; \frac{7}{2}$ **41.** (a) $(-1, 5)$ (b) $(-10, 11)$ **43.** 12
45. $(1, 10), (6, 17), (11, 24), (16, 31)$ **47.** \$1458.03 by road; \$1344.75 by air.

PROBLEM SET 4-2 (Page 140)

1.
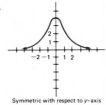

3.
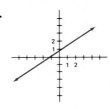
Symmetric with respect to y-axis

5.
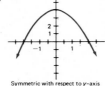

7.
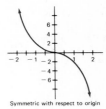
Symmetric with respect to origin

9.
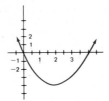
Symmetric with respect to y-axis

11.
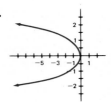

13.
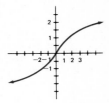

15.

17. $x^2 + y^2 = 36$ **19.** $(x - 4)^2 + (y - 1)^2 = 25$ **21.** $(x + 2)^2 + (y - 1)^2 = 3$ **23.**

25.

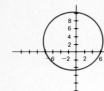

27. $(-1, 5)$; 1   **29.** $(6, 0)$; 1   **31.** $(-\frac{1}{2}, \frac{3}{2})$; $\frac{3}{2}$

33.

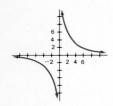

35.

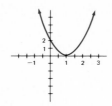

37.

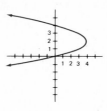

39. $x^2 + y^2 = 41$ **41.** $x^2 + y^2 = 58$ **43.** $(x - 6)^2 + (y - 6)^2 = 36$ **45.** $(x - 4)^2 + (y - 1)^2 = 4$

PROBLEM SET 4-3 (Page 146)

1. $\frac{5}{2}$ **3.** $-\frac{2}{7}$ **5.** $-\frac{5}{3}$ **7.** 0.1920   **9.** $4x - y - 5 = 0$ **11.** $2x + y - 2 = 0$
13. $2x + y - 4 = 0$ **15.** $y - 5 = 0$ **17.** $5x - 2y - 4 = 0$ **19.** $5x + 3y - 15 = 0$
21. $1.56x + y - 5.35 = 0$ **23.** $x - 2 = 0$ **25.** 3; 5   **27.** $\frac{2}{3}$; $-\frac{4}{3}$ **29.** $-\frac{2}{3}$, 2   **31.** -4; 2
33. (a) $y + 3 = 2(x - 3)$ (b) $y + 3 = -\frac{1}{2}(x - 3)$ (c) $y + 3 = -\frac{2}{3}(x - 3)$ (d) $y + 3 = \frac{3}{2}(x - 3)$
(e) $y + 3 = -\frac{3}{4}(x - 3)$ (f) $x = 3$ (g) $y = -3$ **35.** $y + 4 = 2x$ **37.** $(-1, 2)$; $y - 2 = \frac{3}{2}(x + 1)$
39. $(3, 1)$; $y - 1 = -\frac{4}{3}(x - 3)$ **41.** $\frac{7}{5}$ **43.** $\frac{18}{13}$ **45.** $\frac{6}{5}$ **47.** $-\frac{4}{3}$ **49.** $y - 2 = -(x - 4)$
51. (a) Parallel.   (b) Neither.   (c) Perpendicular.   (d) Parallel.   (e) Perpendicular.   **53.** (a) $\frac{9}{5}$ (b) $\frac{9}{5}$; 32
(c) $F = -40$, $C = -40$. At $40°$ below zero, and only then, the two temperature scales give the same reading.
55. Slope is 1600; A-intercept is 20,000.   **57.** -9600. Each year the piece of equipment depreciates $9600.
59. $V = 80,000 - 3900n$ **61.** (a) .75   (b) Increasing production by one item increases the total cost by 75¢.
63. The slope of the line through $(a, 0)$ and $(0, b)$ is $-b/a$. An equation of the line is $y = (-b/a)(x - a)$ which can be rewritten $x/a + y/b = 1$.   **65.** $x + y = 5$

PROBLEM SET 4-4 (Page 154)

1.

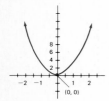

3.

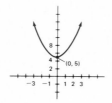

5.

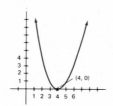

7.

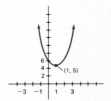

9.

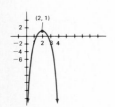

11. $y = 2x^2 - 4x + 9$ **13.** $y = -\frac{1}{2}x^2 + 5x - 10$

15.

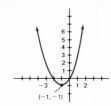

17.

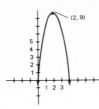

19.

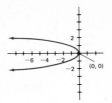

21.

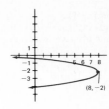

23.

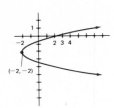

25. $(-3, 4); (0, 1)$ **27.** $(-1, 3); (2, -3)$ **29.** $(-.64, 2.25); (5.04, 10.75)$

31. $y = \frac{1}{12}x^2$ **33.** $y = -\frac{1}{8}(x + 2)^2$ **35.** $y = 2x^2 - 1$ **37.**

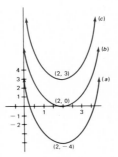

39. 32, 0, −24; two, one, none. **41.** $\frac{5}{4}$ **43.** $(-2, -8); (4, 4)$
45. Eliminating y gives $x^2 - 4x + 5 = 0$, which has no real solutions. **47.** $\frac{45}{4}$ **49.** $y = \frac{7}{6}x^2 - \frac{3}{2}x - \frac{2}{3}$
51. (a) $C = 12,000 + 80x$ (c) 334 (d) 1667; $21,333 **53.** $y = \frac{1}{12}x^2$

PROBLEM SET 4-5 (Page 163)

1. Center: (0,0);
Endpoints of major diameter: $(\pm 5, 0)$
Endpoints of minor diameter: $(0, \pm 3)$

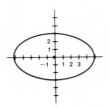

3. Center: (0, 0);
Endpoints of major diameter: $(0, \pm 5)$;
Endpoints of minor diameter: $(\pm 3, 0)$.

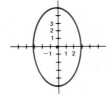

5. Center: (2, −1);
Endpoints of major diameter: (−3, −1), (7, −1);
Endpoints of minor diameter: (2, −4), (2, 2).

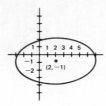

7. Center: (−3, 0);
Endpoints of major diameter: (−3, −4), (−3, 4);
Endpoints of minor diameter: (−6, 0), (0, 0).

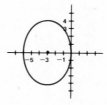

9. (0, 0); $x^2/36 + y^2/9 = 1$ **11.** (0, 0); $x^2/16 + y^2/36 = 1$ **13.** (2, 3); $(x − 2)^2/36 + (y − 3)^2/4 = 1$

15. Center: (0, 0);
Vertices: (±4, 0).

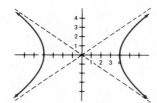

17. Center: (0, 0);
Vertices; (0, ±3).

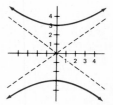

19. Center: (3, −2);
Vertices: (0, −2), (6, −2).

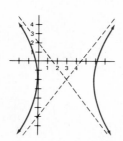

21. Center: (0, −3);
Vertices: (0, −5), (0, −1).

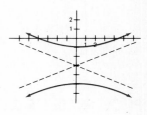

23. $x^2/16 − y^2/25 = 1$ **25.** $(x − 6)^2/4 − (y − 3)^2/4 = 1$ **27.** $(y − 9)^2/36 − (x − 4)^2/16 = 1$
29. $(x + 2)^2/16 + (y − 3)^2/9 = 1$; horizontal ellipse; center: (−2, 3); vertices: (−6, 3), (2, 3).
31. $(x − 2)^2/9 − (y + 1)^2/4 = 1$; horizontal hyperbola; center: (2, −1); vertices: (−1, −1), (5, −1).
33. $x^2/4 + (y − 2)^2/100 = 1$; vertical ellipse; center: (0, 2); vertices: (0, −8), (0, 12).
35. $(y + 2)^2/9 − (x − 4)^2/\frac{9}{4} = 1$; vertical hyperbola; center: (4, −2); vertices: (4, −5), (4, 1).

37. Ellipse.

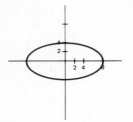

39. Hyperbola.

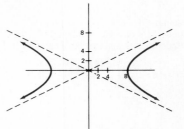

41. Ellipse.

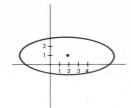

43. Hyperbola.

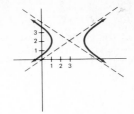

45. $(x - 3)^2/49 + (y - 2)^2/25 = 1$ **47.** $x^2/36 - y^2/16 = 1$ **49.** ± 3.7486 **51.** (a) 27.57 (b) 33.10
53. $x^2/25 + y^2/16 = 1$

CHAPTER 4. REVIEW PROBLEM SET (Page 168)

1. (a) Line. (b) Circle. (c) Line. (d) Parabola. (e) Parabola. (f) Circle. (g) Ellipse. (h) Hyperbola.
(i) Parabola. (j) Ellipse. (k) Hyperbola. (l) Hyperbola. **2.** (a)

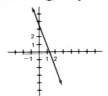

(b)

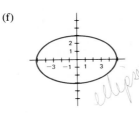

(c)

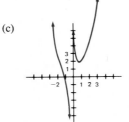

(d)

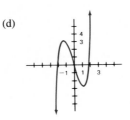

(e)

(f)

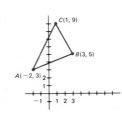

3. (a)

(b) $d(A, B) = \sqrt{29}$; $d(B, C) = 2\sqrt{5}$; $d(C, A) = 3\sqrt{5}$ (c) $\frac{2}{5}$; -2; 2 (d) $y - 3 = \frac{2}{5}(x + 2)$; $y - 5 = -2(x - 3)$;
$y - 3 = 2(x + 2)$ (e) $(\frac{1}{2}, 4)$, $(2, 7)$, $(-\frac{1}{2}, 6)$ (f) $y - 3 = -2(x + 2)$ (g) $y - 3 = \frac{1}{2}(x + 2)$ (h) $12/\sqrt{5} \approx 5.37$ (i) 12
 4. (a) $x - 4 = 0$ (b) $y + 1 = 0$ (c) $3x + 2y - 10 = 0$ (d) $5x - y - 15 = 0$ (e) $2x - 3y + 13 = 0$
(f) $4x - y - 1 = 0$ (g) $x + y - 10 = 0$ (h) $12x + 5y - 169 = 0$
 5. (a) $(0, 0)$; down. (b) $(0, 0)$; to the right. (c) $(-2, 1)$; up. (d) $(\frac{1}{4}, -\frac{1}{2})$; to the left. (e) $(2, -4)$; up.
(f) $(-17, -3)$; to the right. **6.** (a) $y = -\frac{1}{20}x^2$ (b) $x = -\frac{1}{72}y^2 + 3$ (c) $x = 2(y - 5)^2 - 3$ **7.** $(1, 2)$, $(4, 20)$
 8. (a) $(x + 1)^2 + (y - 2)^2 = 7$; circle with center $(-1, 2)$ and radius $\sqrt{7}$. (b) $(x + 1)^2/10 + (y - 2)^2/20 = 1$;
vertical ellipse with center $(-1, 2)$. (c) $(x + 1)^2/6 - (y + 2)^2/12 = 1$; horizontal hyperbola with center $(-1, -2)$.
(d) $y = 2(x + 1)^2 - 16$; vertical parabola opening up with vertex $(-1, -16)$.
 9. (a) $x^2/9 + y^2/25 = 1$ (b) $(x - 2)^2/16 + y^2/4 = 1$ **10.** $(y - 8)^2/9 - (x + 2)^2/1 = 1$

PROBLEM SET 5-1

1. (a) 0   (b) −4   (c) −15/4   (d) −3.99   (e) −2   (f) $a^2 - 4$ (g) $(1 - 4x^2)/x^2$ (h) $x^2 + 2x - 3$
3. (a) $\frac{1}{4}$ (b) $-\frac{1}{2}$ (c) 2   (d) −8   (e) Undefined.   (f) 100   (g) $x/(1 - 4x)$ (h) $1/(x^2 - 4)$ (i) $1/(h - 2)$ (j) $-1/(h + 2)$
5. All real numbers.   **7.** $\{x : x \neq \pm 2\}$ **9.** $\{x : x \neq -2 \text{ and } x \neq 3\}$ **11.** All real numbers.
13. $\{x : x \geq 2\}$ **15.** $\{x : x \geq 0 \text{ and } x \neq 25\}$ **17.** (a) 9   (b) 5   (c) 3; the positive integers.
19. $f(\#) = \boxed{(}\ \#\ \boxed{+}\ 2\ \boxed{)}\ \boxed{x^2}\ \boxed{=}; f(2.9) = 24.01$ **21.** $f(\#) = 3\ \boxed{\times}\ \boxed{(}\ \#\ \boxed{+}\ 2\ \boxed{)}\ \boxed{x^2}\ \boxed{-}\ 4\ \boxed{=}; f(2.9) = 68.03$
23. $f(\#) = \boxed{(}\ 3\ \boxed{\times}\ \#\ \boxed{+}\ 2\ \boxed{\div}\ \#\ \boxed{\sqrt{x}}\ \boxed{)}\ \boxed{y^x}\ 3\ \boxed{=}; f(2.9) = 962.80311$
25. $f(\#) = \boxed{(}\ \#\ \boxed{y^x}\ 5\ \boxed{-}\ 4\ \boxed{)}\ \boxed{\sqrt{x}}\ \boxed{\div}\ \boxed{(}\ 2\ \boxed{+}\ \#\ \boxed{1/x}\ \boxed{)}\ \boxed{=}; f(2.9) = 6.0479407$ **27.** $x + \sqrt{5}$
29. $2x^2 + 3x$ **31.** $\sqrt{3(x-2)^2 + 9}$ **33.** 20   **35.** 5   **37.** 8   **39.** Undefined.   **41.** $18xy - 10x$
43. $3 - 5x$ **45.** $y = 4x$ **47.** $y = 1/x$ **49.** $I = 324s/d^2$ **51.** (a) $R = 2v^2/45$ (b) About 28,444 feet.
53. (a) −6   (b) −15   (c) $\frac{5}{4}$ (d) $-3/x^2 + 12x$ (e) $-9 + 4\sqrt{3}$ (f) $-3(a + b)^2 + 12/(a + b)$
55. (a) 10   (b) 6   (c) $10\sqrt{2}$ **57.** $\{x : x \neq 2\}$ **59.** $\{x : x \geq 0 \text{ and } x \neq \frac{1}{2}\}$ **61.** $(-4 + 1/x^2)^3$
63. $y = 3/x^2; \frac{3}{16}$ **65.** 4000 pounds   **67.** $V(d) = (\pi/12)d^3$
69. $T(x) = 6x - 800 - 20\sqrt{x}; A(x) = (6x - 800 - 20\sqrt{x})/x; \{x : x \text{ is a nonnegative integer}\}; \{x : x \text{ is a positive integer}\}$

PROBLEM SET 5-2   (Page 184)

1.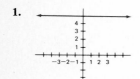

3.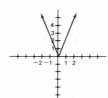

5.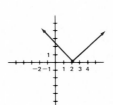

7.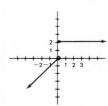

9.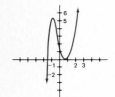

11.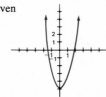

13.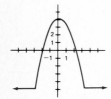

15.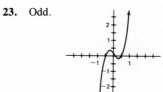

17.

19. Even

21. Neither even nor odd.

23. Odd.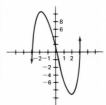

25. Even.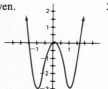

27.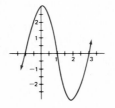

29.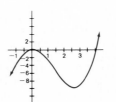

31.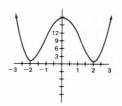

33.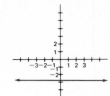

35.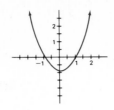

37.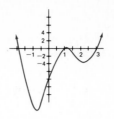

39.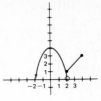

41.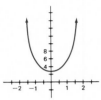

43.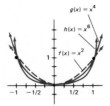

45. 55; 10.1923; 1033.9648

47.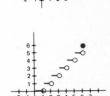

49. $V(x) = 8000 - 541.67x$

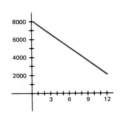

51. $C(x) = \begin{cases} 15 & \text{if } x < 1 \\ 15 + 10[x] & \text{if } x \geq 1 \end{cases}$

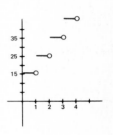

PROBLEM SET 5-3 (Page 192)

1.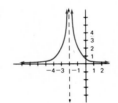

3.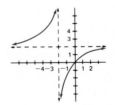

5.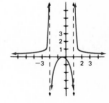

7.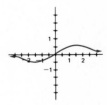

9.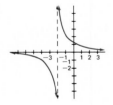

11.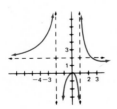

13.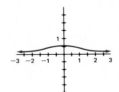

15.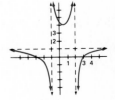

17.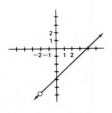

19.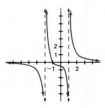

21.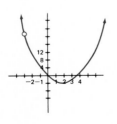

23.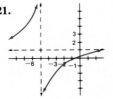

25. 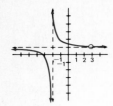   **27.**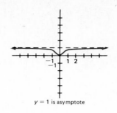

$y = 0$ is asymptote $y = 1$ is asymptote $y = x$ is asymptote

29. $U(x) = (20,000 + 50x)/x$

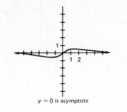

PROBLEM SET 5-4 (Page 198)

1. (a) 3 (b) Undefined. (c) -1 (d) -10 (e) 2 (f) $\frac{1}{2}$ (g) 10 (h) $\frac{2}{5}$ (i) 3
3. $(f + g)(x) = x^2 + x - 2$, all real numbers; $(f - g)(x) = x^2 - x + 2$, all real numbers;
$(f \cdot g)(x) = x^3 - 2x^2$, all real numbers; $(f/g)(x) = x^2/(x - 2)$, $\{x : x \neq 2\}$.
5. $(f + g)(x) = x^2 + \sqrt{x}$, $\{x : x \geq 0)$; $(f - g)(x) = x^2 - \sqrt{x}$, $\{x : x \geq 0\}$;
$(f \cdot g)(x) = x^2\sqrt{x}$, $\{x : x \geq 0\}$; $(f/g)(x) = x^2/\sqrt{x}$, $\{x : x > 0\}$
7. $(f + g)(x) = (x^2 - x - 3)/(x - 2)(x - 3)$, $\{x : x \neq 2$ and $x \neq 3\}$;
$(f - g)(x) = (-x^2 + 3x - 3)/(x - 2)(x - 3)$, $\{x : x \neq 2$ and $x \neq 3\}$;
$(f \cdot g)(x) = x/(x - 2)(x - 3)$, $\{x : x \neq 2$ and $x \neq 3\}$; $(f/g)(x) = (x - 3)/x(x - 2)$, $\{x : x \neq 0, x \neq 2, x \neq 3\}$.
9. $(g \circ f)(x) = x^2 - 2$, all real numbers; $(f \circ g)(x) = (x - 2)^2$, all real numbers.
11. $(g \circ f)(x) = (3x + 1)/x$, $\{x : x \neq 0\}$; $(f \circ g)(x) = 1/(x + 3)$, $\{x : x \neq -3\}$
13. $(g \circ f)(x) = x - 4$, $\{x : x \geq 2\}$; $(f \circ g)(x) = \sqrt{x^2 - 4}$, $\{x : |x| \geq 2\}$
15. $(g \circ f)(x) = x$, all real numbers; $(f \circ g)(x) = x$, all real numbers.   **17.** $g(x) = x^3; f(x) = x + 4$
19. $g(x) = \sqrt{x}; f(x) = x + 2$ **21.** $g(x) = 1/x^3; f(x) = 2x + 5$ **23.** $g(x) = |x|; f(x) = x^3 - 4$

25. 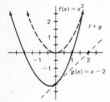   **27.** 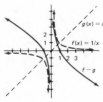   **29.** 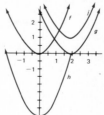   **31.**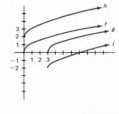

33. (a) $x^3 + 2x + 3$ (b) $x^3 - 2x - 3$ (c) $2x^4 + 3x^3$ (d) $(2x + 3)/x^3$ (e) $2x^3 + 3$ (f) $(2x + 3)^3$ (g) $4x + 9$ (h) x^{27}
35. $1/|x^2 - 4|$; $\{x : x \neq -2$ and $x \neq 2\}$ **37.** (a) $2x + h$ (b) 2 (c) $-1/x(x + h)$ (d) $-2/(x - 2)(x + h - 2)$
39. (a) .7053 (b) 1.6105   **41.** (a) $P = \sqrt{27 + \sqrt{t} + t}$ (b) 6.773

PROBLEM SET 5-5 (Page 206)

1. (a) i, ii; iii; iv; vii; viii (b) i; ii; viii; (c) i; ii; viii   **3.** (a) 1 (b) $-\frac{1}{3}$ (c) $\frac{16}{3}$ **5.** $f^{-1}(x) = (\frac{1}{5})x$
7. $f^{-1}(x) = (\frac{1}{2})(x + 7)$ **9.** $f^{-1}(x) = (x - 2)^2$ **11.** $f^{-1}(x) = 3x/(x - 1)$ **13.** $f^{-1}(x) = 2 + \sqrt[3]{(x - 2)}$

15.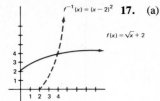
$f^{-1}(x) = (x-2)^2$
$f(x) = \sqrt{x} + 2$

17. (a) 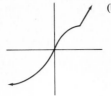 (b) 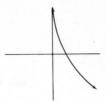 (c)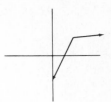

19. $(f \circ g)(x) = 3\left(\dfrac{2x}{3-x}\right) \Big/ \left(\dfrac{2x}{3-x} + 2\right) = \dfrac{6x}{3-x} \Big/ \dfrac{6}{3-x} = x;\ (g \circ f)(x) = 2\left(\dfrac{3x}{x+2}\right) \Big/ \left(3 - \dfrac{3x}{x+2}\right) = \dfrac{6x}{x+2} \Big/ \dfrac{6}{x+2} = x$

21. $\{x : x \ge 1\}; f^{-1}(x) = 1 + \sqrt{x}$ **23.** $\{x : x \ge -1\}; f^{-1}(x) = -1 + \sqrt{x+4}$ **25.** $\{x : x \ge -3\}; f^{-1}(x) = -3 + \sqrt{x+2}$

27. $\{x : x \ge -2\}; f^{-1}(x) = x - 2$ **29.** $\{x : x \ge 1\}; f^{-1}(x) = 1 + \sqrt{2x/(1+x)}$

31.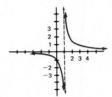

33. $f^{-1}(x) = (x+1)/x$

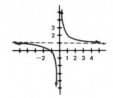

(a) $\frac{1}{2}$ (b) 3   (c) -1 (d) 0   (e) $\frac{4}{3}$ (f) $\frac{1}{2}$

35.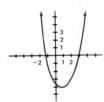

$\{x : x \ge 1\}; f^{-1}(x) = 1 + \sqrt{x+4}$

37. $f^{-1}(x) = \frac{1}{2}(x-5);\ g^{-1}(x) = 1/x;\ (g \circ f)(x) = 1/(2x+5);\ (g \circ f)^{-1}(x) = (1-5x)/2x;\ (f^{-1} \circ g^{-1})(x) = \frac{1}{2}(1/x - 5) = (1-5x)/2x$

39. They are equal.

CHAPTER 5. REVIEW PROBLEM SET   (Page 210)

1. (a) 15   (b) 4   (c) Undefined.   (d) 0   (e) $-\frac{3}{4}$ (f) $\frac{2}{15}$ **2.** $\{x : x \ge -1 \text{ and } x \ne 1\}$ **3.** $y = \frac{1}{8}x^3$ **4.** 12

5. (a) 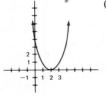 (b) 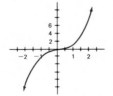 (c) 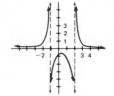 (d)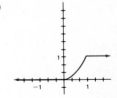

6. 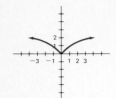 **7.**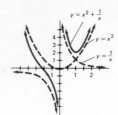

8. (a) $x^3 + 6x^2 + 12x + 9$ (b) $x^3 + 3$ (c) $x^9 + 3x^6 + 3x^3 + 2$ (d) $x + 4$ (e) $\sqrt[3]{x-1}$ (f) $x - 2$ (g) $x + h + 2$
(h) 1 (i) $27x^3 + 1$ **9.** It is moved 2 units to the right and up 3 units.
10. (a) Even. (b) Odd, one-to-one. (c) Even. (d) One-to-one.

11. 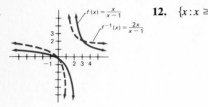 **12.** $\{x : x \geq -2\}$

PROBLEM SET 6-1 (Page 215)

1. 3 **3.** 2 **5.** 7 **7.** $\frac{3}{2}$ **9.** 25 **11.** 9 **13.** 2 **15.** $\frac{1}{100}$ **17.** $\sqrt{2}/2$ **19.** $\sqrt{5}$
21. $3xy\sqrt[3]{2xy^2}$ **23.** $(x+2)y\sqrt[4]{y^3}$ **25.** $x\sqrt{1+y^2}$ **27.** $x\sqrt[3]{x^3-9y}$ **29.** $(xz^2/y^2)\sqrt[3]{x}$
31. $2(\sqrt{x}-3)/(x-9)$ **33.** $2\sqrt{x+3}/(x+3)$ **35.** $\sqrt[4]{2x}/2x$ **37.** $(2y/x)\sqrt[3]{x^2}$ **39.** $\sqrt{2}$ **41.** 26
43. $\frac{9}{2}$ **45.** $-\frac{32}{15}$ **47.** 0 **49.** 4 **51.** $2[(\sqrt{x}-\sqrt{x+h})/\sqrt{x}\sqrt{x+h}]$ **53.** $(x+7)/\sqrt{x+6}$
55. $x/(2\sqrt[3]{x+2})$ **57.** $-9/(x^2\sqrt{x^2+9})$ **59.** 2 **61.** $2y^2$ **63.** $-3x/y$ **65.** $3bc^2\sqrt{6b}$
67. $x\sqrt[4]{1+x^4y^4}$ **69.** $2\sqrt{a-1}/(a-1)$ **71.** $(a^2+1)/a$ **73.** 6 **75.** 13 **77.** 0; 2
79. (a) 2.44289 (b) 5.97209 (c) 1.09648 (d) 5 **81.** (a) $|x|\sqrt{x^2+4}$ (e) $y^2\sqrt[3]{y}/|x|$

83. They are reflections of each other in the line $y = x$. **85.** $0 \leq x \leq 1$

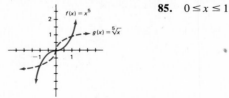

87. (a) 3.875; 3.8730 (b) 6.333; 6.3246 (c) 6.357; 6.3246
89. $\dfrac{1/\sqrt{x+h}-1/\sqrt{x}}{h} = \dfrac{\sqrt{x}-\sqrt{x+h}}{h\sqrt{x}\sqrt{x+h}} \cdot \dfrac{\sqrt{x}+\sqrt{x+h}}{\sqrt{x}+\sqrt{x+h}} = \dfrac{-h}{h\sqrt{x}\sqrt{x+h}(\sqrt{x}+\sqrt{x+h})} = \dfrac{-1}{\sqrt{x}\sqrt{x+h}(\sqrt{x}+\sqrt{x+h})}$

PROBLEM SET 6-2 (Page 222)

1. $7^{1/3}$ **3.** $7^{2/3}$ **5.** $7^{-1/3}$ **7.** $7^{-2/3}$ **9.** $7^{4/3}$ **11.** $x^{2/3}$ **13.** $x^{5/2}$ **15.** $(x+y)^{3/2}$
17. $(x^2+y^2)^{1/2}$ **19.** $\sqrt[9]{16}$ **21.** $1/\sqrt{8^3}=\sqrt{2}/32$ **23.** $\sqrt[4]{x^4+y^4}$ **25.** $y\sqrt[5]{x^4 y}$ **27.** $\sqrt{\sqrt{x}+\sqrt{y}}$
29. 5 **31.** 4 **33.** $\frac{1}{27}$ **35.** .04 **37.** .000125 **39.** $\frac{1}{5}$ **41.** $\frac{1}{16}$ **43.** $\frac{1}{16}$ **45.** $-6a^2$ **47.** $8/x^4$
49. x^4 **51.** $4y^2/x^4$ **53.** y^9/x^{30} **55.** $(2y^3-1)/y$ **57.** $x+y+2\sqrt{xy}$ **59.** $(7x+2)/3(x+2)^{1/5}$
61. $(1-x^2)/(x^2+1)^{2/3}$ **63.** $\sqrt[6]{32}$ **65.** $\sqrt[12]{8x^2}$ **67.** $\sqrt{x}$ **69.** 2.53151 **71.** 4.6364
73. 1.70777 **75.** .0050463

77. **79.** **81.** **83.** $a^{2/3}$ **85.** $(a+2b)^{3/5}$

87. $\sqrt[4]{216}$ **89.** $1/\sqrt{4+\sqrt{x}}$ **91.** 2 **93.** $\frac{1}{4}$ **95.** .0625 **97.** 5 **99.** $(1+x^2)/x$

101. **103.** **105.** $5^{5/3}$; $5^{\sqrt{3}}$; $5^{7/4}$ **107.** 1; 8

PROBLEM SET 6-3 (Page 229)

1. (a) Decays. (b) Grows. (c) Grows. (d) Decays. **3.** (a) 4.66095714 (b) 17.00006441
(c) 4801.02063 (d) 9750.87832 **5.** 1480 **7.** (a) 5.384 billion (b) 6.562 billion (c) 23.772 billion
9. (a) \$185.09 (b) \$247.60 **11.** (a) \$76,035.83 (b) \$325,678.40 **13.** $p(1+r/100)^n$ **15.** \$7401.22
17. \$7102.09 **19.** \$7305.57 **21.** $\$1000(1+.08/12)^{120}=\$2219.64$ **23.** 1600; 1200; 900; 675 **25.** 24,000
27. (a) \$8635.70 (b) \$8832.16 (c) $\$4000(1+.08/12)^{120}=\$8878.56$ (d) $\$4000(1+.08/365)^{3650}\approx\$8901$
29. (a) $2500(1+.1125/4)^{18}=\$4118.76$ (b) $4175(1+.1076/12)^{141}=14{,}699.00$ **31.** 54,000; 18,000; 2000
33. (a) 5000 (b) 7071 (c) 40,000 (d) 14,142 **35.** 1.25 **37.** (a) 842.433 (b) 503.664 (c) 40.43
39. About \$122,390 **41.** (a) \$2621.44 (b) \$4000 (c) No.

PROBLEM SET 6-4 (Page 237)

1. $\log_4 64=3$ **3.** $\log_{27} 3=\frac{1}{3}$ **5.** $\log_4 1=0$ **7.** $\log_{125}(1/25)=-\frac{2}{3}$ **9.** $\log_{10} a=\sqrt{3}$
11. $\log_{10}\sqrt{3}=a$ **13.** $5^4=625$ **15.** $4^{3/2}=8$ **17.** $10^{-2}=.01$ **19.** $c^1=c$ **21.** $c^y=Q$ **23.** 2
25. -1 **27.** 1/3 **29.** -4 **31.** 0 **33.** $\frac{4}{3}$ **35.** 2 **37.** $\frac{1}{27}$ **39.** -2.9 **41.** 49 **43.** .778
45. 1.204 **47.** $-.602$ **49.** 1.380 **51.** $-.051$ **53.** .699 **55.** 1.5314789 **57.** -1.9100949
59. 3.9878003 **61.** $\log_{10}[(x+1)^3(4x+7)]$ **63.** $\log_2[8x(x+2)^3/(x+8)^2]$ **65.** $\log_6(\sqrt{x}\,\sqrt[3]{x^3+3})$
67. 47 **69.** $-\frac{11}{4}$ **71.** $\frac{16}{7}$ **73.** 5; 2 is extraneous. **75.** 7 **77.** 4.0111687 **79.** 2.0446727
81. (a) $\log_{16}(\frac{1}{8})=-\frac{3}{4}$ (b) $\log_{10} y=3.24$ (c) $\log_{1.4} b=a$ **83.** (a) 2 (b) $\frac{1}{2}$ (c) 125 (d) 32 (e) 10 (f) 4
85. $\{x:x>3\}$ **89.** $\log_2 x=c\log_2 10=\log_{10} x\cdot\log_2 10$ **91.** $1=\log_a a=\log_b a\log_a b$

93.

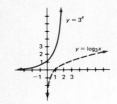

95. (a) 1 (b) $x > 1$ (c) $0 < x < 1$ **97.**

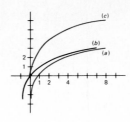

PROBLEM SET 6-5 (Page 246)

1. 1 **3.** 0 **5.** $\frac{1}{2}$ **7.** -3 **9.** 3.5 **11.** $-.2$ **13.** -7.5 **15.** 4.787 **17.** 6.537 **19.** .182
21. 9.1 **23.** .9 **25.** 90 **27.** 1.4609379 **29.** -2.0635682 **31.** -1.8411881 **33.** .50833303
35. 11.818915 **37.** 61.146135 **39.** 8.3311375 **41.** .8824969 **43.** 915.98 **45.** About 2.73.
47. About -0.737. **49.** About 6.84. **51.** Approximately 6.12 years. **53.** Approximately 4.71 years.

55. (a) 5^{10} (b) 9^{10} (c) 10^{20} (d) 10^{1000} **57.**

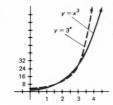

59. $y = ba^x$; $a \approx 1.5$, $b \approx 64$

61. $y = bx^a$; $a \approx 4$, $b \approx 12$ **63.** 3.5 **65.** 2 **67.** 3.303 **69.** 7.844
71. (a) About -1.05. (b) About 26. **73.** (a) 0 (b) $2C \ln r$ (c) $C \ln 2$ **75.** About 0.231 years
77. Approximately 119.2 years. **79.** $x^x = e^{x \ln x}$, $x > 0$ **81.** $(x - 5)^{1/x} = e^{(1/x)\ln(x-5)}$, $x > 5$
83. (a) 6.29209 (b) 10.1148 **85.** (a) $330.04 (b) $331.95 (c) $100e^{(.12)10} \approx 332.01

PROBLEM SET 6-6 (Page 253)

1. 4 **3.** -2 **5.** $\frac{11}{2}$ **7.** -15 **9.** 10,000 **11.** .01 **13.** $10^{3/2}$ **15.** $10^{-3/4}$ **17.** .6355
19. 2.1987 **21.** $.5172 - 2$ **23.** 5.7505 **25.** 8.9652 **27.** 32.8 **29.** .0101 **31.** 3.98×10^8
33. 166 **35.** .838 **37.** .7191 **39.** 3.8593 **41.** $.0913 - 3$ **43.** 7.075 **45.** 8184 **47.** .03985
49. $\frac{5}{4}$ **51.** $\frac{5}{6}$ **53.** 1 **55.** $10^{2/3}$ **57.** .9926 **59.** $.3404 - 3$ **61.** .9833 **63.** 856.0 **65.** 5.370
67. .003570 **69.** 2.75 **71.** .78 **73.** -6

PROBLEM SET 6-7 (Page 257)

1. 128 **3.** .0959 **5.** .0208 **7.** 7.12×10^7 **9.** 3.50 **11.** .983 **13.** 6.05 **15.** 4762
17. 6.143 **19.** 3.530×10^{-6} **21.** 8.90 **23.** 18.2 **25.** 5.19 **27.** $-.5984$ **29.** 2.24
31. .0120 **33.** .00632 **35.** 101 **37.** 1.81 **39.** $\frac{2}{9}$ **41.** -4; 1 **43.** (a) 33.2 grams (b) 391
45. (a) About 6.61 billion. (b) In the year 2013. **47.** 21.7 feet

CHAPTER 6. REVIEW PROBLEM SET (Page 260)

1. (a) $-2y^2\sqrt[3]{z}/z^5$ (b) $2xy^2\sqrt[4]{2x}$ (c) $2\sqrt[6]{5}$ (d) $2(\sqrt{x} + \sqrt{y})/(x - y)$ (e) $5\sqrt{2 + x^2}$ (f) $6\sqrt{2}$ **2.** (a) 12 (b) 4

3. (a) $125a^3$ (b) $1/a^{1/2}$ (c) $1/5^{7/4}$ (d) $3y^{13/6}/x^6$ (e) $x - 2x^{1/2}y^{1/2} + y$ (f) $2^{7/6}$ **4.**

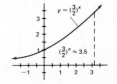

5. $(\frac{1}{2})^{81} \approx 4.14 \times 10^{-25}$ **6.** 16 million   **7.** \$220.80   **8.** (a) 3   (b) $\frac{1}{8}$ (c) 7   (d) 1   (e) $\frac{3}{2}$ (f) 5   (g) 10
(h) 1.14   **9.** $\log_4[(3x+1)^2(x-1)/\sqrt{x}]$ **10.** (a) $\frac{3}{2}$ (b) $\frac{4}{3}$ (c) $\frac{1}{2}$ (d) -2.773 **11.** (a) 1.680   (b) 9.3

(c) 3.517   (d) .9   **12.** 1.807   **13.** 13.9 years   **14.** .1204   **15.** 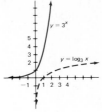   **16.** 3999

PROBLEM SET 7-1   (Page 267)

1. $x + 1; 0$ **3.** $3x - 1; 10$ **5.** $2x^2 + x + 3; 0$ **7.** $x^2 + 4; 0$ **9.** $x + 2 + 5/x^2$
11. $x - 1 + (-x + 3)/(x^2 + x - 2)$ **13.** $2 + (3 - 4x)/(x^2 + 1)$ **15.** $2x^2 + x + 2; -2$ **17.** $3x^2 + 8x + 10; 0$
19. $x^3 + 3x^2 + 7x + 21; 62$ **21.** $x^2 + x - 4; 6$ **23.** $x^3 + x^2 + 3x + 4; 0$ **25.** $x^2 - 3ix - 4 - 6i; 4$
27. $x^3 + 2ix^2 - 4x - 8i; -1$ **29.** $x^2 + 8x + 11; 2x - 48$ **31.** $2x^2 - 4x + 13; -28$
33. $x^4 + 2x^3 + 4x^2 + 8x + 16; 0$ **35.** $Q = x^4 + 3x^3 + 6x^2 + 12x + 8; R = 0$
37. $Q = x^4 + (1 - 2i)x^3 + (-4 - 2i)x^2 + (-4 + 8i)x + 8i; R = 0$ **39.** 11   **41.** 25

PROBLEM SET 7-2   (Page 273)

1. -2 **3.** $-\frac{9}{4}$ **5.** -6 **7.** 14   **9.** 1, -2, and 3, each of multiplicity 1.
11. $\frac{1}{2}$ (multiplicity 1); 2 (multiplicity 2); 0 (multiplicity 3).   **13.** $1 + 2i$ and $-\frac{2}{3}$, each of multiplicity 1.
15. $P(1) = 0$ **17.** $P(3) = 0$ **19.** $(5 \pm \sqrt{7}i)/4$ **21.** 3; 2 (multiplicity 2).   **23.** $(x - 2)(x - 3)$
25. $(x - 1)(x + 1)(x - 2)(x + 2)$ **27.** $(x - 6)(x - 2)(x + 5)$ **29.** $(x + 1)(x + 1 - \sqrt{13})(x + 1 + \sqrt{13})$
31. $x^3 + x^2 - 10x + 8$ **33.** $12x^2 + 4x - 5$ **35.** $x^3 - 2x^2 - 5x + 10$ **37.** $4x^5 + 20x^4 + 25x^3 - 10x^2 - 20x + 8$
39. $x^4 - 3x^2 - 4$ **41.** $x^4 - x^3 - 9x^2 + 79x - 130$ **43.** $-3, -2; (x - 1)^3(x + 3)(x + 2)$ **45.** $\pm i$
47. $(x - 3)(x + 3)(x - 2i)$ **49.** $(x + 1)^2(x - 1 - i)$ **51.** 22   **53.** 1 and 2, each of multiplicity 2.
55. (a) $12x^3 - 25x^2 - 4x + 12$ (b) $x^3 + x^2 - 8x - 12$

57.

59. 3, 2.35   **61.** $16x^4 - 32x^3 + 24x^2 - 8x + 1$

$y = 8$ at $x = -4, -\sqrt{2}, \sqrt{2}$
63. $P(x) \geq 15$ for all real x.
65. If the coefficients are not all zero, then $P(x)$ can have at most n distinct zeros.

PROBLEM SET 7-3   (Page 282)

1. $2 - 3i$ **3.** $-4i$ **5.** $4 + \sqrt{6}$ **7.** $(2 + 3i)^8$ **9.** $2(1 - 2i)^3 - 3(1 - 2i)^2 + 5$ **11.** $5 + i$
13. $3 + 2i; 5 - 4i$ **15.** $-i; \frac{1}{2}$ **17.** $1 - 3i; -2; -1$ **19.** $x^2 - 4x + 29$ **21.** $x^3 + 3x^2 + 4x + 12$
23. $x^5 - 2x^4 + 18x^3 - 36x^2 + 81x - 162$ **25.** $-1; 1; 3$ **27.** $\frac{1}{2}; -1 \pm \sqrt{2}$ **29.** $\frac{1}{2}; (1 \pm \sqrt{5})/2$
31. $3 + 2i; -\frac{1}{2}$ **33.** $-1; -2; 3 \pm \sqrt{13}$ **35.** (a) False.   (b) False.   (c) False.   (d) True.   (e) False.
37. $x - 3i = 0$ has no real solution and is of odd degree.
39. If c/d is a solution in reduced form, then d divides the leading coefficient 1. Thus $d = 1$ and c/d is an integer.
41. Remaining solution: -10.   **43.** $(x + 1)(x - 1)(x^2 + 3x + 4)$
45. (a) $(\sqrt{2} \pm \sqrt{2}i)/2; (-\sqrt{2} \pm \sqrt{2}i)/2$ (b) $2; (\sqrt{2} \pm \sqrt{2}i)/2; (-\sqrt{2} \pm \sqrt{2}i)/2$
47. The zeros are $\pm\frac{1}{2}, \pm i; (-1)^4 a_0/a_4 = -\frac{1}{4} = (\frac{1}{2})(-\frac{1}{2})(i)(-i); -a_3/a_4 = 0 = \frac{1}{2} - \frac{1}{2} + i - i$.

PROBLEM SET 7-4 (Page 290)

1. $4x - 5; -1$ **3.** $4x + 1; 5$ **5.** $10x^4 + 4x^3 - 6x^2 + 8; 16$

7.

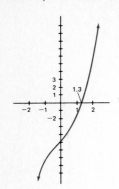 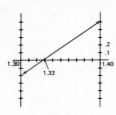 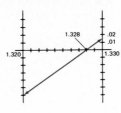

9.

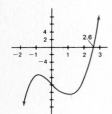 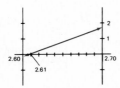 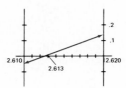

11. $x_1 = 1.3; x_2 = 1.33; x_3 = 1.328; x_4 = 1.3283$ **13.** $x_1 = 2.6; x_2 = 2.61; x_3 = 2.613; x_4 = 2.6129$

15.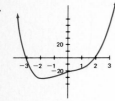

First estimate	Solution to thousandths
-3.2	-3.193
2.2	2.193

17.

First estimate	Solution to thousandths
-1.9	-1.879
$.4$	$.347$
1.5	1.532

19. $-2.11; .25; 1.86$ **21.** $r^3 - 9r^2 + 9r - 3 = 0; r = 7.91$ **23.** 9.701 percent **25.** $y - 12 = 12(x - 2)$

27.

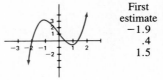

CHAPTER 7. REVIEW PROBLEM SET (Page 294)

1. (a) $2x - 5; 20x - 20$ (b) $x^2 - 3x + 1; -3x + 5$
2. (a) $x^2 - 4; -1$ (b) $2x^3 - 6x^2 + 3x - 5; 12$ (c) $x^2 + 5x + 1; -1$ **3.** $1; 97$

4. (a) -1 (two); 1 (two); i (one); $-i$ (one)   (b) 0 (one); $1 + \sqrt{3}i$ (one); $1 - \sqrt{3}i$ (one); $-\pi$ (three)
5. $2(x - 3)(x - \frac{1}{2})(x + 3)$ **6.** $(x + 4)(x - \sqrt{7})(x + \sqrt{7})$ **7.** $x^4 - 9x^3 + 18x^2 + 32x - 96$
8. $x^4 - 10x^3 + 31x^2 - 38x + 10$ **9.** Remaining zeros: $-2, 4$.   **10.** $6x^3 - 25x^2 + 3x + 4$ **11.** -6
12. $x^3 - 6x^2 + 9x + 50$ **13.** $2 \pm 5i; \pm\sqrt{5}$ **14.** $\frac{3}{2}; 3 \pm 2\sqrt{2}$
15. A cubic equation has 3 solutions (counting multiplicities) and the nonreal solutions for an equation with real coefficients occur in conjugate pairs. The only possible rational solutions are ± 1 and ± 7, but none of them work.

16.

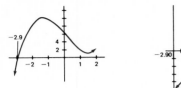 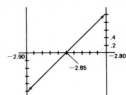 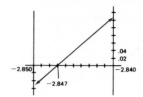

17. $x_1 = -2.9; x_2 = -2.85; x_3 = -2.847$

PROBLEM SET 8-1 (Page 299)

1. $(2, -1)$ **3.** $(-2, 4)$ **5.** $(1, -2)$ **7.** $(0, 0, -2)$ **9.** $(1, 4, -1)$ **11.** $(2, 1, 4)$ **13.** $(0, 0, 0)$
15. $(5, 6, 0, -1)$ **17.** $(15z - 110, 4z - 32, z)$ **19.** $(2y - 3z - 2, y, z)$ **21.** $(-z, 2z, z)$ **23.** Inconsistent.
25. $(-z + \frac{2}{5}, z + \frac{16}{5}, z)$ **27.** $(2, 4); (10, 0)$ **29.** $(5, -7); (6, 0)$ **31.** $(-1, 2); (1, 2)$ **33.** $(\frac{23}{11}, -\frac{9}{11})$
35. $(0, -6)$ **37.** $(0, 2, 1)$ **39.** $a = -6; b = 9$ **41.** 10 nickels; 6 dimes; 12 quarters   **43.** 375
45. 13 inches by 10 inches   **47.** 60; 30; 10

PROBLEM SET 8-2 (Page 308)

1. $\begin{bmatrix} 2 & -1 & 4 \\ 1 & -3 & -2 \end{bmatrix}$ **3.** $\begin{bmatrix} 1 & -2 & 1 & 3 \\ 2 & 1 & 0 & 5 \\ 1 & 1 & 3 & -4 \end{bmatrix}$ **5.** $\begin{bmatrix} 2 & -3 & -4 \\ 3 & 1 & -2 \end{bmatrix}$ **7.** $\begin{bmatrix} 1 & 0 & 0 & 5 \\ 1 & 2 & -1 & 4 \\ 3 & -1 & -5 & -13 \end{bmatrix}$
9. Unique solution.   **11.** No solution.   **13.** Unique solution.   **15.** Infinitely many solutions.
17. No solution.   **19.** $(1, 2)$ **21.** $(x, \frac{3}{2}x - \frac{1}{2})$ **23.** $(1, 4, -1)$ **25.** $(\frac{16}{3}z + \frac{32}{3}, -\frac{7}{3}z - \frac{2}{3}, z)$
27. $(3, 0, 0)$ **29.** $(4.36, 1.26, -.97)$ **31.** $\begin{bmatrix} 2 & -1 & -1 & -1 \\ 4 & -1 & 0 & 3 \\ 2 & -1 & 0 & 7 \end{bmatrix}$ **33.** No solution.
35. Infinitely many solutions.   **37.** No solution.   **39.** $a = -4; b = 8; c = 0$ **41.** $D = -6; E = -4; F = -12$

PROBLEM SET 8-3 (Page 314)

1. $\begin{bmatrix} 8 & 4 \\ 1 & 10 \end{bmatrix}; \begin{bmatrix} -4 & -6 \\ 5 & 4 \end{bmatrix}; \begin{bmatrix} 6 & -3 \\ 9 & 21 \end{bmatrix}$ **3.** $\begin{bmatrix} 5 & 4 & 4 \\ 8 & 3 & -6 \end{bmatrix}; \begin{bmatrix} 1 & -8 & 6 \\ 0 & -3 & 0 \end{bmatrix}; \begin{bmatrix} 9 & -6 & 15 \\ 12 & 0 & -9 \end{bmatrix}$
5. $\begin{bmatrix} 14 & 7 \\ 4 & 36 \end{bmatrix}; \begin{bmatrix} 27 & 29 \\ 5 & 23 \end{bmatrix}$ **7.** $\begin{bmatrix} -3 & -4 & 2 \\ 8 & 22 & -13 \\ -2 & 0 & 9 \end{bmatrix}; \begin{bmatrix} 0 & 5 & -17 \\ 13 & 10 & 3 \\ 1 & -3 & 18 \end{bmatrix}$
9. **AB** not possible; **BA** $= \begin{bmatrix} 7 & 2 & -7 & 6 \\ 15 & 2 & -11 & 16 \end{bmatrix}$
11. **AB** $= \begin{bmatrix} 2 \\ 16 \\ -2 \end{bmatrix}$; **BA** not possible.   **13.** **AB** $=$ **BA** $= \begin{bmatrix} 0 & 0 \\ 0 & 0 \end{bmatrix}$ **15.** $\begin{bmatrix} -4 & 7 & 9 \\ -5 & -5 & 8 \end{bmatrix}$

17. $\mathbf{A}(\mathbf{B} + \mathbf{C}) = \mathbf{AB} + \mathbf{AC} = \begin{bmatrix} -7 & -3 \\ 39 & 34 \end{bmatrix}$; the distributive property. **19.** 93.5917

21. $\begin{bmatrix} 2 & 5 & -1 \\ -8 & 5 & -3 \\ 16 & -2 & -1 \end{bmatrix}$; $\begin{bmatrix} -16 & -6 & 8 \\ 12 & 0 & 22 \\ 11 & -16 & 19 \end{bmatrix}$; $\begin{bmatrix} 32 & -3 & 12 \\ 36 & 29 & 24 \\ 34 & 6 & 25 \end{bmatrix}$

23. $(\mathbf{A} + \mathbf{B})^2 = \mathbf{A}^2 + \mathbf{AB} + \mathbf{BA} + \mathbf{B}^2$; not for matrices. **25.** It consists of zeros.

27. If $\mathbf{A}$ is $m \times n$ and $\mathbf{AB}$ and $\mathbf{BA}$ both make sense, then $\mathbf{B}$ is $n \times m$ and so $\mathbf{AB}$ is $m \times m$, $\mathbf{BA}$ is $n \times n$.

29. (a) $\begin{bmatrix} 16 \\ 13 \\ 14 \end{bmatrix}$ → Art's wages on Monday. (b) $\begin{bmatrix} 18 \\ 14 \\ 16 \end{bmatrix}$ Each man's corresponding wages on Tuesday.
→ Bob's wages on Monday.
→ Curt's wages on Monday.

(c) $\begin{bmatrix} 7 & 9 & 3 \\ 9 & 3 & 4 \\ 8 & 5 & 4 \end{bmatrix}$ The combined output for Monday and Tuesday. (d) $\begin{bmatrix} 34 \\ 27 \\ 30 \end{bmatrix}$ Each man's combined wages for the two days.

31. It multiplies the first row of $\mathbf{B}$ by 3, the second row by 4, and the third by 5. It multiplies the first column of $\mathbf{B}$ by 3, the second column by 4, and the third by 5.

PROBLEM SET 8-4 (Page 322)

1. $\begin{bmatrix} -1 & -3 \\ 1 & 2 \end{bmatrix}$ **3.** $\begin{bmatrix} \frac{1}{6} & \frac{7}{6} \\ 0 & \frac{1}{2} \end{bmatrix}$ **5.** $\begin{bmatrix} 1 & 0 \\ 0 & 1 \end{bmatrix}$ **7.** $\begin{bmatrix} 1/a & 0 \\ 0 & 1/b \end{bmatrix}$ **9.** $\begin{bmatrix} -2 & \frac{3}{2} \\ 1 & -\frac{1}{2} \end{bmatrix}$

11. $\begin{bmatrix} -\frac{4}{7} & \frac{2}{7} & \frac{3}{7} \\ \frac{6}{7} & -\frac{3}{7} & -\frac{1}{7} \\ \frac{5}{7} & \frac{1}{7} & -\frac{2}{7} \end{bmatrix}$ **13.** $\begin{bmatrix} -\frac{1}{9} & \frac{1}{9} & \frac{8}{9} \\ \frac{10}{9} & -\frac{1}{9} & -\frac{26}{9} \\ \frac{1}{9} & -\frac{1}{9} & \frac{1}{9} \end{bmatrix}$ **15.** $\begin{bmatrix} 1 & -1 & 2 & -\frac{5}{4} \\ 0 & \frac{1}{2} & -\frac{3}{2} & \frac{7}{8} \\ 0 & 0 & 1 & -\frac{3}{4} \\ 0 & 0 & 0 & \frac{1}{4} \end{bmatrix}$ **17.** $(\frac{5}{7}, \frac{10}{7}, -\frac{1}{7})$

19. $(\frac{47}{9}, -\frac{128}{9}, \frac{7}{9})$ **23.** Inverse does not exist. **25.** $\begin{bmatrix} -\frac{14}{57} & \frac{22}{57} & -\frac{27}{57} \\ -\frac{9}{57} & \frac{6}{57} & \frac{3}{57} \\ \frac{11}{57} & -\frac{1}{57} & \frac{9}{57} \end{bmatrix}$ **27.** $\begin{bmatrix} \frac{1}{2} & 0 & 0 \\ 0 & \frac{1}{3} & 0 \\ 0 & 0 & -\frac{1}{4} \end{bmatrix}$

29. $x = (-14a + 22b - 27c)/57$; $y = (-9a + 6b + 3c)/57$; $z = (11a - b + 9c)/57$

PROBLEM SET 8-5 (Page 329)

1. −8 **3.** 22 **5.** −50 **7.** 0 **9.** (a) 12 (b) −12 (c) 36 (d) 12 **11.** −30 **13.** −16
15. 7 **17.** 42.3582 **19.** (2, −3) **21.** (2, −2, 1) **23.** 0 **25.** 0 **27.** (1, 2, 3)
29. $\begin{vmatrix} ka & kb \\ c & d \end{vmatrix} = kad - kbc = k(ad - bc) = k\begin{vmatrix} a & b \\ c & d \end{vmatrix}$ **31.** $D = 0$

PROBLEM SET 8-6 (Page 335)

1. −20 **3.** −1 **5.** 39 **7.** 4 **9.** 57 **11.** $x = 2$ **13.** 6 **15.** −72 **17.** −960
19. $aehj$ **21.** 156.8569
23. Add $-r$ times the first row to the third row; then add $-s$ times the second row to the third row. You will get a row of zeros. **27.** They are rational numbers.

PROBLEM SET 8-7 (Page 342)

1.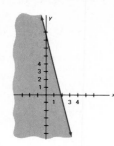

3.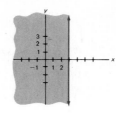

5.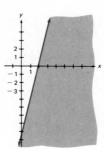

7.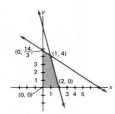

9.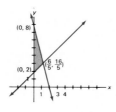

11. Maximum value: 6; minimum value: 0. **13.** Maximum value: $-\frac{4}{5}$; minimum value: -8.
15. Minimum value of 14 at (2, 2). **17.** Minimum value of 4 at $(\frac{3}{2}, 1)$.

19.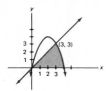

21.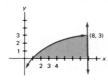

23.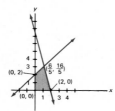

25. Maximum value of 4 at (2, 0); minimum value of -2 at (0, 2).
27. Maximum value of $\frac{11}{2}$ at $(\frac{9}{4}; \frac{13}{4})$; minimum value of 0 at (0, 0).
29. $5700 (1700 barrels of type A, 300 barrels of type B) **31.** $9000 (150 pairs of shoes, 50 pairs of boots)
33. 60 acres of oats and 40 acres of wheat

CHAPTER 8. REVIEW PROBLEM SET (Page 346)

1. (2, 6) **2.** (5, 2, 3) **3.** No solution. **4.** No solution. **5.** $(20w - 36, -9w + 22, -2w + 4, w)$
6. (a) $\begin{bmatrix} -1 & 8 & -2 \\ 8 & -1 & 3 \\ 5 & 13 & 0 \end{bmatrix}$ (b) $\begin{bmatrix} 7 & -11 & 9 \\ -1 & 7 & -11 \\ 0 & -16 & -5 \end{bmatrix}$ (c) $\begin{bmatrix} 0 & 12 & 3 \\ -8 & 6 & -9 \\ -3 & -3 & 0 \end{bmatrix}$ (d) $\begin{bmatrix} 7 & -5 & -5 \\ 3 & 9 & 0 \\ 32 & 14 & -10 \end{bmatrix}$
7. $\begin{bmatrix} \frac{1}{6} & \frac{1}{2} & -\frac{1}{3} \\ \frac{1}{6} & -\frac{1}{2} & \frac{2}{3} \\ \frac{2}{3} & 0 & -\frac{1}{3} \end{bmatrix}$ **8.** $(-1, 2, 3)$ **9.** 2 **10.** 0 **11.** -114 **12.** 0 **13.** -144 **14.** $(1, -2, 4)$

15.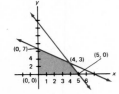

16.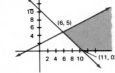

17. 14 **18.** 11 **19.** 24 **20.** 30 suits; 65 dresses

PROBLEM SET 9-1 (Page 354)

1. (a) 9; 11 (b) 5; 2 (c) $\frac{1}{16}$; $\frac{1}{32}$ (d) 81; 121 **3.** (a) 11; 43 (b) $\frac{5}{6}$; $\frac{9}{10}$ (c) 49; 81 (d) -27; 81
5. (a) $a_n = 2n - 1$ (b) $b_n = 20 - 3n$ (c) $c_n = (\frac{1}{2})^{n-1}$ (d) $d_n = (2n - 1)^2$ **7.** (a) 14 (b) 162 (c) $\frac{1}{2}$ (d) 81
9. (a) $a_n = a_{n-1} + 2$ (b) $b_n = b_{n-1} - 3$ (c) $c_n = c_{n-1}/2$ (d) $d_n = d_{n-1} + 8(n - 1)$ **11.** 48 **13.** 42
15. 79 **17.** 69 **19.** (a) 11; 59 (b) 16; 1,048,576 **21.** (a) 27 (b) $\frac{1}{6}$ **23.** (a) 40 (b) 62
25. (a) $a_n = 2 \cdot 3^{n-1}$ (b) $b_n = 2 + 4(n - 1) = -2 + 4n$ (c) $c_n = n(n - 1) + 2 = n^2 - n + 2$ (d) Hard.
(e) $e_n = 2(\frac{1}{2})^{n-1} = 2^{2-n}$ (f) $f_n = n^2 + 1$ **27.** 1; 2; 3; 5; $(1/\sqrt{5})[((1 + \sqrt{5}/2)^{n+1} - ((1 - \sqrt{5})/2)^{n+1}]$
29. $F_n = f_{n+2} - 1$ **31.** (a) 4; 8 (b) There are no primes between $m - 1$ and $m + 20$.
33. (a) $s_n = n^2$ (b) $s_n = s_{n-1} + 2n - 1$ (c) $t_n = t_{n-1} + n$ (d) $t_n = n(n + 1)/2$ **35.** $a_8 = 4$; $a_{27} = 2$; $a_{53} = 5$
37. The number of letters in the English word for n is a_n.

PROBLEM SET 9-2 (Page 360)

1. (a) 13; 16 (b) 3.2; 3.5 (c) 12; 8 **3.** (a) 3; 88 (b) .3; 10.7 (c) -4; -88 **5.** (a) 1335 (b) 190.5
(c) -900 **7.** .5 **9.** (a) 10,100 (b) 10,000 (c) 6,633 **11.** 382.5 **13.** 3.8; 4.6; 5.4; 6.2 **15.** 11; 2,772
17. (a) 90 (b) $\frac{25}{6}$ (c) 15,350 (d) 9,801 **19.** (a) $\Sigma_{i=3}^{20} b_i$ (b) $\Sigma_{i=1}^{19} i^2$ (c) $\Sigma_{i=1}^{n} 1/i$ **21.** (a) 2.5 (b) 130
(c) 3442.5 **23.** 923 **25.** 11,400 **27.** (a) $\Sigma_{i=1}^{112} b_i$ (b) $\Sigma_{i=0}^{100} (19 + 7i)$, or $\Sigma_{i=1}^{101} (12 + 7i)$ **29.** 2419
31. (a) 1; 7; 19; 37; 61 (b) 6; 12; 18; 24; Arithmetic.

PROBLEM SET 9-3 (Page 367)

1. (a) 8; 16 (b) $\frac{1}{2}$; $\frac{1}{4}$ (c) .00003; .000003 **3.** (a) 2; 2^{n-2} (b) $\frac{1}{2}$; $8(\frac{1}{2})^{n-1} = 1/2^{n-4}$ (c) .1; $.3(.1)^{n-1} = 3 \times 10^{-n}$
5. (a) $2^{28} \approx 2.68 \times 10^8$ (b) $(\frac{1}{2})^{26} \approx 1.49 \times 10^{-8}$ (c) 3×10^{-30} **7.** (a) $\frac{31}{2}$ (b) $\frac{31}{2}$ (c) .33333
9. (a) $\frac{1}{2}(2^{30} - 1) \approx 5.3687 \times 10^8$ (b) $16(1 - 1/2^{30}) = 16 - (\frac{1}{2})^{26}$ (c) $.3[1 - (.1)^{30}]/.9 = \frac{1}{3}(1 - (.1)^{30})$
11. $100(2)^{10} = 102,400$ **13.** $\$(2^{32} - 1)$, which is over \$4 billion. **15.** (a) $\frac{1}{2}$ (b) $\frac{4}{15}$
17. 30 feet **19.** $\frac{1}{9}$ **21.** 25/99 **23.** 611/495 **25.** 625; 125; 25; 5; 1
27. $a_n = 120(.2)^{n-1}$; $c_n = 6\pi - 2\pi n$; $e_n = 100(1.08)^n$; $f_n = 106 + 2n$
29. (a) 4429.124 (b) .467 (c) 46.522 (d) 2.414 **31.** $100(1.02)^{40}$; \$220.80; after 11 years and 9 months.
33. $25(1.02)(1.02^{40} - 1)/.02$; \$1540.25
35. If $r \geq 1$, the sum S_n of the first n terms grows large without bound as n increases. If $r \leq -1$, the sum S_n oscillates in value
and does not approach a fixed value. **37.** $\frac{1}{4}$

PROBLEM SET 9-4 (Page 374)

(*Note:* In the text, several proofs by mathematical induction are given in complete detail. To save space, we show only the key
step here, namely, that P_{k+1} is true if P_k is true.)
1. $(1 + 2 + \cdots + k) + (k + 1) = k(k + 1)/2 + k + 1) = [k(k + 1) + 2(k + 1)]/2 = (k + 1)(k + 2)/2$
3. $(3 + 7 + \cdots + (4k - 1)) + (4k + 3) = k(2k + 1) + (4k + 3) = 2k^2 + 5k + 3 = (k + 1)(2k + 3)$
5. $(1 \cdot 2 + 2 \cdot 3 + \cdots + k(k + 1)) + (k + 1)(k + 2) = \frac{1}{3}k(k + 1)(k + 2) + (k + 1)(k + 2) = \frac{1}{3}(k + 1)(k + 2)(k + 3)$
7. $(2 + 2^2 + \cdots + 2^k) + 2^{k+1} = 2(2^k - 1) + 2^{k+1} = 2^{k+1} - 2 + 2^{k+1} = 2(2^{k+1} - 1)$ **9.** P_n is true for $n \geq 8$.
11. P_1 is true. **13.** P_n is true whenever n is odd. **15.** P_n is true for every positive integer n.
17. P_n is true whenever n is a positive integer multiple of 4.
19. $n = 4$. If $k + 5 < 2^k$, then $k + 6 < 2^k + 1 < 2^k + 2^k = 2^{k+1}$.
21. $n = 1$. Since $k + 1 < 10k$, $\log(k + 1) < 1 + \log k < 1 + k$.
23. $n = 1$. Multiply both sides of $(1 + x)^k \geq 1 + kx$ by $(1 + x)$: $(1 + x)^{k+1} \geq (1 + x)(1 + kx) =$
$1 + (k + 1)x + kx^2 > 1 + (k + 1)x$.
25. $x^{2k+2} - y^{2k+2} = x^2(x^2 - y^2) + (x^{2k} - y^{2k})y^2$. Now $(x - y)$ is a factor of both $x^2 - y^2$ and $x^{2k} - y^{2k}$, the latter
by assumption.
27. $(k + 1)^2 - (k + 1) = k^2 + 2k + 1 - k - 1 = (k^2 - k) + 2k$. Now 2 divides $k^2 - k$ by assumption and clearly divides $2k$.
29. $(5 + 15 + \cdots + (10k - 5)) + (10k + 5) = 5k^2 + 10k + 5 = 5(k + 1)^2$
31. $(1 + 2 + \cdots + k) + (k + 1) = (k^2 + k - 6)/2 + (k + 1) = (k^2 + 3k - 4)/2 = [(k + 1)^2 + (k + 1) - 6]/2$. No.
33. $(1^3 + 2^3 + \cdots + k^3) + (k + 1)^3 = [k(k + 1)/2]^2 + (k + 1)^3 = (k + 1)^2(k^2 + 4k + 4)/4 = [(k + 1)(k + 2)/2]^2$

35. (a) $[a + (a + d) + \cdots + (a + (k - 1)d)] + (a + kd) = (k/2)[2a + (k - 1)d] + (a + kd) =$
$\frac{1}{2}[k(2a + kd - d) + 2a + 2kd] = \frac{1}{2}[(k + 1)2a + (k^2 + k)d] = [(k + 1)/2][2a + kd]$ (b) $(a + ar + \cdots + ar^{k-1}) + ar^k =$
$a[(1 - r^k)/(1 - r)] + ar^k = (a - ar^k + ar^k - ar^{k+1})/(1 - r) = a[(1 - r^{k+1})/(1 - r)]$
37. The statement is true when $n = 3$ since it asserts that the angles of a triangle have a sum of $180°$. Now any $(k + 1)$-sided convex polygon can be dissected into a k-sided polygon and a triangle. Its angles add up to $(k - 2)180° + 180° = (k - 1)180°$.
39. $[(1 - \frac{1}{4})(1 - \frac{1}{9}) \cdots (1 - 1/k^2)](1 - 1/(k + 1)^2) = [(k + 1)/2k][((k + 1)^2 - 1)/(k + 1)^2] = (k + 2)/2(k + 1)$
41. $F_{k+1} = F_k + f_{k+1} = f_{k+2} - 1 + f_{k+1} = f_{k+3} - 1$
43. Assume the equality holds for a_k and a_{k+1}. Then $a_{k+2} = (a_k + a_{k+1})/2 = \frac{2}{3}[(1 - (-\frac{1}{2})^k + 1 - (-\frac{1}{2})^{k+1})/2] = \frac{2}{3}[1 - \frac{1}{2}(-\frac{1}{2})^k - \frac{1}{2}(-\frac{1}{2})^{k+1}] = \frac{2}{3}[1 - (-\frac{1}{2})^{k+2}]$.

PROBLEM SET 9-5 (Page 382)

1. (a) 6 (b) 12 (c) 90 **3.** (a) 20 (b) 3024 (c) 720 **5.** 30 **7.** (a) 6 (b) 5 **9.** (a) 36 (b) 72 (c) 6
11. 1680 **13.** 25; 20 **15.** 240 **17.** 300 **19.** (a) 30 (b) 180 (c) 120 (d) 120 (e) 450
21. (a) $12 \cdot 11 \cdot 10 \cdot 9 \cdot 8 \cdot 7 \cdot 6 \cdot 5 \cdot 4$ (b) $2 \cdot 10 \cdot 9 \cdot 8 \cdot 7 \cdot 6 \cdot 5 \cdot 4 \cdot 3$ (c) $2 \cdot 11 \cdot 10 \cdot 9 \cdot 8 \cdot 7 \cdot 6 \cdot 5 \cdot 4$ **23.** 210 **25.** 34,650
27. 840 **29.** 126 **31.** (a) $5!/2!$ (b) $10!/2!$ (c) $8!/6!$ **33.** (a) 720 (b) 120 **35.** (a) 120 (b) 24
37. 1024 **39.** 9,000,000 **41.** 100 **43.** 120

PROBLEM SET 9-6 (Page 389)

1. (a) 720 (b) 120 (c) 120 (d) 1 (e) 6 (f) 6 **3.** (a) 1140 (b) 161,700 **5.** 56 **7.** 495 **9.** 720
11. 63 **13.** (a) 36 (b) 100 (c) 24 **15.** (a) 84 (b) 39 (c) 130 **17.** $_{26}C_{13}$ **19.** 4^{13}
21. $_4C_2 \cdot {}_{48}C_{11} + {}_4C_3 \cdot {}_{48}C_{10} + {}_4C_4 \cdot {}_{48}C_9$ **23.** $_{52}C_5$ **25.** $_{13}C_2 \cdot {}_4C_2 \cdot {}_4C_2 \cdot 44$ **27.** 35 **29.** 126
31. (a) 10 (b) 10 (c) 16 **33.** 1680 **35.** (a) 4 (b) 8 (c) 16; $_nC_0 + {}_nC_1 + \cdots + {}_nC_n = 2^n$ **37.** 5040
39. (a) 45 (b) 90 (c) 100 (d) 55 **41.** (a) 36 (b) 6 (c) 15 **43.** $(6!)^2 = 518,400$

PROBLEM SET 9-7 (Page 396)

1. $x^3 + 3x^2y + 3xy^2 + y^3$ **3.** $x^3 - 6x^2y + 12xy^2 - 8y^3$ **5.** $c^8 - 12c^6d^3 + 54c^4d^6 - 108c^2d^9 + 81d^{12}$
7. $x^{20} + 20x^{19}y + 190x^{18}y^2$ **9.** $x^{20} + 20x^{14} + 190x^8$ **11.** (a) 16 (b) 32 (c) 64 **13.** $405x^2y^{24}$
15. $40x^4$ **17.** 104.076 **19.** 294.4 **21.** 1219 (using 3 terms)
23. $32x^{15} - 80x^{11} + 80x^7 - 40x^3 + 10/x - 1/x^5$ **25.** $b^{36} + 36b^{33}c + 594b^{30}c^2$
27. There are n decisions to make and 2 ways of making each decision; hence there are 2^n ways of selecting a subset.
29. $1 + 10(.0003) + 45(.0003)^2 \approx 1.0030041$ **31.** $2^8 - 1 = 255$ **33.** 105

CHAPTER 9. REVIEW PROBLEM SET (Page 399)

1. (a) and (c) are arithmetic; (b) and (e) are geometric.
2. (a) $a_n = a_{n-1} + 3$ (b) $b_n = 3b_{n-1}$ (c) $c_n = c_{n-1} - .5$ (d) $d_n = d_{n-1} + d_{n-2}$ (e) $e_n = e_{n-1}/3$
3. (a) $a_n = 2 + (n - 1)3 = 3n - 1$ (b) $b_n = 2 \cdot 3^{n-1}$ **4.** 6767 **5.** $3^{100} - 1$ **6.** 3 **7.** 89
8. 40 **9.** 250,500 **10.** $\frac{5}{9}$ **11.** $100(1.02)^{48}$ **13.** P_n is true for n a multiple of 3. **14.** 60 **15.** 64
16. (a) 5040 (b) 210 (c) 30 (d) 1225 (e) 28 (f) 2520 **17.** (a) 504 (b) 224 (c) 84 (d) 40 **18.** 15,120
19. 56 **20.** 256 **21.** $x^{10} + 20x^9y + 180x^8y^2 + 960x^7y^3$ **22.** $-20a^3b^6$ **23.** 1.0408

PROBLEM SET 10-1 (Page 406)

1. (a) $\frac{1}{6}$ (b) $\frac{1}{2}$ (c) $\frac{1}{3}$ (d) $\frac{1}{2}$ (e) $\frac{1}{2}$ **3.** (a) $\frac{1}{8}$ (b) $\frac{3}{8}$ (c) $\frac{1}{2}$ **5.** (a) $\frac{1}{6}$ (b) $\frac{1}{6}$ (c) $\frac{5}{9}$ **7.** (a) 16 (b) $\frac{1}{8}$ (c) $\frac{13}{16}$
9. (a) States have different populations and so are not equally likely as a place of birth. (b) The two events are not disjoint; some people smoke and drink. (c) The two events are complementary; their probabilities should add to 1.
(d) Ties are possible in football. **11.** (a) $\frac{8}{25}$ (b) $\frac{5}{12}$ (c) $\frac{2}{3}$ (d) $\frac{8}{75}$ **13.** $\frac{1}{24}$ **15.** $\frac{9}{11}$ **17.** 1 to 3

19. (a) $\frac{1}{2}$ (b) $\frac{1}{4}$ (c) $\frac{1}{13}$ **21.** (a) $_{26}C_3/_{52}C_3 \approx .118$ (b) $_{13}C_3/_{52}C_3 \approx .013$ (c) $_4C_1 \cdot _{48}C_2/_{52}C_3 \approx .204$
(d) $_4C_3/_{52}C_3 \approx .0002$ **23.** $_8C_5/_{52}C_5 \approx .00002$ **25.** $\frac{1}{8}$; $\frac{7}{8}$ **27.** $_{26}C_5/_{52}C_5 \approx .025$; $2 \cdot _{26}C_5/_{52}C_5 \approx .051$
29. $1 - _{44}C_3/_{52}C_3 \approx .401$ **31.** (a) $\frac{5}{12}$ (b) $\frac{5}{6}$ (c) $\frac{11}{12}$ **33.** $P(A \cup B) = P(A \cap B') + P(A' \cap B) + P(A \cap B) =$
$[P(A \cap B') + P(A \cap B)] + [P(A' \cap B) + P(A \cap B) - P(A \cap B)] = P(A) + P(B) - P(A \cap B)$
35. $4 \cdot 10/_{52}C_5 \approx .000015$; $4/_{52}C_5 \approx .0000015$

PROBLEM SET 10-2   (Page 414)

1. $1/216$ **3.** (a) $\frac{1}{8}$ (b) $\frac{3}{8}$ (c) $\frac{1}{8}$ (d) $\frac{5}{24}$ (e) $\frac{1}{2}$
5. (a) No. Most would agree that doing well in physics is heavily dependent on skill in mathematics.   (b) Yes. Some might disagree, but we think these two events are quite unrelated.   (c) Yes. New shirts and stubbed toes have nothing to do with each other.   (d) No. Check the mathematical condition for independence.   (e) No. It is more likely that a doctor is a man.
(f) Yes. Let old superstitions die.   **7.** $(\frac{9}{10})^3 = .729$ **11.** (a) $\frac{1}{4}$ (b) $\frac{2}{9}$ **13.** (a) $(\frac{3}{10})^3 = .027$ (b) $(\frac{2}{10})^3 = .008$
(c) $.027 + .008 + .125 = .16$ (d) $6(\frac{2}{10})(\frac{3}{10})(\frac{5}{10}) = .18$ **15.** (a) $\frac{1}{5}$ (b) $\frac{1}{5}$ **17.** (a) $.9$ (b) $.5$ (c) $.8$
19. (a) $.27$ (b) $.58$ (c) $.27/.55 \approx .49$ **21.** (a) $25/144$ (b) $5/33$ **23.** (a) $25/324$ (b) $110/324$
(c) $1 - 110/324 = 214/324$ **25.** $_6C_3/_{10}C_3 = \frac{1}{6}$ **27.** (a) $(\frac{2}{3})^4 \approx .198$ (b) $(\frac{1}{3})^4 \approx .012$ (c) $.198 + .012 = .210$
(d) $4(\frac{2}{3})^4(\frac{1}{3}) \approx .263$ (e) $4(\frac{2}{3})^4(\frac{1}{3}) + 4(\frac{1}{3})^4(\frac{2}{3}) \approx .296$

PROBLEM SET 10-3   (Page 421)

1. (a) 6 (b) 15 (c) 20 **3.** (a) $1/64$ (b) $6/64$ (c) $15/64$ (d) $20/64$ (e) $22/64$ **5.** (a) $_{12}C_4(\frac{2}{3})^4(\frac{1}{3})^8$
(b) $_{12}C_6(\frac{2}{3})^6(\frac{1}{3})^6$ **7.** $.59$; $.92$ **9.** (a) $.0162$ (b) $.2522$ **11.** $.7759$ **13.** $.7384$ **15.** $.2617$ **17.** 2
19. $_{20}C_{10}(\frac{1}{2})^{20}$ **21.** $1 - (\frac{5}{6})^{20} - 20(\frac{5}{6})^{19}(\frac{1}{6})$ **23.** $\sum_{k=25}^{100} {}_{100}C_k(\frac{1}{3})^k(\frac{2}{3})^{100-k}$ **25.** (a) $.0563$ (b) $.5256$
27. (a) $6(\frac{1}{6})^5 \approx .00077$ (b) $6 \cdot _5C_4(\frac{1}{6})^4(\frac{5}{6}) \approx .01929$ (c) $6 \cdot _5C_3(\frac{1}{6})^3(\frac{5}{6})^2 \approx .1929$ (d) $6 \cdot 5 \cdot _5C_2(\frac{1}{6})^5 \approx .03858$
(e) $2(5!)(\frac{1}{6})^5 \approx .03086$

PROBLEM SET 10-4   (Page 427)

1. $\$1.25$ **3.** $\$1.00$ **5.** $\$.05$ **7.** (a) $\$25,000$ (b) $\$26,000$
9. Yes, since the mathematical expectation corresponding to the two uncertain positions amounts to only $\$9625$.
11. $\frac{4}{9}$ of a dollar   **13.** $\$3.88$ **15.** $-\$.0263$ **17.** $\$.0825$ **19.** $\$.50$ **21.** $\$7000$
23. $2(\frac{1}{16}) \cdot 4 + 2 \cdot _4C_1(\frac{1}{2})^5 \cdot 5 + 2 \cdot _5C_2(\frac{1}{2})^6 \cdot 6 + 2 \cdot _6C_3(\frac{1}{2})^7 \cdot 7 \approx 5.8$

CHAPTER 10. REVIEW PROBLEM SET   (Page 430)

1. $\frac{2}{7}$; $\frac{4}{7}$ **2.** $1/216$; $3/216$; $212/216$ **3.** $81/100$ **4.** $1 - (5/6)^3 = 91/216$ **5.** $.35$ **6.** (a) $1/15$ (b) $9/100$
7. $\frac{11}{30}$ **8.** (a) $.42$ (b) $.40$ **9.** (a) $1 - (\frac{3}{4})^4 = 175/256$ (b) $_4C_2(\frac{1}{4})^2(\frac{3}{4})^2 \approx 54/256$ **10.** (a) $_{100}C_{30}(\frac{1}{6})^{30}(\frac{5}{6})^{70}$
(b) $\sum_{k=0}^{14} {}_{100}C_k(\frac{1}{6})^k(\frac{5}{6})^{100-k}$ **11.** $\$15.16$ **12.** $\$.10$ **13.** $\$.007435$

PROBLEM SET 11-1   (Page 437)

1. $\$82.25$; $\$317.25$ **3.** $\$900$; $\$2100$ **5.** $\$2598.75$; $\$4848.75$ **7.** $\$4209.70$; $\$2209.70$
9. $\$3835.53$; $\$2035.53$ **11.** $\$4228.17$; $\$1978.17$ **13.** 8.775 percent   **15.** 7.8 percent, compounded quarterly.
17. $\$4746.42$ **19.** $\$5001.10$ **21.** 9.25 percent   **23.** 7.045 percent
25. (a) 8.333 percent (b) 5.95 percent (c) 5.82 percent   **27.** 15.13 years   **29.** 9.83 years
31. (a) 11.11 years (b) 7.73 years (c) 7.70 years   **33.** $\$6168.00$ **35.** $\$1815.29$ **37.** 6.45 years
39. 9.32 percent   **41.** $\$10,782.10$ **43.** $\$1.08329$

PROBLEM SET 11-2 (Page 444)

1. $5098.95; $22,665.98 **3.** $10,740.38; $32,672.74 **5.** $29,290.80 **7.** $29,262.50 **9.** $151,191.15
11. $240.20 **13.** $406.35 **15.** $406.87 **17.** $122.62 **19.** 24 percent, compounded monthly
21. $6725.08; $32,054.16 **23.** $1310.59 **25.** $2457.18 **27.** $3797.40
29. $S_p = R + R[1 - (1 + i)^{-n+1}]/i$, $S_f = R(1 + i)[(1 + i)^n - 1]/i$
31. The first is larger since both represent the present value of the same annuity, but the first is at a lower interest rate. The calculator values are 23.915 and 19.596, respectively.

PROBLEM SET 11-3 (Page 450)

1. (a) $719.46 (b) $5749.29 (c) $247.22 **3.**

Payment Number	Principal Before Payment	Interest Due	Payment	Amount of Principal Repaid
1	12,000.00	516.00	719.46	203.46
2	11,796.54	507.25	719.46	212.21
3	11,584.33	498.13	719.46	221.33
4	11,363.00	488.61	719.46	230.85

5. (a) $58.49 (b) $2744.56 (c) $80.45 **7.**

Deposit Number	Amount at Beginning of Month	Interest Earned	Deposit	Amount at End of Month
1	———	———	58.49	58.49
2	58.49	.47	58.49	117.45
3	117.45	.94	58.49	176.88
4	176.88	1.42	58.49	236.79

9. $1442.03 **11.** (a) $1816.36 (b) $12,327.20 **13.** (a) $6997.40 (b) $4010.40 **15.** $6954.60
17. $6123.91

PROBLEM SET 11-4 (Page 454)

1. $6578.64 **3.** $3624.53 **5.** $506.74 **7.** $99.85 **9.** b **11.** a **13.** b **15.** $138.20
17. $5769.71 **19.** $3201.73 **21.** $827.96 **23.** $4243.86 **25.** $12,428.56
27. Yes. The present value of the three payments is $6440.45; the present value of the proposed payments is $6455.90.
29. $180.17

CHAPTER 11. REVIEW PROBLEM SET (Page 459)

1. (a) $1440 (b) $1727.14 (c) $1775.71 (d) $1780.27 **2.** About 8.33 percent. **3.** $410.65
4. About 4.80 percent. **5.** 3374 days, or 9.24 years **6.** $9202.53; $5703.45 **7.** $400.61 **8.** $271.85
9. $244.85 **10.** $24,396.73 **11.** $3079.26 **12.** $3182.36 **13.** $1384.08

INDEX

FORMULAS FROM ALGEBRA

Exponents

$$a^m a^n = a^{m+n}$$

$$(a^m)^n = a^{mn}$$

$$\frac{a^m}{a^n} = a^{m-n}$$

$$(ab)^n = a^n b^n$$

$$\left(\frac{a}{b}\right)^n = \frac{a^n}{b^n}$$

Radicals

$$(\sqrt[n]{a})^n = a$$

$$\sqrt[n]{a^n} = a, \text{ if } a \geqslant 0$$

$$\sqrt[n]{ab} = \sqrt[n]{a}\sqrt[n]{b}$$

$$\sqrt[n]{\frac{a}{b}} = \frac{\sqrt[n]{a}}{\sqrt[n]{b}}$$

Logarithms

$$\log_a MN = \log_a M + \log_a N$$

$$\log_a (M/N) = \log_a M - \log_a N$$

$$\log_a (N^P) = p \log_a N$$

Factoring Formulas

$$x^2 - y^2 = (x-y)(x+y)$$

$$x^3 - y^3 = (x-y)(x^2 + xy + y^2)$$

$$x^3 + y^3 = (x+y)(x^2 - xy + y^2)$$

$$x^2 + 2xy + y^2 = (x+y)^2$$

$$x^2 - 2xy + y^2 = (x-y)^2$$

$$x^3 + 3x^2 y + 3xy^2 + y^3 = (x+y)^3$$

Quadratic Formula

The solutions to $ax^2 + bx + c = 0$ are $x = \dfrac{-b \pm \sqrt{b^2 - 4ac}}{2a}$

Binomial Formula

$$(x+y)^n = {}_nC_o x^n y^0 + {}_nC_1 x^{n-1} y^1 + \cdots + {}_nC_{n-1} x^1 y^{n-1} + {}_nC_n x^0 y^n$$

$${}_nC_r = \frac{n!}{(n-r)!\, r!} = \frac{n(n-1)\cdots(n-r+1)}{r(r-1)\cdots 3 \cdot 2 \cdot 1} \qquad {}_nC_0 = {}_nC_n = 1$$